AF352813

Intelligent Biosignal Processing in Wearable and Implantable Sensors

Intelligent Biosignal Processing in Wearable and Implantable Sensors

Editors

Hariton-Nicolae Costin
Saeid Sanei

MDPI • Basel • Beijing • Wuhan • Barcelona • Belgrade • Manchester • Tokyo • Cluj • Tianjin

Editors
Hariton-Nicolae Costin
Institute of Computer Science
of the Romanian Academy
Iași, Romania

Saeid Sanei
Nottingham Trent University
UK

Editorial Office
MDPI
St. Alban-Anlage 66
4052 Basel, Switzerland

This is a reprint of articles from the Special Issue published online in the open access journal *Biosensors* (ISSN 2079-6374) (available at: https://www.mdpi.com/journal/biosensors/special_issues/biosignal_processing_wearable).

For citation purposes, cite each article independently as indicated on the article page online and as indicated below:

LastName, A.A.; LastName, B.B.; LastName, C.C. Article Title. *Journal Name* **Year**, *Volume Number*, Page Range.

ISBN 978-3-0365-4601-8 (Hbk)
ISBN 978-3-0365-4602-5 (PDF)

Contents

About the Editors

Hariton-Nicolae Costin

Hariton-Nicolae Costin, **Senior researcher, Institute of Computer Science of the Romanian Academy, Iasi Branch, Romania**. **Scientific interests:** biosignal processing; biomedical image processing; artificial intelligence (neural networks, fuzzy systems, bio-inspired algorithms); (bio)sensors/transducers; e-health and telemedicine; assistive technologies; Internet of Things (IoT). **Current projects:** biomedical image processing and analysis by means of bio-inspired algorithms, machine/deep learning, and other AI methods; gait and human activity and movement analysis; rehabilitation assessment of locomotion injuries and post stroke disabilities using machine and deep learning algorithms applied to patient images; pattern recognition and anomaly detection in fetal morphology using deep learning and statistical learning. Prof. Costin has published or edited 18 books and book chapters and has authored over 200 peer-reviewed scientific publications and 34 annual research reports for the Romanian Academy. **Awards:** Two national prizes. He is Senior member of the IEEE.

Saeid Sanei

Saeid Sanei, **School of Science and Technology, Nottingham Trent University, Clifton campus, Nottingham, UK**. **Scientific interests:** Since 1991, when he received his PhD in Biomedical Signal and Image Processing from Imperial College London, he has been working on speech analysis, brain–computer interfaces, EEG, MEG, joint EEG–fMRI analysis and other brain screening modalities, biosignals and medical imaging, Internet of Things (IoT), body sensor networking and the associated AI, machine learning, and signal processing algorithms. Currently, he is also an Academic Visitor at Imperial College London. Prof Sanei has published or edited 5 books and several book chapters and has authored over 400 peer-reviewed publications. **Current projects:** speech, brain–computer interfacing, EEG, MEG, joint EEG–fMRI and other brain screening modalities, other biosignals, medical imaging, Internet of Things (IoT) and body sensor networking together with the associated AI, machine learning, and signal processing algorithms. **Awards:** Prof. Sanei is a Fellow of British Computer Society (FBCS), has an Honory Diploma from Imperial College London, and is a Senior Member of the IEEE.

Preface to "Intelligent Biosignal Processing in Wearable and Implantable Sensors"

Wearable technology, including sensors, sensor networks, and the associated devices, have led to the development of a variety of applications. Long-term, noninvasive, and nonintrusive monitoring of the human body through the collection of data on as many as possible biometrics and body state indicators as possible is the major goal of healthcare wearable technology developers. For instance, patients suffering diabetes require a simple noninvasive tool to monitor their blood sugar levels on an hourly basis. Those suffering from seizure require the necessary instrumentation to alert them before any seizure onset to prevent fall injuries. Stroke patients need to have their heart rate recorded constantly. These examples show how crucial and necessary wearable healthcare systems can be.

A remote low-cost monitoring strategy significantly promotes social and clinical wellbeing. This can only be achieved if sufficiently reliable recorded information from the human body is available. Such information may be metabolic, biological, physiological, behavioural, psychological, functional, or movement-related.

On the other hand, continuous development of mobile telephones and their improvement up to now, together with availability of large size memory and wideband communication channels, significantly ease achieving the above objectives without hospitalisation or the need to have care takers in hospitals and care units for a long time. This may be considered as a revolution in human welfare. Therefore, more effective and efficient collection of biosignals and biometrics from human body has a tremendous potential for impacting and influencing healthcare and the associated technology.

The state of a patient during rest, walking, working, and sleeping can be well recognised if all the biomarkers of the physiological, biological, and behavioural changes of human body can be measured and processed. This requirement sparks the need for deployment of wearable multi-sensor and multimodal data collection systems. Hence, wearable technology and body sensor networks are central to a complete solution for patient monitoring and healthcare.

The measurable underlying information may not, however, always be visible to the naked eye, and therefore, signal processing, machine learning, and artificial intelligence (AI) techniques have been constantly under research and development for better understanding and recognition of human body states from records of raw data. Although the objective is to have noninvasive and less intrusive sensors, the use of implanted sensors is inevitable particularly for recording of in vivo information, where the human bioindicators need to be monitored for longer times.

Incorporation of AI into medical care leads to the so-called third generation of pervasive health applications. This recent branch of research area aims to combine continuous health monitoring with other sources of medical information and knowledge. Thus, the main objective in third-generation applications is to integrate intelligent agents that implement technologies such as stream and real-time processing, data mining, machine learning, and genetic and multi-omics data. On the other hand, the use of smart sensors paves the path for personalised medicine, which is one of the objectives of future healthcare. With more intelligent systems developed through advanced processing and learning algorithms, the number of sensors can be also reduced, which is another objective of less-intrusive monitoring.

This Special Issue aims to address major advances in the integration and intelligent processing of data coming from wearable, portable, or clinically approved implantable devices. Another aim is to highlight new research opportunities in biomedical informatics and the clinical environment.

Incorporation of on-chip machine learning and AI can lead to the realisation of smart sensors.

In this respect, this Special Issue came as a natural step and attracted the attention of a large number of authors from all over the world enthusiastically working in the related areas. The audience for the resulting book is considered to be numerous researchers, academics, students, and anyone passionate about the synergy between signal processing and AI for patients' benefit.

Hariton-Nicolae Costin and Saeid Sanei
Editors

Editorial

Intelligent Biosignal Processing in Wearable and Implantable Sensors

Hariton-Nicolae Costin [1,*] and Saeid Sanei [2]

[1] Institute of Computer Science, Romanian Academy, 700481 Iasi, Romania
[2] School of Science and Technology, Nottingham Trent University, Nottingham NG11 8NS, UK; saeid.sanei@ntu.ac.uk
* Correspondence: hcostin@gmail.com

Citation: Costin, H.-N.; Sanei, S. Intelligent Biosignal Processing in Wearable and Implantable Sensors. *Biosensors* **2022**, *12*, 396. https://doi.org/10.3390/bios12060396

Received: 8 June 2022
Accepted: 8 June 2022
Published: 9 June 2022

Publisher's Note: MDPI stays neutral with regard to jurisdictional claims in published maps and institutional affiliations.

1. Introduction

Wearable technology including sensors, sensor networks, and the associated devices have opened up space in a variety of applications. Long-term, noninvasive, and nonintrusive monitoring of the human body through collecting as many biometrics and body state indicators as possible is the major goal of healthcare wearable technology developers. Patients suffering diabetes need a simple noninvasive tool to monitor their blood sugar on an hourly basis. Those suffering from seizures require the necessary instrumentation to alarm them before any seizure onset to prevent them from a fall injury. Stroke patients need their heart rate recorded constantly. These are only some examples to show how crucial and necessary the wearable healthcare systems can be.

A remote low-cost monitoring strategy significantly promotes social and clinical wellbeing. This can only be achieved if sufficiently reliable recorded information from the human body is available. Such information may be metabolic, biological, physiological, behavioural, psychological, functional, or movement-related.

On the other hand, the development of mobile telephones since the early 1990's and their improvement till now, together with the availability of large size memory and wideband communication channels, make it significantly easier to achieve the above objectives without hospitalising patients in hospitals and care units for a long time. This may be considered a revolution in human welfare. Therefore, more effective and efficient collection of biosignals and biometrics from the human body has a tremendous impact and influence on healthcare and the technology involved.

The state of a patient during rest, walking, working, and sleeping can be well recognised if all the biomarkers of the physiological, biological, and behavioural changes of human body can be measured and processed. This requirement sparks the need for the deployment of wearable multi-sensor and multimodal data collection systems. Hence, wearable technology and body sensor networks are central to a complete solution for patient monitoring and healthcare.

The measurable underlying information, however, may not be always visualized by the naked eye, and therefore, signal processing, machine learning, and artificial intelligence (AI) techniques have been constantly under research and development in the hope that these techniques can achieve a better understanding and recognition of the human body state from raw data records. Although the objective is to have noninvasive and less intrusive sensors, the use of implanted sensors becomes inevitable for particular in-vivo recordings where the human bioindicators need to be monitored for a longer time.

The incorporation of AI into medical care leads to the so-called third generation of pervasive health applications. This recent branch of research area aims to combine continuous health monitoring with other sources of medical information and knowledge. Thus, the main objective in third-generation applications is to integrate intelligent agents that implement technologies such as stream and real-time processing, data mining, machine learning, and genetic and multi-omics data. On the other hand, the use of smart sensors paves the path for personalized medicine, which is one of the objectives of future healthcare. With more intelligent systems developed through advanced processing and learning algorithms, the number of sensors can also be reduced, as less intrusive monitoring is another objective.

This Special Issue aims to address major advances in the integration and intelligent processing of data coming from wearable, portable, or implantable clinically approved devices. It is also intended to highlight new research opportunities in biomedical informatics and the clinical environment. The incorporation of on-chip machine learning and AI can lead to the realization of smart sensors.

Delightfully, this Special Issue attracted the attention of a large number of authors enthusiastically working in the related areas. Among them the following submission topics successfully achieved the goals of this issue due to their pioneering contributions in the field.

The study by Sawan et al. [1] applies EEG-based brain-machine interfaces during medical rehabilitation, by separating various tasks during motor imagery (MI) and assimilating MI into motor execution (ME). The authors implement intelligent, straightforward, comprehensible, time-efficient, and channel-reduced methods to classify ME versus MI and left- versus right-hand MI. Aside from time-domain information, they map EEG signals to feature space, using extraction methods including statistics, wavelet coefficients, average power, sample entropy, and common spatial patterns. To evaluate their practicability, a support-vector machine as an intelligent classifier model and sparse logistic regression as a feature-selection technique were adopted, and a rate of 79.51% accuracy was obtained. The achieved results make the proposed approach highly suitable to be applied to the rehabilitation of paralyzed limbs.

The paper by Lee et al. [2] analyzes the misalignments and detection errors of quasi-synchronous alignment between echocardiography images and seismocardiogram signals, the latter coming from accelerometer-based devices. Two diagnostic parameters—the ratio of pre-ejection period to left ventricular ejection time (PEP/LVET) and the Tei index—were examined with two statistical verification approaches. In this context, a dynamic time warping (DTW) algorithm was used to align fiducial points. The proposed approach may enable the standardization of the fiducial point detection and the signal template generation. In this way, the program-generated annotation data may serve as the labeled training set for the supervised machine-learning instrument.

The paper by Liu et al. [3] is dedicated to the evaluation of sympathetic nerve activity (SNA), using a skin sympathetic nerve activity (SKNA) signal by means of a Teager-Kaiser energy (TKE) operator, which preprocesses the SKNA signal. The SKNA energy ratio (SKNAER) was proposed for quantifying the SKNA. SKNAER improved the detection accuracy for the burst of SKNA, with 98.2% for detection rate and 91.9% for precision, compared to other approaches. The authors appreciate that the proposed developed feature may play an important role in continuously monitoring of SNA and containing potential for further clinical tests.

COVID-19 could not be missing from this Special Issue. The study by Attallah et al. [4] introduces a novel automated diagnostic tool based on ECG data to diagnose COVID-19, which utilizes 10 deep learning (DL) models of various architectures. It obtains significant features from the last fully connected layer of each DL model and then combines them. Afterward, the tool presents a hybrid feature selection based on the chi-square test and sequential search to select significant features. Finally, it employs several machine-learning classifiers to perform two classification levels: a binary level to differentiate between normal and COVID-19 cases and a multiclass to discriminate COVID-19 cases from normal and other cardiac complications. The proposed method reached an accuracy of 98.2% and 91.6% for binary and multiclass levels, respectively. This performance indicates that the ECG could be used as an alternative means of diagnosis of COVID-19, and perhaps for other diseases.

The study by Rieta et al. [5] proposes a classification model to discriminate between normotensive and hypertensive subjects, employing electrocardiographic and photoplethysmographic (PPG) recordings as an alternative to traditional cuff-based methods. By using 17 discriminatory features extracted from the ECG signal, the main outcome of this research uncovers the relevance of previous calibration to obtain accurate hypertension risk assessment. The k-nearest neighbor classifier provided the best outcomes with an accuracy for new subjects before calibration of 51.48%. The inclusion of just one calibration measurement into the model improved classification accuracy by 30%, reaching gradually more than 96%. Thus, the use of PPG and ECG recordings combined with previous subject calibration can significantly improve discrimination between normotensive and hypertensive individuals.

The paper by Faragó et al. [6] proposes a wearable physiograph for qualitative and quantitative Parkinsonians gait assessment, which performs bilateral tracking of the foot biomechanics and unilateral tracking of arm balance. In this way, the main objective is the monitoring and assessment of gait in Parkinson's disease patients. The novelty is given by the proposed AI-based decisional support procedure for gait assessment, which is validated in a clinical environment. The authors claim that a platform empowering multidisciplinary, AI-evidence-based decision support assessments for optimal dosing between drug and non-drug therapy could lay the foundation for affordable precision medicine.

In [7], the authors analyze the gait signal obtained from an inertial-sensor-based wearable gait device as a tool to manage bone loss and muscle loss in daily life and classify them into seven gait phases. Then, they use explainable AI to analyze the contribution and importance of descriptive statistical parameters on osteopenia and sarcopenia. They confirm high classification accuracy and the statistical significance of gait factors used for osteopenia and sarcopenia management

In [8], the authors propose a comparative analysis of the projection matrices and dictionaries used for compressive sensing (CS) of electrocardiographic (ECG) signals by making compromises between the complexity of preprocessing and the accuracy of reconstruction. They use several types of projection matrices and the reconstructed signals are analyzed quantitatively and qualitatively.

Roy et al. [9] developed an auto-characterization algorithm to leverage the AI-powered auto-signal-enhancing scheme such as denoising autoencoder and adaptive cell characterization technique based on the transfer of learning in deep neural networks. They reported a considerable increase in accuracy and signal enhancement.

In [10], the authors use a carbon nanotube yarn (CNTY) biosensor to chronically record from the vagus nerves of freely moving rats for over 40 continuous hours. Vagal activity is analyzed and spike-cluster-firing rates are found to correlate with food intake. Hence, the neural-firing rates are used to classify eating and other behaviors. This is claimed to be the first chronic recording and decoding of activity in the vagus nerve of freely moving animals enabled by the axon-like properties of the CNTY biosensor in both size and flexibility. This technology is an important step forward in understanding spontaneous vagus-nerve function.

The purpose of the exploratory study by Reuken et al. [11] is to determine whether liver dysfunction can be generally classified with a wearable electronic nose based on semiconductor metal oxide (MOx) gas sensors, and whether the extent of this dysfunction can be quantified. Three sensor modules with a total of nine different MOx layers are used to detect reducible, easily oxidizable, and highly oxidizable gases through non-invasive, rapid, and cost-effective analysis.

Jiang et al. [12] have analyzed surface Electromyography (sEMG) and used it for prosthesis control. They explore how the grasp classification accuracy changes during reaching and grasping, and they identify the period during which the grasp classification accuracy and detection are high. This period has been found suitable for early grasp classification with reduced delay. They also explore the training strategies for better grasp classification in real-time applications.

Chon et al. [13] present an automated arterial fibrillation (AF) prediction algorithm for critically ill sepsis patients, using electrocardiogram (ECG) signals. They extract features from 5-min ECG, using the traditional time, frequency, and nonlinear domain methods.

Different classifiers are then used to classify the existing cardiology dataset. The proposed algorithm achieved 80% accuracy for predicting AF events 10 min earlier than the AF onset.

Faupel et al. [14] use a convolutional neural network (CNN) for epileptic seizure detection capable of running on an ultra-low-power microprocessor, optimised and simulated by MATLAB and implemented on a GAP8 microprocessor with RISC-V architecture. It is claimed that the proposed detector outperforms related approaches in terms of power consumption by a factor of 6. The universal applicability of the proposed CNN based detector is verified with the recording of epileptic rats.

For classification of ECG and EEG signals, Goraș et al. [15] investigate three techniques for reducing dimensionality, namely Laplacian eigenmaps, locality preserving projections, and compressed sensing. The effect of dimensionality reduction is assessed by considering the classification rates for the processed biosignals in the new spaces with different classifiers.

An approach to detect premature ventricular contractions (PVCs) from long-term ECG has been proposed in [16]. The suggested method utilizes deep metric learning to extract features, with compact intra-product variance and separated inter-product differences, from the heartbeat. The use of k-nearest neighbor (KNN), together with the proposed feature extraction method, can extract features by supervised deep-metric learning, which can avoid the bias caused by manual feature engineering. The simulation events show that it is reliable to use deep metric learning and KNN for PVC recognition.

It is our great pleasure to invite you to read this diverse range of papers, as we are hopeful that these submitted works will constitute strong foundations for more research and development in the areas of sensors, wearable technology, and the related signal and data processing techniques by means of AI methods.

Author Contributions: Conceptualization, H.-N.C. and S.S.; methodology, H.-N.C. and S.S.; validation, H.-N.C. and S.S.; formal analysis, H.-N.C. and S.S.; investigation, H.-N.C. and S.S.; resources, H.-N.C. and S.S.; writing—original draft preparation, H.-N.C. and S.S.; writing—review and editing, H.-N.C. and S.S. All authors have read and agreed to the published version of the manuscript.

Funding: This research received no external funding.

Conflicts of Interest: The authors declare no conflict of interest.

References

1. Wang, J.; Chen, Y.-H.; Yang, J.; Sawan, M. Intelligent Classification Technique of Hand Motor Imagery Using EEG Beta Rebound Follow-Up Pattern. *Biosensors* **2022**, *12*, 384. [CrossRef]
2. Chen, C.-H.; Lin, W.-Y.; Lee, M.-Y. Computer-Aided Detection of Fiducial Points in Seismocardiography through Dynamic Time Warping. *Biosensors* **2022**, *12*, 374. [CrossRef]
3. Xing, Y.; Zhang, Y.; Xiao, Z.; Yang, C.; Li, J.; Cui, C.; Wang, J.; Chen, H.; Li, J.; Liu, C. An Artifact-Resistant Feature SKNAER for Quantifying the Burst of Skin Sympathetic Nerve Activity Signal. *Biosensors* **2022**, *12*, 355. [CrossRef] [PubMed]
4. Attallah, O. An Intelligent ECG-Based Tool for Diagnosing COVID-19 via Ensemble Deep Learning Techniques. *Biosensors* **2022**, *12*, 299. [CrossRef] [PubMed]
5. Cano, J.; Fácila, L.; Gracia-Baena, J.M.; Zangróniz, R.; Alcaraz, R.; Rieta, J.J. The Relevance of Calibration in Machine Learning-Based Hypertension Risk Assessment Combining Photoplethysmography and Electrocardiography. *Biosensors* **2022**, *12*, 289. [CrossRef] [PubMed]
6. Ileșan, R.R.; Cordoș, C.-G.; Mihăilă, L.-I.; Fleșar, R.; Popescu, A.-S.; Perju-Dumbravă, L.; Faragó, P. Proof of Concept in Artificial-Intelligence-Based Wearable Gait Monitoring for Parkinson's Disease Management Optimization. *Biosensors* **2022**, *12*, 189. [CrossRef] [PubMed]
7. Kim, J.-K.; Bae, M.-N.; Lee, K.; Kim, J.-C.; Hong, S.G. Explainable Artificial Intelligence and Wearable Sensor-Based Gait Analysis to Identify Patients with Osteopenia and Sarcopenia in Daily Life. *Biosensors* **2022**, *12*, 167. [CrossRef] [PubMed]
8. Fira, M.; Costin, H.-N.; Goraș, L. A Study on Dictionary Selection in Compressive Sensing for ECG Signals Compression and Classification. *Biosensors* **2022**, *12*, 146. [CrossRef] [PubMed]
9. Vaghashiya, R.; Shin, S.; Chauhan, V.; Kapadiya, K.; Sanghavi, S.; Seo, S.; Roy, M. Machine Learning Based Lens-Free Shadow Imaging Technique for Field-Portable Cytometry. *Biosensors* **2022**, *12*, 144. [CrossRef] [PubMed]
10. Marmerstein, J.T.; McCallum, G.A.; Durand, D.M. Decoding Vagus-Nerve Activity with Carbon Nanotube Sensors in Freely Moving Rodents. *Biosensors* **2022**, *12*, 114. [CrossRef] [PubMed]

11.   Voss, A.; Schroeder, R.; Schulz, S.; Haueisen, J.; Vogler, S.; Horn, P.; Stallmach, A.; Reuken, P. Detection of Liver Dysfunction Using a Wearable Electronic Nose System Based on Semiconductor Metal Oxide Sensors. *Biosensors* **2022**, *12*, 70. [CrossRef] [PubMed]

12.   Wang, S.; Zheng, J.; Zheng, B.; Jiang, X. Phase-Based Grasp Classification for Prosthetic Hand Control Using sEMG. *Biosensors* **2022**, *12*, 57. [CrossRef] [PubMed]

13.   Bashar, S.; Ding, E.; Walkey, A.; McManus, D.; Chon, K. Atrial Fibrillation Prediction from Critically Ill Sepsis Patients. *Biosensors* **2021**, *11*, 269. [CrossRef] [PubMed]

14.   Bahr, A.; Schneider, M.; Francis, M.; Lehmann, H.; Barg, I.; Buschhoff, A.-S.; Wulff, P.; Strunskus, T.; Faupel, F. Epileptic Seizure Detection on an Ultra-Low-Power Embedded RISC-V Processor Using a Convolutional Neural Network. *Biosensors* **2021**, *11*, 203. [CrossRef] [PubMed]

15.   Fira, M.; Costin, H.-N.; Goraș, L. On the Classification of ECG and EEG Signals with Various Degrees of Dimensionality Reduction. *Biosensors* **2021**, *11*, 161. [CrossRef] [PubMed]

16.   Yu, J.; Wang, X.; Chen, X.; Guo, J. Automatic Premature Ventricular Contraction Detection Using Deep Metric Learning and KNN. *Biosensors* **2021**, *11*, 69. [CrossRef] [PubMed]

Article

Intelligent Classification Technique of Hand Motor Imagery Using EEG Beta Rebound Follow-Up Pattern

Jiachen Wang [1], Yun-Hsuan Chen [1,2,*], Jie Yang [1,2] and Mohamad Sawan [1,2,*]

[1] Center of Excellence in Biomedical Research on Advanced Integrated-on-Chips Neurotechnologies (CenBRAIN Neurotech), School of Engineering, Westlake University, Hangzhou 310024, China; jiachenwang@utexas.edu (J.W.); yangjie@westlake.edu.cn (J.Y.)
[2] Institute of Advanced Technology, Westlake Institute for Advanced Study, Hangzhou 310024, China
* Correspondence: chenyunxuan@westlake.edu.cn (Y.-H.C.); sawan@westlake.edu.cn (M.S.)

Abstract: To apply EEG-based brain-machine interfaces during rehabilitation, separating various tasks during motor imagery (MI) and assimilating MI into motor execution (ME) are needed. Previous studies were focusing on classifying different MI tasks based on complex algorithms. In this paper, we implement intelligent, straightforward, comprehensible, time-efficient, and channel-reduced methods to classify ME versus MI and left- versus right-hand MI. EEG of 30 healthy participants undertaking motional tasks is recorded to investigate two classification tasks. For the first task, we first propose a "follow-up" pattern based on the beta rebound. This method achieves an average classification accuracy of 59.77% ± 11.95% and can be up to 89.47% for finger-crossing. Aside from time-domain information, we map EEG signals to feature space using extraction methods including statistics, wavelet coefficients, average power, sample entropy, and common spatial patterns. To evaluate their practicability, we adopt a support vector machine as an intelligent classifier model and sparse logistic regression as a feature selection technique and achieve 79.51% accuracy. Similar approaches are taken for the second classification reaching 75.22% accuracy. The classifiers we propose show high accuracy and intelligence. The achieved results make our approach highly suitable to be applied to the rehabilitation of paralyzed limbs.

Keywords: wearable electroencephalography; motor imagery; motor execution; beta rebound; brain–machine interface; feature extraction; EEG classification

---

Citation: Wang, J.; Chen, Y.-H.; Yang, J.; Sawan, M. Intelligent Classification Technique of Hand Motor Imagery Using EEG Beta Rebound Follow-Up Pattern. *Biosensors* **2022**, *12*, 384. https://doi.org/10.3390/bios12060384

Received: 20 April 2022
Accepted: 20 May 2022
Published: 2 June 2022

Publisher's Note: MDPI stays neutral with regard to jurisdictional claims in published maps and institutional affiliations.

1. Introduction

Motor imagery (MI) is a popular method developed to help patients undergoing post-stroke rehabilitation to learn or improve specific motor functions [1]. It is a dynamic state in which patients experience sensations without any actual execution. Studies demonstrate that MI may enhance functional recovery of paralyzed limbs [2], since similar activation sequences occur in the motor cortex during both MI and actual motor execution (ME) [3]. A brain–machine interface (BMI) allows users to interact with the external world through their brain signals instead of their peripheral muscles [4]. Extensive research has been conducted to exploit BMIs for post-stroke rehabilitation, as they assist in the restoration of motional ability [5]. Cincotti et al. demonstrated that, compared with MI alone, rehabilitation training integrated with BMI neurofeedback makes motor areas become more involved, such as by enhancing neuroplasticity in affected regions [6]. Noninvasive electroencephalography (EEG) is a frequently used BMI modality, and one study demonstrated that the majority of stroke patients could use EEG-based MI BMI [7]. One possible application is for evaluating the restoration of physical functions. Until now, the types of assessments commonly used have been time-consuming and can be affected by subjective evaluation.

On the other hand, neurophysiology has revealed that EEG signals experience suppression or enhancement during both MI and ME in the mu and beta frequency bands, which is known as event-related desynchronization (ERD) or event-related synchronization (ERS).

Various EEG-based MI BMIs have been developed to detect this phenomenon. The authors of [8] have concluded that beta rebound (beta ERS occurs shortly after movement) is solid and stable without training, promising fast and universal detection. Leeb et al. applied the beta rebound generated by foot MIs as a feature to detect the user's control intention [9]. Based on beta rebound after foot MIs, Müller-Putz et al. proposed a brain-switch with one-channel EEG [10]. Few studies touch on beta rebound of hand MIs. To the best of our knowledge, no ME versus MI classifications using beta rebound have been reported. Diverse feature extraction methods were proposed to classify left versus right MI [11]. Common spatial pattern (CSP) and its derivatives are proved to result in good accuracy in subsequent classification tasks [12–14]. Wang et al. used SampEn to extract features in MI-EEG data and trained classifier, proving its effectiveness [15]. Other techniques, including statistical, wavelet-based, and power-based, were popular in physiological signal processing. Rajdeep et al. extracted 26 features based on these techniques and finished left versus right hand movements classification [16]. These works have already achieved competitive accuracies, but high-dimensional feature vectors can spoil classifier performance, which calls for feature selection to remove redundant features and retain relevant ones [17]. Gu et al. applied sparse logistic regression (SLR) and its derivatives to select features and to estimate their weight parameters for classification, improving the performance of foot MI and acquiring satisfactory results [18]. However, no prior-art publications were found applying SLR to hand MI classification. Foot MI can generate more observable signals and is, therefore, easier to classify [19], but we cannot overlook hand MI as their deftness and indispensable role in daily life.

To assess the restoration objectively, we investigated the difference between ME and MI, intending to assimilate MI into ME in EEG signals with neurofeedback. It is possible to retrain the brain toward becoming more capable of movement, which improves recovery. While the lateral classification (left versus right hand) has achieved high accuracy in upper and lower limbs, few studies have investigated the difference between ME and MI [20,21]. Focusing on power in different frequency bands, Miller et al. confirmed that spatial distribution of neuronal activity during MI mimics that during ME, and its magnitude is ~25% of ME [22]. More detailed distinction should be drawn to ensure a stable detector. Moreover, existing studies of EEG-based MI BMI share the following limitations: (1) few studies have specified the movement or decoded different motors within the same limb [23]; (2) the multichannel EEG signals in these research activities may reduce processing accuracy and speed, while optimal sets of channels are preferable from a practical point of view [24]; (3) little comparative analysis has been conducted to evaluate different feature extraction methods on experimental data set in parallel to determine which ones are preferable [25]; and (4) they feed large quantities of feature vectors directly into classifiers, which will severely limit the accuracy of classifiers [18].

We built a dataset underlining both ME and MI involving delicate motors to address the above-described limitations. This dataset aimed to explore the feasibility of differentiating between ME versus MI and left- versus right-hand MI by optimizing the feature extraction and classification methods. We put forward a stable and straightforward detector of ME and MI based on beta rebound called "follow-up pattern". We also proposed corresponding methods to address the limitations mentioned before: (1) we reproduced motions that require the engagement of both hands, investigating their application to ME and MI classification; (2) we optimized the number and location of EEG channels to achieve high accuracy with a few channels of EEG-based MI BMI, also proposing a stable and straightforward detector of ME and MI based on beta rebound; (3) we adopted various approaches for extracting features and trained classifiers to validate their utility; and (4) we recognized useful features that improved classification performance with feature-selection techniques. The prepared dataset and analysis methods we proposed can be combined with noninvasive brain stimulation (NIBS) techniques to induce plasticity during post-stroke rehabilitation [26].

The paper is summarized as follows: details of the experiments and the methods of feature selection are described in Section 2; Section 3 illustrates the "follow-up" pattern based on the beta rebound and presents the outcomes of different detection methods; in Section 4 we compare our results with related work; and the conclusions and future work are the subjects of Section 5.

2. Materials and Methods

2.1. Experiments

In our research, 30 healthy individuals (15 males, 15 females; aged 20–35 years, mean $\pm$ SD: 24.26 $\pm$ 3.46; 29 are right-handed) volunteered. All participants provided written informed consent in accordance with the Declaration of Helsinki before the experiment, which was approved by the ethical committee of Westlake University, Hangzhou, China (approval ID: 20191023swan001). All participants received CNY 100 as an inconvenience allowance. Participants were required to make movements based on auditory stimuli, undertaking the following actions: finger tapping, holding a pen, opening a pen, crossing fingers, and moving the arm, as shown in Figure 1. The tasks were set to examine the feasibility (whether joints and hard tissues constrain the freedom of movement) and coordination (all fingers should work in coordination to serve a common purpose, i.e., participants place their hands flat on the table in a comfortable way, while each finger start to orchestrate the required movement after coordination stimuli) of both-hand motion. Each task included five trials for ME and five trials for MI. Each trial was followed by a 2 s rest time. The timing paradigm of a single trial is shown in Figure 2.

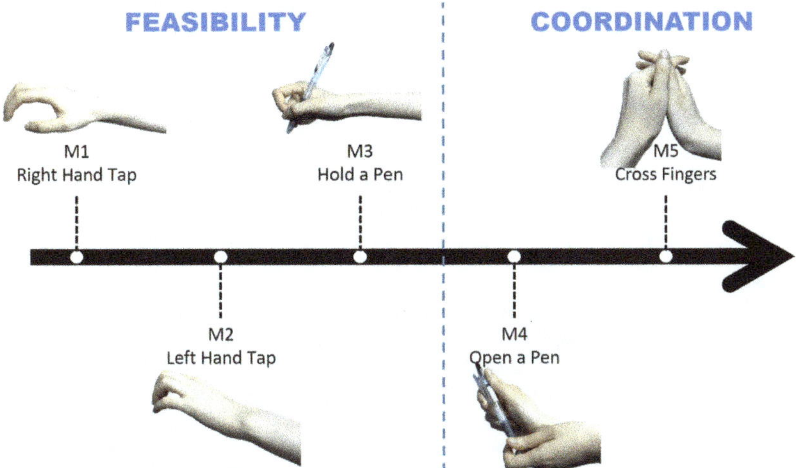

Figure 1. Tasks in the experiment. M1–M3 were to examine the feasibility, and M4–M5 were set for coordination. M1: move specific right fingers according to the auditory code; M2: move specific left fingers according to the auditory code; M3: make the gesture of holding a pen and ready to write; M4: unscrew the pen; M5: fingers of both hands cross over each other.

Figure 2. Timing paradigm of one trial: the duration of motor execution can be 15 s (tapping each finger for 3 s) or 4 s (other tasks); the endpoint of motor imagery depends on the participant's self-regarded "completion". The overall time course is estimated and denoted at the bottom. It can vary between subjects and tasks.

2.2. EEG System

The EEG system examined in this study was the *Brain Products actiCHamp Plus* (EEG signal amplifier) and actiCAP slim (active EEG electrodes) provided by Brain Products GmbH, Munich, Germany, as shown in Figure 3. Thirty-two active electrodes including a reference electrode and a ground electrode were introduced to the system. These electrodes can be placed onto three fabric caps (54–56 cm, 56–58 cm, or 58–60 cm), catering for participants' head circumstances. A chin belt was attached to each cap to achieve better fixation and maintain electrodes' position on the scalp. In total, 32 possible electrode positions arranged under a 10–20 international standard system were marked on each cap.

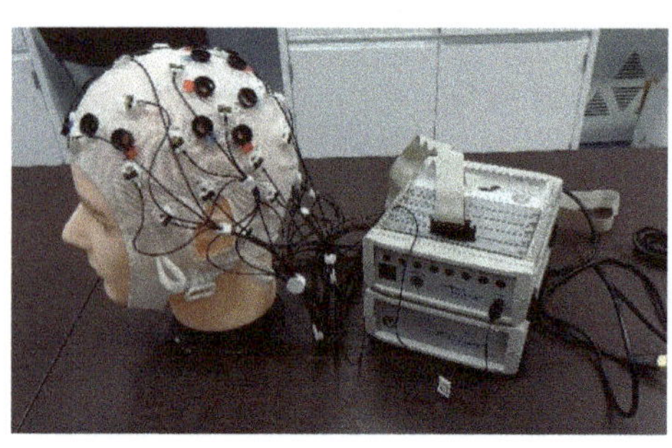
(a)

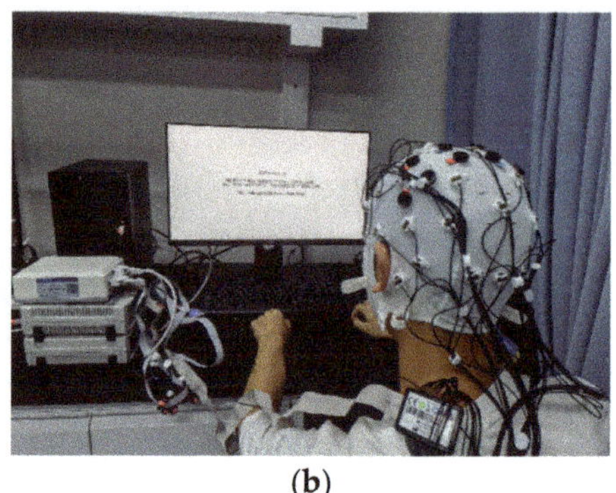
(b)

Figure 3. (**a**) The EEG system used in our study; (**b**) the recording scene: a participant is following the instructions showing on the screen when the EEG signals are recorded.

Before each experiment, a disinfectant wipe was applied to the electrodes. When finished, electrodes and caps were carefully cleaned from gels. These practices can effectively prevent crosstalk between channels induced by resting gels and enhance connectivity by removing dust and particles within the system.

2.3. Data Recording and Preprocessing

EEG signals were recorded with Ag/AgCl electrodes in a 32-channel cap arranged under a 10–20 international standard system (Brain Products, Inc, Gilching, Germany). The central frontal electrode (Fz) served as a reference to a common ground, and the impedance was controlled to be lower than 10 kΩ. The EEG data were recorded with a sampling rate of 1000 Hz. The montage used in our experiment is shown in Figure 4.

Preprocessing included the following procedures: removal of bad channels (channels that coupled noise or had irregular power spectra) or segments, re-referencing to a common average (common average reference is the average electrical activity measured across all scalp channels, re-referencing is conducted by subtracting it from each channel.), filter from 1 to 60 Hz and a 50 Hz notch filter (the interferences from mainline power are removed by the 50 Hz notch filter and EEG signals at 1 to 60 Hz contain most useful information for our applications.), independent component analysis (ICA), epoch extraction, and baseline correction. In the two sets (ME and MI) of preprocessed EEG data, a total of 812 epochs were generated. According to [27], the primary motor cortex (PMC) region, where channels C3, C4, and Cz are located, includes more signals for higher classification performance than other brain areas. We adopted these channels in subsequent analysis to shorten the experiment's preparation time and to reduce the computation load to realize a BMI that requires less input information. We attempted to classify ME and MI with a single channel, Cz, and EEG signals from 19 subjects (with good-quality Cz) were applied. While the classification of left- versus right-hand MI requires more lateral information, 10 participants (with good-quality C3, C4, and Cz) were selected for this task.

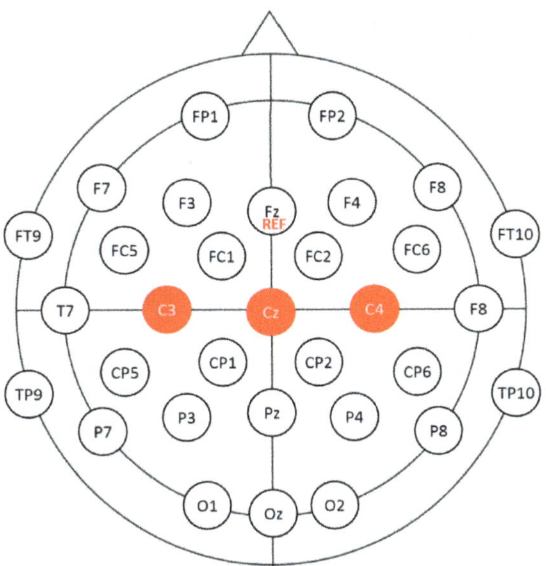

Figure 4. The 32-channel EEG recording montage used in our experiments. Channels C3, C4, and Cz are in the mid-central area, marked as red circles. REF denotes that Fz is the reference electrode.

2.4. Event-Related Desynchronization/Synchronization Analysis

The definition of the ratio ERD/ERS can be formulated as:

$$ERD/ERS_i = \frac{A_i - R}{R} \times 100\%$$

(1)

where A_i is the average power of ith sample over all the trials and R is the average power in the reference interval [28]. The value is defined as ERS when A_i is greater than R.

ERD/ERS values ranging from 13 to 40 Hz were computed to observe beta rebound. ERD/ERS values were considered significant with 95% confidence by adopting a bootstrap *t*-test.

2.5. Feature Extraction

Sample entropy (SampEn) evaluates the complexity and regularity of time-series data, measuring the unpredictability of fluctuations in physiological signals [29]. Let $x(T)$ denote the EEG time series, where T represents the length. To calculate SampEn, we should determine the series of vectors length, m, beforehand:

$$X(i) = [x(i), x(i + m - 1)], \quad i = 1, 2, \ldots, (T - m + 1)$$

(2)

Similar tolerance r controls the number of vector $X(j)$ such that:

$$N^m(i) = card\{X(j)|dist^m\{X(i), X(j)\} < r\}$$

(3)

where $dist^m\{X(i), X(j)\}$ is defined as the most considerable absolute difference between each scalar component.

$$B^m(r) = \frac{1}{(T - m + 1)^2} \sum_{i=1}^{T-m+1} N^m(i)$$

(4)

SampEn is then defined as the negative logarithm of $\frac{B^{m+1}(r)}{B^m(r)}$.

Here, we computed the SampEn of the Cz, C3, and C4 channels from 10 participants, with a series of vector lengths $m = 2$ based on both raw EEG data ($r = 1.0 * SD$, where SD denotes the standard deviation) and ERD/ERS data ($r = 0.1 * SD$). These values were chosen by enumeration and while examining their performance when training classifiers.

The common spatial pattern (CSP) is an advanced algorithm based on principal component analysis (PCA), and it has been successfully applied to brain–computer interfaces [30]. CSP filters EEG signals of two classes to make a clear distinction between them. The feature vectors f_i are defined by Equation (5):

$$f_i = \log\left(\frac{var(Y_i)}{\sum_{k=1}^{k=2} \log(var(Y_k))}\right), \ i = 1, 2 \tag{5}$$

where var represents the variance of a specific sequence and Y_i denotes the corresponding column of CSP-filtered data.

Statistical feature vectors include standard deviation of raw signals and the mean of the absolute values of both the first and second differences of the raw and standardized signal.

We applied Daubechies mother wavelets of order 4 (db4) to analyze the raw EEG data, and the detailed coefficients at level 3 were used to extract features. The related feature vectors were wavelet root mean square (RMS), energy (ENG), and entropy (ENT) [31].

Average power within a specific frequency band was estimated by the average power spectrum density (PSD). The average band power is defined as the power ratio in a specific frequency band to total power. We applied the Welch approach to estimate the PSD with a Hamming window. We performed a PSD estimation on two rhythms, alpha (8–12 Hz) and beta (13–40 Hz).

Details of the feature vectors applied to the classification of ME versus MI, and the classification of laterality in MI, are listed in Tables 1 and 2, respectively.

Table 1. Feature vectors classifying motor execution (ME) versus motor imagery (MI).

Feature Vectors	Size (No. of Trials × No. of Features)
Statistical Features	532 × 6
Wavelet-based Features	532 × 3
Power Features	532 × 4
Total	532 × 13

Table 2. Feature vectors classifying left versus right hand.

Feature Vectors	Size (No. of Trials × No. of Features)
Statistical Features	100 × 18
Wavelet-based Features	100 × 9
Power Features	100 × 24
SampEn	100 × 6
CSP	100 × 2
Total	100 × 59

2.6. Support Vector Machine Classifier

With statistical learning, a support vector machine (SVM) can tackle problems involving small training sets and nonlinear relationships in classification tasks [32]. SVM is used to optimize a hypersurface to separate different classes and to enlarge the distances between them. The MATLAB function *fitcsvm* was applied to train and cross-validates SVM models for our classification tasks.

2.7. Feature Selection

For neuroimaging data, where the training set is small while the feature dimensionality is large, logistic regression is not applicable. In sparse logistic regression (SLR), every weight parameter has its own adjustable variance referred to as relevance parameters, controlling the possible range of the corresponding weight parameters. The weight parameters are estimated as the marginal posterior mean, which can be estimated by variational Bayesian approximation (SLR-VAR) or Laplace approximation (SLR-LAP). The L1-norm-SLR with a Laplace approximation (L1-SLR-LAP) and the component-wise implementation (L1-SLR-COMP) were also investigated in this study [18].

3. Results

3.1. "Follow-Up" Pattern

Beta rebound is a stable phenomenon that occurs several seconds after ME or MI. As shown in Figure 5, the beta rebound is the beta ERS (refer to Formula (1)) that occurs within 1 s after a stimulus (represented as blue lines). It can be observed in participants with little or no training. Taking advantage of this primitive and perceptible reaction, we proposed a method based on the beta rebound in the time-domain signals to discriminate between ME and MI that requires a light computational load and little pre-training. This time-domain "follow-up" pattern helps therapists gain information from the beta rebound in real time, evaluate the performance of paralyzed patients, and then guide and rectify their MI tasks. With proper training, the beta rebound can offer novel targets for therapeutic interventions [33].

Figure 5 demonstrates the difference between ME and MI in both the time and spatial domains. In the time domain, ME and MI have a distinction in amplitude, time delay, and latency. Figure 5a (ME) and Figure 5c (MI) illustrate it as ERD/ERS time courses during the same finger-tapping movement (motion Tap: Right Finger 1), with dashed lines from five different individuals while bold red lines delineating the average time course across these subjects. Beta rebounds are represented as peaks in these lines. "Stimulus" marks the time when subjects hear the auditory instructions. Compared with ME, the beta rebounds of MI have smaller amplitude, appear later after stimulus, and last longer. In the spatial domain, ME and MI have different topographic distributions. Figure 5b (ME) and Figure 5d (MI) demonstrate it with topo-plots (topographic maps of EEG fields in a circular 2D view looking down at the top of the head) depicting ERD/ERS distribution. These topo-plots are from subject S01 for motion Tap: Right Finger 1 at the time when the beta rebound is the most remarkable (ME: 1.624 s; MI: 1.818 s). Black dots mark the locations of electrodes. ERS is in red while ERD is in blue. During the MI task (Figure 5d), the beta rebound was constrained within the mid-central areas (channel Cz). In contrast, the rebound of ME (Figure 5b) had a more enormous scope of influence, affecting adjacent electrodes (channels Cz, FC1, FC2, PC1, PC2, and P3). Cz and the surrounded channels are related to sensorimotor cortex, which accounts for the peak at the mid-central areas. The other peak at channel P3 may attribute to the touch sensation function of parietal lobe, which only occur during ME. To conclude, in the time domain, there is a high probability that beta rebounds of lower intensity, higher latency, and longer duration indicate MIs instead of MEs; in the spatial domain, if beta rebounds mainly affect channel Cz, it will most likely represent MIs.

We computed the difference in the ERD/ERS values between ME and MI by subtracting the signals of each motion recorded from each subject. The results of the subtracted signals during M4 (Figure 1), open a pen, are illustrated by a pseudo-color map in Figure 6, with the x-axis representing post-stimuli time and the y-axis representing subjects. Each pixel indicates the intensity by color, where red denotes the beta rebound of ME, and blue denotes the beta rebound of MI. As marked by black frames (as an example) in Figure 6, most participants' data observes the "follow-up" patterns. The "follow-up" pattern implies that the beta rebound of ME can occur faster than that of MI, following the difference described above in Figure 5. We marked all the peaks in the ERS series and counted all the "follow-up" phenomena across subjects and across motional tasks. The results are shown in Tables 3 and 4. Table 3 defines the percentage as the ratio of motions that displayed "follow-up" patterns. Some subjects, e.g., S06 and S18, achieved high accuracy under this criterion, which reflected the variation across subjects: some subjects are more adapted to imaginary tasks than others. Throughout all the motions listed in Table 4, opening pens and finger-crossing were distinctive compared to the others, and they are both motions designated to examine coordination in movement. The motions that require both hands' involvement and synchronization have more significant potential to be applied in the evaluation system of a rehabilitation process. The parameters of beta rebound (amplitude and time) of ME and MI tasks can be distinguished more obviously.

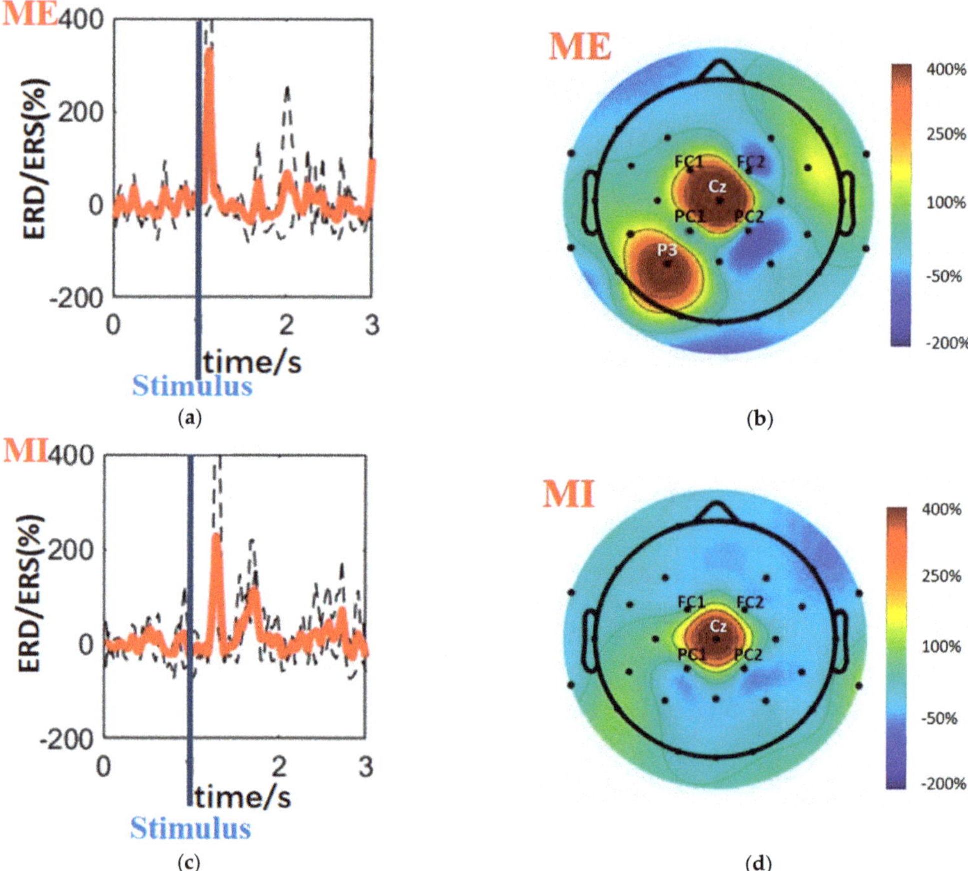

Figure 5. "Follow-up" pattern based on the beta rebound works as an indicator of ME and MI in amplitude, latency, duration, and distribution. An example of the time courses and topo-plots of ME (parts (**a**,**b**)) and MI (parts (**c**,**d**)) event-related desynchronization/synchronization (ERD/ERS) during the motion Tap: Right Finger 1. The bold red lines are the average across five subjects, while the dashed lines are the individuals' ERD/ERS time courses at channel Cz. "Stimulus" marks the time when subjects hear the auditory instructions. Topo-plots are from subject S01 for motion Tap: Right Finger 1 at the time when the beta rebound is the most remarkable (ME: 1.624 s; MI: 1.818 s). Black dots mark the locations of electrodes. ERS is in red while ERD is in blue. Note that parts (**a**,**c**) are based on a part of our whole dataset (26.32%) to make the time courses more explicit for demonstration.

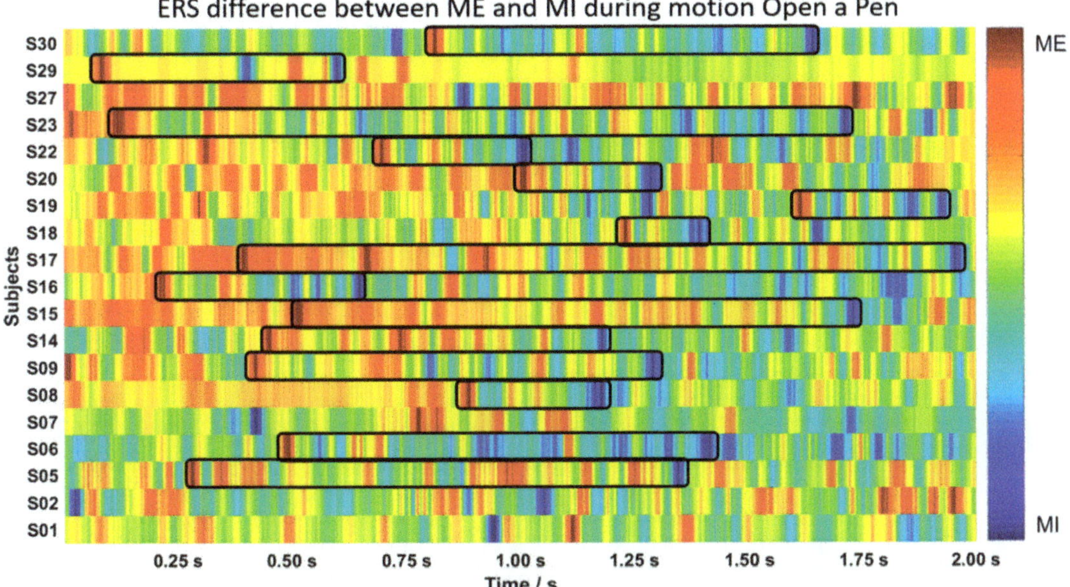

Figure 6. Color map of the differences between ME and MI tasks during the motion, open a pen. Red blocks show ERS during ME, while blue blocks represent ERS during MI. In most cases, a "follow-up" pattern—a red block followed by a blue block—can be observed, marked by the black frames.

Table 3. Percentage of "follow-up" pattern in subjects at Cz among all motions.

Subjects	Percentage (%)	Subjects	Percentage (%)
S01	50.00	S17	57.14
S02	50.00	S18	85.71
S05	57.14	S19	42.86
S06	78.57	S20	57.14
S07	42.86	S22	57.14
S08	64.29	S23	64.29
S09	50.00	S27	64.29
S14	71.43	S29	71.43
S15	71.43	S30	50.00
S16	50.00	Mean ± SD	59.77 ± 11.95

Table 4. Percentage of "follow-up" pattern in motions at Cz among all subjects.

Motions	Percentage (%)	Motions	Percentage (%)
Tap: Right Finger 1	57.89	Tap: Left Finger 4	57.89
Tap: Right Finger 2	36.84	Tap: Left Finger 5	42.10
Tap: Right Finger 3	57.89	Hold a Pen	63.16
Tap: Right Finger 4	63.16	Open a Pen	84.21
Tap: Right Finger 5	52.63	Finger-crossing	89.47
Tap: Left Finger 1	68.42	Arm Movement	52.63
Tap: Left Finger 2	52.63	Mean ± SD	59.77 ± 13.58
Tap: Left Finger 3	57.89		

Based on the findings mentioned above, we can conclude how to identify ME and MI with beta rebound at channel Cz in the time-domain: compared with ME, beta rebounds of MI have smaller amplitude, appear later after stimulus, and last longer.

3.2. ME versus MI Classification

We used feature vectors in Table 1 to train SVM and adopted hyperparameter optimization during training to search for kernel functions and related parameters to induce the best performance. Such procedures achieved a classification accuracy of 78.57%. We drew the scatter plots and found that power-related features may perform better in EEG classification tasks. We selected those four power-based feature vectors to describe the data set and trained the SVM again. The overall accuracy improved slightly to 79.51%, but the dimension of features was reduced, which will mitigate the computational load. Additionally, this indicates that excessive large feature vectors may not necessarily lead to higher accuracy in SVM classification tasks. Feature selection methods can be applied in training classification models, which enlightened us about resorting to SLR in left- versus right-hand MI classification tasks, as described in the following paragraphs.

3.3. Left—Versus Right-Hand Motor Imagery Classification

We adopted features in Table 2 to train a classifier that may facilitate SVM task in a higher dimensionality. The accuracy was only 62%, which was even lower than when sample entropy feature vectors were applied alone. This phenomenon warned us there were some redundant feature vectors in the SVM training data that spoiled the overall result.

We adopted different derivatives of SLR to select features and to calculate weights. The number of features left and the corresponding accuracies are shown in Table 5. Among all the models adopted, L1-SLR-LAP, which applied Laplacian approximation and L1-norm in SLR learning, attained the best performance. The accuracy of L1-SLR-LAP is 75.22%, and the corresponding confusion matrix is displayed in Figure 7. Note that the values here are the average number of 10-fold cross-validation. Higher accuracy was achieved in left-hand MI. Forty-two feature vectors were left after the selection procedure in L1-SLR-LAP. By checking their weights, we found that power features and SampEn displayed distinctive weights in the remaining vectors, which indicated that they were primary factors in the classification task.

Table 5. Accuracy of different classifiers used on our EEG data.

Models	Features Left	Accuracy
SVM	59	62.00%
SLR-LAP	2	57.78%
SLR-VAR	9	50.22%
L1-SLR-LAP	42	75.22%
L1-SLR-COMP	35	58.67%

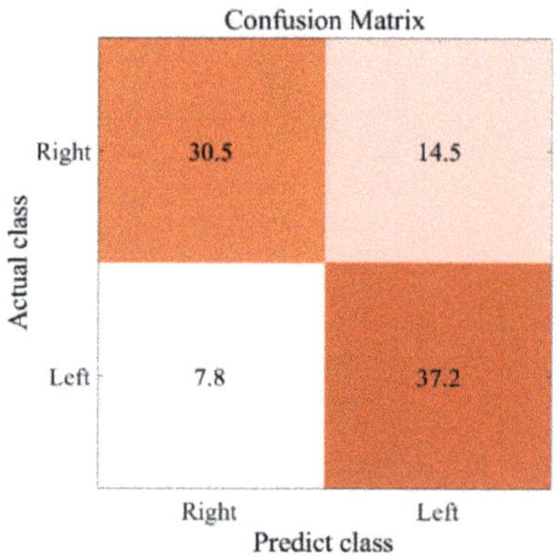

Figure 7. Confusion matrix of the "L1-SLR-LAP" classifier to distinguish the MI of left- and right-hands.

3.4. Comparison and Analyses of Classification Accuracies

Previous studies of MI classification tasks generate interesting classification accuracies, based on different datasets, models and techniques. Table 6 compares the classification results of left- and right-hand MI among the proposed dataset and other datasets. It is important to note that the accuracy of our proposed method is obtained through group-level classification, while in other works, classifiers are trained in a subject-specific manner. Group-level classifications will reduce training sessions and be more applicable to patients, as elucidated in Section 4. Using the same EEG channels and classifier models as the ones we proposed, Malan et al. [34] suggested a novel feature selection algorithm, regularized neighborhood component analysis (RNCA), which outperformed other conventional feature selection techniques. The diverse parameters of RNCA increase its computational burden, while SLR is lighter. The dimension of features in [35] was relatively low, so the accuracy is comparable without feature selection. We achieved a similar accuracy with fewer EEG channels, which can lighten the workload of experiment and computation. Accuracies in [36] seem lower than other studies, which may verify that SVM is more preferable in such contexts.

Table 6. Comparison of classification accuracies among different datasets and methods.

Authors	EEG Channels	Participants	Feature Extraction	Classifiers	Feature Selection	Average Accuracy
This work	3	10	Statistics, Wavelet Coefficients, Average Power, SampEn, CSP	SVM	L1-SLR-LAP	75.2%
Malan et al., 2019 [34]	3	10	DTCWT	SVM	GA	78.9%
					PCA	64.3%
					ReliefF	75.7%
					RNCA	80.7%
Tang et al., 2017 [35]	28	2	Power spectrum	SVM	-	77.2%
Voinas et al., 2022 [36]	16	6	WPD+HOS	RF	-	71.0%
			CSP			66.0%
			Filter Bank CSP			69.0%

CSP: common spatial pattern; DTCWT: dual-tree complex wavelet transform; GA: genetic algorithm; L1-SLR-LAP: L1-norm-SLR with a Laplace approximation; PCA: principal component analysis; RF: Random Forest; RNCA: regularized neighborhood component analysis; SampEn: sample entropy; SVM: support vector machine; WPD+HOS: wavelet packet decomposition combined with higher order statistics.

4. Discussion

We applied a single neuroimaging modality, EEG, in the present study. EEG has a high temporal resolution and can produce good performance in BMI [18]. Other modalities have been explored, e.g., functional magnetic resonance imaging (fMRI) [37], functional near-infrared spectroscopy (fNIRS) [38], magnetoencephalography (MEG) [39], and electrical impedance tomography (EIT) [40]. Due to portability, non-invasiveness, and cost-effectiveness, EEG and fNIRS have an advantage in natural environment applications [41]. In terms of classification accuracy, EEG-based BMI outperforms fNIRS-based BMI [24]. Recent progress of hybrid EEG-fNIRS in BMI demonstrates great potential because data with complementary spatiotemporal resolution can exhibit synergistic effects, bringing about insights into crucial brain processes and structures.

Most reported EEG-based BMI systems can be categorized into one of three paradigms: motor imagery (MI), event-related potential (ERP), and steady-state visually evoked potential (SSVEP). We adopted MI, although successful cases of other paradigms have been proposed, such as P300 ERP [42], SSVEP [43], spatial attention [44], selective attention [45], mental arithmetic [46], action observation [47], late positive potential (LPP) [48], etc. With no need for external stimuli, motor imagery tasks are self-paced, simple, and stable. Our results validate its utility in EEG-based BMI.

The "follow-up" pattern we proposed is based on beta rebound. The mid-centrally located beta rebounds reveal electrophysiological correlates of synchronized "resetting" from overlapping brain networks. The occurrence of beta rebound depends heavily on the types of MI. Our study found that motors with more fingers involved can lead to better results. It can probably be explained by the superposition effect of MI, i.e., the neural activities triggered by hand MI can be interpreted as the summation of the activities invoked by simple finger MIs, which is validated in [49]. The variation is not limited to upper limbs. According to [19], most subjects displayed beta ERS during foot MI, while tongue MI induced no beta rebound in any subject. Luckily, even if there is only a slight laterality difference in the subject, improved BMI control accuracy can be achieved through visual feedback [50].

It is common practice to extract features based on statistical properties, wavelet coefficients, and average power [16]. In this work, we compared the features generated by these above-mentioned principles and SampEn and CSP. Our results show that power features and SampEn play a dominant role in classification tasks. Other innovative methods were proposed for extraction to solve MI classification tasks. Functional brain networks are being widely applied to extract extra features, delineating the interactions between each pair of electrodes [51].

Despite its popularity, SVM is not the only classifier model that can succeed in MI-EEG classification tasks. To evaluate their performance in EEG-based MI BMI, a comparative analysis of five classifiers, SVM, k-NN, naïve Bayes, decision tree, and logistic regression, was conducted in [52] and it concluded that SVM, logistic regression, and naïve Bayes outperformed the others in accuracy. Recently, with automatic end-to-end learning, deep learning (DL) is competent in this context, simplifying processing pipelines; hence, improved performance can be achieved [53].

Instead of feeding large quantities of feature vectors directly into classifiers, a three-feature selection method SLR was applied to lower the dimensionality of features in this work, with the intention of improving classification accuracy. Gu et al. applied a similar method in foot MI classification tasks, with the most remarkable accuracy of 75.33% achieved by SLR-variational approximation (SLR-VAR) [18]. Rejer et al. compared different methods of feature selection on the left- and right-hand MI [54]. Feature selection may also help discover new patterns of brain behavior and invent new explanations for neural pathways. μ-rhythm was suggested to reflect the translation of hearing an instruction into performing the required action, which is well in line with the feature selection results [55].

It is important to note that only a small portion of channels was used in subsequent analysis—to be specific, a single channel (Cz) in ME versus MI classification and three channels (Cz, C3, and C4) in left- versus right-hand MI classification. These channels have been proven to induce better classification results [24,27]. In previous EEG-based MI BMI, a large portion used many EEG channels. Thirty-two EEG electrodes are used in [56] and they achieve a classification accuracy of 59.65%. We successfully classified ME versus MI by 79.51% with one channel (Cz) and left- versus right-hand MI by 75.22% with three channels (Cz, C3, and C4). In general, we achieved better performance with less data, which can alleviate the computation load and reduce experiment preparation time.

Most classifiers in EEG-based BMI studies are trained in a subject-specific manner, which can decode intention from a specific patient based on his own signal features [18,57]. This manner demands laborious training for subjects and repetitive signal processing to ensure solid results. Moreover, it is also infeasible for physically disabled patients to provide these training data. Here, we trained classifiers with population-level features obtained from different subjects and gained competitive performance. It demonstrates excellent potential for simplified application, since real-time EEG signals can be acquired from patients without training and compared with the existing training sets.

Despite the group-level classification, our accuracy is still comparable. A possible merit lies in our carefully selected features for training SVM. In classification between ME and MI, we selected power-based features manually, which improved the accuracy minutely but reduced the feature dimension greatly. In classification between left- and right-hand MI, we adopted SLR to abrogate redundant feature vectors, so the corresponding accuracy increased by 13.22%. Admittedly, we did not reach perfect accuracy, but this appears reasonable given that untrained subjects can be unaffected by BMI protocols. The term "BMI illiteracy" was coined for this non-negligible portion of users, which is estimated at 15% to 30% [58]. The BMI illiteracy rate matches our classification results.

Our research explored the feasibility of EEG for evaluating post-stroke recovery. Previous work cross-validates the efficacy of EEG signals with other assessments, such as motor functions and activities of daily living (ADL), Fugl–Meyer assessment (FMA) scores, and the modified Ashworth scale (MAS). Based upon this fact and our results, we further propose a prospective for EEG assessment, wherein therapists record EEG signals from patients during rehabilitative MI tasks, then label them with classifiers trained by group level training sets and provide real-time feedback to make patients aware of the similarity between their neural activities and the correct ones.

Admittedly, the current study embodies some limitations. The experimental procedure can be modified, allowing subjects to repeat MI within specific time slots. Such modification will not only facilitate analysis but also induce more detectable signals. Moreover, our analysis was based on sensor-level techniques, while the volume conduction effect calls for source-level analysis, which would map EEG signals to cortical areas. The "follow-up" pattern we generalized still improves classification accuracy, so more detailed characteristics can be drawn from it.

5. Conclusions

EEG-based MI BMI has great potential in evaluating post-stroke rehabilitation. However, present assessments suffer from low efficiency and lack of objectiveness, and few related studies underline the difference between ME and MI. In this work, we proposed a dataset and corresponding analysis methods to classify both ME versus MI and left- versus right-hand MI tasks, which can induce plasticity during restoration. This study put forward a stable and straightforward detector of ME and MI based on beta rebound, investigated extracted feature vectors, and applied SVM with SLR to classification. The conclusions are summarized as follows:

"Follow-up" pattern based on the beta rebound is a stable indicator of ME and MI. Compared with ME, the beta rebounds of MI have smaller amplitude, appear later after stimulus, and last longer. The phenomenon is most significant at channel Cz. Such characteristics defined the "follow-up" pattern. Its occurrence is 59.77% $\pm$ 11.95% among all subjects, and motors with more fingers involved can generate better results (finger-crossing: 89.47%).

The ME versus MI classification accuracy is 79.51% with power-based features and SVM. We extracted 13 features with statistic, wavelet-based, and power-based methods. SVM generated a classification accuracy of 78.57% with these feature vectors. After examining the support vectors, features fed into SVM were pruned back to four power-based ones, while the accuracy increased.

The left- versus right-hand MI classification accuracy is 75.22% with SVM and L1-SLR-LAP. We extracted 59 features with statistic, wavelet-based, power-based, SampEn, and CSP methods. We compared the performance of different derivatives of SLR and found out that L1-SLR-LAP win over others with 42 feature vectors left. We concluded that power-based features and SampEn displayed distinctive weights in the remaining vectors.

Therefore, this work demonstrates an innovative approach that can be used for evaluating the rehabilitation results of MI BMI with neurofeedback. In future work, we will focus on the back-end design of the system and explore the addition of NIBS as an adjunct therapy. BMI+NIBS interventions could inform patients and therapists about real-time MI performance and enhance rehabilitation with additional clinical gains.

Author Contributions: Conceptualization, J.W. and Y.-H.C.; methodology, J.W. and Y.-H.C.; software, J.W.; validation, J.W., Y.-H.C. and J.Y.; formal analysis, J.W.; investigation, J.W. and Y.-H.C.; resources, Y.-H.C. and M.S.; data curation, J.W.; writing—original draft preparation, J.W.; writing—review and editing, Y.-H.C., J.Y. and M.S.; visualization, J.W. and Y.-H.C.; supervision, M.S.; project administration, Y.-H.C.; funding acquisition, M.S. All authors have read and agreed to the published version of the manuscript.

Funding: This work was funded by Westlake University [041030080118] and Zhejiang Key R&D Program from Science and Technology Department Zhejiang Province [2021C03002].

Institutional Review Board Statement: The study was conducted in accordance with the Declaration of Helsinki and approved by the Ethical Committee of Westlake University, Hangzhou, China (approval ID: 20191023swan001).

Informed Consent Statement: Informed consent was obtained from all subjects involved in the study.

Data Availability Statement: The data presented in this study are available from the corresponding authors, Y.-H.C. and M.S., upon request. The data are not publicly available because they contain information that could compromise the privacy of research participants.

Acknowledgments: The authors acknowledge the support from Westlake University, Hangzhou, from Zhejiang Key R&D Program No. 2021C03002, and from the Key Laboratory of Child Development and Learning Science of the Ministry of Education, School of Biological Science and Medical Engineering, Southeast University, Nanjing.

Conflicts of Interest: The authors declare that the research was conducted in the absence of any commercial or financial relationships that could be construed as potential conflicts of interest.

References

1. López, N.D.; Pereira, E.M.; Centeno, E.J.; Page, J.C.M. Motor imagery as a complementary technique for functional recovery after stroke: A systematic review. *Top. Stroke Rehabil.* **2019**, *26*, 576–587. [CrossRef] [PubMed]
2. Carrasco, D.G.; Cantalapiedra, J.A. Effectiveness of motor imagery or mental practice in functional recovery after stroke: A systematic review. *Neurología* **2016**, *31*, 43–52. [CrossRef]
3. Chen, M.; Lin, C.-H. What is in your hand influences your purchase intention: Effect of motor fluency on motor simulation. *Curr. Psychol.* **2021**, *40*, 3226–3234. [CrossRef]
4. Deng, X.; Yu, Z.L.; Lin, C.; Gu, Z.; Li, Y. A bayesian shared control approach for wheelchair robot with brain machine interface. *IEEE Trans. Neural Syst. Rehabil. Eng.* **2019**, *28*, 328–338. [CrossRef]
5. López-Larraz, E.; Sarasola-Sanz, A.; Irastorza-Landa, N.; Birbaumer, N.; Ramos-Murguialday, A. Brain-machine interfaces for rehabilitation in stroke: A review. *NeuroRehabilitation* **2018**, *43*, 77–97. [CrossRef]
6. Cincotti, F.; Pichiorri, F.; Aricò, P.; Aloise, F.; Leotta, F.; Fallani, F.D.V.; Millán, J.D.R.; Molinari, M.; Mattia, D. EEG-based Brain-Computer Interface to support post-stroke motor rehabilitation of the upper limb. In Proceedings of the Annual International Conference of the IEEE Engineering in Medicine and Biology Society, San Diego, CA, USA, 28 August–1 September 2012; pp. 4112–4115.
7. Ang, K.K.; Guan, C.T.; Chua, K.S.G.; Ang, B.T.; Kuah, C.W.K.; Wang, C.C.; Phua, K.S.; Chin, Z.Y.; Zhang, H.H. A Large Clinical Study on the Ability of Stroke Patients to Use an EEG-Based Motor Imagery Brain-Computer Interface. *Clin. Eeg Neurosci.* **2011**, *42*, 253–258. [CrossRef]
8. Han, C.-H.; Müller, K.-R.; Hwang, H.-J. Brain-switches for asynchronous brain-computer interfaces: A systematic review. *Electronics* **2020**, *9*, 422. [CrossRef]
9. Leeb, R.; Friedman, D.; Müller-Putz, G.R.; Scherer, R.; Slater, M.; Pfurtscheller, G. Self-paced (asynchronous) BCI control of a wheelchair in virtual environments: A case study with a tetraplegic. *Comput. Intell. Neurosci.* **2007**, *2007*, 79642. [CrossRef]
10. Müller-Putz, G.R.; Kaiser, V.; Solis-Escalante, T.; Pfurtscheller, G. Fast set-up asynchronous brain-switch based on detection of foot motor imagery in 1-channel EEG. *Med. Biol. Eng. Comput.* **2010**, *48*, 229–233. [CrossRef]
11. Lee, S.-B.; Kim, H.-J.; Kim, H.; Jeong, J.-H.; Lee, S.-W.; Kim, D.-J. Comparative analysis of features extracted from EEG spatial, spectral and temporal domains for binary and multiclass motor imagery classification. *Inf. Sci.* **2019**, *502*, 190–200. [CrossRef]
12. Yang, H.; Sakhavi, S.; Ang, K.K.; Guan, C. On the use of convolutional neural networks and augmented CSP features for multi-class motor imagery of EEG signals classification. In Proceedings of the 37th Annual International Conference of the IEEE Engineering in Medicine and Biology Society (EMBC), Milan, Italy, 25–29 August 2015; pp. 2620–2623.
13. Park, S.H.; Lee, D.; Lee, S.G. Filter Bank Regularized Common Spatial Pattern Ensemble for Small Sample Motor Imagery Classification. *IEEE Trans. Neural Syst. Rehabil.* **2018**, *26*, 498–505. [CrossRef] [PubMed]
14. Park, Y.; Chung, W. Frequency-Optimized Local Region Common Spatial Pattern Approach for Motor Imagery Classification. *IEEE Trans. Neural Syst. Rehabil.* **2019**, *27*, 1378–1388. [CrossRef] [PubMed]

15. Wang, L.; Xu, G.; Yang, S.; Wang, J.; Guo, M.; Yan, W. Motor imagery BCI research based on sample entropy and SVM. In Proceedings of the Sixth International Conference on Electromagnetic Field Problems and Applications, Dalian, China, 19–21 June 2012; pp. 1–4.

16. Chatterjee, R.; Bandyopadhyay, T. EEG based Motor Imagery Classification using SVM and MLP. In Proceedings of the 2nd international conference on Computational Intelligence and Networks (CINE), Bhubaneswar, India, 11 January 2016; pp. 84–89.

17. Padfield, N.; Zabalza, J.; Zhao, H.; Masero, V.; Ren, J. EEG-based brain-computer interfaces using motor-imagery: Techniques and challenges. *Sensors* **2019**, *19*, 1423. [CrossRef] [PubMed]

18. Gu, L.; Yu, Z.; Ma, T.; Wang, H.; Li, Z.; Fan, H. EEG-based Classification of Lower Limb Motor Imagery with Brain Network Analysis. *Neuroscience* **2020**, *436*, 93–109. [CrossRef]

19. Pfurtscheller, G.; Neuper, C.; Brunner, C.; Silva, F.L.D. Beta rebound after different types of motor imagery in man. *Neurosci. Lett.* **2005**, *378*, 156–159. [CrossRef]

20. Chen, C.; Zhang, J.; Belkacem, A.N.; Zhang, S.; Xu, R.; Hao, B.; Gao, Q.; Shin, D.; Wang, C.; Ming, D. G-causality brain connectivity differences of finger movements between motor execution and motor imagery. *J. Healthc. Eng.* **2019**, *2019*, 5068283. [CrossRef]

21. Kim, Y.K.; Park, E.; Lee, A.; Im, C.-H.; Kim, Y.-H. Changes in network connectivity during motor imagery and execution. *PLoS ONE* **2018**, *13*, e0190715. [CrossRef]

22. Miller, K.J.; Schalk, G.; Fetz, E.E.; Nijs, M.D.; Ojemann, J.G.; Rao, R.P. Cortical activity during motor execution, motor imagery, and imagery-based online feedback. *Proc. Natl. Acad. Sci. USA* **2010**, *107*, 4430–4435. [CrossRef]

23. Dai, M.; Zheng, D.; Liu, S.; Zhang, P. Transfer kernel common spatial patterns for motor imagery brain-computer interface classification. *Comput. Math. Methods Med.* **2018**, *2018*, 9871603. [CrossRef]

24. Ge, S.; Yang, Q.; Wang, R.; Lin, P.; Gao, J.; Leng, Y.; Yang, Y.; Wang, H. A brain-computer interface based on a few-channel EEG-fNIRS bimodal system. *IEEE Access* **2017**, *5*, 208–218. [CrossRef]

25. Herman, P.; Prasad, G.; McGinnity, T.M.; Coyle, D. Comparative analysis of spectral approaches to feature extraction for EEG-based motor imagery classification. *IEEE Trans. Neural Syst. Rehabil. Eng.* **2008**, *16*, 317–326. [CrossRef] [PubMed]

26. O'Brien, A.; Bertolucci, F.; Torrealba-Acosta, G.; Huerta, R.; Fregni, F.; Thibaut, A. Non-invasive brain stimulation for fine motor improvement after stroke: A meta-analysis. *Eur. J. Neurol.* **2018**, *25*, 1017–1026. [CrossRef] [PubMed]

27. Erdoğan, S.B.; Özsarfati, E.; Dilek, B.; Kadak, K.S.; Hanoğlu, L.; Akın, A. Classification of motor imagery and execution signals with population-level feature sets: Implications for probe design in fNIRS based BCI. *J. Neural Eng.* **2019**, *16*, 026029. [CrossRef] [PubMed]

28. Pfurtscheller, G.; Silva, F.H.L.D. Event-related EEG/MEG synchronization and desynchronization: Basic principles. *Clin. Neurophysiol.* **1999**, *110*, 1842–1857. [CrossRef]

29. Delgado-Bonal, A.; Marshak, A. Approximate entropy and sample entropy: A comprehensive tutorial. *Entropy* **2019**, *21*, 541. [CrossRef] [PubMed]

30. Gubert, P.H.; Costa, M.H.; Silva, C.D.; Trofino-Neto, A. The performance impact of data augmentation in CSP-based motor-imagery systems for BCI applications. *Biomed. Signal Processing Control.* **2020**, *62*, 102152. [CrossRef]

31. Chatterjee, R.; Bandyopadhyay, T.; Sanyal, D.K.; Guha, D. Comparative analysis of feature extraction techniques in motor imagery EEG signal classification. In Proceedings of the First International Conference on Smart System, Innovations and Computing, Jaipur, India, 15–16 April 2017; pp. 73–83.

32. Paul, J.K.; Iype, T.; Dileep, R.; Hagiwara, Y.; Koh, J.W.; Acharya, U.R. Characterization of fibromyalgia using sleep EEG signals with nonlinear dynamical features. *Comput. Biol. Med.* **2019**, *111*, 103331. [CrossRef]

33. Espenhahn, S.; Rossiter, H.E.; Wijk, B.C.V.; Redman, N.; Rondina, J.M.; Diedrichsen, J.; Ward, N.S. Sensorimotor cortex beta oscillations reflect motor skill learning ability after stroke. *Brain Commun.* **2020**, *2*, fcaa161. [CrossRef]

34. Malan, N.S.; Sharma, S. Feature selection using regularized neighbourhood component analysis to enhance the classification performance of motor imagery signals. *Comput. Biol. Med.* **2019**, *107*, 118–126. [CrossRef]

35. Tang, Z.; Li, C.; Sun, S. Single-trial EEG classification of motor imagery using deep convolutional neural networks. *Optik* **2017**, *130*, 11–18. [CrossRef]

36. Voinas, A.E.; Das, R.; Khan, M.A.; Brunner, I.; Puthusserypady, S. Motor Imagery EEG Signal Classification for Stroke Survivors Rehabilitation. In Proceedings of the 10th International Winter Conference on Brain-Computer Interface (BCI), Gangwon-do, Korea, 21–23 February 2022; pp. 1–5.

37. Ge, S.; Liu, H.; Lin, P.; Gao, J.; Xiao, C.; Li, Z. Neural basis of action observation and understanding from first-and third-person perspectives: An fMRI study. *Front. Behav. Neurosci.* **2018**, *12*, 283. [CrossRef] [PubMed]

38. Hong, K.-S.; Ghafoor, U.; Khan, M.J. Brain-machine interfaces using functional near-infrared spectroscopy: A review. *Artif. Life Robot.* **2020**, *25*, 204–218. [CrossRef]

39. Fukuma, R.; Yanagisawa, T.; Yokoi, H.; Hirata, M.; Yoshimine, T.; Saitoh, Y.; Kamitani, Y.; Kishima, H. Training in use of brain–machine Interface-controlled robotic hand improves accuracy decoding two types of hand movements. *Front. Neurosci.* **2018**, *12*, 478. [CrossRef] [PubMed]

40. Wu, Y.; Jiang, D.; Liu, X.; Bayford, R.; Demosthenous, A. A human–machine interface using electrical impedance tomography for hand prosthesis control. *IEEE Trans. Biomed. Circuits Syst.* **2018**, *12*, 1322–1333. [CrossRef] [PubMed]

41. Uchitel, J.; Vidal-Rosas, E.E.; Cooper, R.J.; Zhao, H. Wearable, Integrated EEG–fNIRS Technologies: A Review. *Sensors* **2021**, *21*, 6106. [CrossRef]

42. Jin, J.; Li, S.; Daly, I.; Miao, Y.; Liu, C.; Wang, X.; Cichocki, A. The study of generic model set for reducing calibration time in P300-based brain–computer interface. *IEEE Trans. Neural Syst. Rehabil. Eng.* **2019**, *28*, 3–12. [CrossRef]

43. Ravi, A.; Beni, N.H.; Manuel, J.; Jiang, N. Comparing user-dependent and user-independent training of CNN for SSVEP BCI. *J. Neural Eng.* **2020**, *17*, 026028. [CrossRef]

44. Nazari, M.R.; Nasrabadi, A.M.; Daliri, M.R. Single-trial decoding of motion direction during visual attention from local field potential signals. *IEEE Access* **2021**, *9*, 66450–66461. [CrossRef]

45. Yao, L.; Sheng, X.; Mrachacz-Kersting, N.; Zhu, X.; Farina, D.; Jiang, N. Performance of brain–computer interfacing based on tactile selective sensation and motor imagery. *IEEE Trans. Neural Syst. Rehabil. Eng.* **2017**, *26*, 60–68. [CrossRef]

46. Chin, Z.Y.; Zhang, X.; Wang, C.; Ang, K.K. EEG-based discrimination of different cognitive workload levels from mental arithmetic. In Proceedings of the 40th Annual International Conference of the IEEE Engineering in Medicine and Biology Society (EMBC), Honlulu, HI, USA, 17–21 July 2018; pp. 1984–1987.

47. Ge, S.; Wang, P.; Liu, H.; Lin, P.; Gao, J.; Wang, R.; Iramina, K.; Zhang, Q.; Zheng, W. Neural activity and decoding of action observation using combined EEG and fNIRS measurement. *Front. Hum. Neurosci.* **2019**, *13*, 357. [CrossRef]

48. Leng, Y.; Zhu, Y.; Ge, S.; Qian, X.; Zhang, J. Neural temporal dynamics of social exclusion elicited by averted gaze: An event-related potentials study. *Front. Behav. Neurosci.* **2018**, *12*, 21. [CrossRef] [PubMed]

49. Lindig-León, C.; Bougrain, L. Comparison of sensorimotor rhythms in EEG signals during simple and combined motor imageries over the contra and ipsilateral hemispheres. In Proceedings of the 37th Annual International Conference of the IEEE Engineering in Medicine and Biology Society (EMBC), Milan, Italy, 25–29 August 2015; pp. 3953–3956.

50. Hashimoto, Y.; Ushiba, J.; Kimura, A.; Liu, M.; Tomita, Y. Change in brain activity through virtual reality-based brain-machine communication in a chronic tetraplegic subject with muscular dystrophy. *BMC Neurosci.* **2010**, *11*, 1–9. [CrossRef] [PubMed]

51. Ai, Q.; Chen, A.; Chen, K.; Liu, Q.; Zhou, T.; Xin, S.; Ji, Z. Feature extraction of four-class motor imagery EEG signals based on functional brain network. *J. Neural Eng.* **2019**, *16*, 026032. [CrossRef] [PubMed]

52. Isa, N.M.; Amir, A.; Ilyas, M.; Razalli, M. Motor imagery classification in Brain computer interface (BCI) based on EEG signal by using machine learning technique. *Bull. Electr. Eng. Inform.* **2019**, *8*, 269–275. [CrossRef]

53. Roy, Y.; Banville, H.; Albuquerque, I.; Gramfort, A.; Falk, T.H.; Faubert, J. Deep learning-based electroencephalography analysis: A systematic review. *J. Neural Eng.* **2019**, *16*, 051001. [CrossRef]

54. Rejer, I. EEG feature selection for BCI based on motor imaginary task. *Found. Comput. Decis. Sci.* **2012**, *37*, 283. [CrossRef]

55. Pineda, J.A. The functional significance of mu rhythms: Translating "seeing" and "hearing" into "doing". *Brain Res. Rev.* **2005**, *50*, 57–68. [CrossRef]

56. Yu, J.; Ang, K.K.; Guan, C.; Wang, C. A multimodal fNIRS and EEG-based BCI study on motor imagery and passive movement. In Proceedings of the 6th International IEEE/EMBS Conference on Neural Engineering (NER), San Diego, CA, USA, 6–8 November 2013; pp. 5–8.

57. Selim, S.; Tantawi, M.M.; Shedeed, H.A.; Badr, A. A CSP\AM-BA-SVM Approach for Motor Imagery BCI System. *IEEE Access* **2018**, *6*, 49192–49208. [CrossRef]

58. Vidaurre, C.; Blankertz, B. Towards a cure for BCI illiteracy. *Brain Topogr.* **2010**, *23*, 194–198. [CrossRef]

 *biosensors*

Article

Computer-Aided Detection of Fiducial Points in Seismocardiography through Dynamic Time Warping

Chien-Hung Chen [1], Wen-Yen Lin [2,3,*] and Ming-Yih Lee [1,3]

[1] Graduate Institute of Biomedical Engineering, Chang Gung University, Taoyuan 33302, Taiwan; d0528005@cgu.edu.tw (C.-H.C.); leemiy@mail.cgu.edu.tw (M.-Y.L.)
[2] Center for Biomedical Engineering, Department of Electrical Engineering, Chang Gung University, Taoyuan 33302, Taiwan
[3] Division of Cardiology, Department of Internal Medicine, Chang Gung Memorial Hospital, Taoyuan 33302, Taiwan
* Correspondence: wylin@mail.cgu.edu.tw; Tel.: +886-3-2118-800 (ext. 3675)

Abstract: Accelerometer-based devices have been employed in seismocardiography fiducial point detection with the aid of quasi-synchronous alignment between echocardiography images and seismocardiogram signals. However, signal misalignments have been observed, due to the heartbeat cycle length variation. This paper not only analyzes the misalignments and detection errors but also proposes to mitigate the issues by introducing reference signals and adynamic time warping (DTW) algorithm. Two diagnostic parameters, the ratio of pre-ejection period to left ventricular ejection time (PEP/LVET) and the Tei index, were examined with two statistical verification approaches: (1) the coefficient of determination (R^2) of the parameters versus the left ventricular ejection fraction (LVEF) assessments, and (2) the receiver operating characteristic (ROC) classification to distinguish the heart failure patients with reduced ejection fraction (HFrEF). Favorable R^2 values were obtained, $R^2 = 0.768$ for PEP/LVET versus LVEF and $R^2 = 0.86$ for Tei index versus LVEF. The areas under the ROC curve indicate the parameters that are good predictors to identify HFrEF patients, with an accuracy of more than 92%. The proof-of-concept experiments exhibited the effectiveness of the DTW-based quasi-synchronous alignment in seismocardiography fiducial point detection. The proposed approach may enable the standardization of the fiducial point detection and the signal template generation. Meanwhile, the program-generated annotation data may serve as the labeled training set for the supervised machine learning.

Keywords: cardiac time interval; dynamic time warping; fiducial point detection; heart failure; seismocardiography

Citation: Chen, C.-H.; Lin, W.-Y.; Lee, M.-Y. Computer-Aided Detection of Fiducial Points in Seismocardiography through Dynamic Time Warping. *Biosensors* **2022**, *12*, 374. https://doi.org/10.3390/bios12060374

Received: 23 April 2022
Accepted: 23 May 2022
Published: 30 May 2022

1. Introduction

The emergence of wearable solutions for health monitoring provides opportunities for remote medical surveillance. Biosensing and cloud computing technologies enable physiological parameters to be tracked unobtrusively, accurately, and in real time in everyday life. In most cases, relatively simple, reproducible, and reliable generic physiological parameters are monitored. Cardiac time intervals (CTIs), durations between specific cardiac events, are closely related to cardiac physiology and function. CTIs play a pivotal role in the diagnostic and prognostic assessments of patients with hemodynamic and valve dysfunction, especially with regard to risk stratification [1,2].

In clinical practice, CTIs associated with valve opening and closure, including the pre-ejection period (PEP), left ventricular ejection time (LVET), isovolumic contraction time (IVCT), and isovolumic relaxation time (IVRT), are employed in the calculation of the myocardial health index. These CTIs describe the time periods between the specific cardiac events of mitral valve opening (MO), mitral valve closure (MC), aortic valve opening (AO), aortic valve closure (AC), and Q wave in electrocardiography (ECG). Four frequently used CTIs are defined by the timing differences between cardiac events as follows:

$$PEP = AO - Q, \tag{1}$$

$$LVET = AO - AC, \tag{2}$$

$$IVCT = AO - MC, \tag{3}$$

$$IVRT = MO - AC. \tag{4}$$

Although some of the CTIs are modulated by heart rate, respiration, or even the individual's posture and position [3,4], the cardiac indexes derived from CTIs are considered reliable parameters for predicting and assessing myocardial contractility [2,3,5]. Two frequently used cardiac indexes are the PEP/LVET ratio (also known as the contractility coefficient) [6,7] and the Tei index (also known as the myocardial performance index) [5]. They are defined as follows:

$$\text{Contractility Coefficient} = PEP/LVET, \tag{5}$$

$$\text{Tei index} = (IVCT + IVRT)/LVET. \tag{6}$$

Both the PEP/LVET ratio and the Tei index have been clinically confirmed to be heart rate-independent and negatively correlated with the left ventricular ejection fraction (LVEF) on a beat-to-beat basis [2,3,5]. Carvalho observed that individuals with normal cardiac function (higher LVEF) exhibited a short PEP and a long ejection time. By contrast, patients (lower LVEF) with reduced stroke volume (SV) and cardiac output (CO) had a longer PEP and shorter LVET [8]. Thus, CTIs and the two cardiac indexes mentioned above can supplement the LVEF and serve as proxies in the assessment of myocardial contractility [9–11].

Echocardiography is considered the gold standard for obtaining CTIs [8]. CTIs are typically acquired using ultrasound modalities such as color Doppler flow imaging, tissue Doppler imaging, or M-mode echocardiography [8,12–14]. From echocardiography images, physicians can accurately determine the timings of the start, peak, and end of specific hemodynamic or cardio-mechanical events (e.g., blood ejection, myocardial motion, and valve opening or closure) in various phases of the cardiac cycle. However, acquiring echocardiograms is time consuming and requires the expertise of a well-trained sonographer. Therefore, for long-term, home-based cardiac monitoring, the use of comparatively simple, noninvasive wearable devices to conduct hemodynamic assessment, such as through impedance cardiography, phonocardiography, and seismocardiography (SCG), is preferred [7,10,12].

SCG, a noninvasive approach for the diagnosis of cardiac conditions, is capable of evaluating CTIs through chest wall vibration analysis. Temporal information of cardiac events can be obtained by identifying the specific fiducial points in the SCG signals corresponding to the events. Single-channel and multichannel SCG monitoring systems and SCG–echocardiography hybrid apparatuses have been proposed [7,15,16]. In 1994, Crow examined trimodal screenshots of simultaneous ECG and SCG signals and echocardiography images to investigate the relationship between SCG and echocardiographic images regarding cardiac events [16]. In the trimodal measurements, the SCG signal was routed to the auxiliary input port of the ultrasound machine and was presented synchronously on top of the echocardiogram together with the ECG signal. This enabled the SCG fiducial point detection and the CTI analysis with the same heartbeat cycle on the same ultrasound image.

Home-based cardiac monitoring requires special analytical methods because no sonographers or ultrasound instruments are available. Heterogeneous modality cooperation may serve as the alternative; that is, conducting the diagnostic assessment using ECG and echocardiogram (also assisting SCG fiducial point identification) while home monitoring using ECG and SCG. Lin et al. introduced a quasi-synchronization method for SCG fiducial point detection from the SCG and ECG data collected through the inertial-sensor-based multichannel SCG systems and from several echocardiogram images acquired at different time [7,9]. The ECG signal from SCG measurement and the ECG signal from echocardiogram image were aligned by the manipulation of uniform stretching (or squeezing) and shifting in time axis to align the ECG R peaks to each other. This approach is referred to as the quasi-synchronous alignment (or "conventional" quasi-synchronous alignment) in this paper and is illustrated in Figure 1.

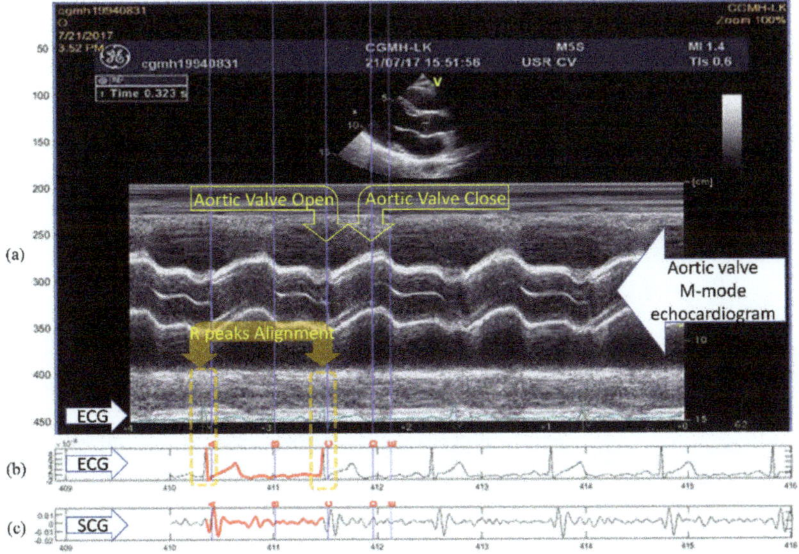

Figure 1. Example of the quasi-synchronous alignment for an echocardiogram image and an SCG measurement: (**a**) M-mode echocardiogram image of the aortic valve (with an ECG signal at the bottom of the image); (**b**) ECG signal simultaneously measured with SCG signal; (**c**) SCG signal.

The development of SCG has been limited by artifact effects, the ambiguity in specific event waveforms and the lack of detection procedures; no standards for fiducial point detection in SCG signals have been established [17]. In this study, the problems of quasi-synchronous alignment were revisited, and an improved SCG fiducial point detection protocol was devised by introducing a reference SCG signal and the dynamic time warping (DTW) algorithm. The new alignment method is referred to as the "DTW-based" quasi-synchronous alignment, so as to distinguish it from the existing quasi-synchronous alignment method (or the "conventional" quasi-synchronous alignment). Compared to the conventional quasi-synchronous alignment, DTW-based quasi-synchronous alignment eliminates the fiducial point detection error due to the stretching (or squeezing) manipulation. Compared to the envelope-based detection methods [14,18–20], DTW-based quasi-synchronous alignment does not limit to any specific fiducial points, as long as the corresponding cardiac events could be identified in an clinical imaging modality (not limited to echocardiogram, that was used in this paper) or cardiac signal that is simultaneously measured with ECG. This method has the potential to untie the knot that limits the development of SCG.

As the deterministic approach in pattern recognition technology, by which machine learning was enabled in many applications [21,22], DTW is known for its capability to align the morphological patterns of two given time series (signals), its flexibility in handling signals of varying length, and its feasibility of implementation through computer programs [23]. DTW has been applied to speech recognition, time series clustering, and protein sequence alignment [24–26]. Herein, DTW was employed to mitigate the waveform variations in SCG signal analysis. In Azad's study, the DTW distance, instead of the Euclidean distance, significantly reduced the morphological variability in clustered SCG signals corresponding to groups of participants in various breathing stages [27]. Based on DTW alignment, an investigation on subject-oriented template generation succeeded to establish the procedures for signal clustering (by using the k-means algorithm) and averaging according to SCG morphological features for the first time [28].

The remainder of this paper is organized as follows. In Section 2, the data acquisition setup, data preprocessing procedure, and the reason for introducing the reference SCG segment are declared. The experimental results are provided in Section 3. The discussions and conclusions are presented in Section 4.

2. Materials and Methods

The study protocol of this work was approved by the Institutional Review Board of Chang Gung Memorial Hospital, Taoyuan, Taiwan (approval number 202100744B0A3). The physiological data (112 echocardiography images and SCG signal clips of 3876 heartbeats) of 56 individuals (30 men, 26 women) collected in the previous study by Lin [9] on a multichannel SCG spectrum system were employed in this study. The data for each individual comprised the diagnostic history, LVEF assessment, echocardiography images, ECG signals, and a set of four-channel SCG signals. In this study, only the first channel SCG data was analyzed, and the signal was recorded from the location of the fifth-left intercostal space in the midclavicular line of the mitral valve with the participant in a supine position. The echocardiogram images captured the parasternal long axis (PLAX) in the M-mode and the Doppler flow images of the mitral and aortic valves, providing the timings of valve opening and closure. Participants were instructed not to exercise before the tests. The raw SCG and ECG data underwent filtering, detrending, wavelet noise reduction, cardiac cycle identification, ECG wave annotation, and cycle segmentation [28].

The following sections introduce concepts and problems of the conventional quasi-synchronous alignment, as well as concepts and advantages of DTW-based quasi-synchronous alignment. Furthermore, a programming flowchart is presented and performed to validate the annotation procedures and the effectiveness of SCG fiducial point detection.

2.1. Conventional Quasi-Synchronous Alignment for Echocardiogram Image and SCG Signal

Although beat-by-beat synchronicity can be achieved in trimodal (echocardiography/ECG/SCG) measurements, the ultrasound probe may interfere with SCG sensors on the chest during such simultaneous measurements. Through simple shifting and rescaling (stretching or squeezing) manipulations on the record signals, the "asynchronously" measured echocardiogram images and SCG signals could be analyzed in the same graph to establish quasi-synchronization [9]. This method, which is as effective as trimodal measurement, caters to the increasing demand for the home-based monitoring for patients with cardiovascular disease and heart failure.

An example of the conventional quasi-synchronous alignment is displayed in Figure 1. The figure consists of an M-mode image of aortic valve motion (with synchronous ECG signal shown overlapped in the lower part of the image) (Figure 1a), an ECG (Figure 1b) and SCG (Figure 1c) signal pair measured at the same time. To align the echocardiogram image and the SCG signal quasi-synchronously, R peaks from the two ECG signals were employed as the beacon targets. The ECG signal in Figure 1b was rescaled and shifted to align the R peaks to the ECG R peaks of the echocardiogram image (in Figure 1a) as indicated in the orange boxes. The vertical blue lines in Figure 1 link the timing information between the echocardiogram image and SCG signals. As the events visually identified in the echocardiogram image, the timing positions of the fiducial points in SCG signal could be easily obtained. Therefore, the fiducial points of the SCG signals could be detected. This technique had also been applied to color Doppler echocardiogram and tissue Doppler echocardiogram images to identify six new fiducial points [9].

2.2. Misalignmenst and Detection Errors in SCG Fiducial Point Detection under Conventional Quasi-Synchronous Alignment

Although the conventional quasi-synchronous alignment is a rapid and intuitive approach in which ECG signals from echocardiogram and SCG measurements are used as the intermedium, the rescaling manipulation can be problematic. Waveform distortion due to rescaling may result in unexpected target shifting as the signals were subjected to the conventional quasi-synchronous alignment. For easier graphic illustration and better clarification, two ECG and SCG signal pairs with different heartbeat lengths were used as examples in Figures 2–4, whereas the mostly concerned targets in this study are echocardiogram image and SCG signal. The issue exists in spite of the target change as long as the end-to-end alignment was achieved by the rescaling manipulation.

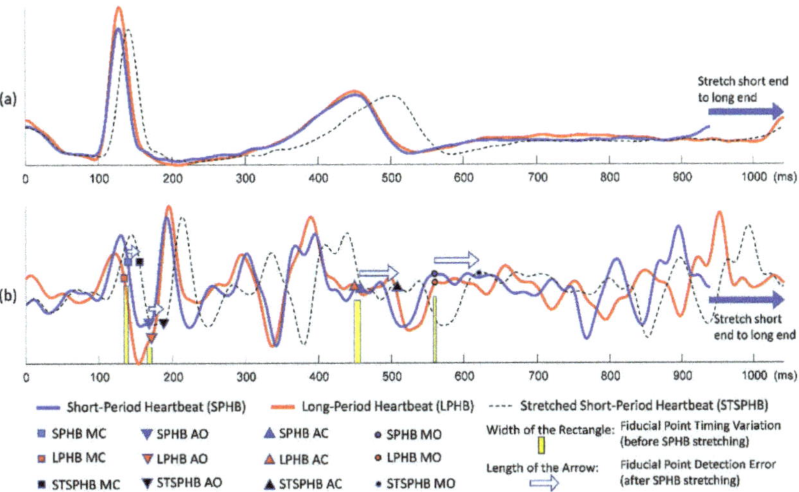

Figure 2. Conventional quasi-synchronous alignment by stretching the short-period signal: (**a**) Long-period (red), short-period (blue), and stretched short-period (black dashed) ECG signals; (**b**) Long-period (red), short-period (blue), and stretched short-period (black dashed) SCG signals with fiducial points and the detection error indicators (white arrows).

Figure 2 demonstrates the conventional quasi-synchronous alignment involving the stretching manipulation of short-period ECG and SCG signals such that they were matched end to end with long-period signals. Figure 3 displays the same alignment approach but performed by squeezing the long-period signals such that they were matched end to end with the short-period signals. The principal problem of the conventional quasi-synchronous alignment is applying a unique rescaling ratio (stretching or squeezing scale ratio) to the entire waveform. The unique rescaling ratio manipulation is suspicious as all the fiducial points (MO, MC, AO, and AC) of the short-period signal (blue line) in Figures 2 and 3 are very close to the corresponding fiducial points in the long-period signal (red line). Over both the short and long period heartbeats, the heart pulsed at a comparable pace in the beginning (before 600 ms). Overall, length and morphology differences were noted in the heartbeat signals; this is attributed to the deviation in the signal length caused by the unequal cardiac paces in the latter portion of the cycle (after 600 ms). The widths of the yellow bars in Figure 2b and b represent the variations in SCG fiducial point timings between the short- and long-period signals before the rescaling manipulation, which are ≤10 ms.

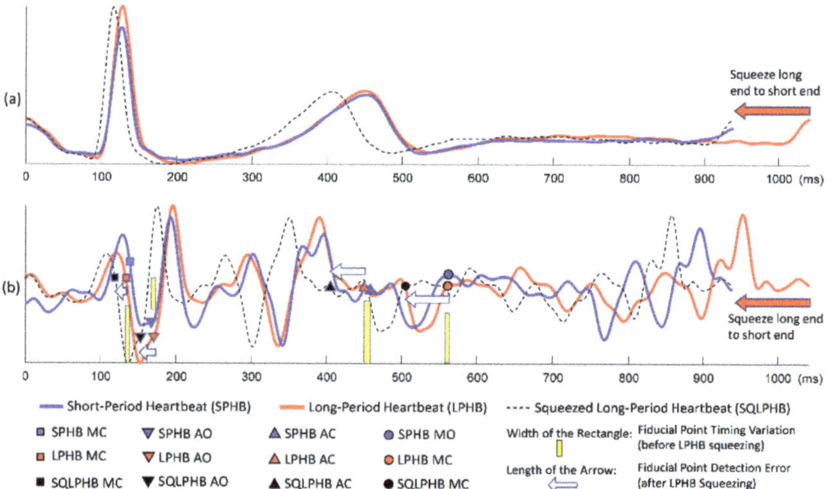

Figure 3. Conventional quasi-synchronous alignment by squeezing the long-period signal: (**a**) Long-period (red), short-period (blue), and squeezed long-period (black dashed) ECG signals; (**b**) Long-period (red), short-period (blue), and squeezed long-period (black dashed) SCG signals with fiducial points and the detection error indicators (white arrows).

Three types of lines (red solid, blue solid, and black dashed) annotated with four types of icons (squares, triangles inverted triangles, and circles) are shown in Figures 2b and 3b. The icons and lines presented in blue and red correspond to the pre-detected fiducial points and SCG signals for the short- and long-period heartbeats, respectively. Figure 2b illustrates the distorted signal (black dashed curve) and timing push off in fiducial points (black icons) ascribable to the uniform stretching process (indicated by the sign of blue rightwards arrow). The amounts of detection error for the four fiducial points are indicated by the lengths of the white arrows. Similar detection errors of the fiducial points are observable in Figure 3b, whereas the long-period signal is squeezed to match the short-period signal. The white arrows are with the lengths (of several tens milliseconds) considerably longer compared to the widths of the yellow bars, and with a trend that the longer arrows appear closer to the right end. The trend reveals that the conventional quasi-synchronous alignment led to different detection errors for different fiducial points. This is owing to the signal morphological distortion as the left end is fixed while applying a unique rescaling ration to the entire signal to make the right end match that of another signal with a different length. Likewise, if the SCG fiducial points detected from the SCG-echocardiogram alignment adopt the conventional quasi-synchronous approach, the same signal distortion and detection errors will occur. That will degrade the CTI assessment and the diagnostic applications of SCG.

2.3. Introduction of Reference Signal in Quasi-Synchronous Alignment

Figure 4 illustrates two ECG and SCG waveforms with the same heart cycle period spontaneously aligning to each other, not only in the former half section but also in the latter half section. This phenomenon (the comparable heart pace and self-alignment in morphology over the entire heart cycle) was leveraged in the proposal of this study (a new quasi-synchronous alignment method) to avoid the signal distortion and to mitigate the misalignments in SCG fiducial point detection. For this reason, a reference SCG signal was introduced as the intermediary to accommodate the fiducial points detected from the alignment to echocardiogram images and also to project the fiducial points to other nonreference SCG signals.

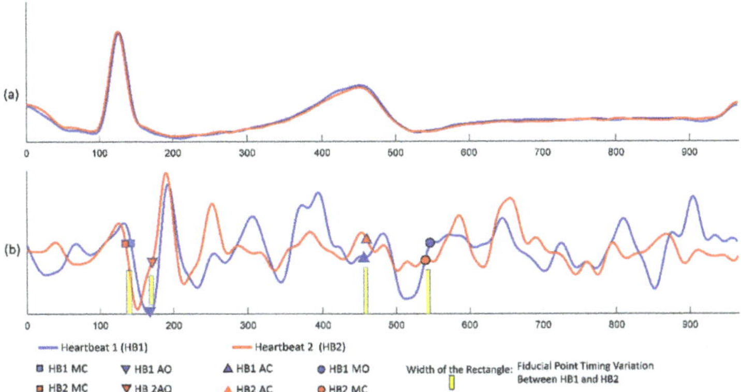

Figure 4. Spontaneous alignment of two ECG and SCG signals with the same heart cycle period: (**a**) ECG signals; (**b**) SCG signals with the detected fiducial points.

To avoid the waveform distortion during the signal alignment, a reference signal pair of ECG and SCG were screened out from the SCG measurements which had the heart cycle period closest to that of the ECG signal in the echocardiogram image. In Figure 5, a reference signal pair example is illustrated, both the selected RR intervals of the ECG signals in echocardiogram (Figure 5a) and the RR intervals in the reference signal pair (Figure 5b) are the same (966 milliseconds). As the RR interval of the reference ECG signal performing end-to-end alignment with the echocardiogram ECG, no rescaling manipulation was needed. In addition, the entire ECG and SCG signals were considered synchronous to the selected cycle in the echocardiogram on the premise of spontaneous alignment discussed in Figure 4. The aortic valve closing event (annotated by "AC" in Figure 5) identified in the echocardiogram, was mapped to the same temporal place in the reference ECG and SCG signals. The reference SCG signal with fiducial points mapped from the echocardiogram was then used as the intermediary template signal for fiducial point projection to other nonreference measurements. The projection process based on the DTW algorithm, which takes signal morphological similarity into consideration. The revised quasi-synchronous alignment method hereinafter referred to as DTW-based quasi-synchronous alignment.

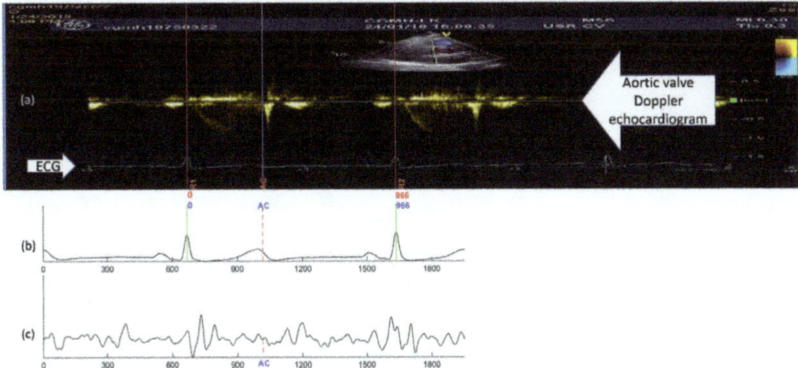

Figure 5. Illustration of aligning a reference signal pair to the selected echocardiogram section and projecting the specific cardiac event (aortic valve closing, AC) from echocardiogram to SCG curve as the detected fiducial point: (**a**) Echocardiogram image and the selected section enclosed within vertical lines marked with R1 and R2; (**b**) The aligned ECG of the reference signal pair; (**c**) The aligned SCG of the reference signal pair.

2.4. DTW

DTW, a dynamic programming base algorithm, has been applied to the speech recognition analysis with the identical voice content but morphologically varied at different speech paces [29]. The flexibility of DTW algorithm in aligning two sequences (not necessarily equal in data length) extends its application from speech recognition to biometrics (e.g., fingerprints), handwriting and many other technological fields [30].

Given two morphologically similar signal sequences, $X = (x_1, x_2,..., x_M)$ of length M ($M \in \mathbb{N}$) and $Y = (y_1, y_2, \dots , y_N)$ of length N ($N \in \mathbb{N}$), the goal of the DTW algorithm is to align X and Y to optimize distance cost function requirements as well as to conform to the warping path constraints of (1) monotonicity, (2) continuity, (3) boundary, (4) warping window, and (5) slope constraints [31,32]. A cost matrix with size of $M \times N$ was constructed to present all possible path points of the dynamic warping. Given a path $W = (w_1, w_2,..., w_K)$ of length K, $max(M,N) \leq K \leq M + N - 1$. Any w_i in W contains two index elements, $w_i = (a_i, b_i)$, with the first and second indexes corresponding to the a_ith and b_ith elements from X and from Y, respectively. Figure 6a illustrates the matrix grids of possible path steps and the aligned warping path for two sequences. Figure 6b shows the point-to-point mapping of the aligned signal sequences. The iterative dynamic planning process drove the warping path W gradually close to the optimized path as the distance cost function was tuning to the minimum scenario. Typically, the DTW distance cost function is defined as follows:

$$\begin{aligned} Distance\ Cost\ Function(W) \\ = DTW(X, Y) \\ = \tfrac{1}{K} \textstyle\sum_{i=1}^{K} D(w_i) \\ = \tfrac{1}{K} \textstyle\sum_{i=1}^{K} \sqrt{\left(x_{a_i} - y_{b_i}\right)^2} \end{aligned} \tag{7}$$

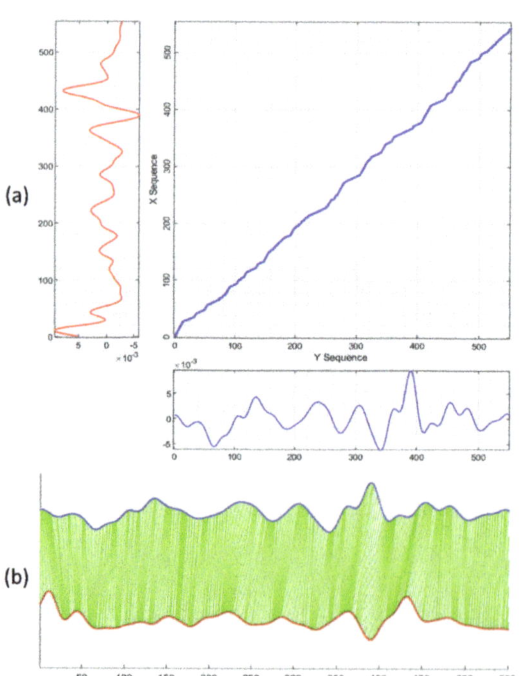

Figure 6. Illustration of the DTW alignment of two signal sequences: (**a**) Matrix grid view of DTW alignment; (**b**) Point-to-point mapping of DTW alignment.

Usually more than one path in the matrix grid can satisfy the requirements of the five constraints, but only one path minimizes the distance cost. A few different types of distance cost function variants have been developed for different application [33,34].

Some DTW applications concern only the Euclidean distance of the aligned points in the signal sequences (as in Equation (7)), whereas in the present study, the morphological similarity was most interested in and critical to fiducial point projection. A hybrid form of the cost function (termed the new cost function) was proposed (as in Equation (8)) because it not only considered the difference in signal values (item led by β) but also evaluates the neighborhood shifting level (item led by α), the difference in signal slope (item led by γ), and the difference in signal concavity (item led by η). The weighting factors (α, β, γ, and η) were tunable and might change across individuals.

$$New\ Cost\ Function(W)$$
$$= \frac{1}{K}\sum_{i=1}^{K}\left(\alpha\sqrt{(a_i - b_i)^2} + \beta\sqrt{(x_{a_i} - y_{b_i})^2} + \gamma\sqrt{\left(\frac{dx_{a_i}}{dt} - \frac{dy_{b_i}}{dt}\right)^2}\gamma + \eta\sqrt{\left(\frac{d^2 x_{a_i}}{dt^2} - \frac{d^2 y_{b_i}}{dt^2}\right)^2}\right) \tag{8}$$

2.5. Fiducial Point Projection with DTW-Based Quasi-Synchronous Alignment

The DTW-base quasi-synchronous alignment introduces two modifications to the conventional quasi-synchronous alignment: (1) the intermediary reference signal pair, and (2) DTW-based point-to-point alignment of two signal sequences. The examples and comparison of SCG fiducial point detections in the nonreference SCG signal using the conventional and DTW-based quasi-synchronous alignment methods are demonstrated in Figure 7. Figure 7a–c illustrate the conventional approach, whereas Figure 7d–g display the DTW-based approach. The echocardiogram shown in Figure 7a is a color Doppler flow measurement at the aortic valve. In Figure 7a, the 966-ms period between R1 and R2 is selected as the synchronization target. The signals in Figure 7b,d represent the same nonreference ECG signal, whereas Figure 7b is rescaled (squeezed to fit signal length of 1003 ms to 966 ms) and shifted such that it can be visually aligned to the ECG R1-R2 section in Figure 7a. The signals in Figure 7c,f represent the same SCG signal (other than the reference SCG signal). Figure 7c has been rescaled and synchronized to Figure 7b.

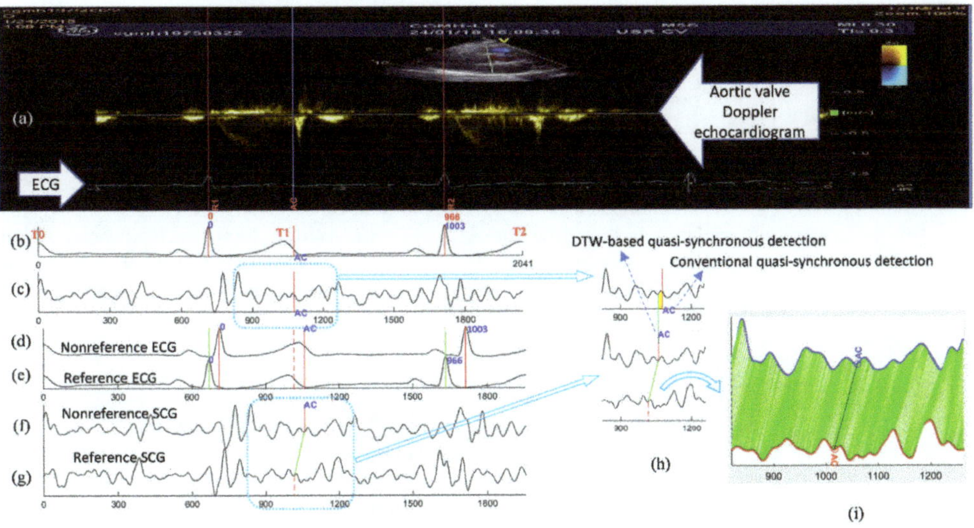

Figure 7. Illustrations and comparison of the fiducial point detection in the nonreference SCG signal between the methods of the conventional and the DTW-based quasi-synchronous alignment: (**a**) Doppler echocardiogram image with the identified ECG R peaks (R1, R2) and aortic valve closing

(AC) event; (**b**) ECG alignment (to the ECG signal in the echocardiogram) under the conventional quasi-synchronization (aligned through shifting and rescaling); (**c**) SCG signal (synchronous to (**b**)) aligned by the conventional approach with the fiducial point AC detected by virtual line extending from the echocardiogram; (**d**) Nonreference ECG signal (no need for R peaks alignment to other ECG signal); (**e**) Reference ECG signal with R peaks aligned to ECG R peaks in the echocardiogram R1 and R2 (aligned by shifting only) under DTW-based quasi-synchronous alignment; (**f**) Nonreference SCG signal (synchronous to (**d**)) with the fiducial point AC projected by DTW-based quasi-synchronous alignment; (**g**) Reference SCG signal (synchronous to (**e**)) aligned to echocardiogram with the fiducial point AC detected by virtual line extending from the echocardiogram under DTW-based approach; (**h**) The detection error (indicated by the width of yellow rectangle in the upper graph) of the conventional method illustrated by comparing the AC points detected from conventional quasi-synchronous alignment (upper trace) and DTW-based (middle trace) approach; (**i**) Point-to-point mapping of the nonreference SCG signal (upper trace) with the reference SCG signal (lower trace) under DTW alignment.

2.5.1. Common Procedures under Both Alignment Methods

An RR interval in the echocardiogram image was selected and the ends were annotated as R1 and R2. A visually recognized AC event was marked by a blue line and labeled as "AC" in the echocardiogram (Figure 7a). The present RR interval (966 ms) was obtained by pixel counting from the timing ticks in the echocardiogram.

2.5.2. Conventional Quasi-Synchronous Alignment

The SCG and ECG signals were shifted and rescaled to align R peaks to R1 and R2 in the ECG of echocardiogram (Figure 7a–c). The fiducial point was obtained by extending the blue line from Figure 7a–c.

2.5.3. DTW-Based Quasi-Synchronous Alignment

A reference signal pair of ECG and SCG was sorted out with the RR interval closest to the R1–R2 period. In this example, the reference signal pair had a heartbeat cycle of 966 ms. Shifting, but not rescaling, was required to align the reference signals until the ECG R peaks matched R1 and R2, as shown in Figure 7e,g. Referring to the condition in Figure 4, the ECG signals in Figure 7a,e, as well as the echocardiogram and SCG signal in Figure 7g, are postulated to align to each other (within the RR interval) spontaneously. The fiducial point AC was detected in the reference SCG signal by extending the blue line from Figure 7a–g. The fiducial point AC was projected from the reference SCG signal (Figure 7g) to nonreference SCG signal (Figure 7f) with the aid of a DTW-based software program, as shown in Figure 7f–i.

As mentioned earlier, a fiducial point detection error resulted from the signal rescaling manipulation in the conventional quasi-synchronous alignment, as indicated by the width of the yellow rectangle in Figure 7h. Figure 7i displays the results of DTW point-to-point alignment with the new cost function between the reference SCG signal (red) and the target nonreference signal (blue).

2.6. Validation of DTW-Based Fiducial Point Detection Approach

An SCG fiducial point detection software tool was developed, using MATLAB R2020a, following the programming flowchart (in Figure 8) which depicts the SCG fiducial point detection procedures with DTW-based quasi-synchronous alignment. Two cardiac diagnostic parameters (PEP/LVET ratio and Tei index) were extracted from a dataset of 56 individuals to assess the clinical application of the proposed SCG fiducial point detection method and to validate the effectiveness of the software. In addition, Figure 8 also reveals that the correlation of averaged diagnostic parameters versus LVEF and ROC classification analysis were conducted at the end of the proof-of-concept experiment.

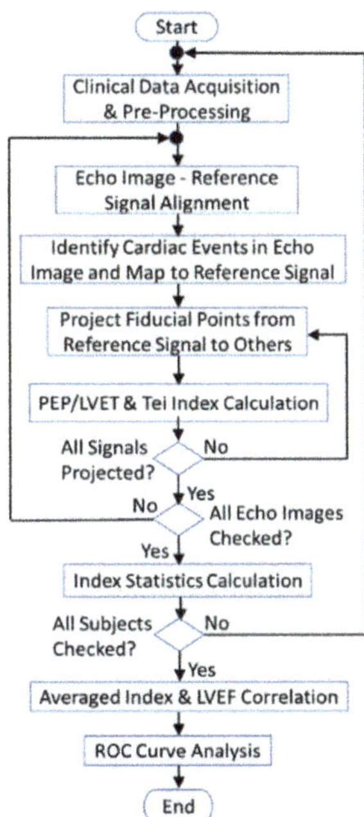

Figure 8. Programming flowchart of the DTW-based fiducial point detection and validation.

2.6.1. Clinical Data Acquisition and Preprocessing

At least two echocardiogram images were required to detect all the interested cardiac events (MO, MC, AO, and AC) because an ultrasonic probe can only examine one cardiac valve at a time. A set of simultaneously measured ECG and SCG signal pairs were clipped from the continuous SCG measurement. As shown in Figure 7b,c, the ECG and SCG signal clips were sectioned according to ECG T waves, from T0 to T2, to include a complete heartbeat section of the centered RR interval and with extra intervals before and after. Prior to the fiducial point detection, signal clips went through the data preprocessing including signal detrending, band-pass filtering, and wavelet denoising.

2.6.2. Echocardiogram–Reference Signal Alignment

An RR interval in the echocardiogram was selected for the interested cardiac event identification as shown in the section confined by R1 and R2 lines in Figure 7a. To minimize the fiducial point detection error, an ECG/SCG signal pair with RR interval period the same as or close to the duration between R1 and R2 lines were screened out from the signal clips as the reference signal pair. The reference signal pair were aligned to the R1–R2 section in the echocardiogram only through shifting, as presented in Figure 5 or in Figure 7e,g.

2.6.3. Identification and Mapping of Cardiac Events in Echocardiogram to Reference Signals

After the alignment of echocardiogram images and reference signals, the valve opening and closure events were identified and annotated in the images, as indicated by the blue line in Figure 5a or Figure 7a. By extending the lines to reference SCG signals, the intersections were annotated as the detected SCG fiducial points in the reference signal.

2.6.4. Projection of Fiducial Points from Reference to Nonreference Signals

The reference SCG signals with fiducial point annotations were used to project the fiducial points to nonreference signals through DTW alignment, under the proposed new cost function optimization. Figure 6 illustrates the results of the point-to-point mapping of the alignment and the aligned path with minimized cost function in the cost matrix.

2.6.5. PEP/LVET Ratio and Tei Index Calculation

To validate the clinical practicability and effectiveness of the DTW-based fiducial point detection approach, the cardiac parameters (PEP/LVET ratio and Tei index) for each nonreference SCG signal clip were calculated according to Equations (5) and (6) from the projected fiducial points.

2.6.6. Index Statistics Calculation

To assess the cardiac health of an individual, the averaged PEP/LVET ratio and Tei index were utilized instead of using the indexes from a single signal clip.

2.6.7. Correlation of the Averaged Indexes to the LVEF

The collection of the averaged indexes (PEP/LVET ratio and Tei index) for 56 individuals were correlated to the individual's clinical LVEF assessment (in Figure 9). The coefficient of determination (R^2) was employed as the indicator of the correlation between the averaged indexes and the clinical LVEF assessment.

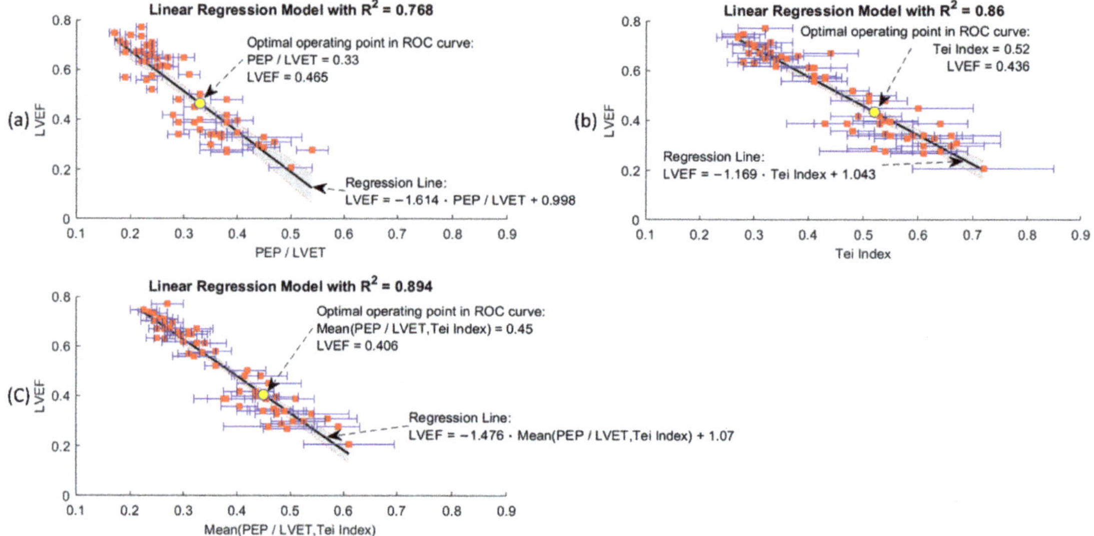

Figure 9. Linear regression models of clinical LVEF assessment versus: (**a**) PEP/LVET ration; (**b**) Tei index; (**c**) mean of PEP/LVET and Tei index. Each graph is annotated with mean (red squares) and standard deviation (blue error bars) of the variable from subjects, the 95% confidence interval of the regression line (shaded area), the coefficient of determination (R2) and the optimal operating point determined in the ROC curve analysis (yellow point).

2.6.8. ROC Curve Analysis

Four predictive models in ROC curve analysis (Figure 10) were used to distinguish the patients of heart failure with reduced ejection fraction (HFrEF), using the predictors of (1) clinical LVEF assessment, (2) PEP/LVEF ratio, (3) Tei index, and (4) the mean PEP/LVEF ratio and Tei index, respectively. The values of area under the ROC curve (AUC) were determined, to evaluate the predictability of the models as well as the effectiveness of these diagnostic parameters derived from the DTW-base quasi-synchronous alignment.

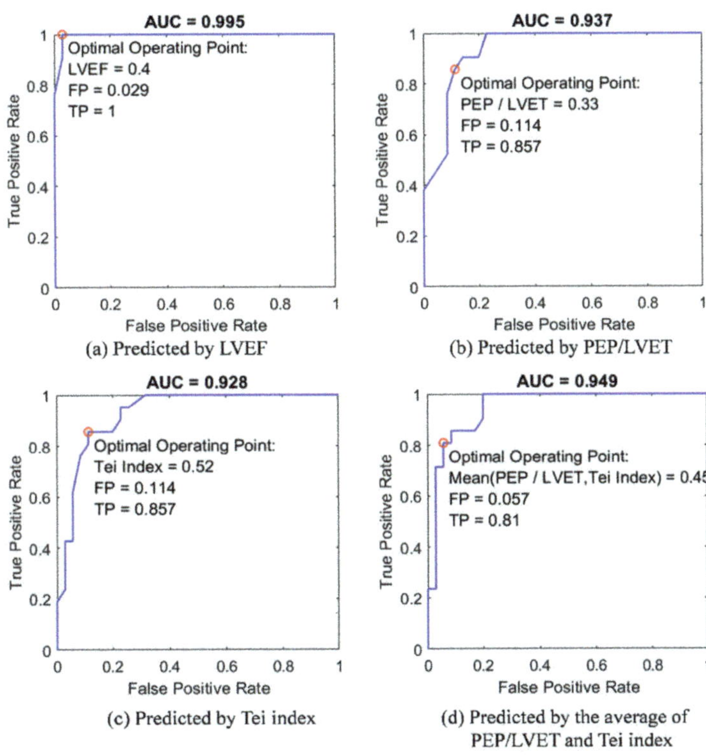

Figure 10. The comparison of ROC model analysis for the classification of patients with HFrEF by using four univariate predictors: (**a**) clinical LVEF assessment; (**b**) the PEP/LVET ratio; (**c**) Tei index; (**d**) the mean of PEP/LVET ration and Tei index.

3. Results

With IRB approval, the SCG–echocardiogram data of the 56 individuals were employed for the proof-of-concept experiments and the effectiveness evaluation of DTW-based fiducial point detection approach. The experimental data were reused from the previous research conducted by Lin et al. on a multichannel SCG system [9].

Table 1 lists the subjects' demographic information and SCG-derived cardiac parameters, including patient ID, sex, age, clinical LVEF assessment, hospital diagnosed cardiovascular diseases, the number of examined SCG clips and the statistics of PEP/LVET ration and Tei index derived by the DTW-based SCG fiducial point detection method. The means ± standard deviations of age and LVEF are 52.1 ± 22.3 years and 50.8% ± 16.3%, respectively. The mean and standard deviation of PEP/LVET ratio and Tei index were calculated from the number of examined SCG clips. The number of data clips used for fiducial point detection ranged from 34 to 106, with a mean ± standard deviation of 69.2 ± 16.2.

Table 1. Clinical assessment and SCG fiducial point derived cardiac parameters for GLM and ROC curve analysis.

Subject ID	Sex	Age	Clinical Assess		Cycles	SCG Fiducial Point Derivative			
			LVEF	Disease		PEP/LVET		Tei Index	
						Avg	SD	Avg	SD
Subject-01	F	22	29.0%	HFrEF	72	0.45	0.03	0.52	0.05
Subject-02	F	22	70.9%	Normal	85	0.24	0.01	0.29	0.04
Subject-03	F	23	77.1%	Normal	56	0.22	0.01	0.32	0.05
Subject-04	F	24	63.6%	Normal	95	0.22	0.01	0.28	0.03
Subject-05	F	26	71.5%	Normal	79	0.18	0.01	0.33	0.04
Subject-06	F	26	74.0%	Normal	78	0.20	0.01	0.27	0.03
Subject-07	F	27	67.5%	Normal	83	0.22	0.01	0.33	0.04
Subject-08	F	34	34.0%	HFrEF	52	0.37	0.04	0.61	0.07
Subject-09	F	37	61.4%	Normal	81	0.27	0.02	0.41	0.06
Subject-10	F	43	61.3%	Normal	34	0.25	0.01	0.40	0.04
Subject-11	F	43	65.0%	Normal	98	0.27	0.02	0.36	0.02
Subject-12	F	47	73.3%	Normal	64	0.22	0.01	0.27	0.04
Subject-13	F	60	34.0%	HFrEF	75	0.35	0.02	0.55	0.08
Subject-14	F	60	42.0%	HFmrEF	65	0.28	0.02	0.53	0.04
Subject-15	F	68	28.0%	HFrEF	67	0.38	0.05	0.54	0.12
Subject-16	F	68	65.0%	MI *	80	0.30	0.02	0.32	0.05
Subject-17	F	70	39.0%	MI	60	0.38	0.02	0.64	0.05
Subject-18	F	70	36.0%	HFrEF	38	0.33	0.03	0.48	0.04
Subject-19	F	72	28.0%	HFrEF	84	0.54	0.03	0.64	0.05
Subject-20	F	80	33.0%	HFrEF	39	0.37	0.02	0.58	0.06
Subject-21	F	82	45.0%	MI	62	0.32	0.02	0.60	0.10
Subject-22	F	84	57.0%	MI	61	0.19	0.01	0.43	0.03
Subject-23	F	85	42.0%	HFmrEF	88	0.38	0.02	0.49	0.05
Subject-24	F	85	27.0%	HFrEF	89	0.38	0.03	0.61	0.06
Subject-25	F	90	35.0%	HFrEF	60	0.40	0.02	0.54	0.05
Subject-26	F	91	39.8%	HFrEF	89	0.40	0.03	0.55	0.05
Subject-27	M	21	70.4%	Normal	62	0.19	0.01	0.30	0.04
Subject-28	M	23	65.9%	Normal	61	0.23	0.01	0.35	0.04
Subject-29	M	23	69.5%	Normal	90	0.24	0.01	0.32	0.03
Subject-30	M	24	66.2%	Normal	76	0.24	0.02	0.38	0.06
Subject-31	M	24	69.9%	Normal	70	0.23	0.01	0.30	0.02
Subject-32	M	24	64.4%	Normal	82	0.25	0.02	0.34	0.04
Subject-33	M	25	62.0%	Normal	69	0.26	0.02	0.34	0.06
Subject-34	M	27	48.0%	HFmrEF	37	0.38	0.03	0.51	0.07
Subject-35	M	28	63.1%	Normal	62	0.23	0.01	0.30	0.03
Subject-36	M	38	74.7%	MI	76	0.17	0.01	0.28	0.04
Subject-37	M	44	58.0%	MI	74	0.31	0.02	0.41	0.04
Subject-38	M	45	39.0%	HFrEF	69	0.38	0.01	0.53	0.03
Subject-39	M	46	67.2%	Normal	73	0.19	0.01	0.31	0.02
Subject-40	M	48	67.3%	MI	75	0.21	0.01	0.44	0.05
Subject-41	M	49	50.3%	HFpEF	70	0.33	0.02	0.51	0.05
Subject-42	M	51	67.2%	Normal	78	0.24	0.01	0.29	0.03
Subject-43	M	48	55.8%	MI	66	0.23	0.02	0.41	0.06
Subject-44	M	58	34.0%	HFrEF	79	0.29	0.02	0.66	0.09
Subject-45	M	62	21.0%	HFrEF	66	0.5	0.04	0.72	0.13
Subject-46	M	62	33.0%	HFrEF	38	0.45	0.07	0.63	0.07
Subject-47	M	63	34.0%	HFrEF	80	0.36	0.02	0.59	0.08
Subject-48	M	63	31.0%	HFrEF	46	0.47	0.03	0.67	0.08
Subject-49	M	65	57.6%	MI	73	0.24	0.02	0.43	0.03
Subject-50	M	66	52.0%	MI	50	0.24	0.01	0.48	0.03
Subject-51	M	68	39.0%	HFrEF	54	0.32	0.04	0.43	0.07
Subject-52	M	69	30.0%	HFrEF	75	0.44	0.03	0.61	0.05
Subject-53	M	70	40.0%	HFrEF	40	0.33	0.03	0.54	0.06
Subject-54	M	79	30.0%	HFrEF	67	0.35	0.02	0.66	0.06
Subject-55	M	79	48.0%	MI	78	0.29	0.01	0.54	0.04
Subject-56	M	88	39.0%	HFrEF	106	0.29	0.01	0.47	0.06

* MI: Myocardial Infarction.

The applicability of SCG fiducial point detection using DTW-based alignment was validated through two experiments: (1) the linear correlation between SCG-derived indexes (PEP/LVET ratio and Tei index) and clinical LVEF assessment and (2) the analysis of ROC classification to distinguish patients with HFrEF or else.

3.1. Linear Correlation Models

The clinical LVEF assessment has been reported to be negatively proportional to the PEP/LVET ratio and Tei index for the patients with cardiac symptoms of varying severity [3,5,7,35,36]. To validate the effectiveness of the SCG fiducial points and the associated CTIs derived using the DTW-based quasi-synchronous alignment method, three general linear models (GLMs) were generated for the correlation analysis to prove the negative proportionality. The trendlines of the 56-subject clinical LVEF assessments were synthesized under three general linear regression models by using the univariate predictors of: (1) the PEP/LVET ratio, (2) Tei index, and (3) the mean PEP/LVET ratio and Tei index (shown in Figure 9). The three general linear regression models are formulated as Equations (9), (10), and (11). Favorable coefficients of determination (R^2) for the three univariate linear models were obtained in this 56-subject experiment, with R^2 = 0.768, 0.86, and 0.894 for (1), (2) and (3), respectively. The negative proportionalities in the graphs of Figure 9 are obvious, whereas the standard deviation (indicated by the blue error bars) and 95% confidence intervals (indicated by the shaded area) are larger for patients assessed as having lower LVEF than normal people (with higher LVEF).

$$\text{LVEF} = -1.614 \times \frac{\text{PEP}}{\text{LVET}} + 0.998 \tag{9}$$

$$\text{LVEF} = -1.169 \,\text{Tei index} \times\ + 1.043 \tag{10}$$

$$\text{LVEF} = -1.476 \times 0.5 \times \left(\frac{\text{PEP}}{\text{LVET}} + \text{Tei index} \right) + 1.070 \tag{11}$$

3.2. ROC Classification

The LVEF assessment has clinical utility for cardiovascular syndrome classification and heart failure diagnosis [37]. According to the 2016 European Society of Cardiology (ESC) guidelines regarding the diagnosis and treatment of acute and chronic heart failure, the class of HFmrEF (heart failure with mid-range ejection fraction) was defined as an LVEF assessment larger than 40% but less than 50%. The other classes of HFrEF and HFpEF (heart failure with preserved ejection fraction) were retained [38].

The second proof-of-concept experiment for the DTW-based fiducial point detection approach was the application of the univariate ROC models to classify the subjects with HFrEF label listed in the disease column of Table 1. In Figure 10, four variables, including (1) LVEF, (2) PEP/LVET ratio, (3) Tei index, and (4) the mean PEP/LVET ratio and Tei index, were employed as the univariate predictors. To build the models, four data pairs, comprising diagnosis disease labels and the values of the predictors, were modeled by the GLMs as with the logit link function and binomial distribution function. The GLM-fitted probability vectors were set as scores and the "HFrEF" tag was set to as the positive label in the ROC curve configurations. Figure 10 exhibits the ROC classification result of four models to predict the diagnosis of HFrEF; extra indicators including the TP and FP rates, AUCs, and the optimal operating points of the predictors are also annotated. The optimal operating point was estimated at the condition that the classifier gave the best trade-off between the costs of failing to detect positives against the costs of raising false alarms.

The AUC of the LVEF model outperformed the other three models, with an AUC of 0.995 and the optimal operating point (LVEF cutoff at 0.4) at TP = 1 and FP = 0.029 (Figure 10a). This result conformed to the lower bound of HFmrEF of 40% suggested in the 2016 ESC guidelines. The AUC of the PEP/LVET ratio model was slightly greater than that of the Tei index model (with AUC = 0.937 > 0.928). In other words, the cardiologists had a 93.7% probability of correctly distinguish a patient with HFrEF from others with the aid of the PEP/LVET ratio model. Using the Tei index as the predictor, HFrEF could be expected to correctly diagnose 92.8% of the cases. As shown in Figure 10b,c, both models has the same TP and FP for the optimal operating points. Using the mean PEP/LVET ratio and Tei index as predictors improved the diagnostic prediction of any original predictors, with AUC = 0.949 (Figure 10d). In general classification evaluations, ROC model predictions with AUC values equal to 0.5 suggest no discriminative ability. AUC values between 0.7 and 0.8, between 0.8 and 0.9, and >0.9 indicate acceptable, excellent, and outstanding discriminative ability, respectively [39].

Through the proof-of-concept experiments of GLM models and ROC curve predictions, the CTIs and the cardiac indexes (PEP/LVET ration and Tei index) derived from the DTW-based quasi-synchronous alignment herein were confirmed to be reasonable in clinical practice. The optimal operating points corresponding to the last three ROC curves are also annotated in Figure 9. Moreover, the corresponding optimal decision points (0.4, 0.465, 0.436 and 0.45) for HFrEF assessment in the ROC curves were in Figure 10 for the comparison of the suggestion in 2016 ESC guidelines, LVEF = 0.4. Hopefully, these findings will provide reference for future study.

4. Discussion and Conclusions

The present study examined the issue of misalignment in the conventional quasi-synchronous alignment method and introduced the intermediary signals (reference signal pair of ECG and SCG) and DTW algorithm to eliminate the timing error in SCG fiducial point detection. The advantages and effects of the collaboration of reference signals and DTW algorithm as well as the new distance cost function were demonstrated in the graphical illustrations. The combination of the intermediary signals and DTW alignment in SCG fiducial point detection was proposed and realized for the first time.

It was known that SCG fiducial point delineation was hindered by artifact effects, the ambiguity in specific event waveforms and the lack of detection procedures. The proposed method abandoned the idea to look for the fiducial point co-occurring waveforms but seek for fusing heterogeneous modalities to allocate the fiducial points in the personal SCG reference signal. DTW algorithm was leveraged afterwards to project the fiducial points to non-reference signals. A merit of aligning signal pair with DTW algorithm is that it does not just align the prominent peaks or valleys but the entire signals. Because the extreme points serve as the anchor points during the alignment, points in between are enforced to be regulated. On the condition that the artifact does not override the signal waveform too much, DTW could overcome the distortion. Therefore, in case of minor signal distortion or featureless points are identified as the fiducial points, the projection can still function correctly with the assist of DTW.

This proposed DTW-based quasi-synchronous alignment is not only dedicated to SCG fiducial point detection but is also applicable to other scenarios. The non-ECG reference signal (SCG reference signal) and the application target (seismocardiography) could be changed to other cardiac signals, such as phonocardiography, ballistocardiography and impedance cardiography. The echocardiogram images could also be changed to other imaging modalities or other cardiac signal templates with the target cardiac events annotated. The application concept of the proposal is not limited in biomedical scenarios; it is possible to extend to the speech recognition or gesture identification and so on.

As the concept was proven clinically in the 56-individual dataset, the software program implemented based on the proposed flowchart achieved high prediction rates (>92.8%) in the experiment of distinguishing patients with HFrEF from others. With the verified diagnostic utility, the proposed quasi-synchronous fiducial point detection procedures could be further refined and standardized to expedite the development in SCG technology and shorten the path from bench to bedside.

Recently, machine learning has been one of the major topics in biomedical engineering; however, building a machine learning model for SCG fiducial point detection requires plenty of labeled training data. The proposed flowchart could be used as a framework and guidelines of the automatic program development for fiducial point detection. Therefore, the program-generated SCG fiducial point annotations could be used as the labeled training data set for the supervised machine learning process in alternative fiducial point detection approaches or other feature identification applications. In addition, by adding new routines, the framework can extend more features to the SCG research, such as signal template generation, signal morphology clustering and multichannel SCG applications. In the future, the framework may be further integrated into the cloud computing services together with the ambulatory ECG/SCG system for home-based real time health monitoring.

Author Contributions: Conceptualization, C.-H.C.; Data curation, W.-Y.L. and M.-Y.L.; Formal analysis, C.-H.C.; Funding acquisition, W.-Y.L. and M.-Y.L.; Investigation, C.-H.C.; Methodology, C.-H.C., W.-Y.L. and M.-Y.L.; Project administration, W.-Y.L. and M.-Y.L.; Resources, W.-Y.L.; Software, C.-H.C.; Supervision, W.-Y.L. and M.-Y.L.; Validation, C.-H.C.; Visualization, C.-H.C.; Writing—original draft, C.-H.C.; Writing—review and editing, C.-H.C., W.-Y.L. and M.-Y.L. All authors have read and agreed to the published version of the manuscript.

Funding: This work was supported in part by the Ministry of Science and Technology, Taiwan (MOST 109-2221-E-182-016, MOST 110-2221-E-182-007), and Chang Gung University Fund (BMRPC50).

Institutional Review Board Statement: The study was conducted in accordance with the Declaration of Helsinki, and approved by the Institutional Review Board of Chang Gung Memorial Hospital, Taoyuan, Taiwan (protocol code: 202100744B0A3, and date of approval: 2021/06/25)." for studies in-volving humans.

Informed Consent Statement: Informed consent was obtained from all subjects involved in the study.

Data Availability Statement: Not applicable.

Acknowledgments: Not applicable.

Conflicts of Interest: All the authors have declared that no conflict of interest exists, no financial or personal relationship with other people or organizations inappropriately influenced this work and the funders had no role in study design, data collection and analysis, decision to publish, or preparation of the manuscript.

References

1. Biering-Sørensen, T. Cardiac Time Intervals by Tissue Doppler Imaging M-Mode Echocardiography: Reproducibility, Reference Values, Association with Clinical Characteristics and Prognostic Implications. Ph.D. Thesis, Faculty of Health and Medical Sciences, University of Copenhagen, Copenhagen, Denmark, 2014.
2. Tavakolian, K. Systolic time intervals and new measurement methods. *Cardiovasc. Eng. Technol.* **2016**, *7*, 118–125. [CrossRef] [PubMed]
3. Lewis, R.P.; Rittogers, S.; Froester, W.; Boudoulas, H. A critical review of the systolic time intervals. *Circulation* **1977**, *56*, 146–158. [CrossRef] [PubMed]
4. Van Leeuwen, P.; Kuemmell, H. Respiratory modulation of cardiac time intervals. *Heart* **1987**, *58*, 129–135. [CrossRef] [PubMed]
5. Chuwa, T.; Rodeheffer, R.J. New index of combined systolic and diastolic myocardial performance: A simple and reproducible measure of cardiac function—A study in normals and dilated cardiomyopathy. *J. Cardiol.* **1995**, *26*, 7–366.
6. Albert, D.E. Cardiac Performance Monitoring System for Use with Mobile Communications Devices. U.S. Patent 8700137B2, 15 April 2014.
7. Lin, W.-Y.; Ke, H.-L.; Chou, W.-C.; Chang, P.-C.; Tsai, T.-H.; Lee, M.-Y. Realization and Technology Acceptance Test of a Wearable Cardiac Health Monitoring and Early Warning System with Multi-Channel MCGs and ECG. *Sensors* **2018**, *18*, 3538. [CrossRef] [PubMed]
8. Carvalho, P.; Paiva, R.; Couceiro, R.; Henriques, J.; Antunes, M.; Quintal, I.; Muehlsteff, J.; Aubert, X. Comparison of Systolic Time Interval Measurement Modalities for Portable Devices. In Proceedings of the 2010 Annual International Conference of the IEEE Engineering in Medicine and Biology, Buenos Aires, Argentina, 31 August–4 September 2010; pp. 606–609.
9. Lin, W.-Y.; Chou, W.-C.; Chang, P.-C.; Chou, C.-C.; Wen, M.-S.; Ho, M.-Y.; Lee, W.-C.; Hsieh, M.-J.; Lin, C.-C.; Tsai, T.-H. Identification of location specific feature points in a cardiac cycle using a novel seismocardiogram spectrum system. *IEEE J. Biomed. Health Inform.* **2016**, *22*, 442–449. [CrossRef]
10. Lee, M.-y.; Lin, W.-Y. Wearable cardiac health monitoring and early warning system. *Impact* **2017**, *2017*, 55–57. [CrossRef]
11. Korzeniowska-Kubacka, I.; Kuśmierczyk-Droszcz, B.; Bilińska, M.; Dobraszkiewicz-Wasilewska, B.; Mazurek, K.; Piotrowicz, R. Seismocardiography-a non-invasive method of assessing systolic and diastolic left ventricular function in ischaemic heart disease. *Cardiol. J.* **2006**, *13*, 319–325.

12. Dehkordi, P.; Khosrow-Khavar, F.; Di Rienzo, M.; Inan, O.T.; Schmidt, S.E.; Blaber, A.P.; Sørensen, K.; Struijk, J.J.; Zakeri, V.; Lombardi, P. Comparison of different methods for estimating cardiac timings: A comprehensive multimodal echocardiography investigation. *Front. Physiol.* **2019**, *10*, 1057. [CrossRef]

13. Su, H.M.; Lin, T.H.; Voon, W.C.; Lee, K.T.; Chu, C.S.; Yen, H.W.; Lai, W.T.; Sheu, S.H. Correlation of Tei index obtained from tissue Doppler echocardiography with invasive measurements of left ventricular performance. *Echocardiography* **2007**, *24*, 252–257. [CrossRef]

14. Khosrow-Khavar, F.; Tavakolian, K.; Blaber, A.P.; Zanetti, J.M.; Fazel-Rezai, R.; Menon, C. Automatic annotation of seismocardiogram with high-frequency precordial accelerations. *IEEE J. Biomed. Health Inform.* **2014**, *19*, 1428–1434. [CrossRef] [PubMed]

15. Wick, C.A.; Inan, O.T.; Bhatti, P.; Tridandapani, S. Relationship between cardiac quiescent periods derived from seismocardiography and echocardiography. In Proceedings of the 37th Annual International Conference of the IEEE Engineering in Medicine and Biology Society (EMBC), Milan, Italy, 25–29 August 2015; pp. 687–690.

16. Crow, R.S.; Hannan, P.; Jacobs, D.; Hedquist, L.; Salerno, D.M. Relationship between seismocardiogram and echocardiogram for events in the cardiac cycle. *Am. J. Noninvasive Cardiol.* **1994**, *8*, 39–46. [CrossRef]

17. Taebi, A.; Solar, B.E.; Bomar, A.J.; Sandler, R.H.; Mansy, H.A. Recent advances in seismocardiography. *Vibration* **2019**, *2*, 64–86. [CrossRef] [PubMed]

18. Rai, D.; Thakkar, H.K.; Rajput, S.S.; Santamaria, J.; Bhatt, C.; Roca, F. A Comprehensive Review on Seismocardiogram: Current Advancements on Acquisition, Annotation, and Applications. *Mathematics* **2021**, *9*, 2243. [CrossRef]

19. Choudhary, T.; Sharma, L.; Bhuyan, M.K. Automatic detection of aortic valve opening using seismocardiography in healthy individuals. *IEEE J. Biomed. Health Inform.* **2018**, *23*, 1032–1040. [CrossRef] [PubMed]

20. Rivero, I.; Valdes, E.; Valdes, F. Robust detection of AO and IM points in the seismocardiogram using CWT. *IEEE Lat. Am. Trans.* **2016**, *14*, 4468–4473. [CrossRef]

21. Karpagavalli, S.; Chandra, E. A review on automatic speech recognition architecture and approaches. *Int. J. Signal Process. Image Process. Pattern Recognit.* **2016**, *9*, 393–404.

22. Jackson, P. Dynamic Time Warping. Available online: http://info.ee.surrey.ac.uk/Teaching/Courses/eem.ssr/eeem034d.pdf (accessed on 22 May 2022).

23. Giorgino, T. Computing and visualizing dynamic time warping alignments in R: The dtw package. *J. Stat. Softw.* **2009**, *31*, 1–24. [CrossRef]

24. Sakoe, H.; Chiba, S. Dynamic programming algorithm optimization for spoken word recognition. *IEEE Trans. Acoust. Speech Signal Processing* **1978**, *26*, 43–49. [CrossRef]

25. Kahveci, T.; Singh, A. Variable Length Queries for Time Series Data. In Proceedings of the 17th International Conference on Data Engineering, Heidelberg, Germany, 2–6 April 2001; pp. 273–282.

26. Lyons, J.; Biswas, N.; Sharma, A.; Dehzangi, A.; Paliwal, K.K. Protein fold recognition by alignment of amino acid residues using kernelized dynamic time warping. *J. Theor. Biol.* **2014**, *354*, 137–145. [CrossRef]

27. Azad, M.K.; Gamage, P.T.; Sandler, R.H.; Raval, N.; Mansy, H.A. Seismocardiographic Signal Variability During Regular Breathing and Breath Hold in Healthy Adults. In Proceedings of the 2019 IEEE Signal Processing in Medicine and Biology Symposium (SPMB), Philadelphia, PA, USA, 7 December 2019; pp. 1–7.

28. Chen, C.-H.; Lin, W.-Y.; Lee, M.-Y. The Applications of K-means Clustering and Dynamic Time Warping Average in Seismocardiography Template Generation. In Proceedings of the 2020 IEEE International Conference on Systems, Man, and Cybernetics (SMC), Virtual Conference, 11–14 October 2020; pp. 1000–1007.

29. Müller, M. Dynamic Time Warping. In *Information Retrieval for Music and Motion*; Springer: Berlin/Heidelberg, Germany, 2007; pp. 69–84.

30. Yadav, M.; Alam, A. Dynamic time warping (DTW) algorithm in speech: A review. *Int. J. Res. Electron. Comput. Eng.* **2018**, *6*, 524–528.

31. Berndt, D.J.; Clifford, J. Using Dynamic Time Warping to Find Patterns in Time Series. In Proceedings of the KDD Workshop, Seattle, WA, USA, 31 July 1994; pp. 359–370.

32. Wang, K.; Gasser, T. Alignment of curves by dynamic time warping. *Ann. Stat.* **1997**, *25*, 1251–1276. [CrossRef]

33. Jeong, Y.-S.; Jeong, M.K.; Omitaomu, O.A. Weighted dynamic time warping for time series classification. *Pattern Recognit.* **2011**, *44*, 2231–2240. [CrossRef]

34. Keogh, E.J.; Pazzani, M.J. Derivative Dynamic Time Warping. In Proceedings of the 2001 SIAM International Conference on Data Mining, Chicago, IL, USA, 5–7 April 2001; pp. 1–11.

35. Reant, P.; Dijos, M.; Donal, E.; Mignot, A.; Ritter, P.; Bordachar, P.; Dos Santos, P.; Leclercq, C.; Roudaut, R.; Habib, G. Systolic time intervals as simple echocardiographic parameters of left ventricular systolic performance: Correlation with ejection fraction and longitudinal two-dimensional strain. *Eur. J. Echocardiogr.* **2010**, *11*, 834–844. [CrossRef] [PubMed]

36. LaCorte, J.C.; Cabreriza, S.E.; Rabkin, D.G.; Printz, B.F.; Coku, L.; Weinberg, A.; Gersony, W.M.; Spotnitz, H.M. Correlation of the Tei index with invasive measurements of ventricular function in a porcine model. *J. Am. Soc. Echocardiogr.* **2003**, *16*, 442–447. [CrossRef]

37. Bishu, K.; Deswal, A.; Chen, H.H.; LeWinter, M.M.; Lewis, G.D.; Semigran, M.J.; Borlaug, B.A.; McNulty, S.; Hernandez, A.F.; Braunwald, E. Biomarkers in acutely decompensated heart failure with preserved or reduced ejection fraction. *Am. Heart J.* **2012**, *164*, 763–770.e763. [CrossRef]
38. Uk, N.A.-A.; Atherton, J.J.; Bauersachs, J.; Uk, A.J.C.; Carerj, S.; Ceconi, C.; Coca, A.; Uk, P.E.; Erol, Ç.; Ezekowitz, J. 2016 ESC Guidelines for the diagnosis and treatment of acute and chronic heart failure. *Eur. Heart J.* **2016**, *37*, 2129–2200.
39. Mandrekar, J.N. Receiver operating characteristic curve in diagnostic test assessment. *J. Thorac. Oncol.* **2010**, *5*, 1315–1316. [CrossRef]

Article

An Artifact-Resistant Feature SKNAER for Quantifying the Burst of Skin Sympathetic Nerve Activity Signal

Yantao Xing [1], Yike Zhang [2], Zhijun Xiao [1], Chenxi Yang [1], Jiayi Li [1], Chang Cui [2], Jing Wang [3], Hongwu Chen [2], Jianqing Li [1,*] and Chengyu Liu [1,*]

[1] School of Instrument Science and Engineering, Southeast University, Nanjing 210096, China; 230198304@seu.edu.cn (Y.X.); zhijunxiao@seu.edu.cn (Z.X.); chenxiyang@seu.edu.cn (C.Y.); lijiayi@seu.edu.cn (J.L.)

[2] Division of Cardiology, The First Affiliated Hospital of Nanjing Medical University, Nanjing 210096, China; zhangyike@njmu.edu.cn (Y.Z.); cuichang@njmu.edu.cn (C.C.); chenhongwu@njmu.edu.cn (H.C.)

[3] Division of Nephrology, The First Affiliated Hospital of Nanjing Medical University, Nanjing 210096, China; wj_flagon@163.com

* Correspondence: ljq@seu.edu.cn (J.L.); chengyu@seu.edu.cn (C.L.)

Citation: Xing, Y.; Zhang, Y.; Xiao, Z.; Yang, C.; Li, J.; Cui, C.; Wang, J.; Chen, H.; Li, J.; Liu, C. An Artifact-Resistant Feature SKNAER for Quantifying the Burst of Skin Sympathetic Nerve Activity Signal. *Biosensors* **2022**, *12*, 355. https://doi.org/10.3390/bios12050355

Received: 6 April 2022
Accepted: 18 May 2022
Published: 20 May 2022

Publisher's Note: MDPI stays neutral with regard to jurisdictional claims in published maps and institutional affiliations.

Abstract: Evaluation of sympathetic nerve activity (SNA) using skin sympathetic nerve activity (SKNA) signal has attracted interest in recent studies. However, signal noises may obstruct the accurate location for the burst of SKNA, leading to the quantification error of the signal. In this study, we use the Teager–Kaiser energy (TKE) operator to preprocess the SKNA signal, and then candidates of burst areas were segmented by an envelope-based method. Since the burst of SKNA can also be discriminated by the high-frequency component in QRS complexes of electrocardiogram (ECG), a strategy was designed to reject their influence. Finally, a feature of the SKNA energy ratio (SKNAER) was proposed for quantifying the SKNA. The method was verified by both sympathetic nerve stimulation and hemodialysis experiments compared with traditional heart rate variability (HRV) and a recently developed integral skin sympathetic nerve activity (iSKNA) method. The results showed that SKNAER correlated well with HRV features (r = 0.60 with the standard deviation of NN intervals, 0.67 with low frequency/high frequency, 0.47 with very low frequency) and the average of iSKNA (r = 0.67). SKNAER improved the detection accuracy for the burst of SKNA, with 98.2% for detection rate and 91.9% for precision, inducing increases of 3.7% and 29.1% compared with iSKNA (detection rate: 94.5% ($p < 0.01$), precision: 62.8% ($p < 0.001$)). The results from the hemodialysis experiment showed that SKNAER had more significant differences than aSKNA in the long-term SNA evaluation ($p < 0.001$ vs. $p = 0.07$ in the fourth period, $p < 0.01$ vs. $p = 0.11$ in the sixth period). The newly developed feature may play an important role in continuously monitoring SNA and keeping potential for further clinical tests.

Keywords: sympathetic activity (SNA); skin sympathetic nerve activity (SKNA); electrodes; electrocardiogram (ECG)

1. Introduction

Cardiovascular diseases (CVDs) are the biggest killer of people globally, accounting for 32.1% of the death cases [1]. With the aggravation of aging [2], the prevention and treatment of CVDs have become a global problem. Some CVDs are manifested as symptomatic cardiac autonomic function neuropathy [3], with sympathetic and vagal innervation disorder or structural damage. Therefore, the evaluation of sympathetic function is of great significance [4]. Microneurography is the gold standard for estimating sympathetic nerve activity (SNA) [5], but it is invasive and rarely used in the clinical scene [6]. Heart rate variability (HRV) is a non-invasive method of assessing SNA [7]. HRV requires proper sinus node function [8], and it is not practicable for patients with atrial fibrillation or premature beat because their rhythm is not sinus rhythm. In addition, HRV cannot reflect the dynamic changes of SNA because of their indirect parameters from segments that last 5 min. Therefore, these methods have some limitations for daily human monitoring applications. The research on electrodermal activity (EDA) provides another possibility for non-invasive SNA evaluation [9], but EDA affected by sweat may lead to some individualized differences in evaluation features [10]. As a non-invasive and real-time method for assessing SNA, skin sympathetic nerve activity (SKNA) has been applied in many clinical events [11,12] and has great potential. It has become a research hotspot in SNA evaluation.

As shown in Figure 1, SKNA is obtained by collecting biopotential signals on the body surface using electrode sensors at high sampling frequency (>2000 Hz) [13]. This method does not require additional inducing conditions and is easy to implement. The amplitude range of SKNA signals is much lower than those of electrocardiogram (ECG), ranging from 0.5 to 80 μV, according to the experiments from our previous work [13] as well as from [14,15]. The sympathetic nerve arises from the spinal cord. In particular, the stellate ganglion sends post-ganglionic sympathetic fibers to the heart and skin. The electrodes measure the subcutaneous sympathetic nerve activity to reflect the cardiac sympathetic nerve activity. Therefore, SKNA is inevitably disturbed by other noises, such as ECG and electromyogram (EMG), resulting in various artifacts in the signal [16]. The electrode-skin interface impedance and skin surface conductivity also affect the signal quality [17]. In addition, power line interference and amplifier saturation distortion have a negative impact on SKNA signal acquisition [18]. Therefore, SKNA has some special characteristics: low amplitude, high noise, and high randomness due to the above factors.

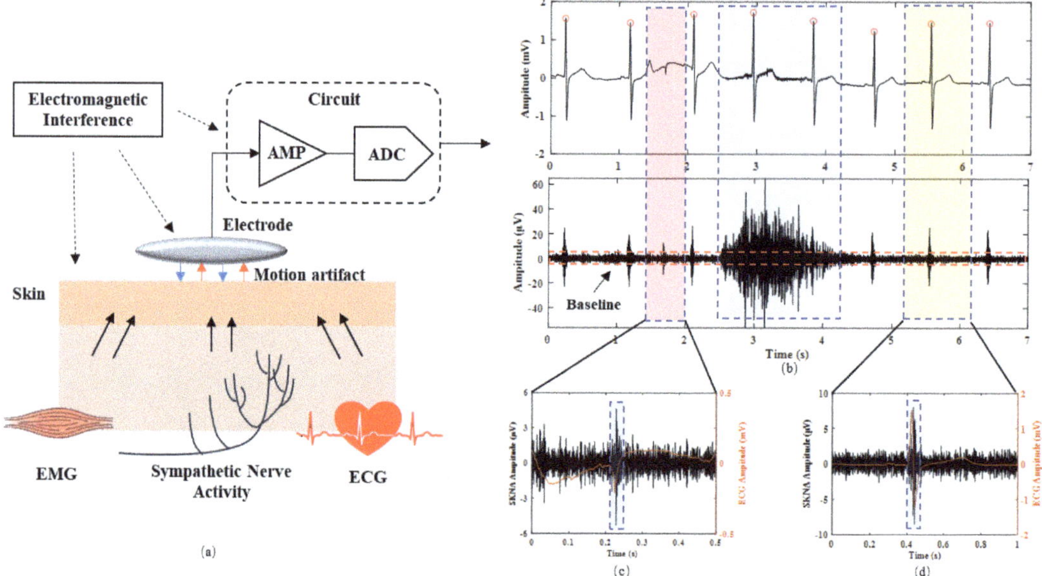

Figure 1. (**a**) The chain of SKNA signal transmission and acquisition. (**b**) Representative examples of acquired signals: the above figure shows the raw signal and the following figure shows SKNA after filtering. (**c**) The step signal artifact. (**d**) The ECG artifact.

There are still some challenges to be addressed in this approach because of the characteristics of SKNA. First, SKNA is aperiodic and has high randomness because it changes with SNA. Therefore, detecting nerve bursts is beneficial to understanding the changes in SKNA. At this point, the nerve bursts are often discriminated with empirical thresholds [19,20]. With the method based on thresholds, it is easy to cause several misjudgments due to the low signal-to-noise ratio of SKNA. The second challenge comes from the quantization of SKNA. Some researchers assess SNA by calculating the average energy of

SKNA [21,22]. Because of noise, signal extraction is easily affected, especially in providing sympathetic-related information. Therefore, it is necessary to further process the obtained signals in order to extract key information from SKNA effectively.

This study aims to develop an artifact-resistant feature to quantify the SKNA signal. We expect this feature to show good performance in short-term and long-term SNA evaluation. The main contributions of this paper can be summarized as follows:

(1) To more accurately detect the burst area, we proposed a method based on TKE operator and envelope and integral signal to detect the burst area. In addition, we proposed to discriminate the ECG artifacts based on QRS complexes.

(2) To accurately quantify the signal, we proposed a new feature, SKNAER, for SNA evaluation based on the detected burst area. We compared SKNAER with aSKNA in the hemodialysis clinical experiments. HRV features related to SNA were calculated simultaneously based on ECG for the comprehensive comparison.

2. Materials and Methods

2.1. Experimental Design

2.1.1. Experimental Setup

PowerLab Data Acquisition Hardware Device (ADInstruments, Lexington Drive Bella Vista New South Wales, Australia) is the signal acquisition system used in the experimental protocol. The data acquired from the signal acquisition system were analyzed by LabChart pro 8 software (ADInstruments, Australia). All the experimental results were imported into MATLAB® (R2019) for further processing and illustrating. The sampling frequency was set to 8 kHz. Two types of experiments were conducted to verify the reliability and effectiveness of the method, including experiments in laboratory environments and clinical experiments. In this work, the signal measurements in the laboratory were carried out in a noise-free sound insulation room, with a temperature of 25 °C and humidity of 50%. The clinical experiments were carried out in the operation room, with a temperature of 25 °C and humidity of 50%. Seven surface electrodes (3M®) were placed on the chest, biceps, forearm, and right abdomen. The positions of electrodes on the body are shown in Figure 2. The sensor positions of the two experiments were the same. Considering the distribution of muscle tissue on the body surface, we assume that ch1 is high signal quality, ch2 is medium signal quality, and ch3 is low signal quality.

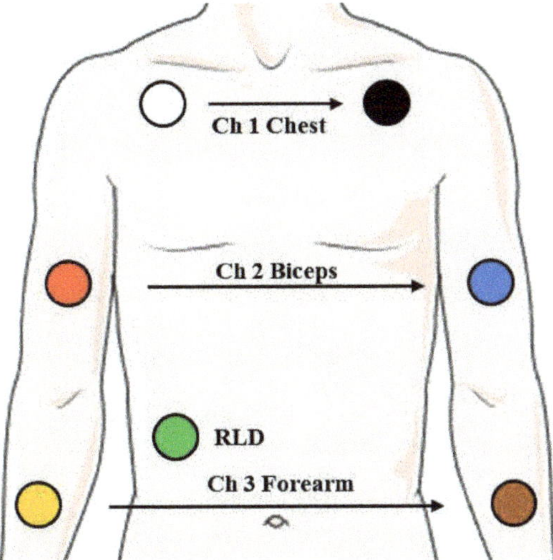

Figure 2. The electrode placement position in the experiment. In Experiment 2, the signal is only collected from channel 1.

2.1.2. Experimental Protocol

Table 1 summarizes the demographic information of the experiments. Ten healthy subjects without CVDs in Experiment 1 and twenty clinical subjects in Experiment 2 from the First Affiliated Hospital of Nanjing Medical University participated in this study. The healthy subjects' average age, height, and weight without CVDs were 25.1 ± 4.6 years old, 173.2 ± 6.5 cm, and 71.0 ± 13.6 kg. The clinical subjects' average age, height, and weight were 58.9 ± 14.6 years old, 170.2 ± 10.3 cm, and 70.5 ± 13.9 kg. Written and informed consent was obtained from each patient. The Ethics Committee has approved the patient experimental protocols of the First Affiliated Hospital of Nanjing Medical University, understudy number 2020-SRFA-183.

Table 1. Summary of demographic information of subjects that participated in experiments.

	Experiment 1		Experiment 2	
	Mean	**Standard Deviation**	**Mean**	**Standard Deviation**
Age/years	25.1	4.6	58.9	14.6
Height/cm	173.2	6.5	170.2	10.3
Weight/kg	71.0	13.6	70.5	13.9
Weight/kg (after dialysis)			67.9	13.6
Cohort size	10		20	

Experiment 1: Standard SKNA signal

We enrolled ten healthy volunteers for signal recording during the cold-water pressor test (CPT) [23] and the Valsalva maneuver (VM) [24]. The two experiments consisted of three steps. The subjects were required to stay in a seated position for a half minute during the first step. This step aimed to eliminate the interference of signal recording at the beginning of the equipment and record the baseline waveform for each subject. The second step was the VM and CPT, the standard procedures for triggering sympathetic discharges. In VM, the subjects were directed to close the glottis after deep inspiration for 30 s. Then, the subjects were monitored for an additional 30 s after directed exhalation. The subjects were guided to use abdominal breathing and avoid unnecessary chest movements. The CPT was performed by placing the subject's left hand up to the wrist in iced water for one minute. The hand was taken out of the iced water after a minute. These experiments were non-invasive and non-drug experiments to change the autonomic nervous system activity. Each subject was required to repeat each task 10 times. A two-minute control and recovery period was recorded for both maneuvers.

Experiment 2: Clinical SKNA signals

The clinical signals were recorded in an operating room. The data of uremic patients during hemodialysis were recorded. Changes in blood volume during hemodialysis can affect the autonomic nervous system. Firstly, the subjects stayed in the supine position before the operation and started recording data simultaneously. After the nurse punctured the arteriovenous fistula of the patients, the patients' blood was drawn out of the body. Then, the blood was exchanged in the dialyzer to remove the toxin. Finally, the processed blood was fed into the patients' bodies. The whole hemodialysis process lasted approximately 4 h. It is worth mentioning that the subjects were patients who needed hemodialysis for their chronic kidney disease. The experiment was an observational data recording experiment and had no additional impact on patients. Data recordings were stopped after hemodialysis. The clinical signal only collected the data from channel 1. On the one hand, the interference of data acquisition on clinical patients should be avoided as much as possible. On the other hand, the emphasis of Experiment 2 is different from that of Experiment 1, and more emphasis is placed on verifying the clinical application effect. The patients were asked to stay supine and avoid unnecessary movement during the recording. Electronic instrument usage, which could produce signal artifacts, was avoided during recordings.

2.2. Burst Detection Method with iSKNA

The burst can be defined as a period of continuous biopotential signal with amplitude higher than the baseline. Therefore, the signal state can be expressed using the following formula:

$$s \sim \begin{cases} 1, & if\ h(x) \geq \tau \\ 0, & if\ h(x) < \tau \end{cases} \tag{1}$$

where s is the state of the signal and $h(x)$ is the statistical feature of the signal. When the statistical characteristic is greater than the threshold τ, it is considered that the signal state at this time is burst. Otherwise, it is the baseline.

As shown in Figure 3a,b, the measured signal was first band-pass filtered (from 500 to 1000 Hz). Then, rectified signal was obtained by full-wave rectification [12].

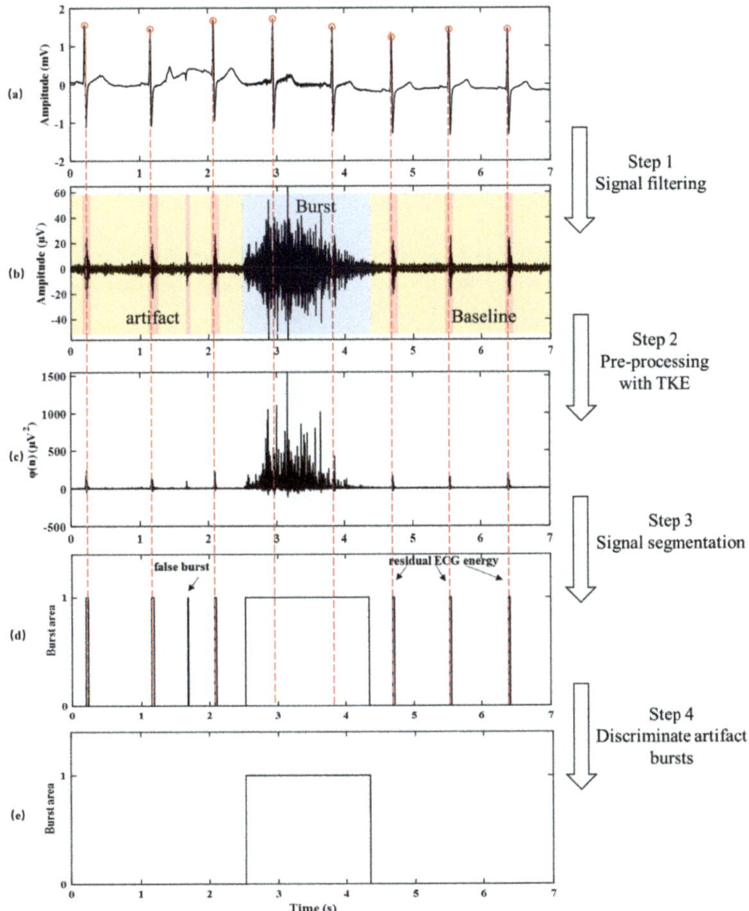

Figure 3. Representative examples of signal preprocessing and segmentation processes using the QRS information complexes and a sensitive threshold. (**a**) The raw signal. (**b**) The filtered SKNA signal. (**c**) The preprocessed signal after TKE operator. (**d**) The segmented burst area based on envelope and integral signal. (**e**) The final segmented burst area. The burst was in the blue box, the baseline was in the yellow box, and the artifact was in the red box.

The envelope of the rectified signal was created with a first-order resistance-capacitance integrating the network with a time constant of 0.1 s. After this step, integral skin sympathetic nerve activity (iSKNA) was obtained. At this point, the baseline and burst are often discriminated with a threshold. The threshold is calculated using the following formula:

$$Threshold = \mu + \rho \times \sigma \tag{2}$$

where μ is the mean of iSKNA, σ is the standard deviation of iSKNA, and ρ is an empirical parameter. Commonly, ρ is set to 3 in previous studies [11,12]. One must decrease ρ to increase sensitivity for SKNA burst. One must increase ρ for higher specificity.

2.3. Optimized Burst Detection Method

2.3.1. Teager–Kaiser Energy Operator

Before burst localization, the TKE operator [25] was introduced to implement the enhancement of the effective signal to highlight the amplitude variation of the SKNA signal. The TKE operator was initially proposed to compute the energy of sound [26] or to detect the onset of sound in the field of non-linear speech signal processing. It can characterize the variation degree of signal in amplitude and frequency domain as shown in Figure 3. Therefore, the amplitude-frequency variation of SKNA can be characterized by preprocessing the signal with the TKE operator.

For a given signal sequence $f(n)$, the TKE operator can be written as:

$$\varphi(n) = f^2(n) - f(n+1)f(n-1) \tag{3}$$

where $f(n)$ is the filtered SKNA signal and $\varphi(n)$ is a new discrete sequence after processing with the TKE operator, as shown in Figure 3b,c. The baseline noise is effectively suppressed. The difference between the baseline and the burst is clearer, which lays a good foundation for the later burst detection algorithm.

2.3.2. Signal Segmentation

The discrete sequence was obtained through the processing of the TKE operator. First, full-wave rectification was applied to the discrete sequence. The time window method was used to obtain the envelope to shield the small fluctuations in the discrete sequence and obtain the overall trend of the signal.

The time window length was set to be n, and the envelope at the time point i was defined as the maximum sequence amplitude in the time window. Since the minimum duration of the nerve action potential is 2 ms, the default value of n is 2 ms if the expert does not have an extra set.

$$Sknae(i) = \max(\varphi(j)), j \in \left[i - \frac{n}{2}, i + \frac{n}{2}\right] \tag{4}$$

The characteristic of burst start/end is that the signal changes from stationary white noise signal to maximum peak value. Therefore, the start/end position of the burst can be preliminarily detected in the light of the magnitude $Sknae(i)$ of the envelope and the derivative $d(i)$ of the point, if:

$$d(i) = Sknae(i) - sknae(i-1) \tag{5}$$

$$Sknae(i) \geq t \text{ and } d(i) > 0 \tag{6}$$

$$eth = \lambda \times Skna_{max} \tag{7}$$

where eth is the envelope threshold, which is defined as an empirical value based on the maximum value of the baseline. λ is an empirical parameter, which is adjusted according to different subjects. Commonly, λ is set to 1. That is, the maximum value of the baseline is used as the threshold of the envelope by default. Similarly, the end of the burst is defined as the envelope value less than t, and the derivative is less than 0.

Through the above operation, the start/end position of the burst was detected. However, the signal rose unsteadily, so the start/end position needs to be adjusted slightly. The significance of using the two methods is that the envelope can be used for rough segmentation. The integral signal is used to adjust the starting point position in more detail to segment the signal more accurately. The integral of $\varphi(n)$ is the integral area of the sequence in unit time. It can reflect the detailed changes of sequence better than the envelope. Therefore, the interference of the signal jitter on the envelope start/end position can be reduced to obtain more accurate burst detection results. The $\varphi_{Integral}(n)$ is defined as:

$$\varphi_{Integral}(n) = \frac{1}{\Delta t} \int_{i-\frac{1}{2\Delta t}}^{i+\frac{1}{2\Delta t}} \varphi(j) \tag{8}$$

where Δt is the length of the time window. In the above process, a larger time window of signal envelope can better reflect the overall trend of the sequence. In this operation, selecting the time window Δt as half of the envelope window n can obtain more accurate segmented positions based on rough segmentation.

The rough segmentation envelope length L can be calculated according to the following formula:

$$L = EP - SP \tag{9}$$

where SP is the starting position of the burst and EP is the ending position of the burst. The range of the moving time window is one-tenth of the envelope length L in the adjustment step. The segmented burst area in this step is shown in Figure 3d. Therefore, the new starting and ending position is:

$$SP_{new} = \min\left(\varphi_{\text{Integral}}(i)\right), \, i \in \left[SP - \frac{1}{10}L, SP + \frac{1}{10}L\right] \tag{10}$$

$$EP_{new} = \min\left(\varphi_{\text{Integral}}(i)\right), \, i \in \left[EP - \frac{1}{10}L, EP + \frac{1}{10}L\right] \tag{11}$$

2.3.3. Discrimination of Artifact Bursts

Burst in the signal was detected through the above process, including true nerve burst and false burst. Since the characteristics of the false burst are similar to the trust burst, they are difficult to distinguish based on the envelope.

SKNA signal was obtained from the body surface by standard lead. Therefore, the ECG artifact was the major noise source in SKNA. Although most energies of ECG were filtered out after a 500–1000 Hz band-pass filter, some residual energy still existed as background noise, especially in QRS complexes. Firstly, the R-peaks detection was performed on ECG using an open-source QRS detector [27]. Then, adaptive thresholds were applied to the length signal to determine the onset and duration of the QRS complexes [28]. In this way, we obtained two pieces of information: the position of the ECG artifact and the maximum width of each ECG artifact [29]. If:

$$SP < Index_R < EP \text{ and } W_{qrs} < L \tag{12}$$

where $Index_R$ is the index of the position of the R wave and W_{qrs} is the duration of the QRS complexes. That is, in case the burst area contains the position of the R wave and the width of the burst area is smaller than the width of the QRS complex, the burst area is supposed to be a false positive.

As shown in Figure 3, false positives resulting from short bursts in SKNA are discriminated by applying time thresholds to the start/end positions. In other words, start/end positions that had time differences shorter than Ts seconds were removed. This discriminating operation was applied after the start/end positions had been detected. This operation allowed the algorithm to have a sensitive threshold for the on-time [30]. Since the minimum duration of the nerve action potential is 2 ms [31], the default value of Ts is 2 ms if the expert does not have an extra set.

2.4. SKNA Energy Ratio

In [32,33], aSKNA was used to assess SKNA over a period. The length of the period depends on the length of time required for clinical analysis, 30 s [11], 5 min [12], and half an hour or more [22]. The feature can be calculated in two steps. First, iSKNA is obtained by calculating the sum of the areas under the SKNA curve in unit time, as shown in Formula (7). The unit time was set to 0.1 s in the previous research [12]. Then, the amount of SKNA can be quantified by calculating the average of iSKNA, which is defined as aSKNA.

$$aSKNA = \frac{\sum iSKNA(i)}{N} \tag{13}$$

where N is the ratio of the time window of the calculated feature to that of $iSKNA$. For example, N is 3000 if the time window of $aSKNA$ is 5 min. Noise has a great impact on this feature, especially the baseline noise and impulse noise. Moreover, due to the differences in sensors and subjects, this feature is unstable in individual comparison. Therefore, according to the concept of signal-to-noise ratio (SNR), we defined a feature to estimate SNA by calculating the ratio of the detected burst sequence energy to the baseline energy. The $SKNA$ energy ratio ($SKNAER$) is defined as:

$$SKNAER = 10\log_{10}\frac{P_{burst}}{P_{baseline}} \tag{14}$$

where P_{burst} is the total energy of the nerve burst area detected after step 4 as show in Figure 3. $P_{baseline}$ is the total energy of the baseline area detected after step 3. It is worth noting that the discriminated artifact burst in step 4 is neither baseline nor nerve burst. It is recognized as noise and is not used to calculate the feature. To avoid error calculation caused by extreme conditions, the signal segment is discriminated as abnormal for secondary processing when it is determined to be all baseline or burst.

2.5. Evaluation Methods

To verify the effectiveness and accuracy of the algorithm, the data were manually labeled by an expert. The expert scrolled through the data using a custom graphical user interface tool and manually placed onsets and offsets of SKNA burst. For the annotation of burst, firstly, we recorded the start/end time of the activation action in the standard signal recording experiment. Based on the start/end time of the activation action and the relative amplitude of the signal, experts trimmed the start/end position of the burst.

The first problem is as follows: what was the detection quality of the proposed algorithm compared with the expert? This was answered with the data of Experiment 1.

To answer the first question, we obtained detection rate and precision. These measures were defined in terms of the following quantities and expressed as percentages. The detection rate was obtained to quantify the difference in the number of detected pairs of onsets/offsets and the number of pairs manually labeled by the expert. It was defined as the percentage of the number of true onset/offset pairs detected by the algorithm to the number of bursts labeled by the expert.

True positives (TPs): numbers of burst areas classified as the burst by both the expert and the algorithm.

False positives (FPs): numbers of burst areas classified as the burst by the algorithm and not by the expert.

False negatives (FNs): numbers of burst areas not classified as the burst by the algorithm and classified as the burst by the expert.

The detection rate and precision were defined as follows.

$$DR = \frac{TP}{TP + FN} \times 100\% \tag{15}$$

$$P_+ = \frac{TP}{TP + FP} \times 100\% \tag{16}$$

In addition, we also calculated the coincidence to answer the first question. The coincidence was defined to compare the coincidence degree of the automatic/manual segmentation results by algorithm/expert. The coincidence was computed as the ratio of the manual segmentation length of the overlapping part of automatic segmentation length and manual segmentation length.

$$CO = \frac{\min(EP_a, EP_e) - \max(SP_a, SP_e)}{EP_e - SP_e} \tag{17}$$

where SP_a and EP_a are the automatic segmentation of starting and ending position by algorithm; SP_e and EP_e are the manual segmentation of starting and ending position by expert.

2.6. Reference Features

The second question is what is the clinical effect of the proposed feature, and whether it is more effective in evaluating SNA than other features? The second question can be answered by using the data from Experiment 1 and 2. In Experiment 1, the CPT and the VM are the standard procedures for triggering sympathetic discharges [32,33], which can increase blood pressure, heart rate, and SNA level. In Experiment 2, hemodialysis is to purify the blood by dispersing and circulating all kinds of harmful and redundant metabolic wastes and excess electrolytes out of the body to achieve the purpose of correcting water-electrolyte and acid-base balance. As a result, pressure on the autonomic nervous system decreases, and SNA changes with the release of toxic substances. In this study, we used two methods to process data to verify the effectiveness of SKNA-based evaluation. One method was to use the proposed method to calculate the SKNAER in different periods. The other was to calculate the aSKNA of different periods without additional processing.

In order to further compare the differences in features, we calculated low frequency/high frequency (LF/HF), the standard deviation of NN intervals (SDNN), and very low frequency (vLF) based on HRV. These indicators can reflect the characteristics of SNA. The increase of LF/HF and vLF represented the increase in SNA [34]. The decrease in SDNN indicated an increase in SNA [35]. In addition, the paired t-test was performed to evaluate the difference in features before and after time.

3. Results

Table 2 shows the statistical results of Experiment 1 in detail. The observed burst number of the signal collected from the arm was less than that from the chest, especially in the biceps position. In addition, the false positives and false negatives of the signal from the arm were also more than those of the signal from the chest. This may be due to more EMG interference on the arm. From this point of view, the signals obtained from the chest position and the forearm may have better signal quality than that of the biceps position. The proposed algorithm was verified on standard datasets. We calculated the burst detection accuracy of signals from three acquisition positions, respectively. The results showed that the detection rate and precision of the proposed algorithm on the acquired signal from the chest were $100.0 \pm 0\%$ and $94.2 \pm 5.0\%$, respectively. Although the detection rate and precision of the algorithm decreased on the acquired signal from the biceps and forearm, the detection rate and precision were still acceptable, $96.4 \pm 5.5\%$ and $87.3 \pm 7.4\%$. In addition, the coincidence area was also calculated in the experiment. The coincidence of the signal from the chest was $96.4 \pm 1.2\%$, and it was higher than that of the signal from the arm. Experimental results showed that the proposed algorithm had a satisfactory performance on the acquired signal from a different position.

Table 2. The mean, standard deviation, and confidence interval (90% CI) of detection rate, coincidence, and precision of the proposed algorithm on the acquired signal of Experiment 1. * $p < 0.05$, ** $p < 0.01$, *** $p < 0.001$.

Position		TP	FN	FP	DR (%)	*p*-Value	CO (%)	*p*-Value	P+ (%)	*p*-Value
Ch1 Chest	Proposed method	159	0	9	100.0 ± 0 [100.0 100.0]	0.18	96.4 ± 1.2 [95.6 97.2]	***	94.2 ± 5.0 [91.9 97.9]	***
	With iSKNA [12]	157	2	103	98.8 ± 2.6 [97.2 100.2]		92.2 ± 1.7 [91.4 92.8]		59.9 ± 3.6 [58.2 62.4]	
Ch2 Biceps	Proposed method	111	3	18	96.4 ± 5.5 [95.6 99.3]	*	92.3 ± 2.1 [91.6 92.9]	***	87.3 ± 7.4 [85.3 89.4]	***
	With iSKNA [12]	100	14	47	87.1 ± 11.0 [84.5 92.5]		87.6 ± 2.4 [87.0 88.6]		67.6 ± 9.3 [63.9 70.5]	
Ch3 Forearm	Proposed method	147	2	10	98.7 ± 3.2 [97.0 100.3]	0.34	94.2 ± 1.3 [93.3 94.9]	***	93.7 ± 2.6 [92.3 95.4]	***
	With iSKNA [12]	146	3	94	97.8 ± 5.8 [94.2 101.0]		91.0 ± 0.9 [90.5 91.5]		60.5 ± 5.8 [57.1 63.8]	
Summary	Proposed method	417	4	46	98.2 ± 3.9 [97.7 99.6]	**	94.3 ± 2.3 [93.6 95.0]	***	91.8 ± 6.2 [89.9 93.9]	***
	With iSKNA [12]	403	18	244	94.5 ± 8.9 [91.7 97.2]		90.3 ± 2.6 [89.5 91.0]		62.8 ± 7.3 [60.4 65.0]	

Compared to the proposed method, there were more false positives and false negatives using the method with iSKNA. On the one hand, the increase of false negatives in the biceps position was more than that in other positions. In the acquisition position with high signal quality, the detection rate of the method with iSKNA was not much different from that of the proposed method. However, the detection rate of the signal collected at the biceps position was reduced to 87.1 ± 11.0%. On the other hand, the difference between the method with iSKNA and the proposed method was mainly reflected in the number of false positives, resulting in the difference in precision. Since there was no additional processing for ECG artifacts, there were a number of false positives using the method with iSKNA. The precision and CO of the method with iSKNA were much lower than those of the proposed method. In general, the proposed method had better performance in burst detection than the method with iSKNA, especially in the signal with low signal quality.

Figure 4a–d indicate the correlation between SKNAER and the other features of the ten patients before and after sympathetic activation in Experiment 1. The results showed that the SKNAER was positively correlated with SDNN (r = 0.60), LF/HF ratio (r = 0.67), vLF power (r = 0.47), and aSKNA (r = 0.67). Figure 5a–e show the box diagram of the features related to the SNA before and after sympathetic activation in Experiment 1. The HRV features and SKNA features of sympathetic activation one minute before and one minute after were calculated. All features showed an upward trend after sympathetic activation. For the HRV features, SDNN increased from 48.75 to 93.19 ms ($p < 0.001$). From the perspective of this feature, SNA showed a downward trend. vLF increased from 5510.96 to 12,213.52 ms^2 ($p < 0.05$), and three outliers occurred in the experiment. LF/HF increased from 1.94 to 4.27 ($p = 0.15$). For the SKNA features, aSKNA increased from 0.91 to 1.42 μV ($p < 0.001$) and SKNAER increased from −13.76 to 8.29 dB ($p < 0.001$). The calculated values of SKNAER before and after sympathetic activation had no overlap. That is, they were almost unaffected by individuals.

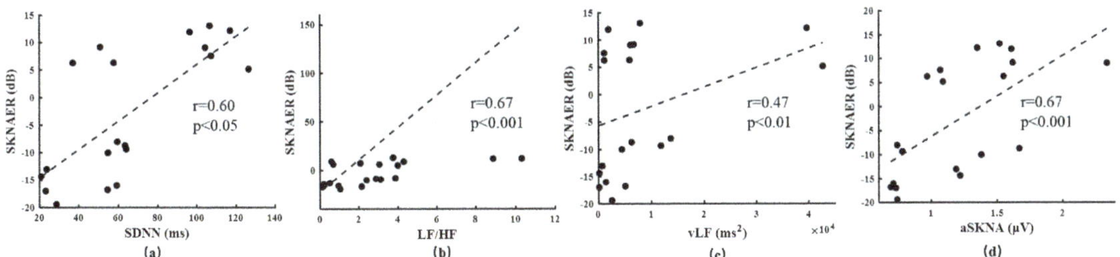

Figure 4. (**a–d**) The correlation between SKNAER and HRV features of the ten patients before and after sympathetic activation in Experiment 1.

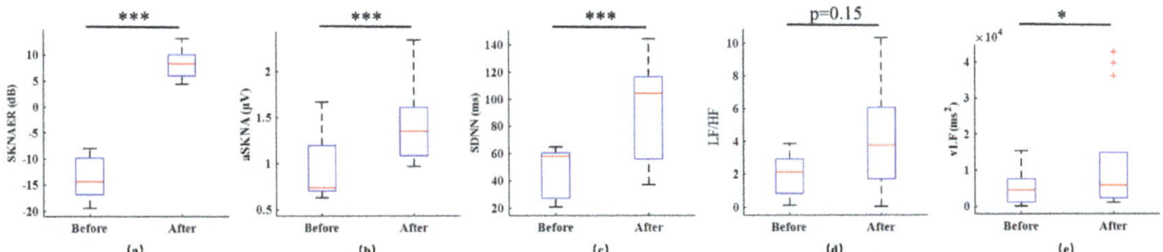

Figure 5. (**a–e**) The statistical results of the features related to the sympathetic nervous activity before and after sympathetic activation in Experiment 1. * $p < 0.05$, *** $p < 0.001$.

The whole data process of renal dialysis for each patient lasted about 4 h. Taking 30 min as a window, data were divided into eight periods. Figure 6 shows the trend of SKNA and HRV features during dialysis in 20 patients with renal failure in the clinical experiment. The SKNA and HRV features of each period were calculated to assess the SNA of renal dialysis patients. For the SKNA features, these two features decreased the second time, aSKNA from 1.19 to 1.05 μV ($p < 0.01$), SKNAER from 1.99 to −3.04 dB ($p < 0.001$). Then, these features increased in the fourth time period, aSKNA from 1.03 to 1.15 μV ($p = 0.07$), SKNAER from −2.45 to 1.94 dB ($p < 0.001$), and began to decline in the sixth time period, aSKNA from 1.24 to 1.12 μV ($p = 0.11$), SKNAER from 2.25 to −0.87 dB ($p < 0.01$), and remained stable until the end of the operation.

For the HRV features, LF/HF maintained an upward trend in the first two hours, especially the fourth period, from 2.13 to 2.70 ($p < 0.05$), and dropped from 2.19 to 1.60 in the sixth period ($p < 0.01$), and then finally increased, reaching 2.01 in the eighth period. vLF fluctuated greatly in the first two hours. It decreased from 1043.50 to 728.80 in the third period ($p < 0.05$) and quickly increased to 918.3 in the fourth period ($p < 0.05$). The trend in the last two hours was similar to that of LF/HF, which decreased first and then increased. SDNN rose continuously in the first three periods and fell from 42.88 to 38.47 ms in the fourth period. It increased in the sixth period, from 38.49 to 40.79 ms, and remained stable. The common trend of the HRV features was that there was a significant difference in the fourth period, indicating the activation of SNA. There was also a significant difference in the sixth period, indicating inhibition of SNA. The variation tendencies of HRV features were consistent with those of aSKNA and SKNAER. However, SKNAER had a more significant difference than aSKNA in evaluating SNA at different times, especially in the fourth and sixth periods.

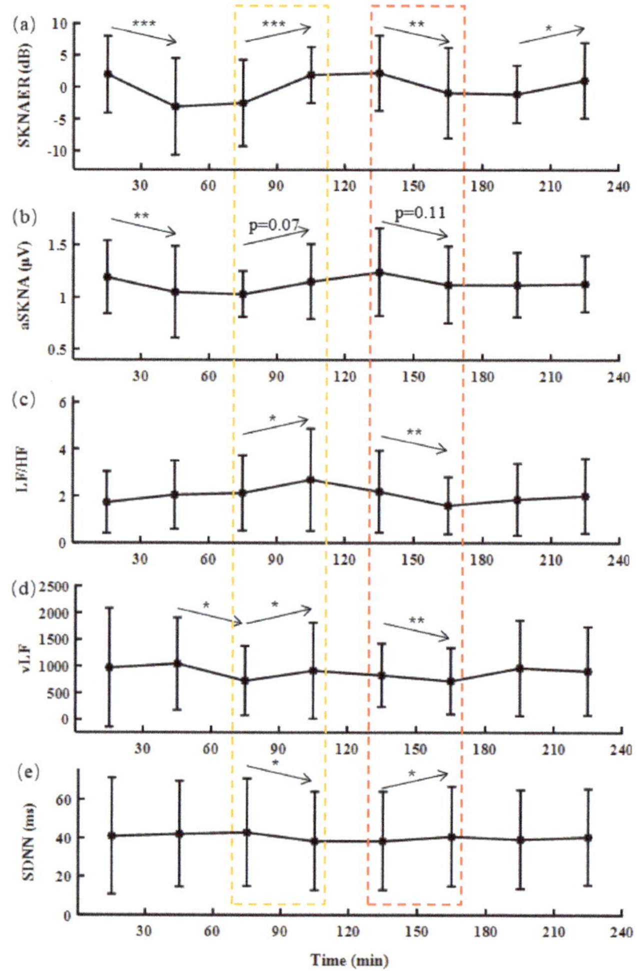

Figure 6. (a–e) The trend of the SKNA and HRV features in the twenty patients during four-hour hemodialysis. * $p < 0.05$, ** $p < 0.01$, *** $p < 0.001$.

4. Discussion

In this work, we developed a burst area detection algorithm and verified the accuracy of the standard sympathetic nerve activation experiments, including the Valsalva experiment and the CPT experiment. We obtained SKNA signals by placing three groups of electrode sensors on the human surface. The data were automatically detected by the proposed algorithm and manually labeled by experts receptively. Experimental results indicated that the consistency of the detected burst area between the algorithm and the expert was high. However, the number of burst areas observed from different positions was different, and the burst number from the biceps was smaller than that of the chest. In other words, the proposed algorithm can effectively locate the burst area, but it cannot further decouple the mixed noise and neural signals. Although the SKNA signal originates from ganglion, the effective information obtained by the signal was different due to the influence of the interference. Preprocessing multichannel signals with principal component analysis and other methods may lay a better foundation before burst segmentation.

The accuracy of burst segmentation was verified on the sympathetic activation dataset. The results showed that the detection rate on the signal of the chest reached 100%. Although the detection rate decreased on the signal of the forearm and biceps, it was also greater than 95%. Therefore, the proposed algorithm showed good performance in different acquisition positions. In addition, the coincidence degree greater than 95% indicated that the sensitivity to detect the burst area was close to the manual label of the expert. Compared with the method with iSKNA, the advantages of the proposed method were the improvement of detection rate on low signal-to-noise ratio signals and the discrimination of false positives. However, due to the complex neural changes and other confounding factors in the experiment, the number of false positives was not satisfactory enough, especially for the acquisition signal at the biceps. This may be affected by the accuracy of the QRS complexes' detected algorithm and EMG artifact, which led to the unrecognized false burst. Further analysis of the difference between the true and false burst may provide a new train of thought to reduce the misjudgment rate.

SKNA-based SNA evaluation has great potential in pathogenesis research such as atrial fibrillation [32] and myocardial infarction [22]. The SNA evaluation is often quantified by the estimation of mean burst amplitude [36], total burst amplitude [37], or burst area [38]. However, direct calculation of the mean burst amplitude, such as aSKNA [22,32], greatly impacted the evaluation of SNA because of the signal noise. The outlier of aSKNA further proved this point during hemodialysis. We discriminated and removed the non-typical burst after detecting the burst area and calculated the burst area energy ratio to assess the SNA. This feature had better significance in reflecting the trend of the SNA during the Hemodialysis experiment. Observational results of the hemodialysis experiment showed that the SNA gradually decreased in the second period, which was the patients' process from dynamic to static. It increased in the fourth period. This is the period when the most malignant clinical events may occur. In the sixth period, the sympathetic nerves returned to calm. The patients gradually completed dialysis for harmful substances in the body during this period. In the eighth period, the SNA increased, and the patients' activity increased at the last moment. From the perspective of aSKNA, except for the second period, the sympathetic nerve of the patient did not change significantly compared with the previous time. However, we observed more time-to-time differences from SKNAER. From this point of view, it can help us better understand the changes in the sympathetic nerves of patients during hemodialysis.

For Experiment 1, VM and CPT are standard sympathetic stimulation procedures. Results showed that there was no significant difference between HRV-based features and SKNA-based features in indicating SNA. It is worth noting that although SDNN showed significant changes before and after sympathetic stimulation procedures, the indicated SNA was actually the opposite. Although SDNN can effectively indicate the changes in the sympathetic nerve in most clinical studies, it is difficult to explain in some specific clinical scenes [39], especially in short-term measurement. It may be because there are many interference factors for RR interval in short-term analysis, such as false heartbeat [40] and algorithm error. Therefore, SKNA-based features were better than HRV-based features in the short-term analysis of SNA. Furthermore, the values of SKNAER before and after sympathetic activation did not overlap, while those of other features overlapped to a certain extent. In other words, SKNAER can better avoid individual differences compared to the other features. This can help us compare the SNA of different individuals, not just individuals themselves. According to the experimental results of this study, the frequency domain index of HRV may have a better correlation and interpretation with SKNAER in short-term analysis, such as LF/HF.

For Experiment 2, the trends of all features were not completely consistent in reflecting changes in SNA. However, these features showed the same change trend in some specific time periods, the fourth and sixth time periods, respectively. In other words, although the two kinds of features were calculated based on different signals, they were consistent in reflecting the trend of SNA at some levels. For the SNA evaluation, many features provided indirect explanations. However, these features sometimes did not always show the same trend [41,42]. This may be because the signal was doped with various interference factors, resulting in inaccurate interpretation. For example, HRV features were calculated based on RR intervals. The RR interval is often affected by the accuracy of the R-wave-detecting algorithm. In addition, patients with atrial fibrillation and premature beats also have a negative impact on the calculation results because of their special heart rhythm. Similarly, SKNAER is also affected by these factors. The wrong R-wave location result may lead to the misjudgment of the SKNA burst, which reduces the precision of the algorithm, especially in the signal with low signal quality. However, the positive outcome is that we discriminated and removed some interference in the calculation to provide a more accurate evaluation of SNA. In addition, SKNAER is calculated based on the energy of burst and baseline, so the influence of R-wave positioning accuracy on SKNA is indirect and less than that of the HRV index. With the advanced QRS positioning algorithm, the influence of R-wave positioning error on the SKNA feature can be further reduced. Furthermore, we established an evaluation method based on SKNA. Since SKNA is the real-time signal transmitted from the sympathetic nerve to the body surface, while ECG is the change of potential cardiac signal caused by the sympathetic nerve affecting cardiac function, thus the method based on SKNA is more direct than that of ECG in evaluating SNA. This also provides an auxiliary basis for the analysis of disease mechanisms.

The findings from this study should be considered in light of several factors. First, the proposed feature assumes that there is sympathetic activation in the measured data segment. The feature may output an error in case there is no burst area in the data segment or the noise covers the burst. Second, the shape and frequency of the burst were not considered for the proposed feature. This more in-depth information should be further processed and characterized to be applied to some specific clinical scenarios. Last, although the TKE factor is used to enhance the amplitude-frequency change of the signal, this operation may also enhance the amplitude-frequency change of some artifacts, resulting in false positives. Research on the conduction mode of nerve signals to the body surface and then distinguishing between effective nerve burst and noise is conducive to further finding effective information from SKNA. In addition, SKNA is greatly affected by motion artifacts The impact comes from the electrodes used to collect signals and the lack of appropriate standards to distinguish effective signals from motion artifacts. This may limit the further application of SKNA. Therefore, SKNA can be used as an effective supplement rather than a substitute for HRV in some specific scenarios, especially in short-term measurement.

5. Conclusions

This paper proposed an SNA evaluation method based on SKNA burst area detection. This method exhibited good performance in terms of detection rate, concordance, and precision on the sympathetic activation dataset manually labeled by the experts. The trend of SNA during hemodialysis was analyzed quantitatively based on the detected burst area. The results showed that SKNAER has a consistent trend in evaluating SNA compared with HRV features. Moreover, it had a more significant difference in the long-term SNA evaluation than aSKNA, consistent with the HRV features. The automatic burst detection algorithm proposed in this work can accurately locate the position of the active SNA, which is helpful to evaluate SNA more accurately. With further development, this new modality could play an important role in continuous monitoring of the autonomic nervous system status, as well as preventing correlated diseases. Future work will explore more useful physiological features for evaluating the autonomic nervous system, such as time-frequency analysis and non-linear dynamic evaluation.

Author Contributions: Conceptualization, C.L. and Y.X.; methodology, Y.X.; software, J.L. (Jiayi Li); validation, Z.X. and J.L. (Jiayi Li); formal analysis, H.C.; investigation, Y.Z.; resources, C.C. and J.W.; data curation, Y.Z.; writing—original draft preparation, Y.X.; writing—review and editing, Y.X. and C.Y.; visualization, Y.X.; funding acquisition, J.L. (Jianqing Li) and. C.L. All authors have read and agreed to the published version of the manuscript.

Funding: This work was supported by the National Key Research and Development Program of China (2019YFE0113800), the National Natural Science Foundation of China (62171123, 81871444, 62001111, and 62071241), and the Natural Science Foundation of Jiangsu Province of China (BK20190014, BK20192004, and BK20200364).

Institutional Review Board Statement: The study was conducted in accordance with the Declaration of Helsinki, and approved by the Ethics Committee of the First Affiliated Hospital of Nanjing Medical University, understudy number 2020-SRFA-183.

Informed Consent Statement: Informed consent was obtained from all subjects involved in the study. Written informed consent was obtained from the patient(s) to publish this paper.

Data Availability Statement: Not applicable.

Conflicts of Interest: The authors declare no conflict of interest.

References

1. World Health Organization. *Cardiovascular Diseases (CVDs)*; World Health Organization: Geneva, Switzerland, 2018.
2. McGill, H.; McMahan, C. Preventing heart disease in the 21st century: Implications of the pathobiological determinants of atherosclerosis in youth (PDAY) study. *Circulation* **2008**, *117*, 1216–1227. [CrossRef] [PubMed]
3. Zhu, C.; Hanna, P.; Rajendran, P.; Shivkumar, K. Neuromodulation for ventricular tachycardia and atrial fibrillation: A clinical scenario-based review. *JACC-Clin. Electrophysiol.* **2019**, *5*, 881–896. [CrossRef] [PubMed]
4. Goldberger, J.; Arora, R.; Buckley, U.; Shivkumar, K. Autonomic nervous system dysfunction. *JACC Focus Semin.* **2019**, *73*, 1189–1206. [CrossRef] [PubMed]
5. Hagbarth, K.; Vallbo, A. Pulse and respiratory grouping of sympathetic impulses in human muscle-nerves. *Acta Physiol. Scand.* **1968**, *74*, 96–108. [CrossRef]
6. Linz, D.; Mahfoud, F.; Schotten, U.; Ukena, C.; Hohl, M.; Neuberger, H.; Wirth, K.; Böhm, M. Renal sympathetic denervation provides ventricular rate control but does not prevent atrial electrical remodeling during atrial fibrillation. *Hypertension* **2013**, *61*, 225–231. [CrossRef]
7. Jung, B.; Dave, A.; Tan, A.; Gholmieh, G.; Zhou, S.; Wang, D.; Akingba, G.; Fishbein, G.; Montemagno, G.; Lin, S.; et al. Circadian variations of stellate ganglion nerve activity in ambulatory dogs. *Heart Rhythm* **2006**, *3*, 78–85. [CrossRef]
8. Robinson, E.; Rhee, K.; Tan, A.; Gholmieh, G.; Zhou, S.; Wang, D.; Akingba, G.; Fishbein, G.; Montemagno, C.; Lin, S.; et al. Estimating sympathetic tone by recording subcutaneous nerve activity in ambulatory dogs. *J. Cardiovasc. Electrophysiol.* **2015**, *26*, 70–78. [CrossRef]
9. Posada-Quintero, H.F.; Florian, J.P.; Orjuela-Cañón, A.D.; Aljama-Corrales, T.; Charleston-Villalobos, S.; Chon, K.H. Power spectral density analysis of electrodermal activity for sympathetic function assessment. *Ann. Biomed. Eng.* **2016**, *44*, 3124–3135. [CrossRef]
10. Posada-Quintero, H.F.; Florian, J.P.; Orjuela-Cañón, A.D.; Chon, K.H. Highly sensitive index of sympathetic activity based on time-frequency spectral analysis of electrodermal activity. *Am. J. Physiol.* **2016**, *311*, 582–591. [CrossRef]
11. Doytchinova, A.; Hassel, J.; Yuan, Y.; Lin, H.; Yin, D.; Adams, D.; Straka, S.; Wright, K.; Smith, K.; Wagner, D.; et al. Simultaneous non-invasive recording of skin sympathetic nerve activity and electrocardiogram. *Heart Rhythm* **2017**, *14*, 25–33. [CrossRef]
12. Kusayama, T.; Wong, J.; Liu, X.; He, W.; Doytchinova, A.; Robinson, E.; Adams, D.; Chen, L.; Lin, S.; Davoren, K.; et al. Simultaneous non-invasive recording of electrocardiogram and skin sympathetic nerve activity (neuECG). *Nat. Protoc.* **2020**, *5*, 1853–1877. [CrossRef] [PubMed]
13. Xing, Y.; Li, J.; Hu, Z.; Li, Y.; Zhang, Y.; Cui, C.; Cai, C.; Liu, C. A portable neuECG monitoring system for cardiac sympathetic nerve activity assessment. In Proceedings of the 2020 International Conference on Sensing, Measurement & Data Analytics in the era of Artificial Intelligence (ICSMD), Xi'an, China, 15–17 October 2020; pp. 407–412.
14. Liu, X.; Rabin, P.; Yuan, Y.; Kumar, A.; Vasallo, P.; Wong, J.; Mitscher, G.; Everett, T.; Chen, P. Effects of anesthetic and sedative agents on sympathetic nerve activity. *Heart Rhythm* **2019**, *16*, 1875–1882. [CrossRef] [PubMed]
15. Jiang, Z.; Zhao, Y.; Doytchinova, A.; Kamp, N.; Tsai, W.; Yuan, Y.; Adams, D.; Wagner, D.; Shen, C.; Chen, L.; et al. Using skin sympathetic nerve activity to estimate stellate ganglion nerve activity in dogs. *Heart Rhythm* **2015**, *12*, 1324–1332. [CrossRef] [PubMed]
16. Zaydens, E.; Taylor, J.A.; Michael, A.C.; Uri, T.E. Characterization and modeling of muscle sympathetic nerve spiking. *IEEE Trans. Biomed. Eng.* **2013**, *60*, 2914–2924. [CrossRef] [PubMed]
17. Beckman, L.; Medrano, G.; Jungbecker, N.; Walter, M.; Gries, T.; Leonhardt, S. Characterization of textile electrodes and conductors using standardized measurement setups. *Physiol. Meas.* **2010**, *31*, 233–247. [CrossRef]
18. Liu, C.; Yang, M.; Di, J.; Xing, Y.; Li, Y.; Li, J. Wearable ECG: History, key technologies, and future challenges. *Chin. J. Biomed. Eng.* **2019**, *38*, 641–652.
19. Diedrich, A.; Charoensuk, W.; Brychta, W.; Ertl, A.; Shiavi, R. Analysis of raw microneurographic recordings based on wavelet de-noising technique and classification algorithm: Wavelet analysis in microneurography. *IEEE Trans. Biomed. Eng.* **2003**, *50*, 41–50. [CrossRef]

20. Okamura, S.; Yonezawa, Y.; Ninomiya, I. A new segmentation method of synchronized sympathetic nerve activity. *Eng. Med. Biol. Soc.* **1995**, *2*, 1003–1004.
21. Brychta, R.J.; Tuntrakool, S.; Appalsamy, M.; Keller, N.; Robertson, D.; Shiavi, R.; Diedrich, A. Wavelet methods for spike detection in mouse renal sympathetic nerve activity. *IEEE Trans. Biomed. Eng.* **2007**, *54*, 82–93. [CrossRef]
22. Liu, C.; Lee, C.; Lin, S.; Tsai, W. Temporal clustering of skin sympathetic nerve activity bursts in acute myocardial infarction patients. *Front. Neurosci.* **2021**, *15*, 1501. [CrossRef]
23. Victor, R.; Leimbach, W.; Seals, D.; Wallin, B.; Mark, A. Effects of the cold pressor test on muscle sympathetic nerve activity in humans. *Hypertension* **1987**, *9*, 429–436. [CrossRef] [PubMed]
24. Levin, A. A simple test of cardiac function based upon the heart rate changes induced by the valsalva maneuver. *Am. J. Cardiol.* **1966**, *18*, 90–99. [CrossRef]
25. Lemyre, C.; Jelinek, M.; Lefebvre, R. New approach to voiced onset detection in speech signal and its application for frame error concealment. In Proceedings of the 2008 IEEE International Conference on Acoustics, Speech and Signal Processing, Las Vegas, NV, USA, 31 March–4 April 2008; pp. 4757–4760.
26. Kaiser, J. A simple algorithm to calculate the energy of a signal. In Proceedings of the IEEE International Conference Acoustics, Speech, and Signal Processing, Albuquerque, NM, USA, 3–6 April 1990; pp. 381–384.
27. Liu, C.; Zhang, X.; Zhao, L.; Liu, F.; Chen, X.; Yao, Y.; Li, J. Signal quality assessment and lightweight QRS detection for wearable ECG SmartVest system. *IEEE Internet Things J.* **2019**, *6*, 1363–1374. [CrossRef]
28. Zong, W.; Moody, G.B.; Jiang, D. A robust open-source algorithm to detect onset and duration of QRS complexes. In Proceedings of the Computers in Cardiology, Thessaloniki, Greece, 21–24 September 2003; pp. 737–740.
29. Liu, C.; Li, P.; Maria, C.; Zhao, L.; Zhang, H.; Chen, Z. A multi-step method with signal quality assessment and fine-tuning procedure to locate maternal and fetal QRS complexes from abdominal ECG recordings. *Physiol. Meas.* **2014**, *35*, 1665–1683. [CrossRef]
30. Rashid, U.; Niazi, I.K.; Signal, N.; Farina, D.; Taylor, D. Optimal automatic detection of muscle activation intervals. *J. Electromyogr. Kinesiol.* **2019**, *48*, 103–111. [CrossRef]
31. Bonato, P.; D'Alessio, T.; Knaflitz, M. A statistical method for the measurement of muscle activation intervals from surface myoelectric signal during gait. *IEEE Trans. Biomed. Eng.* **1998**, *45*, 287–299. [CrossRef] [PubMed]
32. Kusayama, T.; Douglas, A.; Wan, J.; Doytchinova, A.; Wong, J.; Mitscher, G.; Straka, S.; Shen, C.; Everett, T.; Chen, P. Skin sympathetic nerve activity and ventricular rate control during atrial fibrillation. *Heart Rhythm* **2020**, *17*, 544–552. [CrossRef]
33. Kusayama, T.; Wan, J.; Doytchinova, A.; Wong, J.; Kabir, R.; Mitscher, G.; Straka, S.; Shen, C.; Everett, T.; Chen, P. Skin sympathetic nerve activity and the temporal clustering of cardiac arrhythmias. *JCI Insight* **2019**, *4*, e125853. [CrossRef]
34. Vinik, A.; Maser, R.; Mitchell, B.; Freeman, R. Diabetic autonomic neuropathy. *Diabetes Care* **2003**, *26*, 1553–1579. [CrossRef]
35. Moak, J.; Goldstein, D.; Eldadah, B.; Saleem, A.; Holmes, C.; Pechnik, S.; Sharabi, Y. Supine low-frequency power of heart rate variability reflects baroreflex function, not cardiac sympathetic innervation. *Heart Rhythm* **2007**, *4*, 1523–1529. [CrossRef]
36. Sundlöf, G.; Wallin, B. Human muscle nerve sympathetic activity at rest. Relationship to blood pressure and age. *J. Physiol.* **1978**, *274*, 621–637. [CrossRef] [PubMed]
37. Sundlöf, G.; Wallin, B. Muscle-nerve sympathetic activity in man. Relationship to blood pressure in resting normo- and hyper-tensive subjects. *Clin. Sci. Mol. Med.* **1978**, *4*, 387–389. [CrossRef] [PubMed]
38. Sugiyama, Y.; Matsukawa, T.; Suzuki, H.; Iwase, S.; Mano, T. A new method of quantifying human muscle sympathetic nerve activity for frequency domain analysis. *Electroencephalogr. Clin. Neurophysiol.* **1996**, *101*, 121–128. [CrossRef]
39. Federico, L.; Heikki, H.; Scgnudt, G.; Malik, M. Short-term heart rate variability: Easy to measure, difficult to interpret. *Heart Rhythm* **2018**, *15*, 1559–1560.
40. Fred, S.; Zachary, M.; Christopher, L. A critical review of ultra-short-term heart rate variability norms research. *Front. Neurosci.* **2020**, *14*, 594880.
41. Zhang, Y.; Wang, J.; Xing, Y.; Cui, C.; Cheng, H.; Chen, Z.; Chen, H.; Liu, C.; Wang, N.; Chen, M. Dynamics of cardiac autonomic response during hemodialysis measured by heart rate variability and skin sympathetic nerve activity: The impact of interdialytic weight gain. *Front. Physiol.* **2022**, *13*, 890536. [CrossRef]
42. Xing, Y.; Zhang, Y.; Yang, C.; Li, J.; Li, Y.; Cui, C.; Li, J.; Cheng, H.; Fang, Y.; Cai, C.; et al. Design and evaluation of an autonomic nerve monitoring system based on skin sympathetic nerve activity. *Biomed. Signal Process. Control* **2022**, *76*, 103681. [CrossRef]

biosensors

MDPI

Article

An Intelligent ECG-Based Tool for Diagnosing COVID-19 via Ensemble Deep Learning Techniques

Omneya Attallah

Department of Electronics and Communications Engineering, College of Engineering and Technology, Arab Academy for Science, Technology and Maritime Transport, Alexandria 1029, Egypt; o.attallah@aast.edu

Abstract: Diagnosing COVID-19 accurately and rapidly is vital to control its quick spread, lessen lockdown restrictions, and decrease the workload on healthcare structures. The present tools to detect COVID-19 experience numerous shortcomings. Therefore, novel diagnostic tools are to be examined to enhance diagnostic accuracy and avoid the limitations of these tools. Earlier studies indicated multiple structures of cardiovascular alterations in COVID-19 cases which motivated the realization of using ECG data as a tool for diagnosing the novel coronavirus. This study introduced a novel automated diagnostic tool based on ECG data to diagnose COVID-19. The introduced tool utilizes ten deep learning (DL) models of various architectures. It obtains significant features from the last fully connected layer of each DL model and then combines them. Afterward, the tool presents a hybrid feature selection based on the chi-square test and sequential search to select significant features. Finally, it employs several machine learning classifiers to perform two classification levels. A binary level to differentiate between normal and COVID-19 cases, and a multiclass to discriminate COVID-19 cases from normal and other cardiac complications. The proposed tool reached an accuracy of 98.2% and 91.6% for binary and multiclass levels, respectively. This performance indicates that the ECG could be used as an alternative means of diagnosis of COVID-19.

Keywords: deep learning; COVID-19; ECG trace image; transfer learning; Convolutional Neural Networks (CNN); feature selection

Citation: Attallah, O. An Intelligent ECG-Based Tool for Diagnosing COVID-19 via Ensemble Deep Learning Techniques. *Biosensors* 2022, 12, 299. https://doi.org/10.3390/bios12050299

Received: 8 March 2022
Accepted: 24 April 2022
Published: 5 May 2022

Publisher's Note: MDPI stays neutral with regard to jurisdictional claims in published maps and institutional affiliations.

1. Introduction

At the end of December 2019, the world faced a new type of threatening disease called coronavirus, commonly known as COVID-19 [1]. Based on statistics announced by the World Health Organization (WHO) [2], more than 190 million cases of COVID-19 and more than 4 million cases of mortality have been reported worldwide on 31 July 2021. Due to the rapid propagation and the massive increase in the number of new infections of such a disease, the world faced new challenges [3]. These challenges involved travel constraints, countries' lockdown, social distancing, and curfews. Most importantly, healthcare associations of many countries were about to collapse due to the superfluous number of COVID-19 infections that needed beds and deficiencies in vital medical kits and supplies. Consequently, the rapid and precise diagnosis of COVID-19 is important to lower mortality rates and avert the encumbrance on health organizations.

Based on the COVID-19 diagnosis provided by the Chinese government, the real-time reverse transcription-polymerase chain reaction (RT-PCR) test is the gold standard for the diagnosis of COVID-19 [4]. However, late sample acquisition, firm laboratory setting restrictions, and the requirement of qualified experts to perform the RT-PCR exam could lead to a prolonged and inaccurate diagnosis [5]. Therefore, more efficient methods are needed to achieve a more precise and faster diagnosis. Among these approaches are antigen tests and medical imaging, including computed tomography (CT) and X-ray imaging techniques. Although COVID-19 antigen tests are faster and cheaper than the RT-PCR test, they very often produce inaccurate results. The major limitation of antigen tests is their low sensitivity, leading to high false negative outcomes, therefore it is not recommended by the WHO [6]. Moreover, it has been reported to have lower sensitivity compared to RT-PCR tests [7]. On the other hand, CT and X-ray imaging modalities play an important role in the diagnosis of lung-related abnormalities. Numerous research articles have proven the ability of the X-ray and CT modalities to achieve more accurate results than RT-PCR [8,9]. However, these imaging modalities require the presence of a skilled specialized radiologist to perform the diagnosis. Furthermore, the COVID-19 diagnosis procedure is difficult due to the symmetry among the patterns of the new coronavirus and other sorts of similar diseases [10]. Furthermore, the manual investigation requires a long time and thus automatic diagnostic tools are compulsory to decrease observation time and exertion achieved by experts to perform the diagnosis and produce a precise decision.

Artificial intelligence (AI) techniques aim to create automated diagnostic tools capable of analyzing medical data (such as images and bio-signals) simply and fast. They have been utilized successfully to enhance prognosis and diagnosis of various disorders and diseases [11–20]. The ability of AI techniques to facilitate the new coronavirus has been proven in the survey article [21]. Currently, deep learning (DL) approaches are widely used to construct automated diagnostic tools using radiograph images to support the diagnosis of COVID-19 and avoid the challenges of manual inspection [22,23]. Regardless of the achievements of DL methods in diagnosing COVID-19 using radiographic images, these scanning techniques have some limitations. These shortcomings include high cost, immobility, exposure to a large amount of radiation, and the requirement for qualified technicians to acquire these images [24]. Hence, new diagnostic tools based on other modalities are needed to assist in COVID-19 diagnosis whilst the epidemic persists.

It is well-known that COVID-19 primarily affects the respiratory system; however, it also affects the cardiovascular system [25,26]. Numerous research articles have shown various types of cardiovascular alterations in people with COVID-19. These variations involve divergence of the ST segment of the PR interval [27], arrhythmias [28], QRST changes, and conduction disorders [25]. These cardiac variations can be visualized on the electrocardiogram (ECG) of patients with COVID-19. Such cardiovascular modifications [29] have promoted the study of ECG data as a new means of diagnosing the novel coronavirus. Looking at the huge advantages of using ECG, including low cost, mobility, simplicity of use, safe, harmless, and providing real-time monitoring, automated diagnostic tools for COVID-19 based on ECG data could be of significant value in addition to imaging modalities and PCR exams. Thus, further investigation is needed to verify the feasibility of using ECG for the diagnosis

Related Studies

The conventional method to study ECG data by AI is to mine traditional handcrafted features and employ them to train machine learning classifiers. These methods have previously been used to identify cardiac anomalies from ECG records. Numerous research articles used such methods based on 1D ECG signals to detect several cardiac problems [30–34]. However, these methods generally require a trade-off between accuracy and computation load and are subjective to errors [11,35]. Conversely, DL was recently employed to examine ECG by automatically attaining valuable features, thus avoiding the disadvantages of handcrafted methods [36,37]. Many studies have shown that ECG 1D signals converted to 2D demonstrations have better performance and benefits than 1D-based models [38,39]. Several studies analyzed and converted 1D to 2D ECG using transform domains such as short-term frequency transform and wavelet transform [38–44] and used them with DL techniques. Many studies used several forms of DL models to detect abnormalities in ECG signals [45–49]. It is worth mentioning the great efforts that were made by the PhysioNet/Computing in Cardiology Challenge in 2020 and 2021 to stimulate the multitype arrhythmia classification over annotated databases with thousands of 12-lead ECG recordings [50,51]. Despite the success of previous studies in detecting cardiac complications from ECG signals, it could not be easily used in real clinical practice as the above methods mainly rely on ECG signals; however, in real medical practice, this is regularly not the usual scenario. Because the ECG data taken in real clinical practice are acquired and stored as 2D ECG trace images [52]. Unlike the digital ECG signal acquired using wearable sensors, which contain multiple clean and well-detached leads, the ECG trace image data acquired in real practice are ambiguous. Such a trace image has an overlay between ECG waveforms collected from different leads and the rigid surrounding minor axes that raise hardness in mining significant information precisely. Furthermore, in digital ECG signals, that data is collected in hundreds of hertz as a sampling frequency; however, in real medical practice, the ECG data are acquired in few hertz, which results in a huge degradation in the quality of the data which correspondingly impacts the classification performance of AI-based models. One possible solution to resolve that issue is to turn the trace image into a digital ECG signal [53]. However, this conversion is complex, and the converted signal is of low quality due to the extensive noise generated by the conversion [54]. Even with the great capabilities of DL methods, this noise hinders DL techniques in detecting the small variations among different cardiac anomalies, which is the major component of cardiac complications diagnosis.

The abovementioned issues obstruct the digital ECG signals from being used in real-world clinical practice which collects ECG records as trace images. Therefore, some research articles used direct ECG trace images to identify several cardiac complications using AI techniques. The authors in [55] proposed a system to detect myocardial infarctions from ECG trace images. Their system contained multiple divisions based on shallow artificial neural networks (ANN) that used 12-lead ECG, achieving an accuracy of 94.73%. In [56], a discrete wavelet transform (DWT) was used to obtain significant features from the trace images using the 'Haar' wavelets. An ANN was constructed to differentiate between normal and abnormal ECG patterns, obtaining an accuracy of 99%. In [57], five hand-made feature extraction methods along with five classifiers were used to recognize two categories of cardiac arrhythmias. The highest accuracy of 96% was achieved using local binary patterns and ANN. On the other hand, Du et al. [58] proposed a DL pipeline to identify several cardiac diseases. The pipeline determined the prospective distinctive regions and adaptively merged them. Next, a recurrent neural network was employed and attained a sensitivity and precision of 83.59% and 90.42%, respectively. The MobileNet v2-deep DL method was utilized in [59] to identify four cardiac complications with 98% accuracy. In [60], DenseNet was trained with ECG trace images to predict strokes and achieved 85.82% accuracy.

The promising performance achieved using the formerly discussed methods based on ECG trace images triggered the investigation of the possibility of employing this type of ECG data with DL techniques to diagnose COVID-19. An acknowledgment must be made of the recently published public data [61] which has helped to achieve the suggested target. This data has ECG images of patients with COVID-19 and other cardiac findings. To the best of our knowledge, up to today, four studies have utilized this dataset to examine the potential of using ECG trace images in the new diagnosis of coronavirus. This dataset was used in [62] to study the impact of employing various enhancement methods on the diagnosis of COVID-19 using EfficientNet trained with ECG trace images. The paper concluded that augmentation methods are useful to some extent; nevertheless, exceeding this extent will lower the performance. An 81.8% maximum accuracy was achieved. Whereas in [63], six DL approaches were utilized to identify COVID-19 from other cardiac findings in two classification categories. Alternatively, in [64], hexaxial feature and Gray-Level Co-occurrence Matrix (GLCM) approaches were employed to extract considerable features and generate hexaxial mapping images. The created images were fed to DL methods to distinguish COVID-19 from other images as a binary classification category with a precision of 96.2%. The study [65] extracted deep features from two layers of several CNNs to an accuracy of 97.73% and 98.8% for multiclass and binary classification problems.

These previous studies experienced numerous shortcomings. Initially, the tool implemented in [64] performed only using binary classification category; however, the multiclass problem is more complicated and essential but was not considered. Furthermore, the hexaxial feature mapping utilized in it is quite sensitive to image quality, which correspondingly influences the extraction process of the GLCM procedure. The classification results obtained in the tool presented in [62] were considerably low and therefore cannot be reliable. The number of samples and features used in the testing process of the tool introduced in [63] was small, leading to a probable bias. The study [65] used a huge number of features to build their model. On the other hand, the tools proposed in [62,63,66] were based on individual DL models to perform the feature extraction or classification procedure. However, the research articles [67,68] confirmed that the incorporation of features of numerous DL approaches has the capacity to improve the classification results.

This study examined the viability of utilizing ECG information for COVID-19 diagnosis via presenting a novel diagnostic tool using various AI methods. The proposed tool attempts to overcome the limitations of the previous studies by incorporating several DL techniques and using a hybrid feature selection approach to reduce the number of features used to train the classification models. The classification procedure of the proposed tool is performed on two levels. The primary level aims to classify the ECG data to COVID-19 and normal cases (binary class level). The second level is multiclass to distinguish COVID-19 cases from normal and other cardiac complications.

2. Materials and Methods

2.1. ECG Dataset

The proposed diagnostic tool uses a recent dataset that is public [61], including images of ECG records for patients with COVID-19 and other cardiac problems. Until now, to the best of our knowledge, this is the primary and single public dataset for ECG records of COVID-19. ECG images available in the dataset are 1937 of distinct categories. The dataset consists of 250 scans of cases with the novel coronavirus, 300 trace records of cases with a present or former myocardial infarction (MI), 548 ECG records of irregular heartbeats, and 859 normal images without any heart complications as shown in Table 1. Data were acquired using a 12-lead system with a sampling frequency equal to 500 Hz through an EDAN SE-3 series 3-channel electrocardiograph. Table 1 also illustrates the number of images used for the training and validation sets of the proposed tool. The dimension of the images varied from 952 × 1232 to 2213 × 1572. The x-scale is 25 mm/s, and the y-scale is 10 mm/volt. Six ECG electrodes were placed on the chest representing six precordial leads. Another three electrodes were placed on the two arms and left leg representing six limb leads, including augmented voltage right (AVR), augmented voltage left (AVL), augmented voltage foot (AVF), Lead I, II, and III. The images of the dataset were evaluated by medical professionals using a telehealth ECG diagnostic scheme. This evaluation was carried out under the supervision of expert cardiologists who had long experience in ECG annotation and exploration. These medical experts removed all uncertain, ambiguous, and misleading images from the dataset.

In the binary classification level (normal versus COVID-19), 250 normal and 250 novel coronavirus records were utilized. Whereas in the multiclass classification level, a total of 750 images were employed, 250 for cardiac complications, 250 for normal cases, and 250 for COVID-19 cases. To avoid the classification bias that occurs due to the class imbalance structure of the ECG dataset (the number of images per class is not equal) that affects the classification process, an equal number of images was selected and used for each class to train the classification models. An ECG trace record sample for a COVID-19 patient is shown in Figure 1.

Table 1. Description of the ECG dataset including the number of available ECG images per class and the number of images used in the proposed study.

Class	Number of Available Images	Images Used in the Proposed Study	Images Used in Training	Images Used in Validation
COVID-19	250	250	175	75
Normal	859	250	175	75
Cardiac Abnormalities include:	848	250	175	75
• Irregular Heartbeats	548	125	88	37
• Current and Recovered MI	300	125	88	37

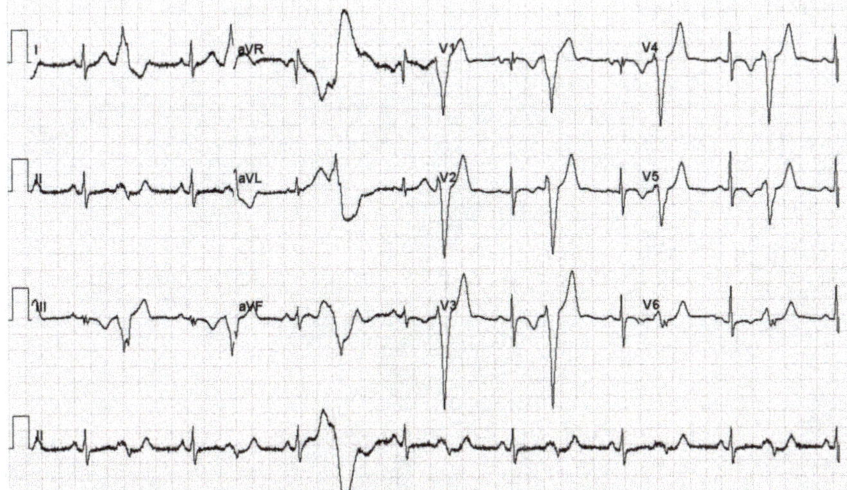

Figure 1. An ECG trace record sample for a COVID-19 patient.

2.2. Proposed Tool

The proposed automated tool consists of four steps: ECG trace image preprocessing, deep feature extraction and feature incorporation, hybrid feature selection, and classification. The proposed method used ten DL approaches. Figure 2 shows a diagram that describes the steps of the proposed diagnostic tool.

DL is an emerging technology that has been widely employed in several fields. DL approaches are the recent class of machine learning (ML). They consist of numerous architectures; however, convolution neural networks (CNNs) are the architectures most widely used for medical images [69]. Therefore, the proposed diagnostic tool utilizes ten CNNs of various architectures. These networks include InceptionResNet, ResNet-18, ResNet-50, ShuffleNet, Inception V3, MobileNet, Xception, DarkNet-19, DarkNet-53, and DenseNet-201.

Inception V3 Google proposed the Inception CNN architecture in 2016 [70]. It is a newer version of GoogleNet [71], but with some modifications. It was first introduced to run well with reduced memory requirements and computational cost. Its principal component is the inception unit which merges numerous filters into a novel filter structure which correspondingly lowers the number of parameters. To expand the information stream into the network, the Inception module considered the depth as well as the width of the layers during the construction of the network [72].

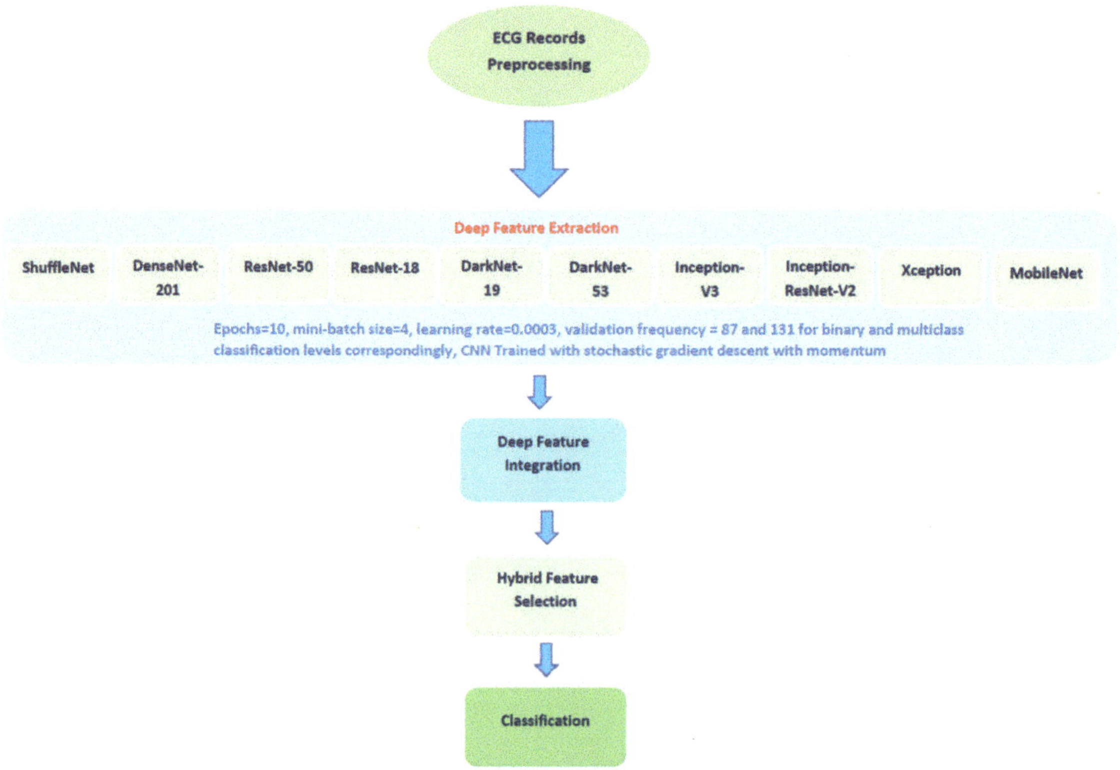

Figure 2. A diagram describing the steps of the proposed diagnostic tool.

ResNet is one of the time-efficient CNNs that gained popularity due to its novel structure created by He et al. in 2015 [73]. ResNet counts on the residual block which embeds crosscuts in the interior layers of a standard CNN to cross several convolution layers which quickens and eases the convergence procedure of the CNN despite the huge number of convolution layers.

Xception is a new version of the Inception network introduced in 2017 [74]. The inception layers contain depthwise convolution layers, followed by a pointwise convolution layer. The Xception structure involves double layers of convolutional, then several depthwise separable convolution layers, and standard layers of convolutional and fully connected. The Xception module is more robust and powerful than the Inception module and can perform cross-channel and spatial interaction correlations while fully dissociated [75].

Inception-ResNet-V2 presented a mixture of residual network architecture and the inception module [76]. It has a number of filters of various dimensions that are merged with residual joints. The main advantage of this fused architecture is enhancing the performance of the network and pace of convergence.

DenseNet was created by Huang et al. [77] in 2017, who extended the idea of shorter connections between layers near the input/output layers. The key building block of this network is the 'dense block'. The major difference between the residual block and dense block is that the latter attaches every layer to each layer having a similar input resolution, whereas the former generates shorter links among adjacent layers. Second, each layer of

DenseNet accomplishes a concatenation of the earlier outputs; in contrast, ResNet performs a summation. DenseNet-201 was utilized in this article, containing 201 layers.

ShuffleNet is an effective CNN primarily designed by Zhang et al. in 2018 [78]. ShuffleNet was initially produced to serve fields that require low computational capability. It contains two key blocks known as pointwise group convolution and channel shuffle. The first block utilizes convolution layers of dimension 1×1 to reduce training speed while attaining adequate precision. The second block supports the data flowing across feature channels by allowing a cluster of layers to control input data belonging to distinct groups, where the output/input channels are connected.

DarkNet is a new DL architecture designed by the authors of [79]. It employs YOLO-V2 as the backbone of its structure. DarkNet uses filters of dimension 3×3 and then doubles the number of channels after every pooling phase. It employs a pooling stage to perform detection and classification as well as 1×1 filters to reduce the feature presentation between 3×3 convolutions. Darknet-19 involves 19 convolutional layers, whereas DarkNet-53 contains 53 convolutional layers.

MobileNet is a fine and time-efficient DL architecture that was originally designed in [80]. It can decrease the complexity of the training model by lowering the number of parameters while maintaining an acceptable performance. These are convolutional layers of dimensions 3×3 and 1×1, respectively. MobileNet has 53 deep layers.

2.2.1. ECG Image Preprocessing

Initially, the dimensions of the ECG images are changed according to the input layer dimension of each CNN model. Then, those ECG records are augmented to increase the amount of records available in the data set and prevent the likelihood of overfitting that could occur in the case of small data. Those augmentation methods included in the proposed diagnostic tool are flipping in both the x- and y-orientations, and translation in both the x- and y-directions where the range of the translation distance is picked randomly within the range $(-30, 30)$. The scaling augmentation method is also applied to the image in the x- and y-directions where the image is scaled with a scale factor chosen randomly from the range $(0.9, 1.1)$. Table 2 demonstrates the dimensions of the input layers of each of the CNN models and the extracted features length. Table 2 shows that the number of features extracted from the last fully connected layer of each CNN for the binary classification and multiclass classification levels is 2 and 3, respectively.

Table 2. The dimensions of the input layers of each of the CNN models and the mined features dimensions.

CNN Construction	Dimension of Input	Length of the Extracted Deep Features
ResNet-50 **ResNet-18** **DenseNet-201** **ShuffleNet** **MobileNet**	$224 \times 224 \times 3$	Binary Classification Level
		2
		Multiclass Classification Level
		3
Inception-V3 **Inception-ResNet** **Xception**	$229 \times 229 \times 3$	Binary Classification Level
		2
		Multiclass Classification Level
		3
DarkNet-19 **DarkNet-53**	$226 \times 226 \times 3$	Binary Classification Level
		2
		Multiclass
		3

2.2.2. Deep Features Extraction and Feature Incorporation

Some complications may occur while CNNs are being trained, including convergence and overfitting. These issues impose the adjustment of a few parameters in the CNNs to guarantee that the weights of the CNN layers are updated at the same rate during the training process. Transfer learning (TL) is a method that can solve this problem. TL re-employs a CNN that was previously learned with a huge dataset like ImageNet for another classification problem [81]. In other words, TL uses a pretrained CNN that has learned feature representations from a large dataset to solve another classification problem dealing with a small dataset (similar to the dataset used in this paper). This process can enhance detection accuracy if used for comparable problems [81]. For that reason, this paper used ten CNNs that were pretrained. Before retraining the ten CNNs, the number of their output layers was changed to 3 or 2 which is equal to the number of classes in the case of the multiclass and binary class classification categories of the proposed diagnostic tool. In other words, the DL models were retrained for the novel classification task. Then, after the retraining process was finished, deep features were extracted from the last fully connected layers of the ten pretrained CNNs. The number of features extracted from each CNN was 2 in the case of the binary classification category and 3 in the multiclass classification category. Afterward, the proposed tool incorporated the deep features extracted from the ten DL models in a concatenation way to form one feature vector consisting of 20 and 30 features in the case of the binary and multiclass classification categories, respectively.

2.2.3. Hybrid Feature Selection

Feature selection (FS) is an essential step to selecting the most valuable features available in the feature space to reduce its dimension, which correspondingly boosts the diagnostic accuracy and avoids overfitting [82,83]. FS methods can be categorized into three categories: filter, wrapper, and hybrid [84]. Hybrid FS merges filter and wrapper methods. This category combines the benefits of previous FS types [84]. Thus, a hybrid FS approach was presented and employed in this study.

The hybrid FS step presented in the diagnostic tool combines the chi-squared test filter FS approach with a wrapper FS approach based on three search strategies. The chi-squared-test is a well-known and commonly used FS method [85]. It attempts to determine the significant features t_k that best differentiate positive and negative sets of instances of class C_i. The chi-squared test score is calculated using Equation (1).

$$\text{Chi} - \text{Squared Test} = \frac{N(AD - CB)^2}{(A + C)(B + D)(A + B)(C + D)} \tag{1}$$

where N is the total number of ECG records (samples in a dataset); A = the number of samples in class c_i that contain the feature t_k; B = the number of samples that contain the feature t_k in other classes; C = the number of samples in class c_i that do not contain the feature t_k; D = the number of samples that do not contain the feature t_k in other classes.

The hybrid FS method initially ranks deep features extracted from the ten CNN models utilizing the chi-squared test filter FS. Then, it employs this ranking to guide the three feature search strategies within the wrapper FS approach. These three search strategies are backward, forward, and bidirectional. The first searching approach starts with all features in the feature space and then ignores features of lower ranks iteratively. Conversely, the forward approach begins with one feature having the greatest rank and then adds the following features one by one. The bidirectional alternates between the forward and backward strategies. Note that for the three strategies, only the features that improve the classification results are kept, while others are deleted.

2.2.4. Classification

The classification phase was performed in two schemes. The first scheme was an end-to-end deep learning classification with ten CNNs, including InceptionResNet, ResNet-18, ResNet-50, ShuffleNet, Inception V3, MobileNet, Xception, DarkNet-19, DarkNet-53, and DenseNet-201. The second scheme used several machine learning classifiers trained with deep features extracted from the last fully connected layers of the ten CNNs. These classifiers involved a support vector machine (SVM), random forest (RF), K-nearest neighbor (KNN), the linear discriminate classifier (LDA), quadratic discriminate analysis (QDA), and decision tree (DT). The classification step included two levels: binary and multiclass. At the former level, classifiers were used to identify COVID-19 and normal patients. The multiclass level classified images into normal, COVID-19, and cardiac complications. The 10-fold cross-validation method was used to validate the results. The classifiers were run 10 times and the average classification performance of all these runs is displayed in the results section. Classification was carried out in two phases. Phase I used the deep features extracted from the ten CNN models to train the classifiers. Phase II employed the hybrid FS approach to select features used to train the classifiers.

LDA is a popular machine learning technique used for both classification and feature reduction. It searches for the linear combinations of features that have a high ability to explain the data. LDA separates class labels of data using hyperplanes. These planes are achieved by looking for the projection of data points that can minimize their variance and maximize the distance between class labels.

K-NN is a commonly used classifier in the field of machine learning due to its simplicity, straightforwardness, and effectiveness even with noisy data. Although it is simplistic, it has the ability to reach good classification accuracy in medical applications. It allocates a label to every instance in the test data equivalent to the label amongst the k nearest neighbors included in the training data. This label is chosen according to the distance measured between the instance being classified and those instances in the training data. This distance shows that instance in the test data to those in the training data. The distance used in our approach was the Euclidean similarity measure and the number of neighbors (k) was equal to 1 and 5 for binary and multiclass classification levels, respectively, with equal distance weights.

Decision Trees are well-known machine learning classifiers that are widely used in medical applications due to several reasons. They are capable of visualizing interactions between extracted features. This visualization process enables a doctor to easily understand how the classifier decision is made. The DT classifier creates instances of data according to conditions. The DT has a tree structure with a root node whose leaves demonstrate class labels, and the branch nodes present the extracted features and reasons that result in this class label. The nodes of a tree are connected by an arc that represents the condition of the feature. The tree is divided into branches and leaves based on a metric such as information gain, gain ratio, or Gini index. The maximum number of splits in this study was 100, and the splitting criterion was the Gini diversity index.

Random Forest is an ensemble classifier that consists of multiple decision trees. RF uses the divide-and-conquer approach (DAC) to perform classification. The DAC method divides the input feature space into several partitions depending on a goodness metric. Subsequently, the classification outputs of all trees are averaged to produce a final decision. The Gain ratio metric was used in the proposed tool. There, the number of trees was 100.

SVM is a robust machine learning classifier. It transforms linear or nonlinear input data points into a new domain that can easily separate between classes of data. A hyperplane is employed to separate between classes of input data to facilitate classification. A kernel function maps the similarity between the input vector and the new higher-dimension feature space. The linear kernel function was employed.

On the other hand, for retraining the CNNs for end-to-end classification, the learning rate, number of epochs, and minimum batch size were adjusted to 0.0003, 10, and 4, respectively. Whereas the validation frequency was modified to 87 and 131 for binary and multiclass classification levels, respectively. The ten CNNs were trained with the stochastic gradient descent with a momentum algorithm. The other hyperparameters were kept unchanged. The proposed diagnostic tool was implemented using the Weka Data Mining Tool [86] and MATLAB R2020a.

2.3. Performance Evaluation

The overall performance of the proposed diagnostic tool was measured using multiple metrics involving the Mathew correlation coefficient (MCC), the F1 score, precision, specificity, and sensitivity calculated using Equations (2)–(7). In addition to confusion, the receiver operating characteristics curve (ROC) and the area under ROC (AUC) were also determined.

$$\text{Accuracy} = \frac{\text{TP} + \text{TN}}{\text{TN} + \text{FP} + \text{FN} + \text{TP}} \tag{2}$$

$$\text{Sensitivity} = \frac{\text{TP}}{\text{TP} + \text{FN}} \tag{3}$$

$$\text{Precision} = \frac{\text{TP}}{\text{TP} + \text{FP}} \tag{4}$$

$$\text{MCC} = \frac{\text{TP} \times \text{TN} - \text{FP} \times \text{FN}}{\sqrt{(\text{TP} + \text{FP})(\text{TP} + \text{FN})(\text{TN} + \text{FP})(\text{TN} + \text{FN})}} \tag{5}$$

$$\text{F1} - \text{Score} = \frac{2 \times \text{TP}}{(2 \times \text{TP}) + \text{FP} + \text{FN}} \tag{6}$$

$$\text{Specificity} = \frac{\text{TN}}{\text{TN} + \text{FP}} \tag{7}$$

where FN refers to the false negative which is the amount of COVID-19 records wrongly categorized as nonCOVID-19, TN is the true negative representing the nonCOVID-19 records correctly recognized. TP is the true positive, which is equal to the number of COVID-19 scans properly identified. Finally, FP is the false positive equivalent to the sum of nonCOVID-19 records improperly classified as COVID-19.

3. Results

3.1. Phase I Classification Results

Phase I represents the use of deep features extracted from the ten CNNs and fused to train the machine learning classifiers. Table 3 illustrates the classification accuracy of phase I for the binary class and multiclass classification levels, respectively. Table 3 shows that the maximum accuracy of 97.78% was achieved for the binary classification level using the RF model. All other classifiers obtained an accuracy that ranged from 97.36% to 97.6%. The highest accuracy of 90.88% was achieved using the RF classifier for multiclass classification. The SVM, LDA, and KNN achieved the next-highest accuracies of 90.43%, 90.35%, and 89.39%. Finally, the DT and QDA classifiers reached the lowest accuracies of 86.56% and 85.6%, respectively. The confusion matrices attained using the LDA and SVM classifiers are shown in Figures 3 and 4 for binary and multiclass classification, respectively. The ROC curve for the SVM and LDA classifiers are shown in Figures 5 and 6 for binary and multiclass, respectively. Figure 7 shows that the AUC for the LDA and SVM classifiers were 0.99 and 0.99 for the binary class classification level. For the multiclass classification level, the AUCs for the LDA and SVM classifiers were 0.97 and 0.98, respectively.

Table 3. Phase I classification accuracy (%) and standard deviations obtained using machine learning classifiers.

Binary Classification Level					
DT	**RF**	**QDA**	**LDA**	**SVM**	**KNN**
97.62 (0.14)	97.78 (0.06)	97.6 (0)	97.6 (0)	97.6 (0)	97.36 (0.23)
Multiclass classification Level					
DT	**RF**	**QDA**	**LDA**	**SVM**	**KNN**
86.56 (0.8)	90.88 (0.19)	85.6 (0.06)	90.35 (0.21)	90.43 (0.28)	89.39 (0.30)

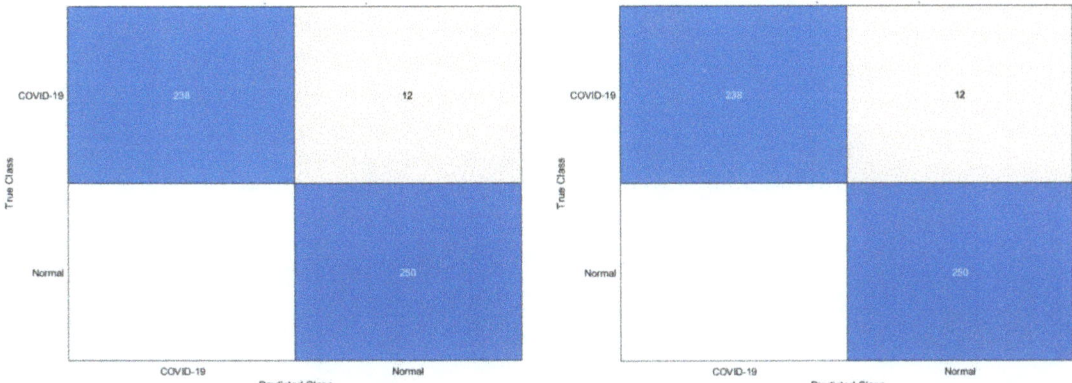

Figure 3. Confusion matrices for the binary class classification level: (**left**) LDA, (**right**) SVM classifiers.

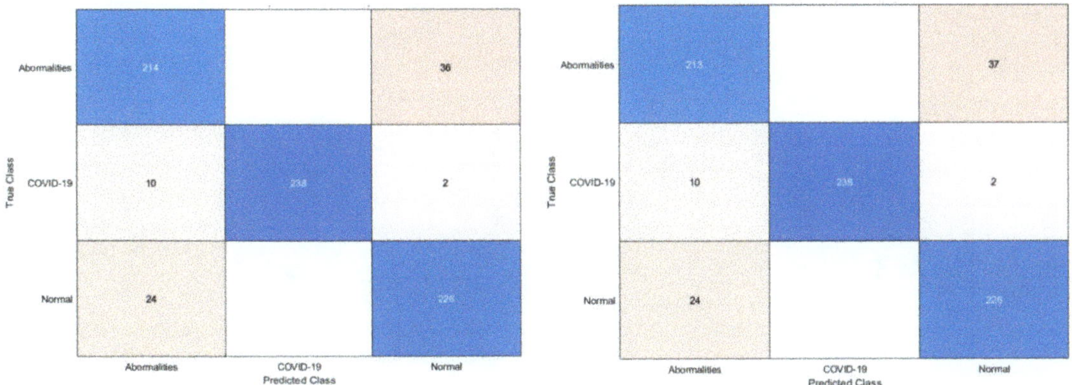

Figure 4. Confusion matrices for the multiclass classification level: (**left**) LDA, (**right**) SVM classifiers.

The two Figures S1 and S2 have been attached to the Supplementary Materials representing a two-dimensional scatter plot of the first two features of the feature space for the binary and multiclass classification levels used as inputs to the classifiers. Also, the two Figures S3 and S4 have been added to the Supplementary Materials representing a two-dimensional scatter plot of the LDA classifier predictions using the first two features of the feature space for the binary and multiclass classification levels. Moreover, the two Figures S5 and S6 have been added to the Supplementary Materials representing a two-dimensional scatter plot of the LDA classifier predictions using the first two features of the feature space for the binary and multiclass classification levels

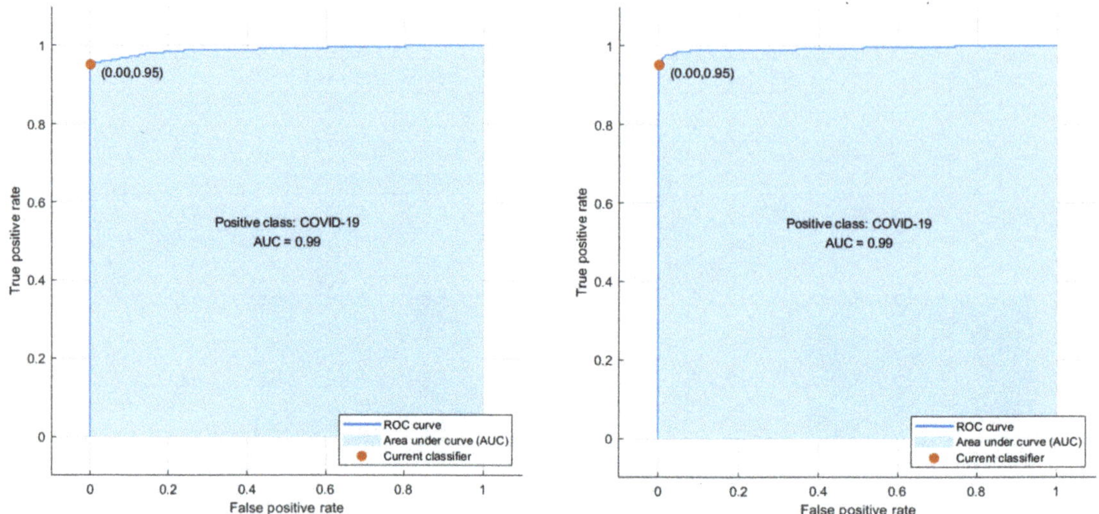

Figure 5. The ROC curves for the binary class classification level: (**left**) LDA, (**right**) SVM classifiers.

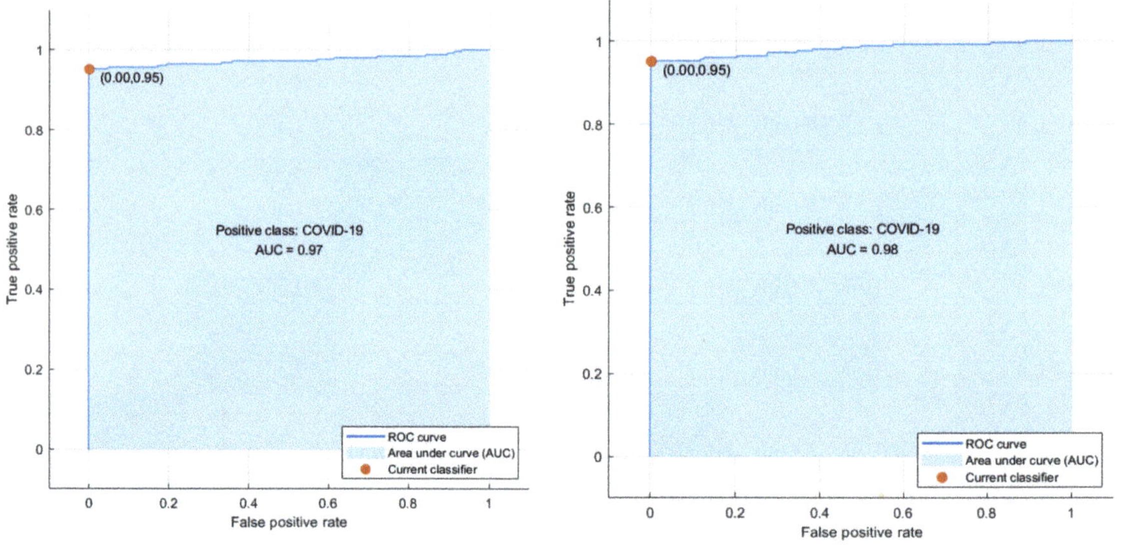

Figure 6. The ROC curves for the multiclass classification level: (**left**) LDA, (**right**) SVM classifiers.

To access and confirm the statistical significance of the performance of the ML classifiers, the one-way analysis of variance (ANOVA) test was applied to the results of the classifiers after a repeated 10-fold cross-validation process. The ANOVA test was performed on the classification accuracy results achieved using the classifiers of the binary classification level to test the statistical significance between them. The results are shown in Table 4. ANOVA was also performed for the results of the multiclass classification problem and the outputs of the test are shown in Table 5. It can be seen in Tables 4 and 5 that the p-values attained from the test were lower than α, where $\alpha = 0.05$. Consequently, it could be concluded that there is a statistically significant difference in the classification accuracies of the classifiers for both the multiclass and binary classification levels.

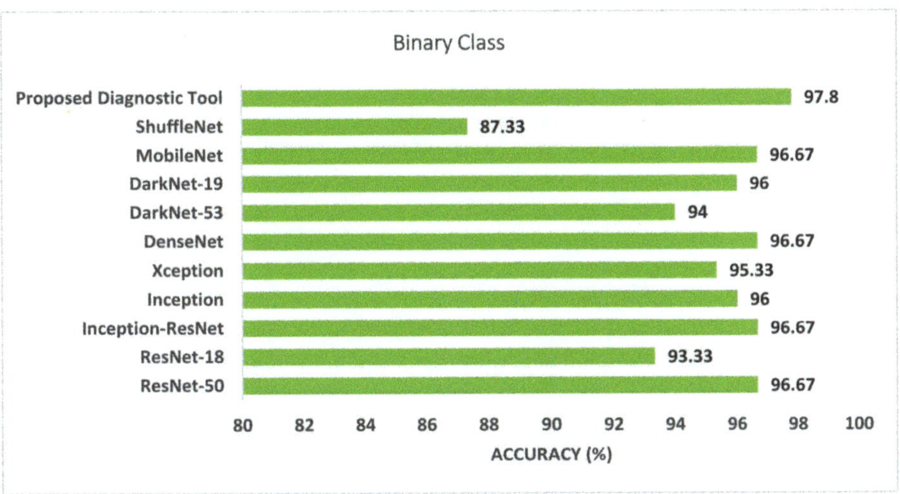

Figure 7. The classification accuracy of the proposed diagnostic tool using the RF classifier of phase I compared to the end-to-end DL classification for the binary level.

Table 4. One-way analysis of variance test details for the binary classification level.

Source of Variation	SS	df	MS	F	p Value
Columns	0.901	5	0.180	12.54	<0.001
Error	0.776	54	0.014		
Total	1.677	59			

Table 5. One-way analysis of variance test details for the multiclass classification level.

Source of Variation	SS	df	MS	F	p Value
Columns	252.148	5	50.429	298.07	<0.001
Error	9.136	54	0.1692		
Total	261.284	59			

Figures 7 and 8 compare the phase I performance of the RF classifier of the proposed diagnostic tool with the end-to-end DL classification for the binary and multiclass levels. Figure 7 proves that the deep features extracted from the last fully connected layers of the ten CNNs had a higher classification accuracy compared to end-to-end pretrained CNNs for the binary classification level. On the other hand, for the multiclass classification level, the RF classifier of the proposed diagnostic tool obtained 90.88% accuracy, which is higher than all other pretrained CNNs. As can be seen in Figure 8, the accuracy of the RF classifier of the proposed diagnostic tool was greater than the 76.44%, 75.56%, 72.89%, 73.33%, 72.89%, 71.59%, 71.11%, 67.11%, 69.33%, and 64.44% achieved using ResNet-50, ResNet-18, Inception-ResNet, Inception, Xception, DenseNet-201, DarkNet-53, DarkNet-19, MobileNet, and ShuffeNet, respectively.

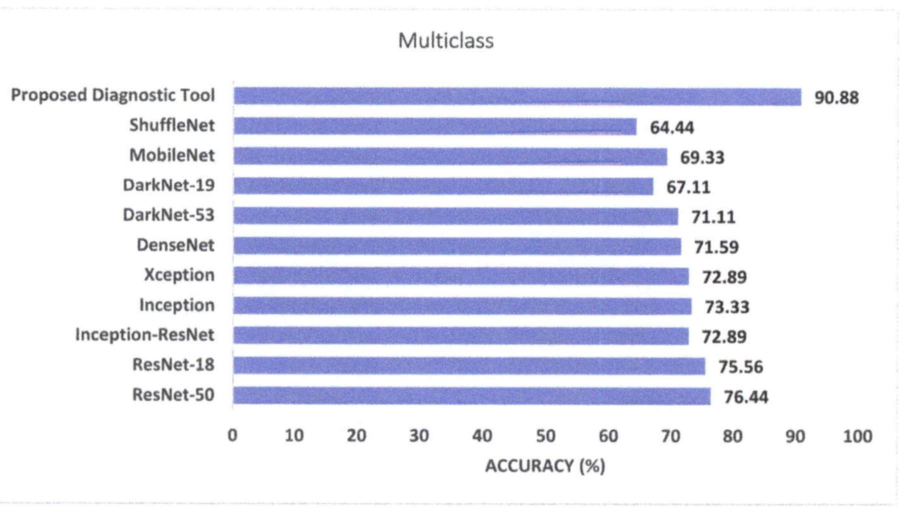

Figure 8. The classification accuracy of the proposed diagnostic tool using the phase I RF classifier compared to the end-to-end DL classification for the multiclass level.

3.2. Phase II Classification Results

Phase II of the proposed diagnostic tool presented the features selected after the hybrid FS approach used them to train the classification models. The following section presents the results of the hybrid feature selection approach based on the three search strategies using three classifiers. First, it shows the rank scores of features using the chi-square test filter FS method. Then, it shows the number of selected features as well as the classification accuracy attained for the binary and multiclass classification levels. Tables 6 and 7 represent the ranking score for each feature attained using the chi-square test FS method for the binary and multiclass classification levels, respectively.

Table 6. The ranking score for each feature attained using chi-square FS along with its order in the feature vector and the name of the feature for the binary classification level.

Rank	Order	Feature Name
461.538	18	Feature 2 of MobileNet
461.538	5	Feature 1 of InceptionResNet
460.68	6	Feature 2 of InceptionResNet
457.854	17	Feature 1 of MobileNet
457.854	7	Feature 1 of Xception
457.592	13	Feature 1 of DarkNet-53
456.608	14	Feature 2 of DarkNet-53
456.397	19	Feature 1 of Shuffle
454.198	3	Feature 1 of Inception
454.198	2	Feature 2 of ResNet-50
454.198	4	Feature 2 of Inception
454.198	20	Feature 2 of Shuffle
454.198	10	Feature 2 of DenseNet

Table 6. *Cont.*

454.198	8	Feature 2 of Xception
454.198	15	Feature 1 of DarkNet-19
454.198	16	Feature 2 of DarkNet-19
454.198	12	Feature 2 of ResNet-18
454.198	9	Feature 1 of DenseNet
454.198	11	Feature 1 of ResNet-18
454.198	1	Feature 1 of ResNet-50

Table 7. The ranking score for each feature attained using chi-square FS along with its order in the feature vector and the name of the feature for the multiclass classification level.

Rank	Order	Feature Name
1021.0997	1	Feature 1 of ResNet-50
980.0815	25	Feature 1 of MobileNet
938.3733	19	Feature 1 of DarkNet-53
932.5696	12	Feature 3 of Xception
926.9128	28	Feature 2 of Shuffle
917.393	6	Feature 3 of Inception
916.3032	18	Feature 3 of ResNet-18
906.3512	13	Feature 1 of DenseNet
898.5766	10	Feature 1 of Xception
894.4025	15	Feature 3 of DenseNet
886.2739	3	Feature 3 of ResNet-50
883.1262	7	Feature 1 of InceptionResNet
877.7686	21	Feature 3 of DarkNet-53
865.6989	22	Feature 1 of DarkNet-19
814.21	2	Feature 2 of ResNet-50
811.4717	24	Feature 3 of DarkNet-53
798.1348	16	Feature 1 of ResNet-18
797.0761	27	Feature 3 of MobileNet
781.2226	4	Feature 1 of Inception
760.9445	8	Feature 2 of InceptionResNet
755.4454	29	Feature 2 of Shuffle
723.6848	11	Feature 2 of Xception
720.4829	23	Feature 1 of DarkNet-19
703.0506	5	Feature 2 of Inception
697.2656	20	Feature 2 of DarkNet-53
697.2656	26	Feature 2 of MobileNet
697.2656	17	Feature 2 of ResNet-18
697.2656	14	Feature 2 of DenseNet
673.7605	9	Feature 3 of InceptionResNet
634.5971	30	Feature 3 of Shuffle

Table 8 shows the binary class classification level after the hybrid FS approach of the proposed diagnostic tool using the three search strategies (phase II) compared to phase I (before FS) for the DT, RF, and QDA classifiers as they achieved the highest accuracies in phase I. Table 8 shows that the hybrid FS approach of the proposed diagnostic tool improved the classification accuracy compared to phase I. This was obvious as the accuracies attained using the forward and bidirectional strategies were 98.2%, 98%, and 97.8%, which are better than those attained before FS. In addition, the accuracies attained using the backward strategy were 98.2%, 98%, 97.6% for the DT, RF, and QDA classifiers, which were higher than those achieved before FS using the same classifier except for the QDA which is equal to that achieved before FS. Some performance measures were calculated for the binary classification level and are illustrated in Table 9. Table 9 reveals the results for the sensitivity (0.968, 0.96, 0.956), specificity (0.996, 1, 1), precision (0.996, 1, 1), F1-score (0.982, 0.961, 0.978), and MCC (0.964, 0.989, 0.957) for the DT, RF, and QDA models, respectively, using the forward search strategy.

Table 8. The accuracy of binary-level classification (%) of the DT, RF, and QDA classifiers that obtained the highest accuracy in phase I compared to after using the three search strategies of the hybrid FS approach (phase II of the proposed diagnostic tool).

Classifier	Before FS	Forward	Backward	Bidirectional
DT	97.62	98.2	98.2	98.2
RF	97.78	98.0	98.0	98.0
QDA	97.6	97.8	97.6	97.8

Table 9. The binary-level performance metrics (%) of the DT, RF, and QDA classifiers that achieved the highest accuracy using the forward search strategies of the hybrid FS approach.

Classifier	Sensitivity	Specificity	Precision	F1-Score	MCC
DT	96.8	99.6	99.6	98.2	96.4
RF	96.0	100	100	96.1	98.9
QDA	95.6	100	100	97.8	95.7

On the other hand, the results of the multiclass classification level of phase II of the proposed diagnostic tool are displayed in Tables 10 and 11. Table 10 shows the multiclass accuracy of the hybrid FS approach of the proposed diagnostic tool using the three search strategies (phase II) compared to phase I (before FS) for the RF, LDA, and SVM classifiers which achieved the highest accuracy in phase I. The accuracies displayed in Table 10 verify that the hybrid FS approach based on the three search methods increased the capacity of the classification model compared to phase I (before FS). This was clear as the forward and bidirectional strategies achieved better accuracies of 91.6% and 90.93% for the RF classifiers, 91.07% and 91.33 for the LDA classifier, and 90.58% and 90.53% for the SVM classifier compared to 90.88%. Using the exact classifiers before FS, 90.35% and 90.43% accuracy was achieved. Similarly, the backward search method reached accuracies of 91.33%, 91.07%, and 90% using the RF, LDA, and SVM classifiers, which were higher than those attained before FS except for the SVM classifier, it remained the same. Table 11 indicates the sensitivity (0.916, 0.911, 0.905), specificity (0.958, 0.955, 0.953), precision (0.918, 0.918, 0.908), F1 score (0.917, 0.917, 0.906), and MCC (0.875, 0.875, 0.859) for the RF, LDA and SVM classifiers, respectively, using the forward search method.

Table 10. The classification accuracy (%) of the RF, LDA, and SVM that obtained the highest accuracy in phase I compared to after using the three search strategies of the hybrid FS approach (phase II of the proposed diagnostic tool).

Classifier	Before FS	Forward	Backward	Bidirectional
RF	90.88	91.6	91.33	90.93
LDA	90.35	91.07	91.07	91.33
SVM	90.43	90.58	90	90.53

Table 11. The multiclass-level performance metrics (%) of the RF, LDA, and SVM that achieved the highest accuracy using the forward search strategies of the hybrid FS approach.

Classifier	Sensitivity	Specificity	Precision	F1-score	MCC
RF	91.6	95.8	91.8	91.7	87.5
LDA	91.1	95.5	91.8	91.7	87.5
SVM	90.5	95.3	90.8	90.6	85.9

Figure 9 shows the number of features selected for the binary and multiclass levels using the forward search strategy that reached maximum accuracy (using the DT classifier for binary and the RF model for multiclass). Figure 9 indicates that the number of features after the hybrid FS for the binary classification problem is three. The three features include Feature 2 of MobileNet, Feature 1 of InceptionResNet, and Feature 2 of ResNet-50. The figure also shows that the number of features after FS for the multiclass classification level is eight. These eight features are Feature 1 of MobileNet, Feature 3 of Inception, Feature 3 of ResNet-18, Feature 1 of Xception, Feature 2 of DarkNet-53, Feature 3 of DarkNet-53, Feature 1 of DarkNet-19, and Feature 3 of InceptionResNet.

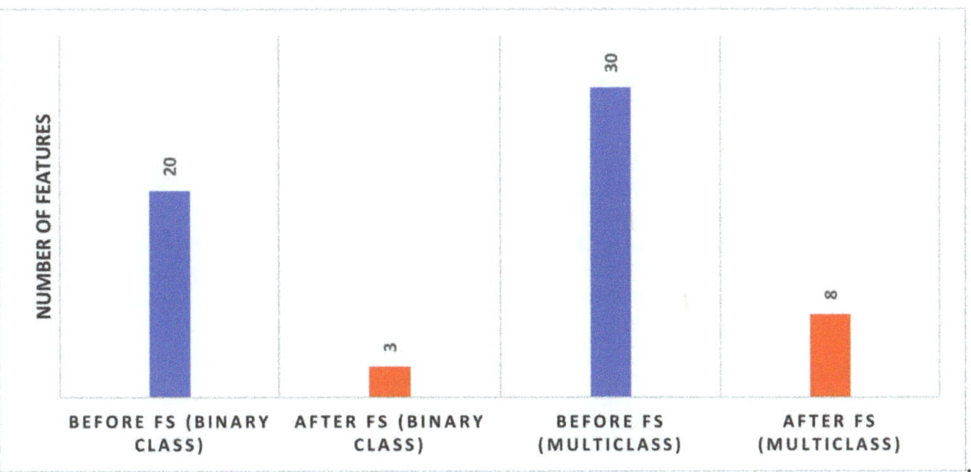

Figure 9. The number of features of phase I and phase II for the binary and multiclass classification levels using forward search strategies (using the classifiers which attained the peak performance).

4. Discussion

Recent relevant studies revealed various forms of cardiovascular variations in ECG data acquired from patients infected by the novel coronavirus as ST-segment changes, QRST irregularities, and arrhythmias. On the other hand, several research articles discussed that COVID-19 could not be the leading reason for these cardiovascular deformities; nevertheless, it should be emphasized that it could reveal the intrinsic conditions or lower them [87]. The entire cardiac findings indicated in the literature have been observed on all the ECG data utilized in this study.

This paper presented a novel diagnostic tool to automatically diagnose COVID-19 by incorporating multiple DL and hybrid FS approaches. This diagnostic tool consists of two classification levels: binary and multiclass. The first level consists of distinguishing COVID-19 and normal cases, while the second level consists of recognizing COVID-19, normal, and other cardiac abnormalities. The proposed tool extracted deep features from the last fully connected layers of ten CNNs models. Next, it fused these features, used several classifiers in the two classification levels, and compared their performance with the end-to-end DL classification. The previous step is known as phase I of the proposed diagnostic tool. Afterward, a hybrid FS method was presented based on three search approaches. This process is called phase II of the proposed diagnostic tool. The results achieved in phase I showed that the deep feature incorporation is better than end-to-end DL classification as shown in Figures 7 and 8. Phase I of the proposed tool attained an accuracy of 97.78% and 90.88% for the binary and multiclass classification levels, respectively. These accuracies are greater than those obtained by the end-to-end deep learning classification, having a range of 87.33–96.67% and 64.44–76.44% for the binary and multiclass classification levels, respectively.

In the second phase of the proposed tool, only classifiers that attained the highest accuracies for either the binary or the multiclass classification levels were employed in the hybrid FS procedure. Table 8 compares the results before and after feature selection for the three classifiers which attained the highest accuracies for the binary classification level. Table 8 shows that the highest accuracy of 98.2% was achieved using DT trained with only three features selected during the FS process of the binary classification level. This accuracy is greater than the 97.62% achieved before FS using the same classifier trained with 20 features. Similarly, Table 10 compares the results before and after FS for the three classifiers which attained the highest accuracies for the multiclass classification level. Table 10 indicates that the maximum accuracy of 91.6% was reached using the RF classifier trained with only eight features chosen during the FS procedure of the multiclass classification level. This accuracy is greater than the 90.56% accomplished before FS using the same classifier learned with 30 features. Thus, the performance of phase II of the proposed tool verifies that the presented hybrid FS method had a further enhancement in classification performance. It also reduced the number of features successfully.

It is worth mentioning that ECG detection requires more physical contact between patients and physicians than the RT-PCR test or CT imaging, which will increase the risk of virus transmission. Therefore, ECG may be more suitable as an auxiliary inspection means of COVID-19 than a primary screening tool.

4.1. Comparison with Related Studies

The performance of phase II of the proposed diagnostic tool versus other relevant tools that are directly copied from published papers is demonstrated in Table 12. The ECG records used to construct the proposed diagnostic tool are added to Appendix A. The results illustrated in Table 12 show that the proposed tool could be used to distinguish between normal and COVID-19 cases. It can also differentiate between normal, COVID-19, and other cardiac abnormalities. The table also indicates that the proposed tool has a performance comparable to those of other related studies. It is worth mentioning that, for the binary classification level, the specificity and precision of the proposed tool are higher than in the other studies [63,64]. However, it has lower sensitivity than other studies. However, for the multiclass classification level, the proposed tool achieved higher sensitivity, specificity, and precision than the studies [62,63]. These results indicate that the proposed tool based on ECG data could be used to diagnose COVID-19. It could be considered a possible novel solution that might be utilized in actual medical scenarios. It can be considered an alternative to current diagnostic tools.

Table 12. The results of phase II of the proposed diagnostic tool versus other related studies that are directly copied from published papers.

Binary Classification Level					
Article	Technique	Sensitivity (%)	Precision (%)	Specificity (%)	Accuracy (%)
[64]	hexaxial feature mapping + GLCM + CNN	98.4	94.3	94	96.2
[63]	ResNet-18	98.6	98.5	96	98.62
Presented diagnostic tool	Fully connected deep features + hybrid FS (forward search with DT classifier)	96.8	99.6	99.6	98.2%
Multiclass Classification Level					
		Sensitivity (%)	Precision (%)	Specificity (%)	Accuracy (%)
[62]	EfficientNet	75.8	80.8	-	81.8
[63]	MobileNet	90.8	91.3	92.8	90.79
Presented diagnostic tool	Fully connected deep features + hybrid FS (forward search with RF classifier)	91.6	91.8	95.8	91.6

4.2. Limitations

This study has several limitations. The first limitation is the small database for training/validation, which is quite insufficient for deep learning of thousands of hyperparameters. In addition, the lack of an independent test dataset is considered another limitation. Furthermore, this study did not consider methods that handle the class imbalance problem. Furthermore, the baseline rhythm of each patient is not available, and the effect of the baseline rhythm is not explored. Additionally, this study did not take into account optimization techniques for the selection of deep learning hyperparameters. In addition, the dataset used in the study is from confirmed COVID-19 patients. The detection of asymptomatic infections may not achieve the same level of sensitivity. Thus, the extension of the scope of the results is to some extent limited. Finally, this study did not consider the uncertainty of the input data.

5. Conclusions

The current study explored the prospect of employing ECG trace images for diagnosing the novel coronavirus. It proposed a novel automated ECG-based diagnostic tool that incorporates deep features from ten DL models. The proposed diagnostic tool used several well-known ML classifiers for classification. The classification procedure was performed on two levels. The primary level aimed to distinguish patients with COVID-19 from normal cases (binary class level). Whereas the second level was multiclass to distinguish cases of COVID-19 from normal and other cardiac complications. The major contributions of the diagnostic tool were, first, the construction of a novel automatic, inexpensive, harmless, susceptible, and quick diagnostic tool as a replacement to the present diagnostic tools to support the automatic detection of COVID-19. In addition, the novel tool relied on 2D ECG trace images to diagnose COVID-19, which is a new approach to achieving a diagnosis. Moreover, in view of the disparities in the performances between DL models, the proposed tool utilized ten DL models of distinctive structures to merge their benefits, not a single architecture. Additionally, it extracted features from the last fully connected layers of the ten DL models instead of end-to-end DL classification (as in previous studies). The proposed tool merged these features to investigate the impact of merging on diagnostic accuracy. Furthermore, it presented a hybrid FS approach based on three search strategies to select the most significant deep features and the lower dimensions of the feature space. Finally, it explored whether the hybrid FS approach boosts the performance of the proposed diagnostic tool. The results achieved using the proposed tool could be evidence that ECG records can be used in diagnosing the new coronavirus. The presented tool may prevent the shortcomings of chest imaging techniques, antigen, and PCR exams. It could be considered an easy, inexpensive, quick, portable, and sensible approach. Therefore, it might help clinicians in diagnosing COVID-19 accurately and automatically. Upcoming experiments will test the efficiency of the proposed tool in actual clinical procedures. Further work will consider using resampling techniques that handle the class imbalance problem. Future work will explore more deep learning techniques as well as hyperparameter optimization approaches. In addition, the uncertainty of the input data will be taken into consideration in future work.

Funding: This research received no external funding.

Supplementary Materials: The following supporting information can be downloaded at: https://www.mdpi.com/article/10.3390/bios12050299/s1, Figure S1: a two-dimensional scatter plot of the first two features of the feature space for the binary classification level; Figure S2: a two-dimensional scatter plot of the first two features of the feature space for the multiclass classification level; Figure S3: a two-dimensional scatter plot of the LDA prediction using the first two features of the feature space for the binary classification level; Figure S4: a two-dimensional scatter plot of the LDA prediction using the first two features of the feature space for the multiclass classification level. Figure S5: a two-dimensional scatter plot of the QDA prediction using the first two features of the feature space for the binary classification level; Figure S6: a two-dimensional scatter plot of the QDA prediction using the first two features of the feature space for the multiclass classification level.

Institutional Review Board Statement: Not applicable.

Informed Consent Statement: Not applicable.

Data Availability Statement: The dataset employed in this paper can be found in the Mendeley [https://data.mendeley.com/datasets/gwbz3fsgp8/2]. (accessed on 1 September 2021).

Conflicts of Interest: The author declares that they have no conflict of interest.

Abbreviations

AI	Artificial intelligence
ANOVA	Analysis of variance
ANN	Artificial neural networks
AVF	Augmented voltage foot
AVL	Augmented voltage left
AVR	Augmented voltage right
CNN	Convolutional Neural Network
CT	Computed Temography
DL	Deep learning
DT	Decision Tree
DWT	Discrete wavelet transform
ECG	Electrocardiogram
FN	False negative
FP	False positive
FS	Feature Selection
GLCM	Gray-Level Co-Occurrence Matrix
KNN	K-nearest neighbor

LDA	Linear discriminate analysis
MI	Myocardial infarction
ML	Machine learning
QDA	Quadratic discriminate analysis
RF	Random Forest
RT-PCR	Real-time reverse transcription-polymerase chain reaction
SVM	Support vector machine
TL	Transfer learning
TN	True negative
TR	True positive
TML	Traditional machine learning techniques
WHO	World Health Organization

Appendix A

The patients' records that have been used are as follows:

For the normal class: records utilized are normal (1) to normal (250).

For the COVID-19 class: all records available in the dataset are utilized

For cardiac abnormalities class: abnormal heartbeats(HB) records utilized are HB(1) to HB (88), myocardial infarction (MI) records used are MB (1) to MB (77), previous myocardial infarction (PMI) records used are PMI (1) to PMI (88).

References

1. Pascarella, G.; Strumia, A.; Piliego, C.; Bruno, F.; Del Buono, R.; Costa, F.; Scarlata, S.; Agrò, F.E. COVID-19 Diagnosis and Management: A Comprehensive Review. *J. Intern. Med.* **2020**, *288*, 192–206. [CrossRef] [PubMed]
2. Attallah, O. A Computer-Aided Diagnostic Framework for Coronavirus Diagnosis Using Texture-Based Radiomics Images. *Digital Health* **2022**, *8*, 20552076221092544. [CrossRef] [PubMed]
3. Alballa, N.; Al-Turaiki, I. Machine Learning Approaches in COVID-19 Diagnosis, Mortality, and Severity Risk Prediction: A Review. *Inform. Med. Unlocked* **2021**, *24*, 100564. [CrossRef] [PubMed]
4. Alsharif, W.; Qurashi, A. Effectiveness of COVID-19 Diagnosis and Management Tools: A Review. *Radiography* **2021**, *27*, 682–687. [CrossRef] [PubMed]
5. Alanagreh, L.; Alzoughool, F.; Atoum, M. The Human Coronavirus Disease COVID-19: Its Origin, Characteristics, and Insights into Potential Drugs and Its Mechanisms. *Pathogens* **2020**, *9*, 331. [CrossRef] [PubMed]
6. Kyosei, Y.; Yamura, S.; Namba, M.; Yoshimura, T.; Watabe, S.; Ito, E. Antigen Tests for COVID-19. *Biophys. Phys.* **2021**, *18*, 28–39. [CrossRef]
7. Scohy, A.; Anantharajah, A.; Bodéus, M.; Kabamba-Mukadi, B.; Verroken, A.; Rodriguez-Villalobos, H. Low Performance of Rapid Antigen Detection Test as Frontline Testing for COVID-19 Diagnosis. *J. Clin. Virol.* **2020**, *129*, 104455. [CrossRef]
8. Raptis, C.A.; Hammer, M.M.; Short, R.G.; Shah, A.; Bhalla, S.; Bierhals, A.J.; Filev, P.D.; Hope, M.D.; Jeudy, J.; Kligerman, S.J. Chest CT and Coronavirus Disease (COVID-19): A Critical Review of the Literature to Date. *Am. J. Roentgenol.* **2020**, *215*, 839–842. [CrossRef]
9. Rousan, L.A.; Elobeid, E.; Karrar, M.; Khader, Y. Chest X-Ray Findings and Temporal Lung Changes in Patients with COVID-19 Pneumonia. *BMC Pulm. Med.* **2020**, *20*, 245. [CrossRef]
10. Xie, X.; Zhong, Z.; Zhao, W.; Zheng, C.; Wang, F.; Liu, J. Chest CT for Typical 2019-NCoV Pneumonia: Relationship to Negative RT-PCR Testing. *Radiology* **2020**, *296*, E41–E45. [CrossRef]
11. Attallah, O.; Anwar, F.; Ghanem, N.M.; Ismail, M.A. Histo-CADx: Duo Cascaded Fusion Stages for Breast Cancer Diagnosis from Histopathological Images. *PeerJ Comput. Sci.* **2021**, *7*, e493. [CrossRef] [PubMed]
12. Attallah, O. DIAROP: Automated Deep Learning-Based Diagnostic Tool for Retinopathy of Prematurity. *Diagnostics* **2021**, *11*, 2034. [CrossRef] [PubMed]
13. Attallah, O.; Sharkas, M. GASTRO-CADx: A Three Stages Framework for Diagnosing Gastrointestinal Diseases. *PeerJ Comput. Sci.* **2021**, *7*, e423. [CrossRef]
14. Karthikesalingam, A.; Attallah, O.; Ma, X.; Bahia, S.S.; Thompson, L.; Vidal-Diez, A.; Choke, E.C.; Bown, M.J.; Sayers, R.D.; Thompson, M.M. An Artificial Neural Network Stratifies the Risks of Reintervention and Mortality after Endovascular Aneurysm Repair; a Retrospective Observational Study. *PLoS ONE* **2015**, *10*, e0129024. [CrossRef]
15. Attallah, O. An Effective Mental Stress State Detection and Evaluation System Using Minimum Number of Frontal Brain Electrodes. *Diagnostics* **2020**, *10*, 292. [CrossRef] [PubMed]
16. Attallah, O.; Karthikesalingam, A.; Holt, P.J.; Thompson, M.M.; Sayers, R.; Bown, M.J.; Choke, E.C.; Ma, X. Using Multiple Classifiers for Predicting the Risk of Endovascular Aortic Aneurysm Repair Re-Intervention through Hybrid Feature Selection. *Proc. Inst. Mech. Eng. Part H J. Eng. Med.* **2017**, *231*, 1048–1063. [CrossRef] [PubMed]

17. Ragab, D.A.; Sharkas, M.; Attallah, O. Breast Cancer Diagnosis Using an Efficient CAD System Based on Multiple Classifiers. *Diagnostics* **2019**, *9*, 165. [CrossRef] [PubMed]

18. Attallah, O. MB-AI-His: Histopathological Diagnosis of Pediatric Medulloblastoma and Its Subtypes via AI. *Diagnostics* **2021**, *11*, 359–384. [CrossRef]

19. Attallah, O. CoMB-Deep: Composite Deep Learning-Based Pipeline for Classifying Childhood Medulloblastoma and Its Classes. *Front. Neuroinform.* **2021**, *15*, 663592. [CrossRef]

20. Attallah, O.; Zaghlool, S. AI-Based Pipeline for Classifying Pediatric Medulloblastoma Using Histopathological and Textural Images. *Life* **2022**, *12*, 232. [CrossRef]

21. Alyasseri, Z.A.A.; Al-Betar, M.A.; Doush, I.A.; Awadallah, M.A.; Abasi, A.K.; Makhadmeh, S.N.; Alomari, O.A.; Abdulkareem, K.H.; Adam, A.; Damasevicius, R. Review on COVID-19 Diagnosis Models Based on Machine Learning and Deep Learning Approaches. *Expert syst.* **2021**, *39*, e12759. [CrossRef] [PubMed]

22. Attallah, O.; Ragab, D.A.; Sharkas, M. MULTI-DEEP: A Novel CAD System for Coronavirus (COVID-19) Diagnosis from CT Images Using Multiple Convolution Neural Networks. *PeerJ* **2020**, *8*, e10086. [CrossRef] [PubMed]

23. Ragab, D.A.; Attallah, O. FUSI-CAD: Coronavirus (COVID-19) Diagnosis Based on the Fusion of CNNs and Handcrafted Features. *PeerJ Comput. Sci.* **2020**, *6*, e306. [CrossRef] [PubMed]

24. Pawlak, A.; Ręka, G.; Olszewska, A.; Warchulińska, J.; Piecewicz-Szczęsna, H. Methods of Assessing Body Composition and Anthropometric Measurements–a Review of the Literature. *J. Educ. Health Sport* **2021**, *11*, 18–27. [CrossRef]

25. Khawaja, S.A.; Mohan, P.; Jabbour, R.; Bampouri, T.; Bowsher, G.; Hassan, A.M.; Huq, F.; Baghdasaryan, L.; Wang, B.; Sethi, A. COVID-19 and Its Impact on the Cardiovascular System. *Open Heart* **2021**, *8*, e001472. [CrossRef] [PubMed]

26. Buja, L.M.; Stone, J.R. A Novel Coronavirus Meets the Cardiovascular System: Society for Cardiovascular Pathology Symposium 2021. *Cardiovasc. Pathol.* **2021**, *53*, 107336. [CrossRef]

27. Barman, H.A.; Atici, A.; Alici, G.; Sit, O.; Tugrul, S.; Gungor, B.; Okuyan, E.; Sahin, I. The Effect of the Severity COVID-19 Infection on Electrocardiography. *Am. J. Emerg. Med.* **2021**, *46*, 317–322. [CrossRef]

28. Babapoor-Farrokhran, S.; Rasekhi, R.T.; Gill, D.; Babapoor, S.; Amanullah, A. Arrhythmia in COVID-19. *SN Compr. Clin. Med.* **2020**, *2*, 1430–1435. [CrossRef]

29. Predabon, B.; Souza, A.Z.M.; Bertoldi, G.H.S.; Sales, R.L.; Luciano, K.S.; de March Ronsoni, R. The Electrocardiogram in the Differential Diagnosis of Cardiologic Conditions Related to the Covid-19 Pandemic. *J. Card. Arrhythm.* **2020**, *33*, 133–141. [CrossRef]

30. Andrysiak, T. Machine Learning Techniques Applied to Data Analysis and Anomaly Detection in ECG Signals. *Appl. Artif. Intell.* **2016**, *30*, 610–634. [CrossRef]

31. Tuncer, T.; Dogan, S.; Pławiak, P.; Acharya, U.R. Automated Arrhythmia Detection Using Novel Hexadecimal Local Pattern and Multilevel Wavelet Transform with ECG Signals. *Knowl.-Based Syst.* **2019**, *186*, 104923. [CrossRef]

32. Hernandez-Matamoros, A.; Fujita, H.; Escamilla-Hernandez, E.; Perez-Meana, H.; Nakano-Miyatake, M. Recognition of ECG Signals Using Wavelet Based on Atomic Functions. *Biocybern. Biomed. Eng.* **2020**, *40*, 803–814. [CrossRef]

33. Pandey, S.K.; Janghel, R.R.; Vani, V. Patient Specific Machine Learning Models for ECG Signal Classification. *Procedia Comput. Sci.* **2020**, *167*, 2181–2190. [CrossRef]

34. Rouhi, R.; Clausel, M.; Oster, J.; Lauer, F. An Interpretable Hand-Crafted Feature-Based Model for Atrial Fibrillation Detection. *Front. Physiol.* **2021**, *12*, 581. [CrossRef] [PubMed]

35. Nanni, L.; Ghidoni, S.; Brahnam, S. Handcrafted vs. Non-Handcrafted Features for Computer Vision Classification. *Pattern Recognit.* **2017**, *71*, 158–172. [CrossRef]

36. Huang, J.; Chen, B.; Yao, B.; He, W. ECG Arrhythmia Classification Using STFT-Based Spectrogram and Convolutional Neural Network. *IEEE Access* **2019**, *7*, 92871–92880. [CrossRef]

37. Zhai, X.; Tin, C. Automated ECG Classification Using Dual Heartbeat Coupling Based on Convolutional Neural Network. *IEEE Access* **2018**, *6*, 27465–27472. [CrossRef]

38. Kiranyaz, S.; Ince, T.; Gabbouj, M. Real-Time Patient-Specific ECG Classification by 1-D Convolutional Neural Networks. *IEEE Trans. Biomed. Eng.* **2015**, *63*, 664–675. [CrossRef]

39. Attallah, O.; Sharkas, M.A.; Gadelkarim, H. Deep Learning Techniques for Automatic Detection of Embryonic Neurodevelopmental Disorders. *Diagnostics* **2020**, *10*, 27. [CrossRef]

40. Izci, E.; Ozdemir, M.A.; Degirmenci, M.; Akan, A. Cardiac Arrhythmia Detection from 2d Ecg Images by Using Deep Learning Technique. In Proceedings of the 2019 Medical Technologies Congress (TIPTEKNO), Izmir, Turkey, 3–5 October 2019; pp. 1–4.

41. Huang, J.-S.; Chen, B.-Q.; Zeng, N.-Y.; Cao, X.-C.; Li, Y. Accurate Classification of ECG Arrhythmia Using MOWPT Enhanced Fast Compression Deep Learning Networks. *J. Ambient Intell. Humaniz. Comput.* **2020**, 1–18. [CrossRef]

42. Singh, S.A.; Majumder, S. A Novel Approach Osa Detection Using Single-Lead ECG Scalogram Based on Deep Neural Network. *J. Mech. Med. Biol.* **2019**, *19*, 1950026. [CrossRef]

43. Bortolan, G.; Christov, I.; Simova, I. Potential of Rule-Based Methods and Deep Learning Architectures for ECG Diagnostics. *Diagnostics* **2021**, *11*, 1678. [CrossRef] [PubMed]

44. Zhu, Z.; Lan, X.; Zhao, T.; Guo, Y.; Kojodjojo, P.; Xu, Z.; Liu, Z.; Liu, S.; Wang, H.; Sun, X. Identification of 27 Abnormalities from Multi-Lead ECG Signals: An Ensembled SE_ResNet Framework with Sign Loss Function. *Physiol. Meas.* **2021**, *42*, 065008. [CrossRef] [PubMed]

45. Jo, Y.-Y.; Kwon, J.; Jeon, K.-H.; Cho, Y.-H.; Shin, J.-H.; Lee, Y.-J.; Jung, M.-S.; Ban, J.-H.; Kim, K.-H.; Lee, S.Y. Detection and Classification of Arrhythmia Using an Explainable Deep Learning Model. *J. Electrocardiol.* **2021**, *67*, 124–132. [CrossRef]
46. Yang, X.; Zhang, X.; Yang, M.; Zhang, L. 12-Lead ECG Arrhythmia Classification Using Cascaded Convolutional Neural Network and Expert Feature. *J. Electrocardiol.* **2021**, *67*, 56–62. [CrossRef]
47. Zhang, H.; Liu, C.; Zhang, Z.; Xing, Y.; Liu, X.; Dong, R.; He, Y.; Xia, L.; Liu, F. Recurrence Plot-Based Approach for Cardiac Arrhythmia Classification Using Inception-ResNet-V2. *Front. Physiol.* **2021**, *12*, 648950. [CrossRef]
48. Krasteva, V.; Christov, I.; Naydenov, S.; Stoyanov, T.; Jekova, I. Application of Dense Neural Networks for Detection of Atrial Fibrillation and Ranking of Augmented ECG Feature Set. *Sensors* **2021**, *21*, 6848. [CrossRef]
49. Dai, H.; Hwang, H.-G.; Tseng, V.S. Convolutional Neural Network Based Automatic Screening Tool for Cardiovascular Diseases Using Different Intervals of ECG Signals. *Comput. Methods Programs Biomed.* **2021**, *203*, 106035. [CrossRef]
50. Alday, E.A.P.; Gu, A.; Robichaux, C.; Ian Wong, A.K.; Liu, C.; Liu, F.; Elola, A.; Seyedi, S.; Li, Q.; Sharma, A. Classification of 12-Lead ECGs: The PhysioNet/Computing in Cardiology Challenge 2020. *Physiol. Meas.* **2021**, *41*, 124003. [CrossRef]
51. Reyna, M.A.; Sadr, N.; Alday, E.A.P.; Gu, A.; Shah, A.J.; Robichaux, C.; Rad, A.B.; Elola, A.; Seyedi, S.; Ansari, S. Issues in the Automated Classification of Multilead ECGs Using Heterogeneous Labels and Populations. *Personnel* **2021**, *4*, 5.
52. Badilini, F.; Erdem, T.; Zareba, W.; Moss, A.J. ECGscan: A Method for Digitizing Paper ECG Printouts. *J. Electrocardiol.* **2003**, *36*, 39. [CrossRef]
53. Mishra, S.; Khatwani, G.; Patil, R.; Sapariya, D.; Shah, V.; Parmar, D.; Dinesh, S.; Daphal, P.; Mehendale, N. ECG Paper Record Digitization and Diagnosis Using Deep Learning. *J. Med. Biol. Eng.* **2021**, *41*, 422–432. [CrossRef] [PubMed]
54. Widman, L.E.; Freeman, G.L. A-to-D Conversion from Paper Records with a Desktop Scanner and a Microcomputer. *Comput. Biomed. Res.* **1989**, *22*, 393–404. [CrossRef]
55. Hao, P.; Gao, X.; Li, Z.; Zhang, J.; Wu, F.; Bai, C. Multi-Branch Fusion Network for Myocardial Infarction Screening from 12-Lead ECG Images. *Comput. Methods Programs Biomed.* **2020**, *184*, 105286. [CrossRef]
56. Mohamed, B.; Issam, A.; Mohamed, A.; Abdellatif, B. ECG Image Classification in Real Time Based on the Haar-like Features and Artificial Neural Networks. *Procedia Comput. Sci.* **2015**, *73*, 32–39. [CrossRef]
57. Ferreira, M.A.A.; Gurgel, M.V.; Marinho, L.B.; Nascimento, N.M.M.; da Silva, S.P.P.; Alves, S.S.A.; Ramalho, G.L.B.; Rebouças Filho, P.P. Evaluation of Heart Disease Diagnosis Approach Using ECG Images. In Proceedings of the 2019 International Joint Conference on Neural Networks (IJCNN), Budapest, Hungary, 14–19 July 2019; pp. 1–7.
58. Du, N.; Cao, Q.; Yu, L.; Liu, N.; Zhong, E.; Liu, Z.; Shen, Y.; Chen, K. FM-ECG: A Fine-Grained Multi-Label Framework for ECG Image Classification. *Inf. Sci.* **2021**, *549*, 164–177. [CrossRef]
59. Khan, A.H.; Hussain, M.; Malik, M.K. Cardiac Disorder Classification by Electrocardiogram Sensing Using Deep Neural Network. *Complexity* **2021**, *2021*, 5512243. [CrossRef]
60. Xie, Y.; Yang, H.; Yuan, X.; He, Q.; Zhang, R.; Zhu, Q.; Chu, Z.; Yang, C.; Qin, P.; Yan, C. Stroke Prediction from Electrocardiograms by Deep Neural Network. *Multimed. Tools Appl.* **2021**, *80*, 17291–17297. [CrossRef]
61. Khan, A.H.; Hussain, M.; Malik, M.K. ECG Images Dataset of Cardiac and COVID-19 Patients. *Data Brief* **2021**, *34*, 106762. [CrossRef]
62. Anwar, T.; Zakir, S. Effect of Image Augmentation on ECG Image Classification Using Deep Learning. In Proceedings of the 2021 International Conference on Artificial Intelligence (ICAI), Islamabad, Pakistan, 5–7 April 2021; pp. 182–186.
63. Rahman, T.; Akinbi, A.; Chowdhury, M.E.; Rashid, T.A.; Şengür, A.; Khandakar, A.; Islam, K.R.; Ismael, A.M. COV-ECGNET: COVID-19 Detection Using ECG Trace Images with Deep Convolutional Neural Network. *arXiv* **2021**, arXiv:2106.00436. [CrossRef]
64. Ozdemir, M.A.; Ozdemir, G.D.; Guren, O. Classification of COVID-19 Electrocardiograms by Using Hexaxial Feature Mapping and Deep Learning. *BMC Med. Inform. Decis. Mak.* **2021**, *21*, 170. [CrossRef] [PubMed]
65. Attallah, O. ECG-BiCoNet: An ECG-Based Pipeline for COVID-19 Diagnosis Using Bi-Layers of Deep Features Integration. *Comput. Biol. Med.* **2022**, *142*, 105210. [CrossRef] [PubMed]
66. Shorten, C.; Khoshgoftaar, T.M. A Survey on Image Data Augmentation for Deep Learning. *J. Big Data* **2019**, *6*, 60. [CrossRef]
67. Amin, S.U.; Alsulaiman, M.; Muhammad, G.; Mekhtiche, M.A.; Hossain, M.S. Deep Learning for EEG Motor Imagery Classification Based on Multi-Layer CNNs Feature Fusion. *Future Gener. Comput. Syst.* **2019**, *101*, 542–554. [CrossRef]
68. Xu, Q.; Wang, Z.; Wang, F.; Gong, Y. Multi-Feature Fusion CNNs for Drosophila Embryo of Interest Detection. *Phys. A Stat. Mech. Its Appl.* **2019**, *531*, 121808. [CrossRef]
69. Ravì, D.; Wong, C.; Deligianni, F.; Berthelot, M.; Andreu-Perez, J.; Lo, B.; Yang, G.-Z. Deep Learning for Health Informatics. *IEEE J. Biomed. Health Inform.* **2016**, *21*, 4–21. [CrossRef]
70. Szegedy, C.; Vanhoucke, V.; Ioffe, S.; Shlens, J.; Wojna, Z. Rethinking the Inception Architecture for Computer Vision. In Proceedings of the IEEE Conference on Computer Vision and Pattern Recognition, Las Vegas, NV, USA, 27–30 June 2016; pp. 2818–2826.
71. Szegedy, C.; Liu, W.; Jia, Y.; Sermanet, P.; Reed, S.; Anguelov, D.; Erhan, D.; Vanhoucke, V.; Rabinovich, A. Going Deeper with Convolutions. In Proceedings of the IEEE Conference on Computer Vision and Pattern Recognition, Boston, Ma, USA, 7–12 June 2015; pp. 1–9.
72. Radhika, K.; Devika, K.; Aswathi, T.; Sreevidya, P.; Sowmya, V.; Soman, K.P. Performance Analysis of NASNet on Unconstrained Ear Recognition. In *Nature Inspired Computing for Data Science*; Springer: Berlin/Heidelberg, Germany, 2020; pp. 57–82.

73. He, K.; Zhang, X.; Ren, S.; Sun, J. Deep Residual Learning for Image Recognition. In Proceedings of the IEEE Conference on Computer Vision and Pattern Recognition, Las Vegas, NV, USA, 27–30 June 2016.
74. Chollet, F. Xception: Deep Learning with Depthwise Separable Convolutions. In Proceedings of the IEEE Conference on Computer Vision and Pattern Recognition, Honolulu, HI, USA, 21–26 July 2017; pp. 1251–1258.
75. Jinsakul, N.; Tsai, C.-F.; Tsai, C.-E.; Wu, P. Enhancement of Deep Learning in Image Classification Performance Using Xception with the Swish Activation Function for Colorectal Polyp Preliminary Screening. *Mathematics* **2019**, *7*, 1170. [CrossRef]
76. Szegedy, C.; Ioffe, S.; Vanhoucke, V.; Alemi, A. Inception-v4, Inception-Resnet and the Impact of Residual Connections on Learning. *arXiv* **2016**, arXiv:1602.07261.
77. Huang, G.; Liu, Z.; Van Der Maaten, L.; Weinberger, K.Q. Densely Connected Convolutional Networks. In Proceedings of the IEEE Conference on Computer Vision and Pattern Recognition, Honolulu, HI, USA, 21–26 July 2017; pp. 4700–4708.
78. Zhang, X.; Zhou, X.; Lin, M.; Sun, J. Shufflenet: An Extremely Efficient Convolutional Neural Network for Mobile Devices. In Proceedings of the IEEE Conference on Computer Vision and Pattern Recognition, Salt Lake City, UT, USA, 18–23 June 2018; pp. 6848–6856.
79. Redmon, J.; Farhadi, A. YOLO9000: Better, Faster, Stronger. In Proceedings of the IEEE Conference on Computer Vision and Pattern Recognition, Honolulu, HI, USA, 21–26 July 2017; pp. 7263–7271.
80. Howard, A.G.; Zhu, M.; Chen, B.; Kalenichenko, D.; Wang, W.; Weyand, T.; Andreetto, M.; Adam, H. Mobilenets: Efficient Convolutional Neural Networks for Mobile Vision Applications. *arXiv* **2017**, arXiv:1704.04861.
81. Tan, C.; Sun, F.; Kong, T.; Zhang, W.; Yang, C.; Liu, C. A Survey on Deep Transfer Learning. In *International Conference on Artificial Neural Networks*; Springer: Berlin/Heidelberg, Germany, 2018; pp. 270–279.
82. Hancer, E.; Xue, B.; Zhang, M. A Survey on Feature Selection Approaches for Clustering. *Artif. Intell. Rev.* **2020**, *53*, 4519–4545. [CrossRef]
83. Chizi, B.; Rokach, L.; Maimon, O. A Survey of Feature Selection Techniques. In *Encyclopedia of Data Warehousing and Mining, Second Edition*; IGI Global: Hershey, PA, USA, 2009; pp. 1888–1895.
84. Attallah, O.; Karthikesalingam, A.; Holt, P.J.; Thompson, M.M.; Sayers, R.; Bown, M.J.; Choke, E.C.; Ma, X. Feature Selection through Validation and Un-Censoring of Endovascular Repair Survival Data for Predicting the Risk of Re-Intervention. *BMC Med. Inform. Decis. Mak.* **2017**, *17*, 115–133. [CrossRef] [PubMed]
85. Bahassine, S.; Madani, A.; Al-Sarem, M.; Kissi, M. Feature Selection Using an Improved Chi-Square for Arabic Text Classification. *J. King Saud Univ.-Comput. Inf. Sci.* **2020**, *32*, 225–231. [CrossRef]
86. Hall, M.; Frank, E.; Holmes, G.; Pfahringer, B.; Reutemann, P.; Witten, I.H. The WEKA Data Mining Software: An Update. *ACM SIGKDD Explor. Newsl.* **2009**, *11*, 10–18. [CrossRef]
87. Haseeb, S.; Gul, E.E.; Çinier, G.; Bazoukis, G.; Alvarez-Garcia, J.; Garcia-Zamora, S.; Lee, S.; Yeung, C.; Liu, T.; Tse, G. Value of Electrocardiography in Coronavirus Disease 2019 (COVID-19). *J. Electrocardiol.* **2020**, *62*, 39–45. [CrossRef] [PubMed]

Article

The Relevance of Calibration in Machine Learning-Based Hypertension Risk Assessment Combining Photoplethysmography and Electrocardiography

Jesús Cano [1], Lorenzo Fácila [2], Juan M. Gracia-Baena [3], Roberto Zangróniz [4], Raúl Alcaraz [4] and José J. Rieta [1,*]

[1] BioMIT.org, Electronic Engineering Department, Universitat Politecnica de Valencia, 46022 Valencia, Spain; jecaser@etsii.upv.es
[2] Cardiology Department, General University Hospital Consortium of Valencia, 46014 Valencia, Spain; lorenzo.facila@uv.es
[3] Cardiovascular Surgery Department, Hospital Clínico Universitario de Valencia, 46010 Valencia, Spain; gracia_juabae@gva.es
[4] Research Group in Electronic, Biomedical and Telecommunication Engineering, University of Castilla-La Mancha, 16071 Cuenca, Spain; roberto.zangroniz@uclm.es (R.Z.); raul.alcaraz@uclm.es (R.A.)
* Correspondence: jjrieta@upv.es

Abstract: The detection of hypertension (HT) is of great importance for the early diagnosis of cardiovascular diseases (CVDs), as subjects with high blood pressure (BP) are asymptomatic until advanced stages of the disease. The present study proposes a classification model to discriminate between normotensive (NTS) and hypertensive (HTS) subjects employing electrocardiographic (ECG) and photoplethysmographic (PPG) recordings as an alternative to traditional cuff-based methods. A total of 913 ECG, PPG and BP recordings from 69 subjects were analyzed. Then, signal preprocessing, fiducial points extraction and feature selection were performed, providing 17 discriminatory features, such as pulse arrival and transit times, that fed machine-learning-based classifiers. The main innovation proposed in this research uncovers the relevance of previous calibration to obtain accurate HT risk assessment. This aspect has been assessed using both close and distant time test measurements with respect to calibration. The k-nearest neighbors-classifier provided the best outcomes with an accuracy for new subjects before calibration of 51.48%. The inclusion of just one calibration measurement into the model improved classification accuracy by 30%, reaching gradually more than 96% with more than six calibration measurements. Accuracy decreased with distance to calibration, but remained outstanding even days after calibration. Thus, the use of PPG and ECG recordings combined with previous subject calibration can significantly improve discrimination between NTS and HTS individuals. This strategy could be implemented in wearable devices for HT risk assessment as well as to prevent CVDs.

Keywords: high blood pressure; hypertension; photoplethysmography; electrocardiography; calibration; classification models; machine learning

Citation: Cano, J.; Fácila, L.; Gracia-Baena, J.M.; Zangróniz, R.; Alcaraz, R.; Rieta, J.J. The Relevance of Calibration in Machine Learning-Based Hypertension Risk Assessment Combining Photoplethysmography and Electrocardiography. *Biosensors* **2022**, *12*, 289. https://doi.org/10.3390/bios12050289

Received: 25 March 2022
Accepted: 28 April 2022
Published: 1 May 2022

Publisher's Note: MDPI stays neutral with regard to jurisdictional claims in published maps and institutional affiliations.

1. Introduction

High blood pressure or hypertension (HT) is the most significant risk factor for many cardiovascular diseases (CVDs) including cardiac arrhythmias, coronary disease, renal failure and stroke [1]. To this, it must be added that most patients with HT are undiagnosed, as in the early stages and even in the elevated blood pressure stage, HT rarely causes symptoms. For these reasons, regular blood pressure monitoring and the assessment of blood pressure levels is crucial for the prevention and early diagnosis of asymptomatic HT and the study of its evolution over time for diagnosed subjects [2].

Arterial blood pressure (BP) values have two components: systolic blood pressure (SBP), determined by the impulse generated by the contractions of the left ventricle, which indicates how much pressure the blood is exerting against the arterial walls when the heart contracts, and the diastolic blood pressure (DBP), which depends on the resistance of the arteries to the passage of blood and indicates the pressure exerted against the walls when the heart relaxes [3]. BP depends mainly on two variables: the volume propelled by the heart in a unit of time and the resistances offered by the arteries to the passage of blood [4]. In turn, these variables depend on the activity of the autonomic nervous system (ANS), which governs heart rate and the resistance of the arterioles, and, on the other hand, the balance of water and salt filtered through the kidneys, which modulates blood volume.

Traditionally, BP has been measured through invasive as well as non-invasive strategies. Invasive BP measurement has been usually reserved for patient hospitalization, especially in Intensive Care Units (ICUs), where the availability of precise and time-continuous BP measurements is relevant [5]. For non-invasive BP estimation, conventional cuff-based measurement devices, which use oscillometric and auscultation methods, are known to be able to offer adequate accuracy. However, they are not designed to be wearable and only offer a one-off measure. Therefore, they are not compatible with continuous measurement throughout the day due to mobility limitations caused by the device, they are uncomfortable, and their measurement procedure, with the repeated inflation and deflation of the cuff, is somewhat tedious, cumbersome and requires patient attention [6].

Machine Learning classifiers provide many advantages to clinical medicine in general and to biosignal-based HT risk assessment in particular over non-invasive traditional measures, as they can be embedded in wearable devices such as smartwatches, facilitating uninterrupted monitoring throughout the day. This allows both the detection of asymptomatic hypertensive patients and the monitoring of diagnosed patients in their daily lives outside the clinical setting by screening changes in blood pressure.

As a consequence of the above factors, work in this field is focused on the development of cuff-less systems that can provide the user with information about the BP condition in near real time [7]. New wearable devices, such as wristbands or smartwatches capable of monitoring physiological signals that change according to BP level, as the electrocardiogram (ECG) and photoplethysmogram (PPG) do, may facilitate the development of these BP measurement systems [8,9]. The most promising signal is the PPG, an optical measurement technique that can be used to detect changes in blood volume in the micro vascular bed of tissues as a result of cardiac pumping. This technique is based on illumination of the skin measuring changes in light absorption [10]. It is typically implemented with a light-emitting diode (LED) to illuminate the skin and a photodetector to measure the amount of light transmitted or reflected through the skin. The change in tissue light absorption is governed by the amount of protein and hemoglobin in blood and the hemodynamic and physiological condition caused by the change in the properties of the artery [11].

In recent years, many studies have investigated methods to estimate BP using PPG signals. The first work that studied the correlation between the PPG and BP was conducted by Teng and Zhang [12], where a linear regression model was used to evaluate the relationship between four PPG features an BP. Once this relationship was known and established, later studies focused on the use of propagation theory, which extracted key features from ECG and PPG signals, simultaneously collected, for BP estimation.

Propagation features, as pulse transit time (PTT) and pulse arrival time (PAT), have been extensively used in previous works [13,14]. PTT was defined as the time taken for the pressure wave to travel between two arterial sites. Thus, it could be estimated as the time delay between a PPG wavefront measured by two separate sensors located in two distal sites of the body. For its part, PAT was defined as the delay between the electrical activation of the heart (R peak of ECG) and the PPG wavefront at the foot, maximum slope point and peak of the PPG signal, which represents the arrival of the pulse at the measurement location. Cavalcante et al. [15] applied this methodology for the first time using the start and end pulse points of these signals as well as PTT, PAT and pulse wave velocity (PWV) to determine the cardiovascular condition. Furthermore, Chen et al. [16] used ear and toe sensors to determinate PTT and its strong relationship with BP.

Other methodologies used the changes in PPG morphology to estimate BP. In this way, Kurylyak et al. [17] extracted 21 features from the PPG waveform, and demonstrated that PPG features could significantly decrease BP estimation error. Li et al. [18] and Kachuee et al. [19] also combined PAT and morphological parameters of the PPG, improving the accuracy of estimation of BP in comparison to only PAT-based features. After analyzing the different proposed methods to estimate BP, this work introduces a combination of both approaches, propagation theory features and morphological PPG features for enhanced HT risk assessment.

In the studies providing a BP value from PPG recordings, this value was just an estimation, so these methods need medical supervision. Thus, the present work introduces an alternative way to solve the problem of BP classification models with reliability, so that they can automatically provide in a continuous and non-invasive way the subject's blood pressure condition and can trigger alarms in case of an asymptomatic hypertensive condition. In this same way, Visvanathan et al. [20] used a support vector machine (SVM) to classify BP and Liang et al. [21] used PAT and PPG features and four distinctive classifiers, these being logistic regression, AdaBoost tree, Bagged tree and K-nearest neighbors, for the classification of subjects as a function of BP estimated values.

However, it has been demonstrated that the relationship between the aforementioned PPG-based propagation parameters and BP depends on many physiological factors, such as arterial walls' thickness and elasticity, age and gender, posture and risk factors of CVDs. Thus, calibration is needed when BP levels from a new subject are going to be evaluated by an automated classification method [22]. Moreover, calibration before measurement is essential to adapt the algorithms to the variations on PPG waveforms, as they are easily corrupted by fluctuations in blood circulation state, affecting the connection between BP and peripheral pulses [23].

The aim of the present study is to develop a classification system for discriminating between normotensive (NTS) and hypertensive (HTS) subjects and to evaluate the need and relevance of per-subject calibration. For this purpose, PPG and ECG simultaneous recordings have been analyzed and processed and propagation features, such as PTT and PAT, combined with other PPG morphological features have been extracted and used to train advanced classification models. The manuscript is organized as follows. Section 2 presents the database, the Machine Learning (ML) method procedure and preprocessing, the analysis techniques and the methods to evaluate the need for calibration. Section 3 presents the results, which will be analyzed in Section 4. Finally, in Section 5, the main scientific contributions of this study are remarked upon.

2. Materials And Methods

2.1. Materials

In this study, the recordings used were obtained from the MIMIC database, which contains information from ICU patients admitted to Beth Israel Deaconess Medical Center in Boston, USA [24]. This database was chosen as it contains ECG, PPG and invasive BP signals recorded simultaneously in ICU. BP signals in which the systolic or diastolic waves were indistinguishable, ECG signals where QRS morphology was distorted or PPG signals in which the systolic and diastolic waves were indistinguishable and the morphology was distorted were dismissed due to the presence of artifacts.

The BP values were labelled according to the report of the Joint National Committee on the prevention, detection, evaluation and treatment of high blood pressure [25]: as normotensive (NTS) for SBP lower than 120 mmHg, prehypertensive (PHT) for SBP between 120 and 140 mmHg and hypertensive (HTS) for SBP higher than 140 mmHg.

After labelling MIMIC recordings according to SBP values, it was observed that several subjects had stable stretches with different labels. One reason that explained these alterations in SBP values was that all patients were in an ICU, so they may have received treatment or medication that significantly altered SBP levels. Moreover, there were subjects with distant stretches, at different time points, whose SBP values were on the borderline between two labels, so that they had a different label across time even though the changes in SBP were only a few mmHg.

Because of the aforementioned reasons, those subjects with huge alterations of their SBP values (labels including NTS and HTS across time) were dismissed, as they were not suitable to train a classification model aimed at assessing the risk of HT. As a result, subjects maintaining the same label across the recording time were selected. A total of 913 recordings from 69 subjects, 45 being NTS and 24 being HTS, with acceptable signal quality conditions were selected from the MIMIC database. The signals were all recorded simultaneously with a duration of 120 s, a common sampling frequency of 125 Hz and a resolution of 8–10 bits [26].

2.2. Signal Preprocessing

The PPG signals were processed by a fourth-order Chebyshev II bandpass filter with cutoff frequencies between 0.5 and 10 Hz [27] to remove minor noises and artifacts caused by sensors' bad contacts, patient movements or any other interfering physiological activity, such as the respiratory activity, that did not provoke signal dismissing in the previous selection stage of minimum signal quality. Furthermore, the mean value of the filtered PPG was removed to prevent drifts and to allow a better comparison between different signals.

Since the waveform of the PPG signal itself is rather simple and not very informative, the derivatives of the signal were also used to better assess the changes in the signals caused by BP. They represent the velocity plethysmogram (VPG) and the acceleration plethysmogram (APG) and were obtained by applying the first and the second order derivatives, respectively, to the processed PPG signal [28].

The ABP signals which reflected the change in BP over the cardiac cycle were clear and did not require any processing to be applied. For its part, standard preprocessing was applied to each ECG [29]. Thus, they were high-pass filtered with cutoff frequency of 0.5 Hz to remove the baseline, and then low-pass filtered with a cutoff frequency of 50 Hz to reduce high-frequency muscle noise and power line interference, in this case, 60 Hz [29].

2.3. Fiducial Points Identification

After signal preprocessing, fiducial points from PPG, VPG and APG were extracted as illustrated in Figure 1. The systolic peaks of the three signals (S, W, a), the onset point of the PPG signal (O), and two local maxima and minimum of the APG signal (b, c, d, e) were extracted [28,30]. Fiducial points in the precessed signals were obtained based on searching local minima and maxima, calculated by establishing threshold and slope criteria in each of the pulses composing every signal.

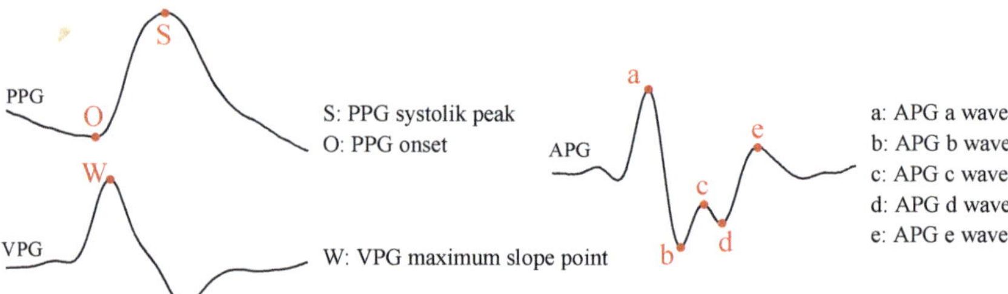

Figure 1. Graphical definition of fiducial points detected from photoplethysmogram (PPG), velocity plethysmogram (VPG) and acceleration plethysmogram (APG) signals.

The maximum systolic blood pressure (SBP) was extracted as the maximum point of each ABP pulse. SBP was used to label every subject as NTS or HTS. Subjects whose selected segments had SBP < 130 mmHg were labeled as NTS, and subjects whose selected segments had SBP > 130 mmHg as HTS. Finally, for each ECG recording, an R-peak detector based on the phasor transform was applied to the processed ECG signal to obtain the position of each beat [30].

2.4. Definition of Discriminatory Features

After the detection of R-peaks in ECG recordings and the fiducial points for each PPG, VPG and APG signals, discriminatory features were defined based on the pulse wave propagation models, such as pulse arrival times (PAT) or pulse transit time (PTT), and other morphological features from the signals that are listed below [28,31,32]. Figure 2 illustrates the definition of the features.

- *PAT*: time interval between R peak and: the O-notch (PAT_{foot}), the maximum slope of PPG signal or W peak of VPG signal ($PAT_{derivate}$) and S-peak (PAT_{peak}).
- *PTT*: time interval between SBP peak in BP signal to S-peak.
- *Sistolic peak amplitude in PPG*: amplitude from the baseline to S-peaks.
- *Sistolic peak amplitude in VPG*: amplitude from the baseline to W-peaks.
- *TPP*: time interval between two consecutive S-peaks.
- *Time pulse interval (TPI)*: time interval between two consecutive O-notches.
- *Rising time*: time interval between O-notch and systolic peak in PPG signal.
- *Width*: pulse width at half the height of systolic peak height.
- *Pulse area*: integral of the signal between two consecutive O-notches.
- *Area 1*: trapezoidal integration of PPG signal from O-notch to S-peak.
- *Area 2*: trapezoidal integration of PPG signal from S-peak to O-notch.
- *Inflection Point Area (IPA)*: ratio of both areas (A2/A1)
- *a-a*: time interval between two consecutive *a*-peaks in APG signal.
- *Ratios between APG waves with the a-wave*: $b/a, c/a, d/a, e/a$.
- *Complex APG ratios*: $(b - c - d - e)/a, (b - e)/a, (b - c - d)/a, (c + d - b)/a$.

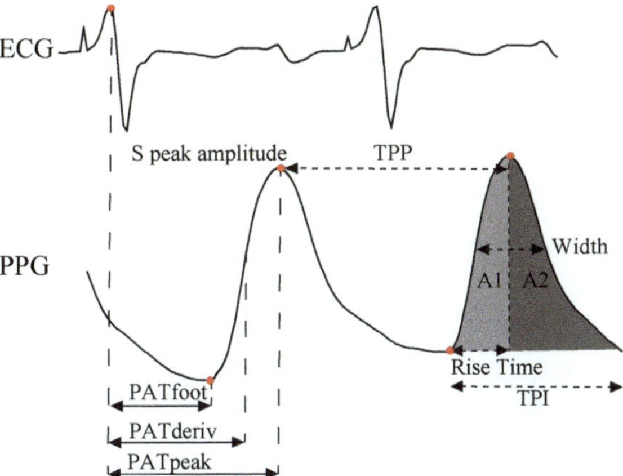

Figure 2. Representation of PAT_{foot}, $PAT_{derivate}$ and PAT_{peak} features obtained by the time interval between ECG R-peak and fiducial points of PPG signals as well as PPG morphological parameters: Systolic peak amplitude, TPP, rise time, areas under the pulse, width and TPI.

2.5. Feature Selection

The aim of the feature-selection stage was to select only those features, from the original 23 discriminating parameters, that presented relevant information for solving the classification problem optimally.

Firstly, since all the features were continuous quantitative variables, it was necessary to carry out a normalization, since each one could take on a different range of values and more weight would be given to the variables with higher values, not necessarily being more important. The normalization was carried out using "zscore" centering the variables so that they had zero mean and scaling so that they had unit standard deviation, as represented in the following equation

$$z = \frac{x - \overline{X}}{S},$$ (1)

where x is a concrete value of a given feature, $\overline{X}$ is the mean of all values of that feature and S the standard deviation.

Once the variables were normalized, *ReliefF* algorithm was applied to rank predictors by importance, determining which ones had the best discriminatory power. The key idea of this method is to estimate the quality of predictors according to how well instances near to each other are distinguished, rewarding predictors that give different values to neighbors of a different class [33]. Furthermore, by means of positive and negative correlation, the independence between pairs of variables was analyzed in order to discard those that did not provide new information for the classification task. Figure 3 illustrates the matrix whose entries are the correlation coefficients obtained by matching pairs of variables, so that highly correlated features can be discarded.

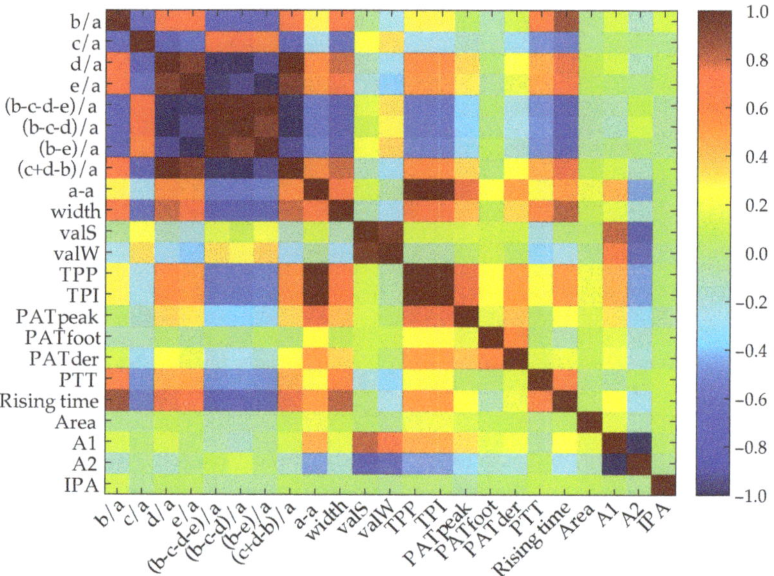

Figure 3. Correlation matrix of the 23 initial discriminatory features used in the study. Dark red values represent higher correlation coefficients and dark blue values represent lower correlation coefficients.

After analyzing the correlation matrix and *ReliefF* results, it was decided to remove three complex APG ratios $(b - c - d - e)/a$, $(b - e)/a$ and $(c + d - b)/a$ as they had high correlation coefficients with other features, as well as the last three *ReliefF* ranked features (TPP, TPI and pulse area), as the deletion of more features worsened classification performance. Finally, after the feature selection, a matrix of 17 normalized features was obtained, which will be used as inputs to train the classification models with ML techniques.

2.6. Implementation Details

The experiment was executed under MATLAB (MathWorks, Natick, MA, USA), a scientific and engineering computing software, running on a computer equipped with an Intel i7-8700 CPU @ 3.2 GHz, 16 GB of memory. The implementation for HT risk assessment combining PPG and ECG signals has been based on testing ML classification strategies such as logistic regression, Naive Bayes, discriminant analysis, support vector machines (SVM), k-nearest neighbors (KNN), ensemble classifiers and various types of decision trees [34]. Finally, SVM, Bagging Ensemble classifier and KNN were selected as they provided the highest percentages of classificatory accuracy.

SVM aims at finding the optimal separating hyper-plane between classes by focusing on the training cases that lie at the edge of the class distributions, the support vectors, so only training samples that lie on class boundaries are needed for discrimination [35]. The Bagging technique builds multiple classifiers based on a number of bootstrap samples. The outputs are decided by majority voting [36]. Finally, the KNN classifier obtains the k-nearest neighbors of the data to be classified and, as the Bagging technique, majority voting among the neighborhood is used to decide the output classification [37].

As stated before, the main objective of this study was testing whether HT risk assessment of new subjects could be improved with previous calibration. However, before addressing this goal, the classification of subjects as NTS or HTS, based on discriminant features extracted from PPG and ECG signals, was tested. In so doing, comparison with previous studies without subject-based calibration could be made. The experiment employed a leave-one-out cross-validation strategy. The classification algorithm was applied as many times as segments in the database, using each segment of 2 min in length as a single validation set and all other segments from the same subject, together with the other subjects, as a training set.

Classification performance was assessed with statistical tests for accuracy (*Acc*), sensitivity (*Se*), specificity (*Sp*) and *F1-Score*. *Acc* represented the percentage of correctly assessed PPG segments. *Se* was defined as the ability to detect as positive HTS subjects, whereas *Sp* was defined as the ability to detect as NTS healthy subjects. Finally, *F1-Score* was considered to be the harmonic mean of *Se* and *Acc*. These statistical tests were mathematically computed as

$$Acc = \frac{TP + TN}{TP + TN + FP + FN} \tag{2}$$

$$Se = \frac{TP}{TP + FN} \tag{3}$$

$$Sp = \frac{TN}{TN + FP} \tag{4}$$

$$F1\text{-}Score = \frac{2 \cdot Se \cdot Acc}{Se + Acc} = \frac{2 \cdot TP}{2 \cdot TP + FP + FN} \tag{5}$$

where TN was the number of correctly classified NTS segments, TP the number of correctly classified HTS segments, FN the number of segments that the model predicted as NTS and were actually HTS and FP the number of patients that the model predicted as HTS and were actually NTS.

2.7. Need for Calibration of New Subjects

Calibration was defined here as the inclusion of at least one previous measurement of the subject under study in the training set. Aimed at studying the importance of calibration in the classification of new subjects as NTS or HTS, three approaches were taken:

1. Classification performance of new subjects without prior subject-based calibration was studied employing those models providing the best classification results with leave-one-out cross-validation strategy. For this purpose, the analyzed segments of the new subject were only used for validation and the remaining segments from subjects other than the one under study were used to train the model.

2. As a second approach, the effectiveness of calibration to improve classification of new measurements performed later and close in time was studied following the sequel procedure:

    a. Signal segments with a duration of 2 min were divided into 12 sub-segments of 10 s in length.

b. The sub-segments of all subjects except the one to be analyzed and the first sub-segment of the analyzed subject, acting as the calibration measurement, were used as training dataset and the next sub-segment of the same subject was used for validation.

c. After its classification, this second sub-segment was introduced in the training dataset, using the next sub-segment for validation. This step was repeated until the 12 consecutive sub-segments of the subject were processed.

d. This procedure was repeated for all 2 min segments of all subjects in the database.

This way, a sequential calibration and validation was performed with the idea being to analyze the improvement in classification as the model was gradually calibrated by introducing previous measurements of the same subject very close in time.

3. Finally, the effectiveness of calibration for the classification of distant measurements was studied. To control the distance between measurements, groups of segments of the same patient that were less than 1 h, between 1 h and 6 h, between 6 h and 24 h and more than one day apart were selected. A sequential validation similar to the described for consecutive sub-segments was also followed in this approach in order to study whether classification results improved as the model was calibrated by introducing previous measurements of the same patient far away in time.

The aforesaid three approaches were developed employing the ML classification model that provided the best classification result in a leave-one-out cross-validation strategy.

3. Results

Statistical results of classification from the cross-validation strategy to discriminate between NTS and HTS segments are shown in Table 1. As can be seen, all three models provided outstanding classification results, with KNN being the model that obtained the best classification performance with a total accuracy of 93.54%, sensitivity of 92.31%, specificity of 94.35% and F1 score of 91.93%.

Table 1. Classification performance to distinguish between NTS and HTS individuals for the best models analyzed with the selected features.

Model	Accuracy	Sensitivity	Specificity	F1-Score
KNN	93.54%	92.31%	94.35%	91.93%
SVM	91.35%	90.93%	91.62%	89.34%
Ensemble	90.69%	82.97%	95.81%	87.66%

Regarding results about the need to calibrate each model to provide the best classification outcome with new subjects, the KNN model was chosen as provided the best classification results with leave-one-out cross-validation. First of all, following the first approach detailed back in Section 2.7, the segments were classified without any previous calibration, in other words, with the training dataset only consisting of segments from other subjects, with each analyzed segment from the subject under study being tested for validation. In this case, classification accuracy with no previous calibration was 51.48%. This proved that hypertension risk assessment of subjects without a prior calibration provided low accuracy results, as has been also reported by previous studies [22,23].

Next, applying the second approach of Section 2.7, aimed at demonstrating whether poor classification results could be improved with calibration, Figure 4 shows the classification accuracy between NTS and HTS individuals, in the form of box-and-whisker plots, employing sequential validation of consecutive sub-segments. Moreover, the Figure indicates in each square the mean accuracy for all selected subjects according to the number of consecutive sub-segments in the training dataset acting as subject calibration. It can be seen that, with the sole incorporation of one prior close in time sub-segment in the training dataset for calibration, the classification performance increased by 30% with respect to the case without calibration. Furthermore, accuracy improved progressively until it was stabilized above 96%, when more than six prior and close in time sub-segments from the same subject were present in the training dataset.

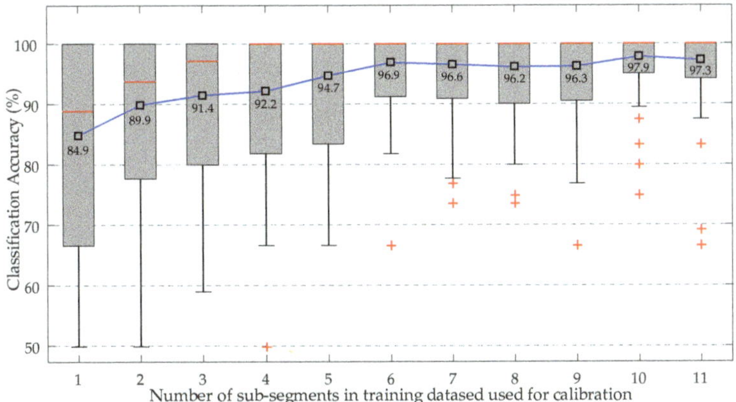

Figure 4. Classification performance provided by the KNN classifier in the discrimination between NTS and HTS individuals. Results obtained for sequential validation of consecutive sub-segments for each of the selected subject segments. In each box, the red line indicates the median, and the bottom and top edges indicate the 15th and 85th percentiles, respectively. The whiskers cover the most extreme data points not considered outliers, and the red symbol (+) stands for outliers. Black squares inside each box indicate mean accuracies.

Finally, for the third option defined in Section 2.7, performing calibration distant from measurements, Figure 5 shows classification outcomes of sequential validation with different distances between segments. It was demonstrated that calibration improved the classification task discriminating between NTS and HTS subjects because, as the number of measurements of the same subject in the model increased, so did the accuracy rate. Figure 5 also shows that with calibration and measurement separated by less than 1 h, the model was able to classify with an accuracy beyond 94% from the sixth calibration measurement onwards. As expected, these outcomes decreased as the distance between calibrations and test measurement increased, thus requiring up to five calibration measurements with distances between 6 h and 24 h to obtain classification accuracies above 75%. In any case, the need to perform several calibration measurements to achieve very good classification accuracy with test measurements, which could be many hours or even days away from calibration, does not seem to be a serious limitation. On the other hand, it is worth mentioning that only five subjects had recording lengths longer than five days, so that less stable results in Figure 5d can be considered as normal, because any misclassification would significantly affect the final accuracy. Although the number of subjects was not quite elevated in this last case, the obtained results demonstrate that classification performance was very good even with distances between calibration and test measurement of several days, which is very promising for real-world applications based on embedding these methodologies into wearable devices.

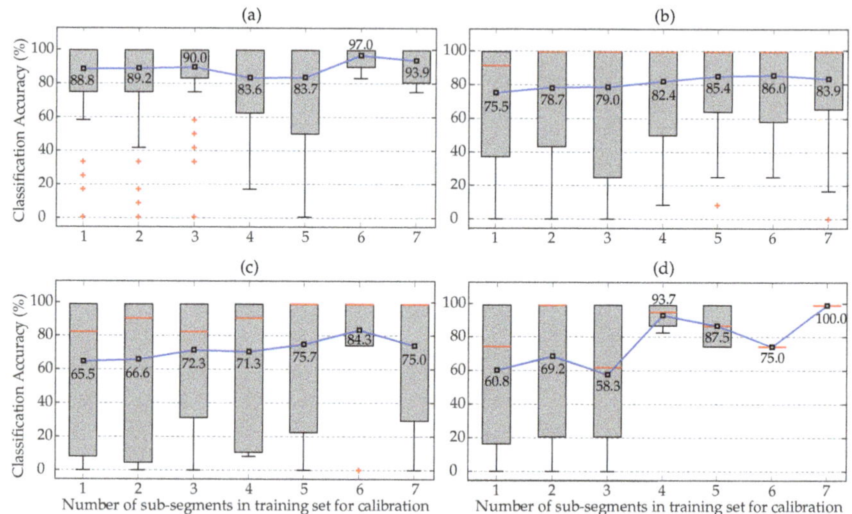

Figure 5. Classification accuracy to distinguish between NTS and HTS individuals using KNN sequential validation of segments with calibration distant from test measurements. (**a**) Distance below 1 h. (**b**) Distance between 1 and 6 h. (**c**) Distance between 6 and 24 h. (**d**) Distance above 24 h. In each box, the red line indicates the median, and the bottom and top edges indicate the 15th and 85th percentiles, respectively. The whiskers extend to the most extreme data points not considered outliers, and the red symbol (+) stands for outliers. Black squares inside each box indicate mean accuracies.

4. Discussion

The continuous measurement of BP is of great importance as it facilitates the early detection and prevention of hypertension, being the main risk factor for many CVDs. With the eruption in recent years of the Internet of Things [38] and cuff-less devices that are able to continuously measure and process physiological signals applying artificial intelligence techniques, such as ML and Deep Learning (DL), alternatives to traditional cuff-based single-time BP measurement methods have been proposed. The main signal used in related studies has been the PPG, as its morphological variations are related to the heart's activity and vascular walls condition, being similar to BP morphology both in frequency and time domains [39]. Furthermore, PPG signal can be acquired by non-invasive low-cost devices as smart watches, obtaining a continuous and real time measurement.

The monitoring of BP through PPG has mainly been studied by two different approaches: (i) addressing the problem of monitoring BP as a regression task estimating systolic and diastolic values; and (ii) addressing the problem of detecting hypertensive subjects as a classification task. In this study, the second approach has been developed, as estimations from the first approach still have serious limitations, so that it is more clinically beneficial to alert hypertensive subjects, acting as a support for clinical decision making.

Tjahjadi et al. [23] proposed the use of KNN technique and PPG signal without ECG, requiring the extraction of 2100 PPG feature points from 2.1 s of data. Their classification results achieved an F1-score of 100% for NTS and PHT patients and 90.80% for HTS patients. Although the authors affirm that this method achieved higher classification performance than other ML and DL methods, obtaining 2100 PPG feature points in such a short period of time required a sampling frequency of 1 kHz, which is a serious drawback for embedding this method in wearable devices, as it significantly increases the amount of data sampled, saved and transmitted which, unavoidably, will involve a considerably high power consumption.

Most studies for HT risk classification use both PPG and ECG signals, as PAT value is directly related to BP value. Although previous works have studied the efficiency of employing PAT as the only parameter to estimate BP [14,40], Liang et al. [21] reported a higher correlation with BP levels by combining PAT with additional PPG features. Dividing the dataset into 70% for training and 30% for validation, the KNN classification model obtained the best performance compared to bagged tree, logistic regression and AdaBoost tree. The F1 scores comparing NTS vs. PHT, NTS vs. HTS and NTS + PHT vs. HTS were 84.34%, 94.84% and 88.49%, respectively. As a consequence, the HT risk-discrimination performance between NTS and HTS was similar to the one achieved in the present study, where F1 score with the KNN classification model was 91.93%, employing leave-one-out cross-validation strategy. Therefore, both in previous studies and in the present work, the KNN classifier has been the best model to assess HT risk, combining PPG recordings and Machine Learning techniques. However, until now, there is no agreement about the discriminant features to be used, since it depends considerably on patient selection and database, mode of acquisition and signal quality.

In recent years, DL approaches have obtained outstanding performance extracting information from images [41,42]. Liang et al. [43] used the continuous wavelet transform of PPG signals and convolutional neural networks to classify BP. The dataset was divided in 80% for training and 20% for testing. The F1 scores for the binary classification comparing NTS vs. PHT were 80.52%, NTS vs. HTS were 92.55% and (NT + PHT) vs. HT trials were 82.95%. The main disadvantages of DL approaches are the requirement of a high computational cost, the extra duration of the training stage and the need for a large number of recordings.

One important consideration introduced by this work, that was not specified in related studies, has been the study of subjects with stable labels of BP. Usually, BP levels vary slightly throughout the day depending on the activities carried out by each person and many other factors, however, each subject would have to be labelled with a single and stable label. For example, any subject cannot be diagnosed as HTS at certain moments of the day, PHT at others and NTS at others. This is a problem when using databases such as the MIMIC, as it consists of recordings from ICU patients that, as a consequence of their unstable condition or the administration of drugs, may have altered and variable SBP levels.

Furthermore, any previous study about hypertension risk assessment has taken into account the relevance of calibration as a factor improving significantly classification results. In this respect, calibration has been only considered in other studies addressing BP estimation in combination with other patient data such as age [44], distance and area of arteries between measure sites or other factors that increase BP as exercise or postural changes [40]. Recently, Schlesinger et al. [45] used convolutional neural networks and PPG signals for BP estimation, achieving a reduction in mean absolute difference of 2.54 mmHg after calibration, using a single 30 s window of PPG signal and the associated BP reading.

The present study has proposed two calibration approaches, trying to improve the poor initial classification accuracy of 51.48% when a new subject entered the method without any previous calibration. The first approach investigated if the method improved classification accuracy when consecutive sub-segments of each subject were used both for calibration and classification, employing sequential validation. The assumption here was the supposed high similarity between calibration measurements and test measurement. Figure 4 showed that the presence of just one calibration sample was enough to increase classification performance more than 30%, which was enhanced even more as the number of calibration measurements raised.

The second approach studied the benefit of calibration for distances between calibration time and measurement time varying from less than 1 h to more than 24 h. This way, it was considered if PPG signal properties from the same patient were kept across time or changed along the day or week. For distances to calibration below 1 h, classification accuracy improved by 30% with just one calibration, keeping these results until more than 6 segments from the same patient were in the training dataset. The improvement of hypertension risk classification decreased slightly as the distance between calibration and measurement increased, although the use of calibration always improved classification results compared to classifying a new uncalibrated subject. Thus, after the fifth calibration, all the experiments provided high accuracy.

These approaches have demonstrated that the properties of each patient's PPG features were variable over time, as worse results were obtained with measurements distant from calibration than with those very close in time to calibration. Therefore, in order to ensure high classification accuracy, several recalibrations performed at distant recording times and, if possible, in different situations are recommended to accurately asses the risk of HT with PPG and ECG recordings, which can be obtained in a simple way through wearable devices.

Finally, this study has certain limitations that are worth considering. Even though more than 900 recordings were analyzed, the total number of patients was not too large and there was no information available on factors that may imply a higher risk of HT such as age, sex or physical condition. In this respect, Mukkamala et al. [46] studied the age factor in calibration predicting a maximum calibration interval of 1 year for subjects of 30 years of age, that declined linearly to 6 months for subjects at the age of 70, using the PTT as discriminatory feature. In addition, this study has only applied the artificial intelligence technique of ML. Future works will address the application of DL classifiers in order to discern whether they are able to improve hypertension risk assessment of current ML classifiers employing calibrated PPG recordings.

5. Conclusions

The combined extraction of discriminant features from PPG and ECG recordings, together with the use of machine learning classification models such as KNN, has been able to perform outstanding hypertension risk assessment in the discrimination between NTS or HTS subjects. The application of per-subject calibration, both in close and distant measurements, has proved its relevance for accurate classification. The implementation of these artificial intelligence techniques in wearable devices would improve the early diagnosis and prevention of cardiovascular diseases associated to hypertension.

Author Contributions: Conceptualization, J.C., R.A. and J.J.R.; methodology, J.C., R.A. and J.J.R.; software, J.C. and J.J.R.; validation, J.C., L.F., J.M.G.-B., R.Z., R.A. and J.J.R.; resources, L.F., J.M.G.-B., R.Z. and J.J.R.; data curation, J.C., L.F., J.M.G.-B. and R.Z.; original draft preparation, J.C.; review and editing, J.C., L.F., J.M.G.-B., R.Z., R.A. and J.J.R. All authors have read and agreed to the published version of the manuscript.

Funding: This research received financial support from grants PID2021-00X128525-IV0, PID2021-123804OB-I00 and TED2021-129996B-I00 of the Spanish Government 10.13039/501100011033 jointly with the European Regional Development Fund (EU), SBPLY/17/180501/000411 from Junta de Comunidades de Castilla-La Mancha and AICO/2021/286 from Generalitat Valenciana.

Institutional Review Board Statement: Data is available upon request and through MIMIC approval. MIMIC project was approved by the Institutional Review Boards of Beth Israel Deaconess Medical Center (Boston, MA, USA) and the Massachusetts Institute of Technology (Cambridge, MA, USA).

Informed Consent Statement: Requirement for individual patient consent was waived as the study did not impact clinical care and all data were de-identified.

Data Availability Statement: The data supporting reported results and presented in this study are available on request from the corresponding author.

Conflicts of Interest: The authors have no association with commercial entities that could be viewed as having an interest in the general area of the submitted manuscript. The funders had no role in the design of the study; in the collection, analyses, or interpretation of data; in the writing of the manuscript, or in the decision to publish the results.

References

1. Kjeldsen, S.E. Hypertension and cardiovascular risk: General aspects. *Pharmacol. Res.* **2018**, *129*, 95–99. [CrossRef] [PubMed]
2. World Health Organization. *A Global Brief on Hypertension: Silent Killer, Global Public Health Crisis: World Health Day 2013*; Technical Report; World Health Organization: Geneva, Switzerland, 2013.
3. Laurent, S.; Cockcroft, J.; Van Bortel, L.; Boutouyrie, P.; Giannattasio, C.; Hayoz, D.; Pannier, B.; Vlachopoulos, C.; Wilkinson, I.; Struijker-Boudier, H. Expert consensus document on arterial stiffness: Methodological issues and clinical applications. *Eur. Heart J.* **2006**, *27*, 2588–2605. [CrossRef] [PubMed]
4. Magder, S.A. The Highs and Lows of Blood Pressure. *Crit. Care Med.* **2014**, *42*, 1241–1251. [CrossRef] [PubMed]
5. Piepoli, M.F.; Hoes, A.W.; Agewall, S.; Albus, C.; Brotons, C.; Catapano, A.L.; Cooney, M.T.; Corrà, U.; Cosyns, B.; Deaton, C.; et al. 2016 European Guidelines on cardiovascular disease prevention in clinical practice. *Eur. Heart J.* **2016**, *37*, 2315–2381. [CrossRef]
6. Frese, E.M.; Fick, A.; Sadowsky, S.H. Blood Pressure Measurement Guidelines for Physical Therapists. *Cardiopulm. Phys. Ther. J.* **2011**, *22*, 5–12. [CrossRef]
7. Kario, K. Management of Hypertension in the Digital Era. *Hypertension* **2020**, *76*, 640–650. [CrossRef]
8. Panula, T.; Sirkia, J.P.; Wong, D.; Kaisti, M. Advances in non-invasive blood pressure measurement techniques. *IEEE Rev. Biomed. Eng.* **2022**. [CrossRef]
9. Quan, X.; Liu, J.; Roxlo, T.; Siddharth, S.; Leong, W.; Muir, A.; Cheong, S.M.; Rao, A. Advances in Non-Invasive Blood Pressure Monitoring. *Sensors* **2021**, *21*, 4273. [CrossRef]
10. Elgendi, M.; Fletcher, R.; Liang, Y.; Howard, N.; Lovell, N.H.; Abbott, D.; Lim, K.; Ward, R. The use of photoplethysmography for assessing hypertension. *npj Digit. Med.* **2019**, *2*, 60. [CrossRef]
11. Allen, J. Photoplethysmography and its application in clinical physiological measurement. *Physiol. Meas.* **2007**, *28*, R1–R39. [CrossRef]
12. Teng, X.; Zhang, Y. Continuous and noninvasive estimation of arterial blood pressure using a photoplethysmographic approach. In Proceedings of the 25th Annual International Conference of the IEEE Engineering in Medicine and Biology Society (IEEE Cat. No.03CH37439), Cancun, Mexico, 17–21 September 2003; Volume 4, pp. 3153–3156. [CrossRef]
13. Cattivelli, F.S.; Garudadri, H. Noninvasive Cuffless Estimation of Blood Pressure from Pulse Arrival Time and Heart Rate with Adaptive Calibration. In Proceedings of the 2009 Sixth International Workshop on Wearable and Implantable Body Sensor Networks, Berkeley, CA, USA, 3–5 June 2009; pp. 114–119. [CrossRef]
14. Liang, Y.; Abbott, D.; Howard, N.; Lim, K.; Ward, R.; Elgendi, M. How Effective Is Pulse Arrival Time for Evaluating Blood Pressure? Challenges and Recommendations from a Study Using the MIMIC Database. *J. Clin. Med.* **2019**, *8*, 337. [CrossRef] [PubMed]
15. Cavalcante, J.L.; Lima, J.A.; Redheuil, A.; Al-Mallah, M.H. Aortic Stiffness. *J. Am. Coll. Cardiol.* **2011**, *57*, 1511–1522. [CrossRef]
16. Chen, Y.; Wen, C.; Tao, G.; Bi, M.; Li, G. Continuous and Noninvasive Blood Pressure Measurement: A Novel Modeling Methodology of the Relationship Between Blood Pressure and Pulse Wave Velocity. *Ann. Biomed. Eng.* **2009**, *37*, 2222–2233. [CrossRef] [PubMed]
17. Kurylyak, Y.; Lamonaca, F.; Grimaldi, D. A Neural Network-based method for continuous blood pressure estimation from a PPG signal. In Proceedings of the 2013 IEEE International Instrumentation and Measurement Technology Conference (I2MTC), Minneapolis, MN, USA, 6–9 May 2013; pp. 280–283. [CrossRef]
18. Li, Y.; Wang, Z.; Zhang, L.; Yang, X.; Song, J. Characters available in photoplethysmogram for blood pressure estimation: Beyond the pulse transit time. *Australas. Phys. Eng. Sci. Med.* **2014**, *37*, 367–376. [CrossRef]
19. Kachuee, M.; Kiani, M.M.; Mohammadzade, H.; Shabany, M. Cuff-less high-accuracy calibration-free blood pressure estimation using pulse transit time. In Proceedings of the 2015 IEEE International Symposium on Circuits and Systems (ISCAS), Lisbon, Portugal, 24–27 May 2015; pp. 1006–1009. [CrossRef]
20. Visvanathan, A.; Banerjee, R.; Dutta Choudhury, A.; Sinha, A.; Kundu, S. Smart phone based blood pressure indicator. In Proceedings of the 4th ACM MobiHoc Workshop on Pervasive Wireless Healthcare—MobileHealth '14, Philadelphia, PA, USA, 11 August 2014; ACM Press: New York, NY, USA, 2014; pp. 19–24. [CrossRef]
21. Liang, Y.; Chen, Z.; Ward, R.; Elgendi, M. Hypertension Assessment via ECG and PPG Signals: An Evaluation Using MIMIC Database. *Diagnostics* **2018**, *8*, 65. [CrossRef] [PubMed]
22. Attarpour, A.; Mahnam, A.; Aminitabar, A.; Samani, H. Cuff-less continuous measurement of blood pressure using wrist and fingertip photo-plethysmograms: Evaluation and feature analysis. *Biomed. Signal Process. Control* **2019**, *49*, 212–220. [CrossRef]
23. Tjahjadi, H.; Ramli, K. Noninvasive Blood Pressure Classification Based on Photoplethysmography Using K-Nearest Neighbors Algorithm: A Feasibility Study. *Information* **2020**, *11*, 93. [CrossRef]
24. Johnson, A.E.; Pollard, T.J.; Shen, L.; Lehman, L.W.H.; Feng, M.; Ghassemi, M.; Moody, B.; Szolovits, P.; Anthony Celi, L.; Mark, R.G. MIMIC-III, a Freely Accessible Critical Care Database. *Sci. Data* **2016**, *3*, 160035. [CrossRef]
25. Chobanian, A.V.; Bakris, G.L.; Black, H.R.; Cushman, W.C.; Green, L.A.; Izzo, J.L.; Jones, D.W.; Materson, B.J.; Oparil, S.; Wright, J.T.; et al. Seventh Report of the Joint National Committee on Prevention, Detection, Evaluation, and Treatment of High Blood Pressure. *Hypertension* **2003**, *42*, 1206–1252. [CrossRef]
26. Goldberger, A.L.; Amaral, L.A.N.; Glass, L.; Hausdorff, J.M.; Ivanov, P.C.; Mark, R.G.; Mietus, J.E.; Moody, G.B.; Peng, C.K.; Stanley, H.E. PhysioBank, PhysioToolkit, and PhysioNet: Components of a New Research Resource for Complex Physiologic Signals. *Circulation* **2000**, *101*, e215–e220. [CrossRef]

27. Liang, Y.; Elgendi, M.; Chen, Z.; Ward, R. An optimal filter for short photoplethysmogram signals. *Sci. Data* **2018**, *5*, 180076. [CrossRef] [PubMed]

28. Elgendi, M. On the Analysis of Fingertip Photoplethysmogram Signals. *Curr. Cardiol. Rev.* **2012**, *8*, 14–25. [CrossRef] [PubMed]

29. Sörnmo, L.; Laguna, P. *Bioelectrical Signal Processing in Cardiac and Neurological Applications*; Elsevier: Amsterdam, The Netherlands, 2005. [CrossRef]

30. Martínez, A.; Alcaraz, R.; Rieta, J.J. Application of the phasor transform for automatic delineation of single-lead ECG fiducial points. *Physiol. Meas.* **2010**, *31*, 1467–1485. [CrossRef] [PubMed]

31. Mousavi, S.S.; Firouzmand, M.; Charmi, M.; Hemmati, M.; Moghadam, M.; Ghorbani, Y. Blood pressure estimation from appropriate and inappropriate PPG signals using A whole-based method. *Biomed. Signal Process. Control* **2019**, *47*, 196–206. [CrossRef]

32. Tanveer, M.S.; Hasan, M.K. Cuffless blood pressure estimation from electrocardiogram and photoplethysmogram using waveform based ANN-LSTM network. *Biomed. Signal Process. Control* **2019**, *51*, 382–392. [CrossRef]

33. Kononenko, I. Estimating attributes: Analysis and extensions of RELIEF. In *Lecture Notes in Computer Science (Including Subseries Lecture Notes in Artificial Intelligence and Lecture Notes in Bioinformatics)*; Springer: Berlin/Heidelberg, Germany, 1994; Volume 784, pp. 171–182. [CrossRef]

34. Shalev-Shwartz, S.; Ben-David, S. *Understanding Machine Learning*; Cambridge University Press: Cambridge, UK, 2014; pp. 1–397, ISBN 9781107298019. [CrossRef]

35. Mathur, A.; Foody, G. Multiclass and Binary SVM Classification: Implications for Training and Classification Users. *IEEE Geosci. Remote Sens. Lett.* **2008**, *5*, 241–245. [CrossRef]

36. Breiman, L. Bagging predictors. *Mach. Learn.* **1996**, *24*, 123–140. [CrossRef]

37. Laaksonen, J.; Oja, E. Classification with learning k-nearest neighbors. In Proceedings of the International Conference on Neural Networks (ICNN'96), Washington, DC, USA, 3–6 June 1996; Volume 3, pp. 1480–1483. [CrossRef]

38. Gao, H.; Qiu, B.; Duran Barroso, R.J.; Hussain, W.; Xu, Y.; Wang, X. TSMAE: A Novel Anomaly Detection Approach for Internet of Things Time Series Data Using Memory-Augmented Autoencoder. *IEEE Trans. Netw. Sci. Eng.* **2022**. [CrossRef]

39. Martínez, G.; Howard, N.; Abbott, D.; Lim, K.; Ward, R.; Elgendi, M. Can Photoplethysmography Replace Arterial Blood Pressure in the Assessment of Blood Pressure? *J. Clin. Med.* **2018**, *7*, 316. [CrossRef]

40. Mukkamala, R.; Hahn, J.O.; Inan, O.T.; Mestha, L.K.; Kim, C.S.; Toreyin, H.; Kyal, S. Toward Ubiquitous Blood Pressure Monitoring via Pulse Transit Time: Theory and Practice. *IEEE Trans. Biomed. Eng.* **2015**, *62*, 1879–1901. [CrossRef]

41. Gao, H.; Xu, K.; Cao, M.; Xiao, J.; Xu, Q.; Yin, Y. The Deep Features and Attention Mechanism-Based Method to Dish Healthcare Under Social IoT Systems: An Empirical Study With a Hand-Deep Local–Global Net. *IEEE Trans. Comput. Soc. Syst.* **2022**, *9*, 336–347. [CrossRef]

42. Gao, H.; Xiao, J.; Yin, Y.; Liu, T.; Shi, J. A Mutually Supervised Graph Attention Network for Few-Shot Segmentation: The Perspective of Fully Utilizing Limited Samples. *IEEE Trans. Neural Netw. Learn. Syst.* **2022**. [CrossRef] [PubMed]

43. Liang, Y.; Chen, Z.; Ward, R.; Elgendi, M. Photoplethysmography and Deep Learning: Enhancing Hypertension Risk Stratification. *Biosensors* **2018**, *8*, 101. [CrossRef] [PubMed]

44. Suzuki, S.; Oguri, K. Cuffless and non-invasive Systolic Blood Pressure estimation for aged class by using a Photoplethysmograph. In Proceedings of the 2008 30th Annual International Conference of the IEEE Engineering in Medicine and Biology Society, Vancouver, BC, Canada, 20–25 August 2008; pp. 1327–1330. [CrossRef]

45. Schlesinger, O.; Vigderhouse, N.; Eytan, D.; Moshe, Y. Blood Pressure Estimation from PPG Signals Using Convolutional Neural Networks And Siamese Network. In Proceedings of the ICASSP 2020—2020 IEEE International Conference on Acoustics, Speech and Signal Processing (ICASSP), Barcelona, Spain, 4–8 May 2020; pp. 1135–1139. [CrossRef]

46. Mukkamala, R.; Hahn, J.O. Toward Ubiquitous Blood Pressure Monitoring via Pulse Transit Time: Predictions on Maximum Calibration Period and Acceptable Error Limits. *IEEE Trans. Biomed. Eng.* **2018**, *65*, 1410–1420. [CrossRef] [PubMed]

 *biosensors*

Article

Proof of Concept in Artificial-Intelligence-Based Wearable Gait Monitoring for Parkinson's Disease Management Optimization

Robert Radu Ileşan [1,2], Claudia-Georgiana Cordoş [3], Laura-Ioana Mihăilă [3], Radu Fleşar [4], Ana-Sorina Popescu [1], Lăcrămioara Perju-Dumbravă [1] and Paul Faragó [3,*]

[1] Department of Neurology and Pediatric Neurology, Faculty of Medicine, University of Medicine and Pharmacy "Iuliu Hatieganu" Cluj-Napoca, 400012 Cluj-Napoca, Romania; robert.ilesan@usb.ch (R.R.I.); popescu.ana.sorina@elearn.umfcluj.ro (A.-S.P.); lperjud@elearn.umfcluj.ro (L.P.-D.)

[2] Clinic of Oral and Cranio-Maxillofacial Surgery, University Hospital Basel, CH-4031 Basel, Switzerland

[3] Bases of Electronics Department, Faculty of Electronics, Telecommunications and Information Technology, Technical University of Cluj-Napoca, 400114 Cluj-Napoca, Romania; claudia.cordos@bel.utcluj.ro (C.-G.C.); laura.mihaila@bel.utcluj.ro (L.-I.M.)

[4] Computer Science, Faculty of Mathematics and Computer Science, West University of Timişoara, 300223 Timişoara, Romania; radu.flesar02@e-uvt.ro

* Correspondence: paul.farago@bel.utcluj.ro

Abstract: Parkinson's disease (PD) is the second most common progressive neurodegenerative disorder, affecting 6.2 million patients and causing disability and decreased quality of life. The research is oriented nowadays toward artificial intelligence (AI)-based wearables for early diagnosis and long-term PD monitoring. Our primary objective is the monitoring and assessment of gait in PD patients. We propose a wearable physiograph for qualitative and quantitative gait assessment, which performs bilateral tracking of the foot biomechanics and unilateral tracking of arm balance. Gait patterns are assessed by means of correlation. The surface plot of a correlation coefficient matrix, generated from the recorded signals, is classified using convolutional neural networks into physiological or PD-specific gait. The novelty is given by the proposed AI-based decisional support procedure for gait assessment. A proof of concept of the proposed physiograph is validated in a clinical environment on five patients and five healthy controls, proving to be a feasible solution for ubiquitous gait monitoring and assessment in PD. PD management demonstrates the complexity of the human body. A platform empowering multidisciplinary, AI-evidence-based decision support assessments for optimal dosing between drug and non-drug therapy could lay the foundation for affordable precision medicine.

Keywords: artificial intelligence; sensors; convolutional neural networks; Parkinson's disease; biomedical monitoring; accelerometer; pressure sensor; disease management; electromyography; correlation

Citation: Ileşan, R.R.; Cordoş, C.-G.; Mihăilă, L.-I.; Fleşar, R.; Popescu, A.-S.; Perju-Dumbravă, L.; Faragó, P. Proof of Concept in Artificial-Intelligence-Based Wearable Gait Monitoring for Parkinson's Disease Management Optimization. *Biosensors* **2022**, *12*, 189. https://doi.org/10.3390/bios12040189

Received: 31 January 2022
Accepted: 17 March 2022
Published: 23 March 2022

Publisher's Note: MDPI stays neutral with regard to jurisdictional claims in published maps and institutional affiliations.

1. Introduction

More than 200 years ago, in 1817, Dr. James Parkinson published a scientific work entitled "The Essay on Shaking Palsy" [1] and with it the foundation of the disease that bears his name. After two centuries, we still struggle to understand and treat neurodegenerative diseases, such as Parkinson's disease (PD), which are growing exponentially, especially in industrialized regions, and given that no one is immune to them, specialists are concerned about "the Parkinson pandemic" [2].

In 2017, we started the PDxOne project, intending to optimize the management of PD by offering sustainable automated or semiautomated solutions.

1.1. Considerations on Parkinson's Disease

Parkinson's disease (PD) is a common progressive neurodegenerative disorder that can cause significant disability and decreased quality of life [3]. PD is today more present than ever, being the second most common neurodegenerative disease after Alzheimer's disease and affecting 0.3% of the population [4]. It is estimated that there are 6.2 million people diagnosed with PD disease worldwide, and the disease caused the death of 117,000 people in 2015 alone [5]. The American Parkinson Disease Association estimates that there are already 1 million people with PD living in the U.S. alone and over ten million worldwide, giving some researchers the evidence of a Parkinson pandemic [2]. While studies are divided on the prevalence by gender, the affinity of the pathology for the aging tissue, particularly neural tissue, is much clearer. Statistically, PD occurs in people over 60 years old, affecting 1% of this age group and increasing to 4% for people over 80 [4]. Unfortunately, PD can also occur in younger people under 50, known as Young Onset Parkinson's disease (YOPD). Studies show that 5–10% of patients diagnosed with PD are between 20 and 50 years old [6]. It is becoming even more complex with the literature showing cases of patients diagnosed with PD who were younger than 20—even some rare cases with patients under ten years of age, and the first symptoms appeared as early as two years of age [7]. This form is known as the juvenile form of PD and was first described in 1875. It has a substantial genetic component that can be diagnosed today with genetic testing [7].

In 2014, Mary Ann Thenganalt et al. performed a systematic review of articles cited in PubMed between 1980 and 2013. They concluded that PD could be divided into several subtypes, the most representative being tremor-dominant and postural instability gait difficulty form (PIGD) [8]. The clinical presentation of PD can take the form of many symptoms, from which the easiest to notice are the motor impairments [9]. These symptoms are caused by the degeneration of the dopaminergic neurons located in the substantia nigra from the ventral midbrain [10]. According to the Movement Disorder Society (MDS), the clinical diagnosis of PD is based on the presence of bradykinesia, along with either rest tremor or rigidity [11].

Motor impairment in PD has long been the focus of researchers, with significant advances being made in diagnostic accuracy, implementation, improvement of more accurate assessment scales, and better management of therapeutic strategies [12]. Unfortunately, at the moment of the diagnosis, a significant number of neurons that produce dopamine are already dysfunctional [6]. During the 15–20 years before the onset of motor symptoms, the patients experience a phase called "prodromal PD", during which the neurodegeneration starts and progresses [13]. It has been proved that olfaction impairment, constipation, depression, rapid eye movement (REM) and sleep behavior disorder (RBD) can be present in the prodromal period of PD [14].

Although the focus of research has been on motor symptoms, clinical studies have shown that non-motor symptoms in PD, such as depression, pain, psychosis and sleep disturbances, should be regarded as equally important when analyzed using quality-of-life questionnaires as well as economic and health indicators [12]. Therefore, we are dealing with a pathology that is increasingly present in our lives, that will double its number globally by 2030 [15], and that will have a significant socio-economic impact. Socially, PD patients become isolated, stigmatized [16], and even discriminated, with severe implications for the course of the disease and the clinical picture. To diagnose PD in a patient means to put a verdict that irreversibly alters their lifestyle and that of their families. Once diagnosed, the patient will also undergo a lifelong treatment, which aims not to cure but to improve or stagnate the symptoms' evolution. Studies, therefore, show a mortality rate double that of the healthy population, which increases and presents a more aggressive clinical picture in patients with YOPD or juvenile PD [17]. Therefore, the aspect of early onset raises many questions about our current ability to understand the pathophysiology of PD and treatment errors that may result in individual and global socio-economic consequences.

Indeed, medical assistance for patients with PD is a major drain on the healthcare budget. This financial amount is given by the complexity of PD, which affects the patient on several levels at once. To understand the real-life implications that affect every one of us and why we should spend the necessary resources, a quantification of socio-economic indicators is necessary. One figure we could start with is the financial effort that countries in Europe make to treat patients with PD, i.e., almost EUR 14 billion per year [18]. Interestingly, this figure also represents the amount the USA is spending per year treating PD patients, namely USD 14.4 billion [19]. The average annual cost per patient for PD in Germany is EUR 20,095 [20]. On average, direct costs represent 65.5% and indirect costs 34.5% [21]. Of the total direct costs (EUR 13,158), EUR 3526 is spent on medication, which is also the largest expenditure, and EUR 3789 is spent annually on hospitalization and home care costs [21]. Costs for home care by the family amount to 20% of the direct costs [21]. The same study shows a directly proportional relation between costs and disease progression, i.e., EUR 18,660 annual costs for stage 1–2, increase to EUR 31,660 annual costs for stage 2–5 (according to Hoehn and Yahr) [21]. Annual costs for PD differ quite a lot between European countries. While in Russia, EUR 5240 is spent per patient per year [21], in England, the annual costs for a patient with advanced stage PD (3–5 according to Hoehn and Yahr) can reach EUR 72,277 [22]. Attention is required, especially regarding the increase in treatment costs if patients are taken out of their environment, out of their home, and moved to a nursing home. In this case, studies show a 500% increase in costs for PD treatment [23]. All these costs are strictly related to PD, but a patient may also have other associated pathologies aggravated by PD and vice versa. Moreover, the financial impact on society is difficult to quantify because it should be taken on a patient-by-patient basis.

A staggering EUR 798 billion is spent annually at the European level to treat brain diseases, according to a study carried out in 2010 [19]. This amount is almost four times higher than Romania's GDP. It shows the importance of continuous research to provide a sustainable medical system for Europe and beyond in a growing and aging population that wants to maintain its standard of living in old age.

These figures motivate the ongoing research on wearables for the early detection of PD-specific symptomatology and prediagnosis of PD in incipient stages, as well as long-term monitoring of the disease in a ubiquitous healthcare environment which provides intelligent decision support algorithms for assessment and patient-specific treatment plans in PD.

1.2. Related Work—Wearables in PD Monitoring

For exemplification, Boroojerdi et al. report on the employment of the NIMBLE wearable biosensor patches, composed of an accelerometer and an electromyography (EMG) sensor, for motor evaluation in PD [24]. As for another example, Jauhiainen et al. report on the employment of a Movesense sensor and a Forciot insole to observe walking patterns in PD [25]. Phan et al. report on the use of BioKin devices to assess daily tasks: pointing, pouring, walking, and walking around a chair [26]. Lonini et al. report on employing BioStampIRC flexible wearable sensors, consisting of a tri-axial accelerometer and gyroscope, to record motion data [27].

Continuous long-term monitoring of motor symptoms in PD using inertial sensors is described by Borzì et al., aiming for the identification of bradykinesia and FOG [28], or by Powers et al., aiming for the identification of tremors and dyskinesia [29]. The employment of built-in smartphone sensors, with dedicated smartphone applications, is described by Heijmans et al. in [30] or by Motolese et al. in [31], for the remote monitoring of the PD patients during daily activities.

1.3. Related Work—AI-Based Decisional Support in PD Assessment

Some noteworthy examples regarding AI-based decisional support are presented as follows: Lonini et al. report on the employment of Random Forest classifiers to identify bradykinesia and tremor [27]. Random Forest for classification in PD was also studied by Aich et al. in [32], along with support vector machine, K-Nearest Neighbor, and Naïve Bayes. PD-specific symptomatology detection and classification using convolutional neural networks (CNN) was reported by Taewoong et al., who assessed daily activities based on 3D acceleration and angular velocity data measured with a Microsoft Band 2 [33]. Further employment of CNN was reported by Lonini et al. for the identification of bradykinesia and tremor [27], and by Steinmetzer et al. for the arm oscillation monitored with Mbientlab portable Motion Rectangle sensor bracelets under a Timed Up and Go (TUG) test scenario [34].

1.4. This Work

The PDxOne research project desires to develop and use the latest technologies to collect and, above all, interpret medical data. With a world population reaching 8 billion people and with today's medical requirements and demands, these tasks simply cannot be hand-operated anymore. Given the sheer medical data volume that must be collected every day, the request for economical and sustainable solutions forces healthcare systems to embrace a worldwide digitalized implementation.

Thus, our work is placed in the context of today's demand for ubiquitous monitoring and intelligent decisional support in healthcare. We target to develop, at the end of our project, a small-size wearable and portable monitoring system for patients diagnosed with PD, aiming for long-term quantitative and qualitative assessment of the PD symptomatology in a continuous fashion, and intended for AI-based decisional support in the discrimination of the pathology and formulation of dedicated treatment plans, which constitutes a novelty in the field.

This article targets the monitoring and assessment of gait in patients diagnosed with PD. Gait disorders are a hallmark of the condition and are associated with a loss of independence and an increased risk of falls. Disturbances of the gait, even if hardly noticeable, are described from the earliest stages of the disease [35] and include shuffling gate, shortened stride length, reduced overall velocity, and increased stance phase (up to doubling), along with reduced or absent arm swing, reduced trunk rotation, and decreased amplitude of motion in the hips, knees, and ankles [3]. In advanced stages, gait disorders often become increasingly complex, including motor blocks, festination, and imbalance [36].

Multiple studies have also been conducted for the early detection of motor deficient behaviors to apply proper therapeutic interventions, which are proved to slow down the motor dysfunction and maintain functional independency (in patients with preserved cognitive function) [3,37–39]. Symptoms such as dyskinesia, which is induced by therapy and manifests as involuntary movement of any body parts, appear in advanced stages of PD [9]. The symptomatic therapy for the classic motor features is usually satisfactory, but antiparkinsonian therapy that does not induce motor complications is still needed [40].

The solution proposed for ubiquitous gait monitoring and AI-based decisional support in gait assessment is envisioned in the shape of a wearable physiograph. The proposed physiograph performs bilateral tracking of the foot biomechanics assessed by means of plantar pressure distribution and lower-limb EMG, in correlation to upper limb balance, which is evaluated by means of arm balance magnitude of acceleration (MA) and variation of acceleration (VA). The recorded signals are transmitted over a Bluetooth radio link to a mobile device, e.g., smartphone or tablet. They are uploaded and stored into an online database and made available for future access, either in real-time or offline, for processing and interpretation.

The proposed physiograph enables both qualitative and quantitative assessment of gait. As such, we perform gait evaluation based on biomechanical parameters, expressed in terms of arm balance, heel strike, and lift-off, and temporal parameters, expressed in terms of cadence, single support, double support, single support to double support ratio, and stride time variability. Next, we evaluate the physiological interdependencies involved during the gait cycle by applying the cross-correlation function to each recorded signal pair. We illustrate that PD-specific gait is identifiable based on the evaluated gait assessment parameters following the evaluation results. Consequently, the biomechanical and temporal parameters and the cross-correlation results are applicable as inputs to an expert system for identifying and discriminating PD-specific gait pathology.

The novelty of the proposed physiograph consists of the underlying AI-based decisional support procedure for gait assessment. We generate a correlation coefficient matrix from the gait monitoring signals to visually represent the gait pattern. Gait assessment using the biomechanical and temporal parameters and the cross-correlation function is contained in the correlation coefficient matrix. Then, we apply the surface plot of the correlation coefficient matrix to a convolutional neural network (CNN) for gait classification.

A proof of concept of the proposed physiograph with AI-based decisional support is validated in the clinical environment on a group of ten subjects consisting of five PD patients and five healthy controls. As such, the proposed solution provides a feasible method for AI-based support for gait monitoring and assessment in a ubiquitous healthcare environment.

2. Materials and Methods

This paper proposes a wearable miniature physiograph with AI-based decisional support for gait monitoring and assessment in PD. Gait evaluation is performed in accordance with the Unified Parkinson's Disease Rating Scale (UPDRS)—motor subscale, and the Movement Disorder Society UPDRS (MDS-UPDRS) [41,42].

The proof of concept of the proposed wearable gait monitoring physiograph was tested extensively in the laboratory and validated indoors in the clinical environment, with a study group consisting of five patients diagnosed with PD and five healthy controls. The PD group includes three males and two females. The healthy control group includes four males and one female. The healthy controls do not have any previously diagnosed neurodegenerative disorder or podiatric condition.

All procedures performed in this study involving human participants were following the ethical standards of the institutional and/or national research committee. Informed consent was obtained from all individual participants involved in the study.

2.1. The Proposed Physiograph for Gait Monitroing in PD

The proposed gait monitoring physiograph is presented in the block diagram from Figure 1a and the practical realization from Figure 1b.

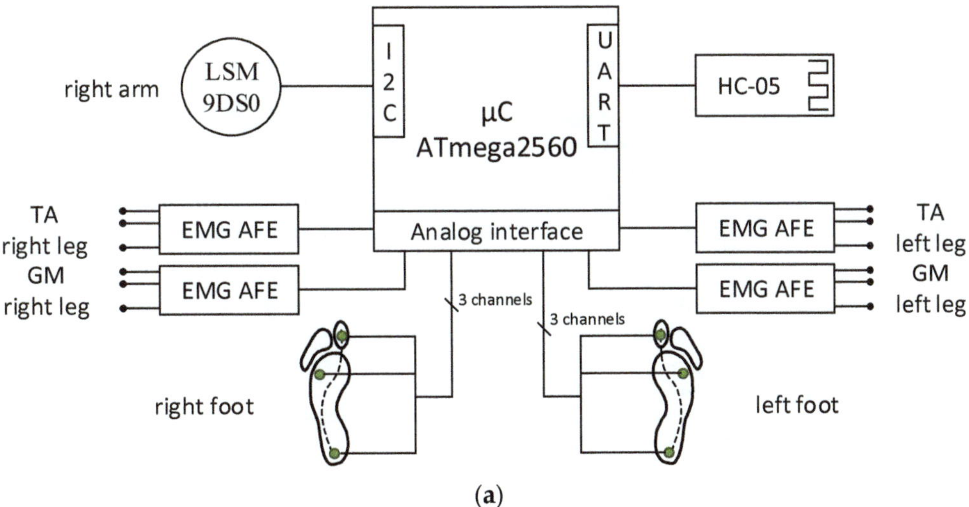

(**a**)

Figure 1. *Cont.*

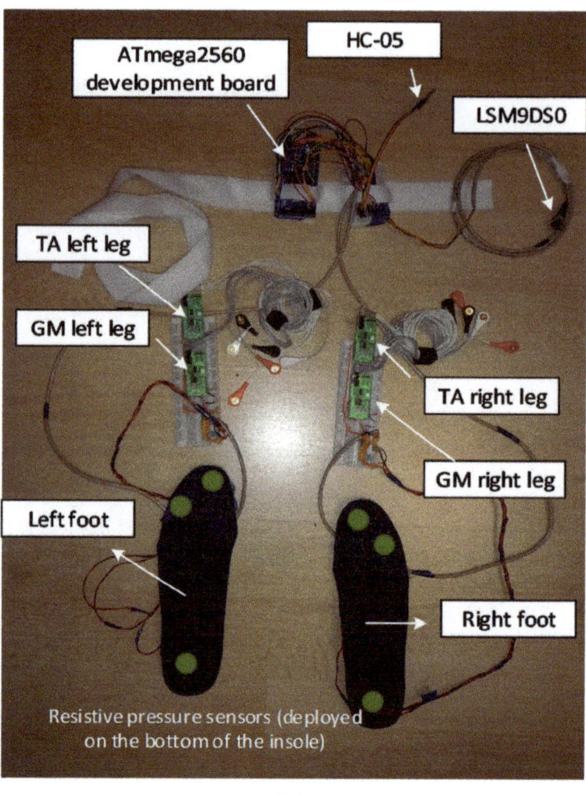

(**b**)

Figure 1. The proposed wearable physiograph for gait monitoring in PD: (**a**) block diagram and (**b**) practical realization.

The proposed wearable physiograph is developed around an ATmega2560 microcontroller (μC), which reads six Aidong IMS C20B thin-film resistive pressure sensors and four EMG channels over the analog ports and a LSM9DS0 module over the I2C interface. Signal acquisition is performed with synchronized sampling, with an fs = 100 Hz sampling frequency and an on-chip 10-bit analog-to-digital converter (ADC). Under this setup, the proposed physiograph performs bilateral tracking of the foot biomechanics through the plantar pressure progression pattern, lower-limb muscular activation, and unilateral monitoring of the arm balance.

The μC development board is attached to a Velcro strip and is worn around the user's waist, as illustrated in Figure 2.

Foot biomechanics is assessed using three pressure sensors and two EMG channels, clustered into a foot biomechanics assessment module [43]. Two such modules are considered for bilateral monitoring.

Operation of the foot biomechanics assessment module is described as follows: bilateral tracking of the plantar pressure progression pattern during the gait cycle is performed using three pressure sensors, attached onto an insole below the toe (FSR_0), metatarsal arch (FSR_1), and heel area (FSR_2), respectively, following the center of pressure (COP) progression line, as illustrated in Figure 3.

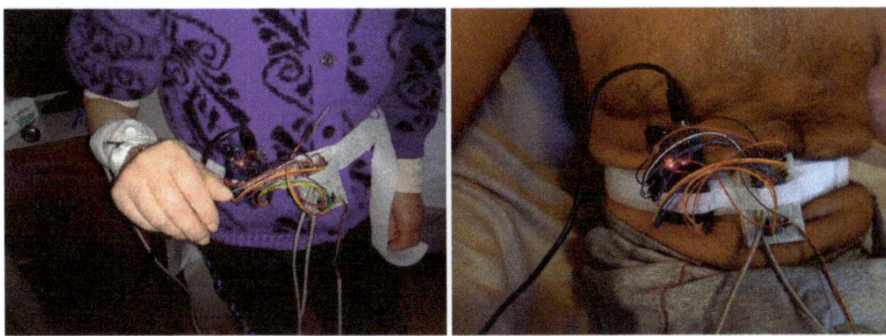

Figure 2. Illustration of the waist-worn ATmega2560 microcontroller development board, attached to a Velcro strip, which reads the peripherals involved in the bilateral monitoring of foot biomechanics and unilateral tracking of the arm balance.

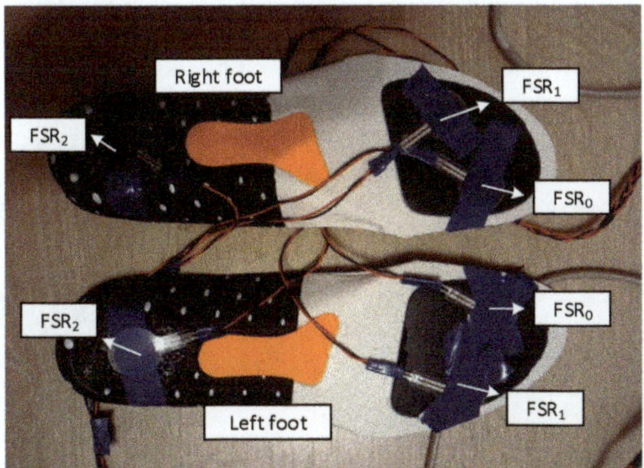

Figure 3. Illustration of the resistive pressure sensor placement onto the shoe insoles, under the toe, metatarsal arch, and heel area, respectively, for bilateral tracking of the plantar pressure progression pattern along the center of pressure progression line.

The sensors are deployed into a resistive divider topology with a $R = 1$ MΩ resistance, as illustrated in Figure 4a, and operate as force sense resistors (FSR) with the sensor resistance value derived as

$$FSR = \frac{V_{FSR}}{V_{DD} - V_{FSR}} \cdot R, \tag{1}$$

where $V_{DD} = 5$ V is the supply voltage and V_{FSR} is the FSR voltage drop. The sensor resistance (kΩ) can be converted to mass (kg) according to the mass vs. resistance characteristics provided in the sensor datasheet and plotted in blue in Figure 4b. Mass can further be converted to pressure (kg/cm^2) by dividing the mass to the sensor area.

In this work, we target gait pattern assessment rather than podiatric assessment. As such, the FSR resistance derived with (1) is sufficient to indicate the application of plantar pressure. In addition to (1), we have changed the polarity of the FSR signal,

$$FSR = \max(FSR) - FSR, \tag{2}$$

to have the "HIGH" signal level indicating pressure, and the "LOW" signal level indicating absence of pressure. This also changes the mass vs. resistance characteristics as plotted in red in Figure 4b.

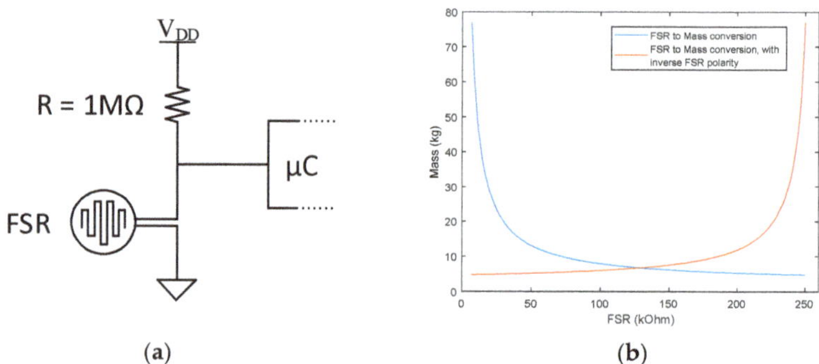

(a) (b)

Figure 4. Deployment of the resistive pressure sensors for operation as force sense resistors: (**a**) schematics, (**b**) mass vs. resistance conversion characteristics.

Tracking of the lower-limb muscular activation pattern during the gait cycle is performed with two EMG channels which acquire the EMG of the Tibialis anterior (TA) and Gastrocnemius medialis (GM) muscles. Off-the-shelf MikroElektronika EMG Click boards were used for each EMG channel analog front end (AFE), respectively. Wet Ag/AgCl electrodes were employed for EMG acquisition. Electrode placement is illustrated in Figure 5, with the active electrodes (white and red) placed onto the TA and GM muscles and the reference electrodes (black) placed onto the lateral and medial malleolus, respectively.

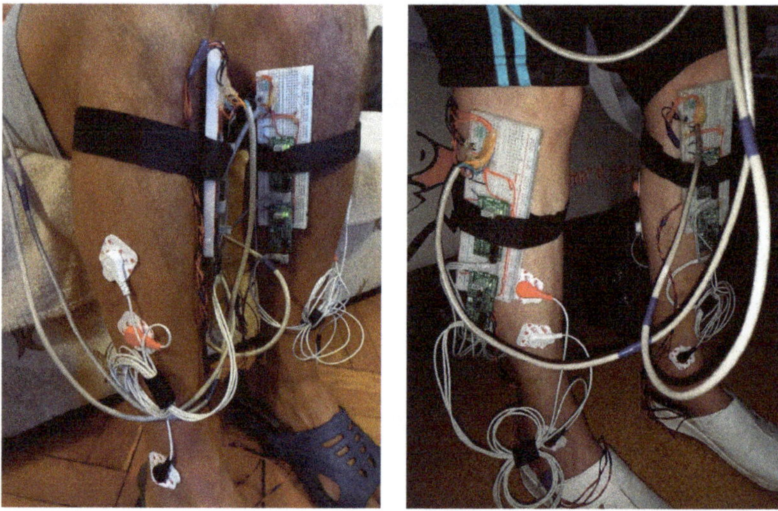

Figure 5. Illustration of the electrode placement for EMG acquisition of the Tibialis anterior and Gastrocnemius medialis, with the reference electrodes placed onto the lateral and medial malleolus, respectively.

A 10× AFE gain is set from the on-board potentiometer to accommodate the prescribed 1 μV–10 mV EMG amplitude range, accounting for 2 mV motor unit action potential (MUAP) amplitude of the healthy muscle, 0.5 mV MUAP amplitude for primary muscular disease, as well as 10 mV MUAP amplitude of intramuscular sprouting and chronic partial denervation [43–46]. On-board filtering is performed with three analog filter stages: two high-pass filters with the cutoff frequencies set to 1.6 Hz and 0.16 Hz, respectively, and a low-pass filter with the cutoff frequency set to 60 Hz.

After acquisition, EMG signal processing accounts for averaging with an 8-sample rectangular window with 50% overlap, a 4th order Butterworth approximation high-pass filter with $f_{cL} = 0.5$ Hz to suppress the DC component, and then a 4th order Butterworth approximation low-pass filter with $f_{cH} = 10$ Hz. To be noted is that the low-pass frequency of 10 Hz was considered as we were interested in the identification rather than evaluation of muscular activity [47].

The accelerometer from a LSM9DS0 module is employed to perform arm balance monitoring. The sensor is attached to the patient's right-hand wrist using a Velcro strip and, if necessary, tightened with an adhesive band, as illustrated in Figure 6.

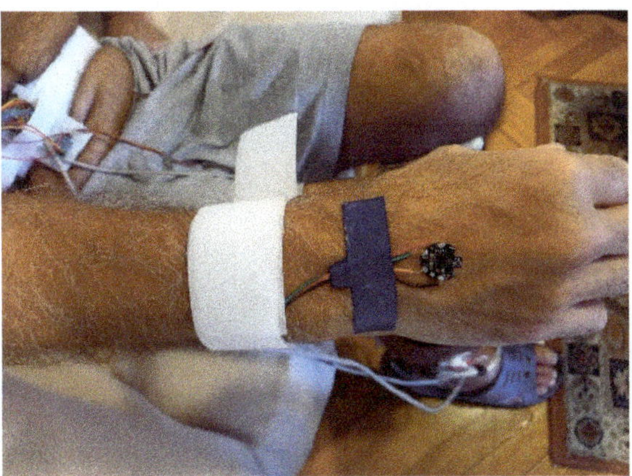

Figure 6. Illustration of the attachment of the LSM9DS0 accelerometer to the user's right wrist, for arm balance tracking during the gait cycle.

The accelerometer was configured for a 2G acceleration range, and the sensor data were read using the Adafruit LSM9DS0 library. Accelerometer signal processing assumes averaging with an 8-sample rectangular window with 50% overlap, a 4th order Butterworth approximation high-pass filter with $f_{cL} = 0.5$ Hz to suppress the DC component standing for the accelerometer initial position [48], and then a 4th order Butterworth approximation low-pass filter with $f_{cH} = 30$ Hz. These filter specifications cover the targeted 1 Hz–10 Hz frequency range of gait-related informational content (most relevant information is available up to 4 Hz) [49], as well as the 4 Hz–6Hz frequency range of tremor [50]. Additionally, the low-pass filter suppresses higher frequency components due to, for example, vibrations as well as noise. The raw signals on the three axes are then converted to acceleration and expressed in m/s^2.

Two metrics are employed for arm motion tracking based on the dynamic acceleration, defined as follows: the magnitude of acceleration (MA) is determined by applying the Pythagorean theorem to the readings on the three axes, respectively [51], according to equation:

$$MA = \sqrt{x^2 + y^2 + z^2}, \tag{3}$$

and defines the absolute acceleration value. The variation of acceleration (VA) has the average of the past readings subtracted from each axis, respectively [47], as defined by equation:

$$VA(k) = \sqrt{\left(x(k) - x_{avg}(k)\right)^2 + \left(y(k) - y_{avg}(k)\right)^2 + \left(z(k) - z_{avg}(k)\right)^2}, \tag{4}$$

where k is the current index and

$$x_{avg}(k) = \frac{1}{k-1} \cdot \sum_{i=1}^{k-1} x_i, \tag{5}$$

$$y_{avg}(k) = \frac{1}{k-1} \cdot \sum_{i=1}^{k-1} y_i, \qquad (6)$$

$$z_{avg}(k) = \frac{1}{k-1} \cdot \sum_{i=1}^{k-1} z_i, \qquad (7)$$

are the average of the past readings, i.e., up to index $k-1$ on each axis, respectively.

The proposed physiograph is aimed at long-term monitoring in a ubiquitous healthcare environment. Wireless connectivity is achieved by deploying the proposed physiograph with a HC-05 Bluetooth module. After acquisition, the raw data are sent over UART to the HC-05 module and transferred to an Android mobile terminal, e.g., a smartphone or tablet. The mobile terminal collects the user data from the physiograph over Bluetooth and assembles it into JavaScript Object Notation (JSON) files. The JSON files are sent one by one over the Internet to the server via REST API and stored in an online database. From this database, the signals are available for later retrieval to a desktop computer. A diagram of the application is presented in Figure 7.

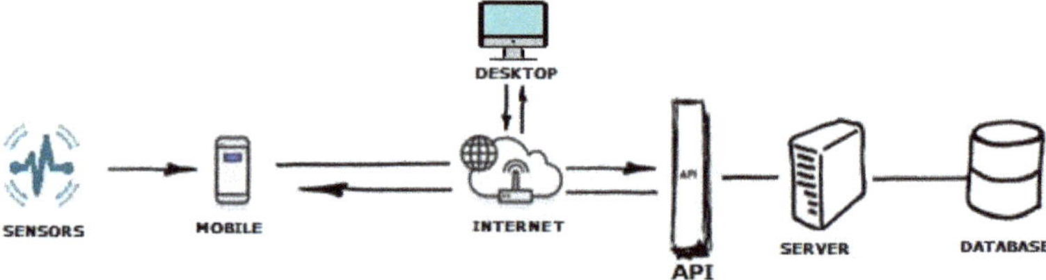

Figure 7. Diagram of the implemented solution to have the signals recorded from the users wearing the gait monitoring physiograph transmitted to a mobile terminal over a Bluetooth radio link and stored into an online database for future retrieval.

An example of a message sent to the API via the mobile phone is provided in Figure 8. The data will be stored in an online database. An example of the database content in Figure 9 illustrates that the data dictionary holds all the data acquired with the gait monitoring physiograph and is made available for later retrieval onto a laptop or personal computer for processing and interpretation.

```
POST /api/server/user/raw-data HTTP/1.1
HOST: my-server
X-Tableau-Auth: AuthKey
Content-Type: application/json
{
  "user": {
    "name": "UserName"
  }
  "data": {
    "acc_x": "12609.00"
    "acc_y": "9771.00"
    "acc_z": "5846.00"
    "acc_medie": "16989.26"
    "stg_TA": "335"
    "stg_GM": "339"
    "stg_presiune_deget_mnare": "541"
    "stg_presiune_metatarsiene": "431"
    "stg_presiune_calcai": "259"
    "dr_TA": "247"
    "dr_GM": "225"
    "dr_presiune_deget_mare": "527"
    "dr_presiune_metatarsiene": "458"
    "dr_presiune_calcai": "685"
  }
}
```

Figure 8. Example of a message sent to the API via the mobile phone.

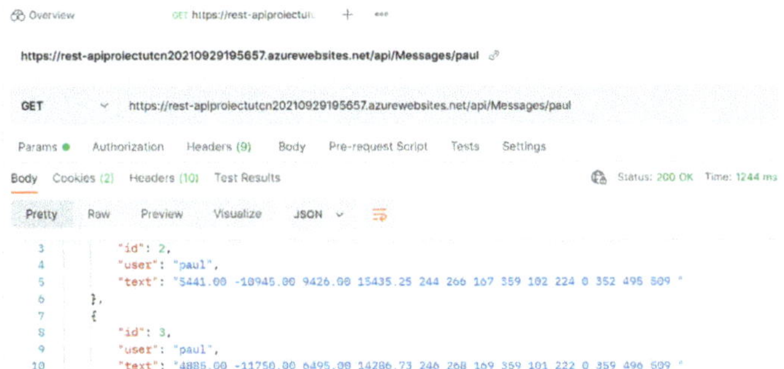

Figure 9. The online database content, resembling the data acquired with gait monitoring physiograph.

2.2. Gait Assessment—Correlation

The study group was instructed to undertake steady-state walking, at a pace of their own choice, and walk around the room. One or two walking trials were performed to make sure that the users were comfortable with the wearable devices and that they understood the requirements of the exercise. The subsequent steady-state walking activity was then recorded for the proposed gait monitoring and assessment procedure. Photographs taken during the trials of gait assessment are illustrated in Figure 10.

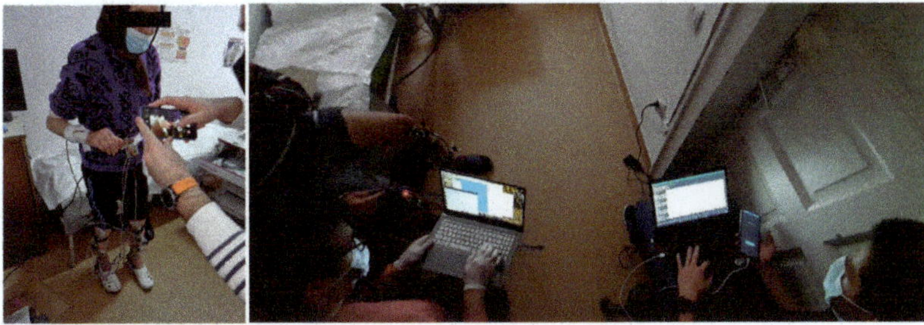

Figure 10. Photographs taken during the trials of gait assessment.

A physiological gait cycle is considered from one heel strike to the next heel strike of the same foot and consists of a stance and a swing phase, respectively [52,53]. The stance phase is further split into:

1. Heel strike—with pressure applied onto the heel area;
2. Support (or foot flat);
3. Midstance;
4. Heel off—with pressure applied onto the metatarsal arch and hallux;
5. Toe off—with pressure applied only onto the hallux before lift-off [43].

The physiological gait cycle is associated with arm balance, which describes a forward sway during the stance phase (from heel strike to lift-off) and a backwards sway during the swing phase (from lift-off to the next heel strike) [43]. This is visible on the arm balance MA waveform which describes a U-shaped pattern. The MA maxima account for the arm sway direction changes, constituting a good indicator for the stance phase initiation and ending.

The gait pattern in PD differs from the physiological gait. The literature describes a flat-foot strike for the PD gait pattern, or toe-to-heel plantar pressure progression in more advanced stages [54], associated with reduced lifting of the foot after lift-off [55] and limited or no arm balance along the gait cycle [56]. As such, the MA waveform exhibits a larger number of peaks, corresponding to the oscillations of the body's center of mass and tremor.

In this case, the VA waveform exhibits a larger variability corresponding to the increased number of MA peaks.

Due to the large number of peaks, the MA waveform cannot be employed to provide indication regarding stance initiation and ending in PD. In this work, we have rather employed the plantar pressures for stance identification. Plantar pressure detection was performed by comparing the FSR value to an empirical threshold level FSR_{th} computed as a fraction of the FSR signals. Accordingly, one stance phase ranges from the first to the last occurrence of plantar pressure, regardless of which pressure point, as illustrated in Figure 11. The stance phases identified in this manner account for the signal frames applied for cross-correlation in the gait assessment procedure described further on.

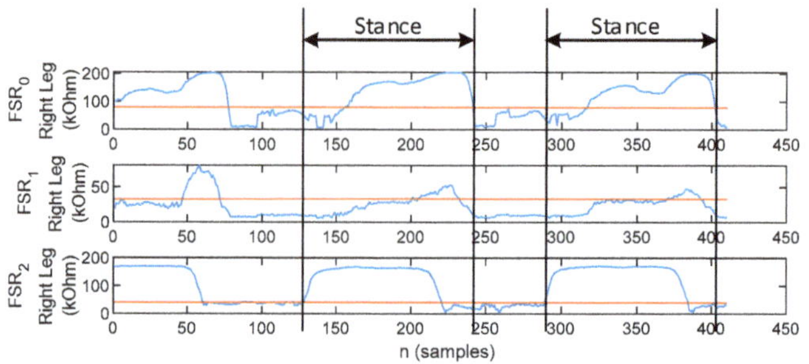

Figure 11. Illustration of the procedure for stance identification: the stance phase lasts from the first to the last occurrence of plantar pressure, identified by having the FSR signals compared to an empirical threshold level.

Physiological gait assumes a precisely defined interdependency between arm balance, plantar pressure, and lower-limb muscular activation. We evaluate signal interdependency in the time domain using the cross-correlation function given by

$$R_{sig_1,sig_2}(m) = \begin{cases} \sum_{n=0}^{N-m+1} sig_1(n+m)\cdot sig_2(n), & m \geq 0 \\ R_{sig_1,sig_2}(-m), & m < 0 \end{cases}, \tag{8}$$

where sig_1 and sig_2 are the signal frames being correlated, m is the cross-correlation index, and N is the frame length [57]. The cross-correlation function defined in (8) provides a measure of the similarity between the two signals sig_1 and sig_2 as a function of m. As such, cross-correlation maxima in the origin account for the identification of signal interdependencies. Provided the cross-correlation maxima are situated outside the origin, the index of cross-correlation peaks accounts for the displacement between the signals.

First, we assess whether the arm balance MA peaks, corresponding to arm balance initiation and ending, are synchronous with the stance initiation and ending determined from the plantar pressure progression pattern. This should be the case for physiological gait. Next, we evaluate the cross-correlation functions for:

1. arm balance MA vs. lower-limb muscular activation signals, i.e., TA and GM respectively,
2. lower limb muscular activation signals vs. FSR signals respectively.

As we move forward, we employ the correlation coefficient matrix to quantify the interdependency between either signal pair. A generic correlation coefficient matrix for M signals is expressed as:

$$R = \begin{pmatrix} 1 & \rho(sig_1,sig_2) & \cdots & \rho(sig_1,sig_M) \\ \rho(sig_2,sig_1) & 1 & \cdots & \rho(sig_2,sig_M) \\ \cdots & \cdots & \cdots & \cdots \\ \rho(sig_M,sig_1) & \rho(sig_M,sig_2) & \cdots & 1 \end{pmatrix}, \tag{9}$$

where

$$\rho\left(sig_i, sig_j\right) = \frac{1}{N-1} \sum_{n=1}^{N} \left(\frac{sig_i(n) - \mu_{sig_i}}{\sigma_{sig_i}}\right) \left(\frac{sig_j(n) - \mu_{sig_j}}{\sigma_{sig_j}}\right), \ i,j = \overline{1,M}, \qquad (10)$$

is the Pearson correlation coefficient of two signals sig_i and sig_j; μ and σ are the mean and standard deviation of the signals indicated in the signal subscripts, respectively; and N is the signal length [58,59]. We aim to generate the correlation coefficient matrix for the signal space

$$SIG = [MA, \ VA, TA_{left}, GM_{left}, FSR_{0,left}, FSR_{1,left} FSR_{2,left},$$
$$TA_{right}, GM_{right}, FSR_{0,right}, FSR_{1,right} FSR_{2,right}], \qquad (11)$$

consisting of the arm balance MA and VA (determined from the accelerometer signals), lower-limb EMG and plantar pressures. To be noted is that the main diagonal of the cross-correlation matrix consists of unity elements and accounts for the fact that the signals are correlated to themselves.

In contrast to (8), the definition of the correlation coefficients given in (10) does not account for the displacement between the signals, but only provides a quantification for signal similarity. Physiological delays originating from the biomechanical processes involved during gait, e.g., arm balance initiated before or after heel strike, arm balance terminated before or after heel strike, etc., which are determined using (8) as shifting of the cross-correlation peak form the origin, are missed using the correlation coefficients in (10).

To address the displacement of the signals in between one another and visualize them on the correlation coefficient plot, we have generated 10 shifted versions of the signal frame. Consequently, we extended the signal space to 120 signals which are to be correlated, resulting in a 120×120 correlation coefficient matrix. A 10×10 section from this matrix illustrates the interdependency between either shifted versions of the signals, rather than the signals themselves. Then, the largest coefficient value, accounting for the best similarity, determines the lag between the signals.

2.3. Gait Assessment—AI-Based Decisional Support

AI-based decisional support for the identification of PD gait pattern is implemented in this work using convolutional neural networks (CNN). The CNN is a deep learning algorithm that takes an image as input, assigns importance to the features in the image, and can differentiate them from each other. Thus, this type of network has the ability to extract local features based on the convolution operation between the original bidimensional data and certain series of the convolution kernels. The preprocessing required in a CNN is much lower compared to other classification algorithms, and such networks are used in applications for image recognition.

One of the benefits of deep learning is the ability to generalize and to learn massive amounts of data. Good network generalization capacities are obtained by accounting for the relationship between the size of the learning database and the complexity of the network architecture. The higher this ratio, the better the network performance on the test dataset. Furthermore, a big advantage of CNN networks is the weight sharing feature, which reduces the number of trainable network parameters and in turn helps the network to enhance generalization and avoid overfitting.

In this study, we used several architectures such as MobileNet, EfficientNetB0, and Xception. MobileNet is a CNN architecture model used for image classification and mobile vision. The advantage of this network is the very low computing power to apply transfer learning, because the model is based on depthwise separable convolution that has the effect of reducing the calculations and the size of the model. MobileNet uses 3×3 depthwise separable convolutions, using 8 to 9 times fewer calculations than standard convolutions with only a small reduction in accuracy. Counting the deep convolutions as separate layers,

MobileNet has 28 layers [60]. The EfficientNet model is based on the uniform scaling of the network width, depth, and resolution to improve performance. This network has been extended to a family of deep learning architectures with very good accuracy and efficiency [61]. Xception is built on two main points: depthwise separable convolution, i.e., a depthwise convolution followed by a pointwise convolution, and shortcuts between convolution blocks.

The surface plot of the correlation coefficient matrix is saved as a jpeg image and is applied to the CNN for classification into physiological and pathological gait. The flowchart of the proposed solution is shown in Figure 12.

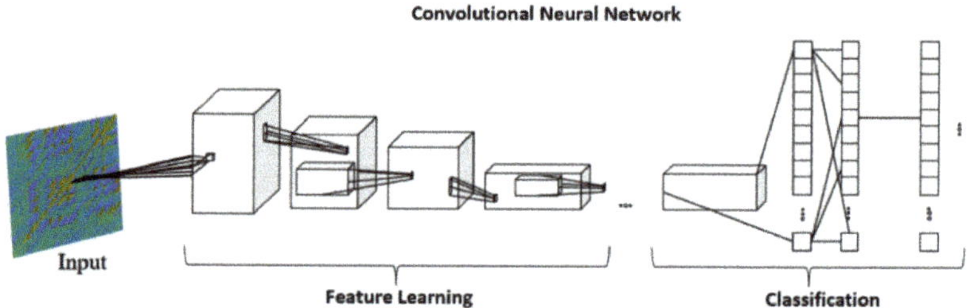

Figure 12. Flowchart of the CNN employed for the classification of the gait pattern into physiological and PD classes.

The parameters used to train the models are listed in Table 1. A very important parameter is the learning rate, which was chosen to be 0.05 for all models and has the role of controlling the model in response to the estimated error each time the model weights are updated. To reduce the nonlinearity of the output, the Softmax output layer activation function is used for all models. This function determines the type of predictions that the model can make. At the same time, the loss is the prediction error of the network, and the loss function has the role of determining the error. In the proposed binary classification system, the binary cross-entropy compares the predicted probability of the model with the actual result, which can be 0 or 1.

Table 1. CNN training parameters.

CNN Parameters	MobileNet	EfficientNetB0	Xception
Image dimension	160 × 160 × 3	160 × 160 × 3	160 × 160 × 3
Learning rate	0.05	0.05	0.05
Epochs	150	150	150
Batch size	32	64	128
Optimizer	Adam	Adam	Adam
Loss function	Binary Cross Entropy	Binary Cross Entropy	Binary Cross Entropy
Output layer activation function	Softmax	Softmax	Softmax

Models were trained on the graphic processing unit (GPU) in Google Colab using Keras. The motivation for Keras is ease of use and extension as neural layers, cost functions, optimizers, initialization schemes, and activation functions. As such, the activation functions are standalone modules that can be combined to create new models defined in Python. Keras offers scalability because it can run on tensor processing units (TPU) or large groups of GPUs, and the model can be exported to run in the browser or on a mobile device.

We have generated a total number of 236 images of correlation coefficient matrix plots, corresponding to the 10-subject database. This constitutes the data set for the CNN which aims to discriminate the walking pattern between physiological gait and PD. Since the data set is small, an augmentation was performed. The Adam optimizer was used to optimize the neural network. The RMSProp optimizer was used in the optimization of the EfficientNetB0 convolutional neural network. Model training was performed with 150 epochs for the MobileNet, EfficientNetB0, and Xception models. The batch size for MobileNet is 32, for EfficientNet it is 64, and for Xception it is 128. To evaluate the performance of each model, the data set was divided as follows: 70% for the training set, 10% for the validation set, and 20% for the test set.

3. Results

3.1. Gait Assessment—Time Domain

A section consisting of three gait cycles acquired during the walking trials is illustrated in Figure 13 for one healthy control and three PD patients. The arm balance MA and bilateral EMG of the TA and GM as well as the FSR resistance values, respectively, are plotted. Red markers are placed to indicate the MA maxima.

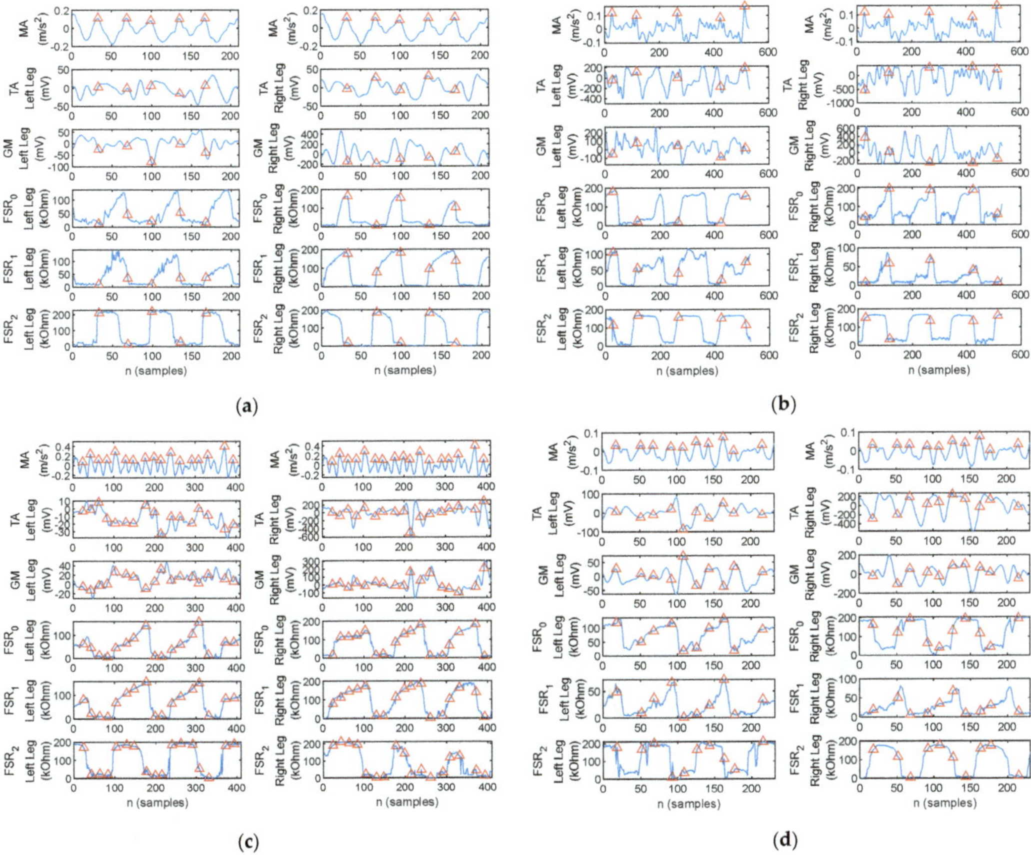

Figure 13. Three-cycle gait section plot of the arm balance MA, bilateral EMG of the TA and GM, and FSR signals, with the red triangles indicating the MA maxima, acquired with the proposed physiograph during walking trials on (**a**) healthy control, (**b**) PD Patient 5, (**c**) PD Patient 1, and (**d**) PD Patient 3.

Figure 13a exhibits the physiological gait pattern (Healthy control 1). Indeed, the peaks of the arm balance MA correspond to the initiation and ending of the stance phases, and the plantar pressures follow the physiological heel→metatarsal arch→hallux progression pattern. Moreover, TA activity can be observed simultaneously with pressure under the heel area and GM activity simultaneously with pressure under the hallux.

In contrast to Figure 13a, the three-cycle gait section plotted in Figure 13b illustrates the gait pattern of a patient with PD (i.e., Patient 5). The walking pattern exhibits bilateral flat-foot strike, which is typical in incipient and mid-stage PD. Muscular activation is present in a direct link to the plantar pressure points. The magnitude of acceleration also exhibits the U-shaped variation due to arm balance, although it is more pronounced during the stance than the swing phase of the gait cycle. However, a series of small-amplitude local peaks originated by tremor are also visible.

The three-cycle gait section recorded on two other patients with PD (i.e., Patient 1 and 3) is plotted in Figure 13c,d. Both patients exhibited flat-foot strike. What stands out, however, on the MA waveform is the absence of arm balance, in which case the peaks are originated by the tremor.

One thing that stands out in Figure 13d for Patient 3 is an asymmetry between the left and right foot during gait. The left foot exhibits a flat-foot strike, typical for PD, although with reduced plantar pressure on the metatarsal arch. The right foot on the other hand exhibits a plantar progression pattern which has toe pressure exerted until the next heel strike with a slight overlap. This accounts for the fact that the patient does not lift the right foot during the sway phase, but rather pulls the right foot with continuous contact between the toe and the ground. According to the FSR plots, the stance phase is terminated when pressure is reduced under the metatarsal arch and tow, i.e., the spike on FSR_0, and the forefoot is only lifted after heel contact.

Tremor on the accelerometer signals can be also assessed on the VA waveform. Regarding arm balance, the gait cycle in PD exhibits limited to no arm balance whatsoever. This is clearly visible on the signals acquired from the accelerometers. The arm balance MA of a healthy control, plotted in Figure 14a, exhibits signal peaks (indicated with a red triangle) which correspond to the stance phase initiation and termination, respectively, and a U-shaped waveform corresponding to arm sway. Consequently, the arm balance VA, plotted in Figure 14a, exhibits peaks during the sway.

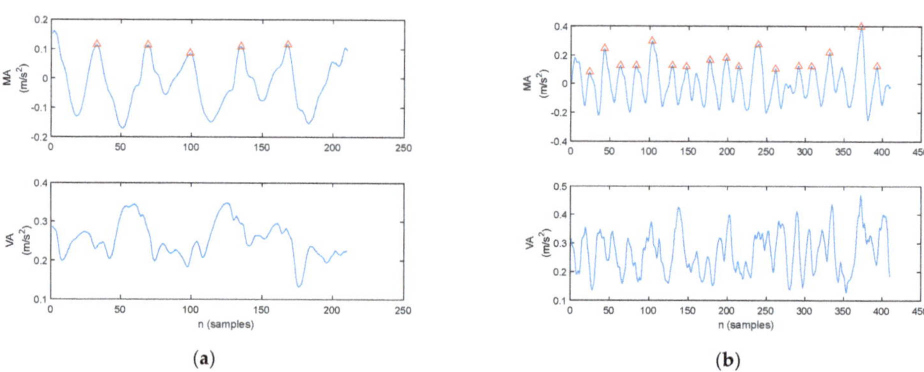

(a)   (b)

Figure 14. The arm balance MA and VA waveforms during the walking trials, with the red triangles indicating the MA maxima. (**a**) Healthy control—the MA exhibits a U-shaped variation and the VA accounts for smaller variation, and (**b**) PD patient—the MA exhibits a larger number of local peaks with reduced amplitude and the VA accounts for larger variation.

In contrast, the arm balance MA of a patient with PD exhibits a larger number of peaks, as illustrated in Figure 14b. Although some MA peaks with larger amplitudes, originated by the oscillation of the body center of mass during gait, can be identified as stance initiation and termination (see Figure 13 for a clear identification), most peaks are rather tremor-induced local peaks. As such, the VA waveform reveals a larger variation.

The time-domain plots further enable the assessment of the biomechanical and temporal parameters, used for motor evaluation in accordance with the UPDRS and MDS-UPDRS rating scales for Parkinson's Disease.

Heel strike accounts for initial contact with the ground at the stance initiation phase under a physiological gait pattern. Heel strike identification is performed by following the plantar pressure progression pattern. The identification of a heel strike is provided by a heel→metatarsal arch→toe pressure progression detection. For illustration, the FSR signals during gait initiation, with heel strike and flat-foot strike, are plotted in Figure 15.

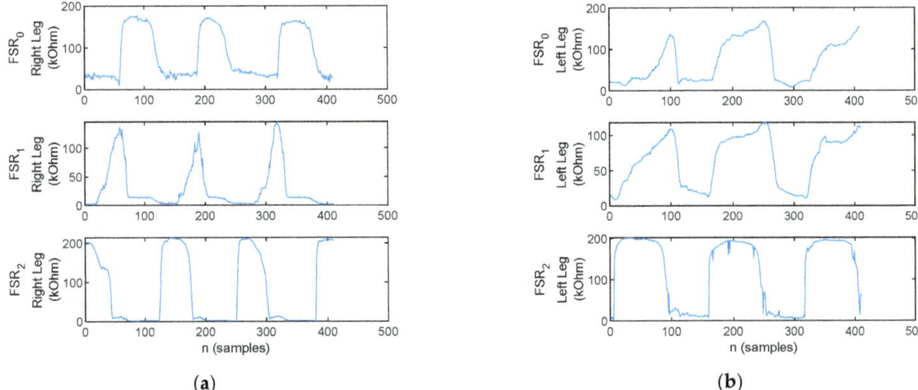

Figure 15. The FSR waveforms (FSR_2—heel, FSR_1—metatarsal arch, FSR_0—toe) during the walking trials for (**a**) healthy control—exhibits the physiological heel→metatarsal arch→toe pressure progression pattern, and (**b**) Patient 5—exhibits flat-foot strike as pressure is applied simultaneously on all three FSRs.

Lift-off accounts for the fact that the foot is lifted from the ground during the swing phase, and consequently stance phase with the opposite foot. Accordingly, no plantar pressure should be recorded during the swing phase, provided the foot is fully lifted from the ground, accounting for the identification of lift-off. For illustration, the FSR signals during sway, with lift-off and without lift-off, are plotted in Figure 16.

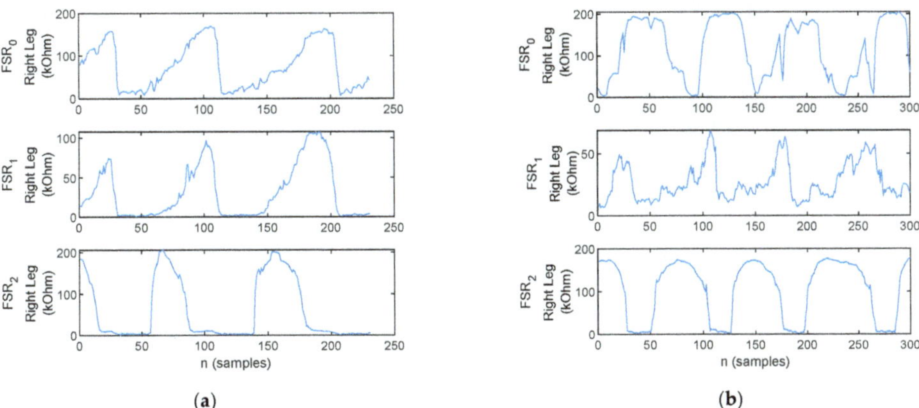

Figure 16. The FSR waveforms (FSR_2—heel, FSR_1—metatarsal arch, FSR_0—toe) during the walking trials for (**a**) healthy control—exhibits a section when no pressure is applied, thus indicating lift-off, and (**b**) Patient 3—exhibits no section when no pressure is applied, indicating that the patient does not lift the foot from the ground but rather pulls the foot during swing.

The biomechanical parameters of the test group are listed in Table 2.

Table 2. Biomechanical parameters of the test group.

Test Group	Arm Balance	Tremor	Heel Strike	Lift-Off
Healthy control 1	Present	Absent	Left: present Right: present	Left: present Right: present
Healthy control 2	Present	Absent	Left: absent (flat-foot strike) Right: absent (flat-foot strike)	Left: present Right: present
Healthy control 3	Present	Absent	Left: present Right: present	Left: present Right: present
Healthy control 4	Present	Absent	Left: present Right: present	Left: present Right: present
Healthy control 5	Present	Absent	Left: present Right: present	Left: present Right: present
Patient 1	Present	Large tremor	Left: absent (flat-foot strike) Right: absent (flat-foot strike)	Left: present Right: present
Patient 2	Absent	Large tremor	Left: absent (flat-foot strike) Right: present	Left: present Right: present
Patient 3	Absent	Large tremor	Left: absent (flat-foot strike) Right: absent	Left: present Right: absent
Patient 4	Present	Small tremor	Left: present Right: present	Left: present Right: present
Patient 5	Present	Small tremor	Left: absent (flat-foot strike) Right: present	Left: present Right: present

The biomechanical parameters of gait, assessed based on the signals recorded with the proposed physiograph, point out the asymmetry between the stance phases with the left and right foot, respectively. The patients exhibit flat-foot strike, which is typical for PD in incipient and middle stages. Only one instance of absent lift-off was observed for Patient 3. Arm balance was identifiable for Patients 1, 4, and 5 who exhibit a small-magnitude tremor.

The healthy control group was not previously diagnosed with any podiatric condition. As illustrated in Table 2, the healthy control group exhibits a physiological gait pattern, except for Healthy Control 2 who exhibits flat-foot strike.

The temporal gait analysis parameters are determined from the time-domain plots of the gait monitoring signals as follows [38,62]:

- Cadence, i.e., the number of steps per minute, is determined as the number of gait cycles in a 10 s segment's gait signals and extrapolated to one minute.
- Single support, i.e., pressure is exerted by only one foot and not the other, determined from a 10 s segment of the gait signals, and expressed in percentages.
- Double support, i.e., pressure is exerted by both feet during double limb support, determined from a 10 s segment of the gait signals, and expressed in percentages.
- Ratio of the single support to double support (s/d) is the ratio for single support and double support.
- Stride time variability is expressed as a coefficient of variation (CoV) computed from the mean (μ_{stride}) and standard deviation (σ_{stride}) of the strides over a 10 s segment of the gait signals [62–64], as given in the following equation

$$CoV = \frac{\sigma_{stride}}{\mu_{stride}} \cdot 100, \tag{12}$$

and is expressed in percentages.

To be noted is that a 10 s interval is sufficient for the assessment considering that the walking trials assume steady-state walking. The temporal parameters of the test group are listed in Table 3.

Table 3. Temporal parameters of the test group.

	Cadence (Steps/min)	Single Support (%)	Double Support (%)	S/D	Stride Time Variability (%)
Healthy Control 1	62	Left: 41.8 Right: 43.3	14.9	Left: 2.8 Right: 2.9	3.8
Healthy Control 2	27	Left: 36.2 Right: 34	29.8	Left: 1.2 Right: 1.1	9.8
Healthy Control 3	24	Left: 35.9 Right: 40.8	23.4	Left: 1.5 Right: 1.7	4.8
Healthy Control 4	56	Left: 38.5 Right: 44.3	17.1	Left: 2.2 Right: 2.6	3.1
Healthy Control 5	55	Left: 39 Right: 41	19.7	Left: 2 Right: 2.1	5.7
Statistics for healthy control group	44.8 ± 17.9	Left: 38.3 ± 2.4 Right: 40.7 ± 4	21 ± 5.9	Left: 1.9 ± 0.6 Right: 2.1 ± 0.7	5.4 ± 2.6
Patient 1	39	Left: 38.4 Right: 36	25.6	Left: 1.5 Right: 1.4	6.3
Patient 2	44	Left: 29.1 Right: 39.2	31.6	Left: 0.9 Right: 1.2	3.4
Patient 3	48	Left: 47 Right: 32.3	20.6	Left: 2.3 Right: 1.6	9
Patient 4	32	Left: 48.5 Right: 42.5	8.9	Left: 5.4 Right: 4.8	13.1
Patient 5	24	Left: 35 Right: 27.4	37.5	Left: 0.9 Right: 0.7	8.2
Statistics for PD group	37.4 ± 9.6	Left: 39.6 ± 8.1 Right: 35.5 ± 5.9	24.9 ± 10.9	Left: 2.2 ± 1.9 Right: 1.9 ± 1.6	8 ± 3.6

What stands out in both Healthy control group and PD group is that the cadence is considerably smaller than the nominal 60 steps/min [52,53]. Both healthy controls and PD patients expressed that their pace was reduced because of physiograph wiring.

The PD group exhibits reduced percentage of single support and S/D in comparison to the healthy control group, while the percentage of double support is not significantly different. On the other hand, stride time variability is increased for the PD group in contrast to the healthy control group.

What stands out for the PD group is that Patient 4 exhibits a very small percentage of single support with the left leg. This is because Patient 4 has a very small right leg lift-off during the swing phase of the gait cycle.

3.2. Gait Assessment—Cross-Correlation

As discussed in Section 2, a physiological gait pattern assumes some specifically defined interdependencies between the plantar pressure, lower-limb muscular activation, and arm balance, assessed using the cross-correlation function. Illustration of the cross-correlation between the signals acquired with the proposed physiograph is presented next.

The arm balance magnitude of acceleration, the TA and GM activation, and the FSR signals for a stance-phase, acquired on a healthy control for both the left foot and the right foot, are plotted in Figure 17. Red markers indicate stance initiation and termination, respectively.

The cross-correlation of the arm balance MA to the lower-limb muscular activation is plotted in Figure 18. As illustrated for the left foot, $R_{MA,EMG\text{-}TA}$ exhibits a peak to the left of the origin, indicating a delay of 70 ms of the TA activation vs. the MA stance initiation peak. Similarly, $R_{MA,EMG\text{-}GM}$ exhibits a peak to the left of the origin, indicating an advancement of 70 ms of the GM activation vs. the MA stance termination peak. Nevertheless, both delays are accounted for as physiological. A similar reasoning can be formulated for the right foot, indicating a delay of 20 ms of the TA vs. the MA stance initiation peak and an advancement of 110 ms of the GM activation vs. the MA stance termination peak.

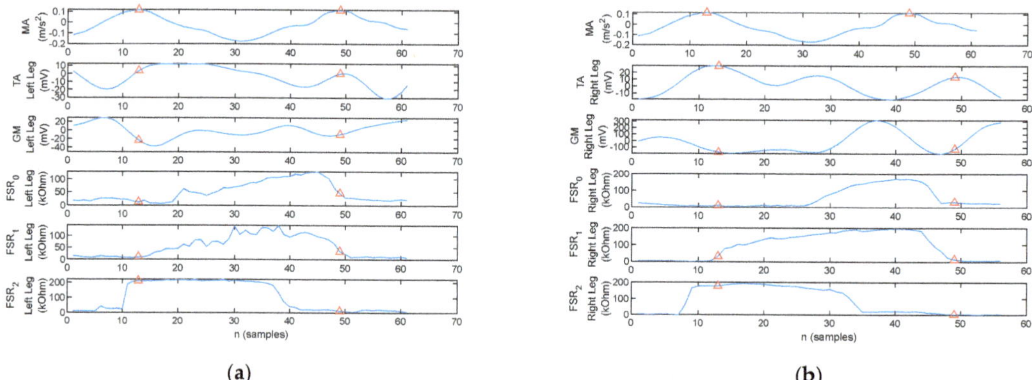

(**a**) (**b**)

Figure 17. One-stance gait section plot of the arm balance MA, EMG of the TA and GM, and FSR signals for a healthy control, with the red triangles indicating the MA maxima: (**a**) left foot and (**b**) right foot.

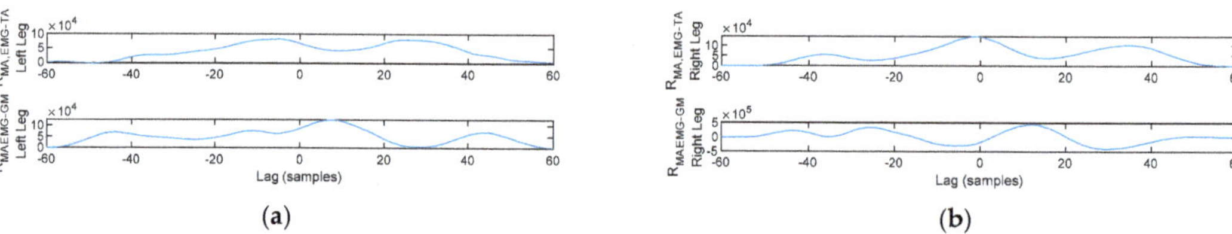

(**a**) (**b**)

Figure 18. Cross-correlation of the MA and EMG signals plotted in Figure 17 for the (**a**) left foot and (**b**) right foot.

Next, the cross-correlation of the TA activation signal to the FSR signals is plotted in Figure 19. As illustrated, $R_{EMG\text{-}TA,FSR2}$ exhibits a cross-correlation peak in the origin, indicating a direct interdependency between TA contraction and plantar pressure under the heel area. In contrast, the peaks of $R_{EMG\text{-}TA,FSR1}$ and $R_{EMG\text{-}TA,FSR0}$ are shifted from the origin, indicating that plantar pressure under the metatarsal arch and the hallux are not directly linked to the TA. The lag indicates the displacement, i.e., delay, between heel strike (TA contraction), heel off (FSR_1 and FSR_0), and toe off (only FSR_0). This reasoning holds for both feet.

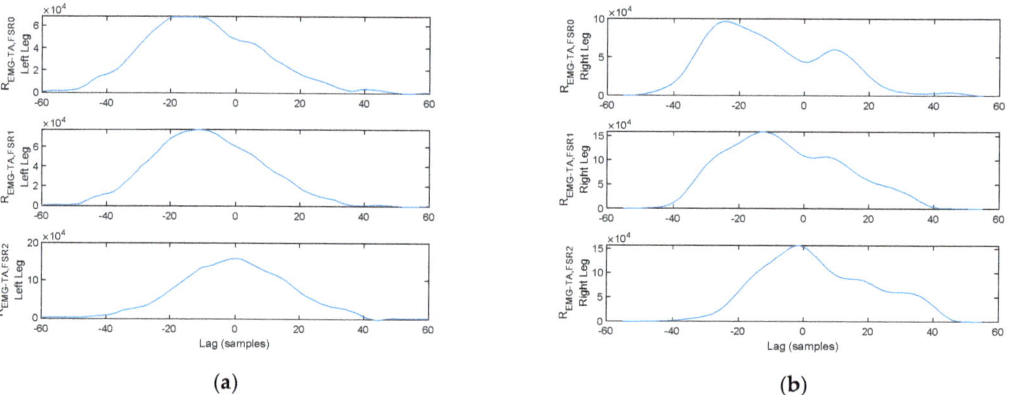

(**a**) (**b**)

Figure 19. Cross-correlation of the TA and FSR signals plotted in Figure 17 for the (**a**) left foot and (**b**) right foot.

A similar reasoning is formulated regarding the cross-correlation of the GM activation signal to the FSR signals plotted in Figure 20. As illustrated, $R_{EMG\text{-}TA,FSR0}$ exhibits a cross-correlation peak in the origin, indicating a direct interdependency between GM contraction and plantar pressure under the hallux. In contrast, the peaks of $R_{EMG\text{-}TA,FSR2}$ and $R_{EMG\text{-}TA,FSR1}$ are shifted from the origin indicating that plantar pressure under the heel area and metatarsal arch are not directly linked to the TA. The lag indicates the displacement, i.e., delay, between heel strike (FSR_2), heel off (FSR_1 and FSR_0), and toe off (GM activation). This reasoning holds for both feet.

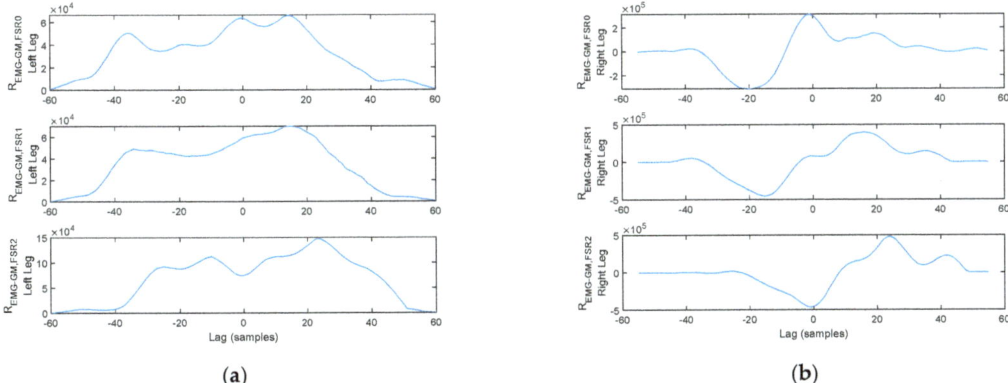

(**a**) (**b**)

Figure 20. Cross-correlation of the GM and FSR signals plotted in Figure 17 for the (**a**) left foot and (**b**) right foot.

In contrast, the arm balance MA, the TA and GM activation, and the plantar pressures for a stance-phase acquired under a bilateral monitoring scenario from a PD patient are plotted in Figure 21.

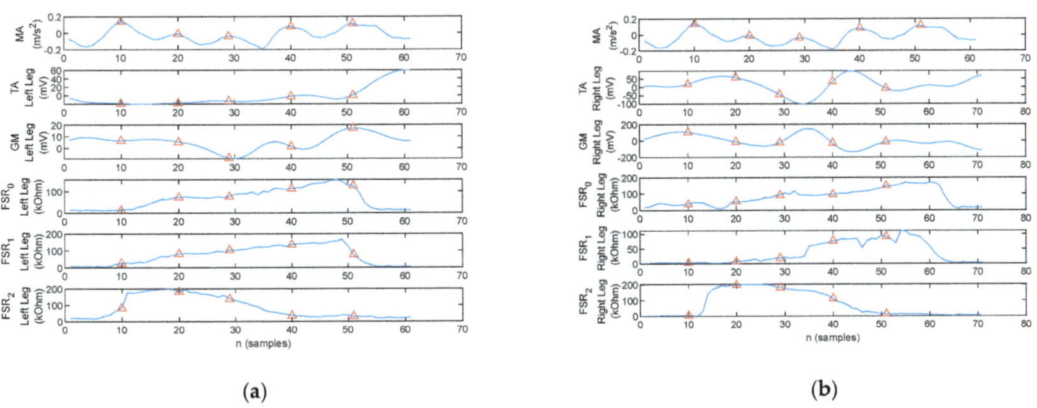

(**a**) (**b**)

Figure 21. One-stance gait section plot of the arm balance MA, EMG of the TA and GM, and FSR signals for a PD patient, with the red triangles indicating the MA maxima: (**a**) left foot and (**b**) right foot.

Two aspects stand out in Figure 21. On one hand, the MA peaks are not in a direct correspondence with the onset and termination of stance. With this consideration, the cross-correlation of the MA to either one of the lower-limb muscles becomes irrelevant, as the cross-correlation peaks indicate correspondence of the muscular activation to MA maxima originating from tremor rather than arm balance. On the other hand, the gait pattern exhibits flat foot strike for the left leg, identified by having pressure exerted simultaneously by the heel, metatarsal arch, and hallux, respectively, rather than the physiological heel→metatarsal arch→toe progression. This latter remark results in cross-correlation maxima in the origin, or close to the origin, for all signal pairs, as plotted in Figures 22 and 23.

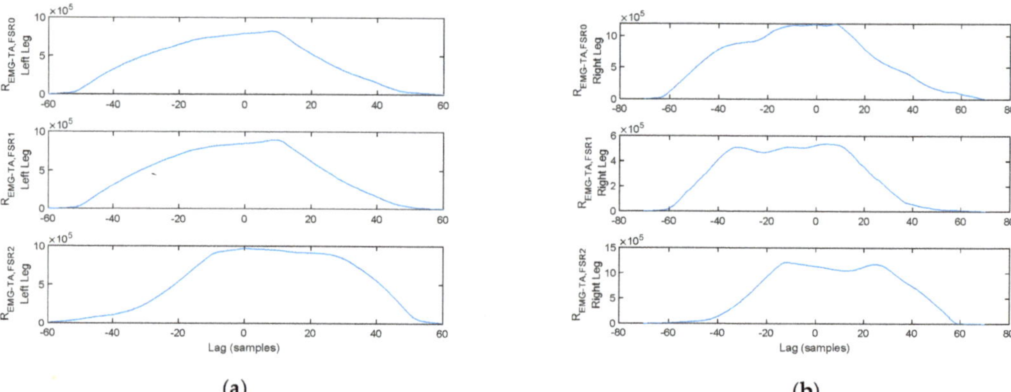

(a)   (b)

Figure 22. Cross-correlation of the TA and FSR signals plotted in Figure 21 for the (**a**) left foot and (**b**) right foot.

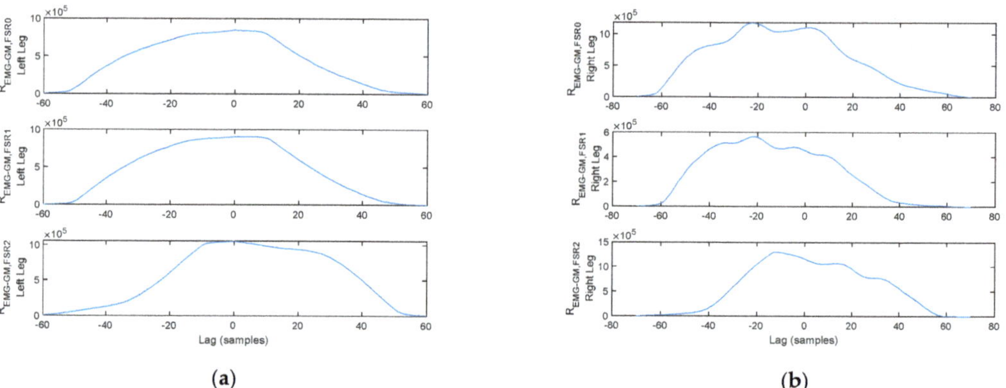

(a)   (b)

Figure 23. Cross-correlation of the GM and FSR signals plotted in Figure 21 for the (**a**) left foot and (**b**) right foot.

The surface plot of the correlation coefficient matrix defined in (9) was employed in this work to visualize the biomechanical and temporal parameters of gait. Consider, for exemplification, the surface plot of the correlation coefficient matrix illustrated in Figure 24. Each 10×10 matrix section, delimited by the squares, displays 100 correlation coefficients for the shifted versions of the signal pairs indicated in the matrix header. Yellow corresponds to large coefficient values, i.e., good similarity between the signals, whereas blue corresponds to small coefficient values. As such, the largest correlation coefficient values, i.e., the brightest colors in the correlation coefficient matrix surface plot, yield the displacement between the signals given in the matrix header. As shown, the displacement determined from the correlation coefficient matrix is equal to the one determined using the cross-correlation function in Figure 19.

For illustration, the surface plot of the correlation coefficient matrix computed for the healthy control and the three PD patients from Figure 13 is illustrated in Figure 25. Interpretation of the correlation coefficient matrices follows in Section 4—Discussion.

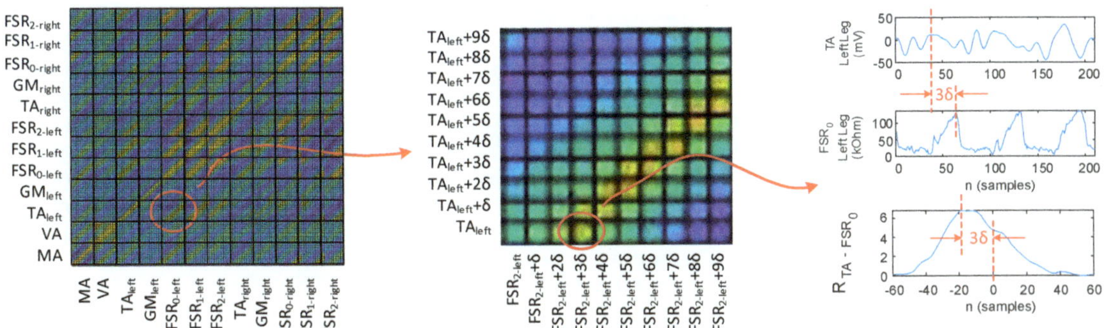

Figure 24. The surface plot of a correlation coefficient matrix split in 12×12 sections corresponding to the monitored signal pairs, with a 10×10 section providing the quantification of interdependency on shifted versions of the corresponding signals.

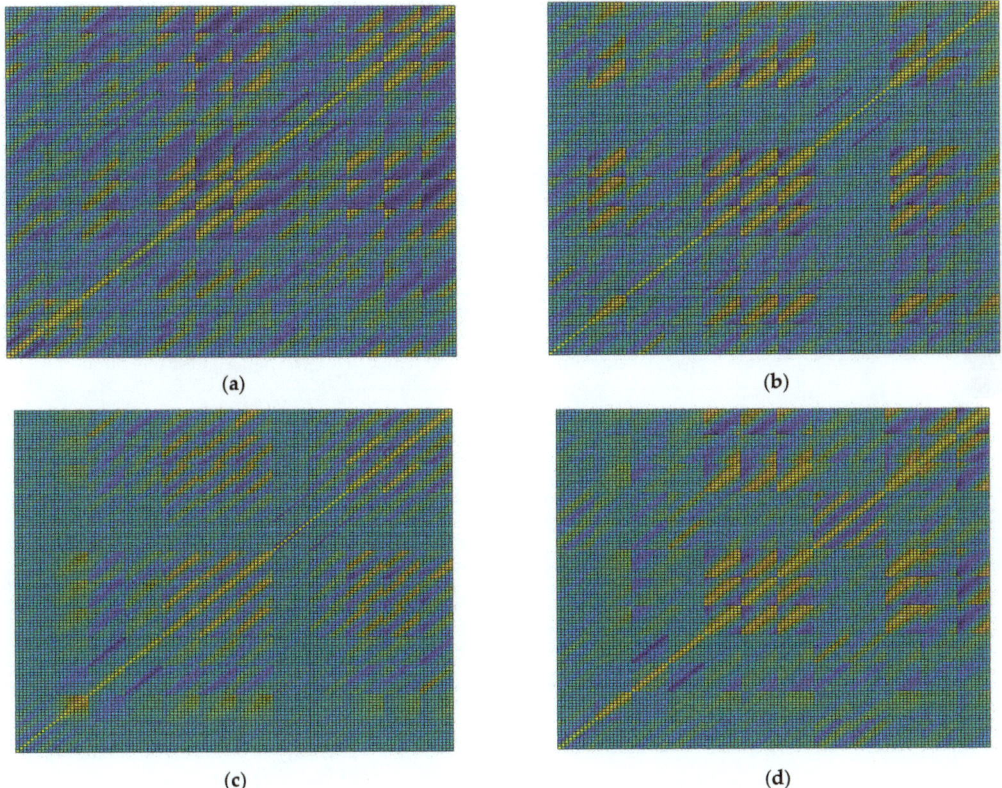

Figure 25. The correlation coefficient matrix surface plots, generated from the signals acquired with the proposed physiograph during the walking trials, for (**a**) healthy control, (**b**) Patient 5, (**c**) Patient 1, and (**d**) Patient 3.

3.3. Gait Assessment—AI-Based Decisional Support

AI-based decisional support for the discrimination of the PD-specific gait pattern was implemented using CNNs. The CNN takes the jpeg of the correlation coefficient matrix surface plot to classify the gait into physiological or PD pattern.

The CNN model performances were evaluated based on accuracy (Acc), sensitivity (Se), specificity (Sp), and precision, i.e., positive predicted values (PPV). These are defined using the true positive (TP), false positive (FP), true negative (TN), and false negative (FN) counts, respectively, as indicated in Equations (13)–(16). TP and TN count the number of correct classifications of Parkinsonian and physiological walking, respectively, whereas FP and FN count the number of incorrect classifications. According to Equations (13)–(16), Acc shows the probability of correct classification, Se evaluates the model's ability to identify the true positive samples, Sp evaluates the model's ability to identify the true negative samples, and PPV indicates the probability of the model to correctly classify a sample as positive. Finally, the error is then computed as the difference between the model training result and the desired result.

$$Acc = \frac{TN + TP}{TP + TN + FP + FN}, \tag{13}$$

$$Se = \frac{TP}{TP + FN}, \tag{14}$$

$$Sp = \frac{TN}{TN + FP}, \tag{15}$$

$$PPV = \frac{TP}{TP + FP} \tag{16}$$

At the end of the training and data validation process, the performance metrics were calculated and listed in Table 4, suggesting that this algorithm for training a CNN network achieves efficient identification. As indicated, the best results were obtained on the MobileNet model with 95% accuracy, 90% sensitivity, and 95% precision. The best results were obtained on the MobileNet model.

The best performance metrics which we achieved in this work are listed in Table 5 for further comparison with the classification performances achieved by some classifiers reported in the literature, including Random Forest and Support Vector Machines (SVM).

Table 4. Performance metrics of the neural network for gait pattern discrimination.

Performance Metrics	CNN Model		
	MobilNet	**EfficientNetB0**	**Xception**
Accuracy	0.95	0.85	0.85
Error	0.19	0.39	0.25
Sensitivity	0.90	0.85	0.9
Specificity	0.96	0.80	0.73
Precision	0.95	0.85	0.74

Table 5. Classification performances of the proposed work in comparison to other solutions reported in the literature.

Performance Metrics	Classifier				
	This Work	**[32]**	**[33]**	**[34]**	**[65]**
Classifier	CNN MobilNet	Random Forest	CNN	CNN	SVM
Accuracy	0.95	0.96	0.63	0.856	0.917
Error	0.19	-	-	-	-
Sensitivity	0.90	0.982	-	0.88	0.752
Specificity	0.96	0.96	-	0.84	0.99
Precision	0.95	0.972	-	0.86	-

4. Discussion

The work described in this article is developed in the context of the PDxOne research project, which aims to implement AI-based support for the collection and interpretation of medical data in PD, in the framework of ubiquitous healthcare.

The application described in this paper targets gait monitoring and assessment, based on the foot biomechanics, i.e., plantar pressure and lower-limb EMG, in correlation to upper-limb balance. To evaluate the gait problems that characterize PD, clinicians use semiquantitative rating scales such as the unified Parkinson's disease rating scale (UP-DRS) [66] or the movement disorders society unified Parkinson's disease rating scale (MDS-UPDRS) [42]. Objective gait evaluation was performed in this work accordingly. For qualitative gait assessment, we have evaluated biomechanical parameters expressed in terms of arm balance, heel strike, and foot lift-off, with the results listed in Table 2. For quantitative gait assessment, we have evaluated temporal parameters expressed in terms of cadence, single support, double support, single support to double support ratio, and stride time variability, with the results listed in Table 3.

Gait impairment is evolving throughout the progression of the disease, and the patterns of gait disturbances that are detected can differ from early to mild/moderate and advanced stages of PD [67], but the relationship between gait features and disease progression is not completely explained [68].

In PD, non-motor symptoms such as anxiety, depression, and cognitive impairment develop along with the motor symptoms, influencing the subjects' ability to perform motor tasks [9]. In the presence of such non-motor symptoms, motor tasks performed under trial conditions with a device attached to the body become a real challenge to the patient. Indeed, having the physiograph modules attached to the body produces an unusual sensorial stimulation to the patient. On the other hand, the presentation to the doctor's office or the medical laboratory is a stress factor itself for many patients, which strikes the emotional component. Consequently, we expected the gait analysis results of the study group to be influenced: smaller stride length, slower gait velocity, and smaller activity motor unit recruitment, although gait is an activity of daily living. Indeed, the results reported in Table 3 illustrate that both PD patients and healthy controls exhibit a smaller cadence compared to the nominal 60 steps/min [43,52,53]. As a natural consequence of the reduced cadence, Table 3 also shows longer single-support and shorter double-support durations compared to the nominal 30% of the gait cycle [43,52,53].

Gait physiology was further presented in a visual manner using the correlation coefficient matrix defined in Section 2.2. Identification of the gait pattern on the correlation coefficient matrix is discussed as follows: physiological gait, as monitored with the proposed physiograph and illustrated in the correlation coefficient matrix surface plot from Figure 25a, accounts for a heel→metatarsal arch→hallux plantar pressure progression pattern in direct correlation to the lower-limb muscular activation. The plantar pressure progression pattern is visualized in the plantar pressure correlation sections by the shift of the in-section yellow diagonal toward the matrix main diagonal. Shifting of the in-section diagonal away from the matrix main diagonal would account for an inverse plantar pressure progression pattern, namely hallux→metatarsal arch→heel.

Activation of the TA accounts for eccentric contraction during heel strike and initial double limb support [43,52,69], which is visualized by the yellow diagonal in the TA-FSR correlation sections, for both left and right feet, respectively. Next, activation of the TA accounts for concentric contraction during the swing phase [43,52,69], which is visualized by the yellow areas in the top left and bottom right, i.e., dark blue diagonal, in the TA-FSR correlation sections for opposing feet.

Activation of the GM describes eccentric contraction during midstance and concentric contraction during heel off and toe off [43,52,69], which is visualized in the GM-FSR correlation sections for both left and right foot, respectively. The GM is inactive during the swing phase, which is visualized in the GM-FSR correlation sections for opposing feet.

Physiological gait assumes a complete arm balance during the gait cycle, with the forward swing accounting for right foot stance and the backward swing accounting for left foot stance. As such, the arm balance is fully correlated to the lower-limb EMG and plantar pressure signals, as visualized in the MA-EMG and MA-FSR correlation sections. To be noted is that only the right arm was considered for assessment in the present work, which explains the yellow diagonal in the correlation sections of the MA with the opposite, i.e., left foot, and the yellow corner areas (dark blue diagonal) in the correlation sections of the MA with the same side, i.e., right foot. Indeed, the forward swing of the right arm—left foot stance produces a larger magnitude of the U-shaped MA signal, in contrast to the backward arm swing—right foot stance, which produces a smaller magnitude (see Figure 14). The VA signal on the other hand exhibits peaks during midstance, thus being correlated to the MA, which explains the yellow diagonal on the MA-VA correlation section. Furthermore, the VA is correlated to shifted versions of the EMG and FSR signals, which explains the yellow areas in the VA-EMG and VA-FSR correlation sections.

Parkinsonian gait is clearly distinguishable from the physiological gait. One of the most representative but non-specific early features of Parkinsonian gait is reduced speed [67]. It has been demonstrated that early PD subjects exhibit a reduced amplitude of arm swing and smoothness of locomotion, as well as increased interlimb asymmetry, all of these being more specific to PD and often the first motor symptoms [70].

Such features of the Parkinsonian gait pattern are identifiable on the correlation coefficient matrix. Some gait features attributable to PD, which we were able to identify and assess during the clinical test of the proposed physiograph, are presented as follows: the correlation coefficient matrix surface plot from Figure 25b corresponds to Patient 5, whose walking pattern described in Table 2 consists of flat-foot strike, bilateral lift-off, presence of arm balance, and small tremor. In contrast to the healthy control, the flat-foot strike is visualized in the left-foot FSR—FSR correlation sections as a yellow main diagonal. Plantar pressure is applied simultaneously to all sensors during flat foot strike, and consequently the FSR signals are correlated to one another (see Figure 13b). The patient keeps a regular lower-limb muscular response during the stance phases, visualized by the TA-FSR and GM-FSR correlation sections. The arm balance is also present, visualized by the MA-EMG and MA-FSR correlation sections. Tremor, although existent, is small in magnitude and consequently allows for the visualization of the yellow diagonals in the VA-EMG and VA-FSR correlation sections.

The correlation coefficient matrix surface plot from Figure 25c corresponds to Patient 1, whose walking pattern described in Table 2 exhibits bilateral flat-foot strike, bilateral lift-off, presence of arm balance, and large tremor. The correlation coefficient matrix follows the same pattern regarding foot biomechanics as for Patient 5. Regarding tremor, however, Figure 13c shows that the tremor and balance magnitudes in the MA and VA signals are comparable. Consequently, both MA and VA signals are uncorrelated to the EMG and FSR signals, respectively. This is visualized by the correlation sections of both MA and VA to the other monitoring sections, which exhibit a rather uniform coloring.

A different walking pattern was identified for Patient 3, with the correlation coefficient matrix surface plot from Figure 25d, who according to Table 2 exhibits absent heel strike and absent lift-off for the right foot. In this regard, the patient pulls the right foot during the swing phase of the gait cycle. This is visualized on the correlation coefficient plot by the right-foot FSR—FSR correlation sections and the left FSR—right FSR correlation sections, which deviate from the yellow diagonal pattern.

As the disease progresses bilaterally, the asymmetry might decrease, and movement becomes more bradykinetic [68,71]. At the same time, along with the neurodegeneration, the movement of the limbs becomes more impaired, and the patients develop shuffling steps with the increased need of double-limb support [72,73]. The further decline in gait is also caused by the postural changes, which are altering the kinematics of the gait, as is the case of the stooped posture [74].

The changes in gait worsen and motor fluctuations, dyskinesias, and freezing of gait become frequent and are accompanied by reduced balance and postural control, all of these exposing the patient to a severe risk of falling. More than that, the decline in the motor capacity can lead, in advanced stages, to the need of wheelchair use [75–77]. We expect to see these changes in the correlation coefficient matrix to facilitate gait pattern interpretation. This could be a game changer for stage-related personalized medicine therapy options.

The proposed physiograph is also envisioned as a portable monitoring solution. For this purpose, the recorded signals are transmitted over a Bluetooth radio link to a mobile device, e.g., smartphone or tablet, and stored in an online database for future retrieval. As such, the proposed solution is applicable for gait monitoring in PD outside the hospital environment. In this scenario, the monitoring protocol can be extended to include the assessment and quantification of influences exercised by daily activities—both domestic and in the community [36], environmental demands [78], environmental manipulation [37], and dual/multiple tasking [36,37].

The proposed physiograph is in a proof-of-concept phase, which was validated in clinical environment. As such, it was to be expected that the performance of both patients and healthy controls was influenced as reported, due to their subjective perception on the worn-in discomfort. There are two reasons for this. Firstly, the proposed physiograph is still in proof-of-concept phase. In accordance with this, the subjects complained of worn-in discomfort and discomfort from wires interfering with their mobility. Secondly, the gait measurements were conducted in a relatively small room, which likely inhibited the automaticity of walking at a normal pace [79]. The clinical environment created conditions for directed attention during the walking trial. As such, the subjects are aware of the motor task and concentrate on performing the task, which may have led to the abovementioned deviations from standard.

The solution to overcome such issues is the repetition of the motor task for a prolonged period (several days) as well as removing the patient from the clinical environment and placing them into their usual living environment, both inside and outside the home. This is enabled by the wearability and portability features of the proposed physiograph.

In continuation of our research, we aim to optimize the system design toward miniaturization, targeting a wireless body area network topology. One step further, the external environment brings multiple sensory information that can divert the patient's attention from the motor task. Thus, monitoring in an external environment provides multiple functional pieces of information with high value in the selection of the therapeutic and recovery interventions.

The solution proposed in this article was validated in a clinical environment for specific applicability in PD. Nevertheless, we envision that it can be easily extrapolated to further neurodegenerative conditions with gait affection.

On the other hand, PD does not exclude the presence of age-specific concurrent diseases such as (degenerative) osteoarthritis and/or cardiovascular diseases and their sequelae; comorbidities are common among patients with PD [80]. For example, one of the study group patients (Patient 3) also presents post-AVC hemiplegia with mild upper arm sequelae. In addition, inherent age-induced changes which affect joint function, e.g., sarcopenia, loss of proprioception and balance, and increased joint laxity, cannot be neglected [80]. All are associated with loss of independence and an increased incidence of falls.

Along with the clinical, biological, and imaging exam, the functional exam is also very important (including the functional exam performed with our device) for diagnostic purposes, as well as for monitoring the pharmacological disease management and rehabilitation outcome. It can be also used to guide kinetic trainings, to impose gait rhythms and velocities, and even to aid in dual tasks.

Along with the paraclinical examinations, the clinical testing, including functional performance, is key for diagnostic purposes. Part of this examination could be performed with our device and could be useful to provide guidance in kinetic trainings, to impose gait rhythms and velocities, and to aid in dual tasks. With further improvements, the pharmacological monitoring, disease management, and rehabilitation outcomes are in reach.

5. Conclusions

PD clearly demonstrates the complexity of the human body and the difficulty in choosing the appropriate treatment. In this paper, we focused on one of the main motor symptoms in PD, the gait. To develop a foundation for sustainable decision-support medical devices, we aim to include other important motor symptoms, as well as the non-motor ones, in future studies. Non-motor symptoms (e.g., cognitive impairment, depression, anxiety, sleep disorders, pain, and other autonomic disturbances) correlate with advanced age and disease severity [81], so they can be considered suggestive for the prognosis of disease progression. Some of these non-motor symptoms may appear much earlier in the course of disease progression [12]. AI-based assessment of these data could raise the importance of new technologies in prediagnosis and digital monitoring of patients. By pre-diagnostic technologies, we refer to those technologies that have the probabilistic ability to provide data interpretation at a high level of certitude for a given diagnosis, often before it is identified by a specialist without computational aid. In our experience, an additional interesting feature of AI systems is that they can identify non-specific symptoms, provided enough data and solid input are available. Regardless of the future development and promising results, the final diagnosis should be reserved, in our opinion, exclusively for the human medical practitioner, following established ethical/legal rules and evidence-based practice.

Back in 2017 when the PDxOne started, the main background idea was to find solutions for PD management optimization by bridging the gap between medicine and computational engineering. We used this approach in our study. Taking into consideration everything mentioned above, we see the necessity for an increased commitment in multidisciplinary treatment by interdisciplinary teams. In our opinion, this interdisciplinary team should include, at least, the following specialists in addition to the neurologist: general practitioner and dentist, geriatrician, trained medical caregiver for patients with PD, speech therapist, occupational therapist, nutritionist, psychologist/psychotherapist ideally specialized in patients with PD, and pharmacologists. To the abovementioned list, we would add new kinds of professionals that have medical degrees such as in a biomedical field (e.g., bioengineer, bioprinting expert, etc.) combined with knowledge of AI and augmented intelligence, as well as lifestyle strategists (advise patients with their health data), telemedicine, and/or health data analyst/biostatistician. Considering that by 2030 we will lack over 18 million healthcare workers worldwide [82], these future professions will be an asset in bridging interdisciplinary activities, they will provide sustainability in healthcare, and they could serve as good choice for internationally trained medical professionals toward alternative career pathways [83] because of labor migration in the era of globalization. We see today this new professional category in its early phase, with clinics having incorporated bioengineers, deep learning developers, bioprinting experts, and so on on their teams. New research projects are already hard to imagine without the congregation of such a team. As with most advancements, this one will also be led by necessity. In the time of Big Data, such an interdisciplinary team could provide personalized treatment for hospitalized patients and those at home. Perhaps the most important aspect of this interdisciplinary team is the medical data, and how interdisciplinary teams could correlate assessments to identify common patterns for optimizing patient-specific treatments. A good line of action would be the development of an international research platform, with standardized parameters, for automated or semiautomated data input. The assessment algorithms for this platform would benefit from the interdisciplinary approach and the fact that coding and treatment could be performed "at the patient's bed". A possible output would be an evidence-based decision support software for optimal dosing between drug and non-drug treatments. With such an effort, adequate treatment could be provided for this complex and multi-faceted disease, as PD is scripted [18]. Such an approach, once established, could offer individualized treatment plans and monitoring programs that would further improve PD management, and with it, socio-economic implications. The ultimate goal of our research community should be to offer affordable precision medicine for everyone.

Author Contributions: Conceptualization, R.R.I. and P.F.; methodology, R.R.I. and P.F.; software, C.-G.C., L.-I.M., R.F. and P.F.; validation, R.R.I., L.P.-D. and P.F.; formal analysis, P.F.; investigation, R.R.I. and P.F.; resources, R.R.I., A.-S.P. and L.P.-D.; data curation, P.F.; writing—original draft preparation, R.R.I., C.-G.C., L.-I.M., R.F., A.-S.P. and P.F.; writing—review and editing, R.R.I., L.P.-D. and P.F.; visualization, C.-G.C., L.-I.M., R.F. and P.F.; supervision, R.R.I. and P.F.; project administration, P.F.; funding acquisition, P.F. All authors have read and agreed to the published version of the manuscript.

Funding: This research was funded by ELYSEUM S.R.L., grant number 21738/30.07.2021.

Institutional Review Board Statement: The study was conducted in accordance with the Declaration of Helsinki and approved by the Ethics Committee of the University of Medicine and Pharmacy "Iuliu Hatieganu" Cluj-Napoca, Romania (protocol code 86 and date of approval 1 February 2018).

Informed Consent Statement: Informed consent was obtained from all subjects involved in the study.

Acknowledgments: The authors would like to thank Florian Thieringer from the Clinic of Oral and Cranio-Maxillofacial Surgery, University Hospital Basel, Basel, Switzerland, for his valuable support in the final review of the paper. The authors would also like to thank Martin Douglas-Jones from the Department of Maxillofacial and Oral Surgery, Frere Provincial Hospital, Eastern Cape, South Africa, for his valuable support in the final review of the paper. The authors would also like to thank Monica-Adriana Farago from the Psycho-Neurosciences and Rehabilitation Department, Faculty of Medicine and Pharmacy, University of Oradea, Oradea, Romania, for her valuable support in the final review of the paper.

Conflicts of Interest: Ileșan Robert Radu is also CEO of Elyseum SRL, the company which funded the grant for the present research. As a physician, maxillofacial surgeon, researcher, and professional, he declares that the outcome of this paper was not influenced from a conflict-of-interest point of view. We hereby declare that the conduit of the research activity is straight and follows the international educational and research values, Dr. med. Dr. med. dent. Ileșan Robert Radu.

References

1. Parkinson, J. An essay on the shaking palsy. 1817. *J. Neuropsychiatry Clin. Neurosci.* **2002**, *14*, 223–236. [CrossRef] [PubMed]
2. Dorsey, E.R.; Sherer, T.; Okun, M.S.; Bloem, B.R. The Emerging Evidence of the Parkinson Pandemic. *J. Parkinson's Dis.* **2018**, *8*, S3–S8. [CrossRef] [PubMed]
3. Svehlík, M.; Zwick, E.B.; Steinwender, G.; Linhart, W.E.; Schwingenschuh, P.; Katschnig, P.; Ott, E.; Enzinger, C. Gait analysis in patients with Parkinson's disease off dopaminergic therapy. *Arch. Phys. Med. Rehabil.* **2009**, *90*, 1880–1886. [CrossRef] [PubMed]
4. De Lau, L.M.; Breteler, M.M. Epidemiology of Parkinson's disease. *Lancet. Neurol.* **2006**, *5*, 525–535. [CrossRef]
5. Lipton, R.B.; Schwedt, T.J.; Friedman, B.W. GBD 2015 Disease and Injury Incidence and Prevalence, Collaborators. Global, regional, and national incidence, prevalence, and years lived with disability for 310 diseases and injuries, 1990–2015: A systematic analysis for the Global Burden of Disease Study 2015. *Lancet* **2016**, *388*, 1545–1602.
6. Samii, A.; Nutt, J.G.; Ransom, B.R. Parkinson's Disease. *Lancet* **2004**, *363*, 1783–1793. [CrossRef]
7. Kreutzer, J.S.; John, D.; Bruce, C. *Encyclopedia of Clinical Neuropsychology*; Springer: New York, NY, USA, 2011; pp. 1380–1383.
8. Thenganatt, M.A.; Jankovic, J. Parkinson disease subtypes. *JAMA Neurol.* **2014**, *71*, 499–504. [CrossRef]
9. Jankovic, J. Parkinson's disease: Clinical features and diagnosis. *J. Neurol. Neurosurg. Psychiatry* **2008**, *79*, 368–376. [CrossRef]
10. Triarhou, L.C. Dopamine and Parkinson's Disease. In *Madame Curie Bioscience Database*; Landes Bioscience: Austin, TX, USA, 2013. Available online: https://www.ncbi.nlm.nih.gov/books/NBK6271/ (accessed on 19 March 2022).
11. Postuma, R.B.; Berg, D.; Stern, M.; Poewe, W.; Olanow, C.W.; Oertel, W.; Obeso, J.; Marek, K.; Litvan, I.; Lang, A.E.; et al. MDS clinical diagnostic criteria for Parkinson's disease. *Mov. Disord. Off. J. Mov. Disord. Soc.* **2015**, *30*, 1591–1601. [CrossRef]
12. Chaudhuri, K.R.; Healy, D.G.; Schapira, A.H. Non-motor symptoms of Parkinson's disease: Diagnosis and management. *Lancet. Neurol.* **2006**, *5*, 235–245. [CrossRef]
13. Mantri, S.; Morley, J.F.; Siderowf, A.D. The importance of preclinical diagnostics in Parkinson disease. *Parkinsonism Relat. Disord.* **2019**, *64*, 20–28. [CrossRef] [PubMed]
14. Postuma, R.B.; Aarsland, D.; Barone, P.; Burn, D.J.; Hawkes, C.H.; Oertel, W.; Ziemssen, T. Identifying prodromal Parkinson's disease: Pre-motor disorders in Parkinson's disease. *Mov. Disord. Off. J. Mov. Disord. Soc.* **2012**, *27*, 617–626. [CrossRef] [PubMed]

15. Dorsey, E.R.; Constantinescu, R.; Thompson, J.P.; Biglan, K.M.; Holloway, R.G.; Kieburtz, K.; Marshall, F.J.; Ravina, B.M.; Schifitto, G.; Siderowf, A.; et al. Projected number of people with Parkinson disease in the most populous nations, 2005 through 2030. *Neurology* **2007**, *68*, 384–386. [CrossRef] [PubMed]
16. Tickle, D.L.; Zebrowitz, L.A.; Ma, H.I. Culture, gender and health care stigma: Practitioners' response to facial masking experienced by people with Parkinson's disease. *Soc. Sci. Med.* **2011**, *73*, 95–102. [CrossRef] [PubMed]
17. Poewe, W. The natural history of Parkinson's disease. *J. Neurol.* **2006**, *253*, VII2–VII6. [CrossRef] [PubMed]
18. Gustavsson, A. Cost of disorders of the brain in Europe 2010. *Eur. Neuropsychopharmacol.* **2011**, *21*, 718–779. [CrossRef]
19. Kowal, S.L.; Stacey, L. The current and projected economic burden of Parkinson's disease in the United States. *Mov. Disord.* **2013**, *28*, 311–318. [CrossRef]
20. Winter, Y. Longitudinal study of the socioeconomic burden of Parkinson's disease in Germany. *Eur. J. Neurol.* **2010**, *17*, 1156–1163. [CrossRef]
21. Kosten von Parkinson Nach Land Sehr Unterschiedlich. Available online: https://healthcare-in-europe.com/de/news/kosten-von-parkinson-nach-land-sehr-unterschiedlich.html (accessed on 24 June 2010).
22. Findley, L.J. The economic burden of advanced Parkinson's disease: An analysis of a UK patient dataset. *J. Med. Econ.* **2011**, *14*, 130–139. [CrossRef]
23. Findley, L. Direct economic impact of Parkinson's disease: A research survey in the United Kingdom. *Mov. Disord.* **2003**, *18*, 1139–1145. [CrossRef]
24. Boroojerdi, B.; Ghaffari, R.; Mahadevan, N.; Markowitz, M.; Melton, K.; Morey, B.; Sheth, N. Clinical feasibility of a wearable, conformable sensor patch to monitor motor symptoms in Parkinson's disease. *Parkinsonism Relat. Disord.* **2019**, *61*, 70–76. [CrossRef] [PubMed]
25. Jauhiainen, M.; Puustinen, J.; Mehrang, S.; Ruokolainen, J.; Holm, A.; Vehkaoja, A.; Nieminen, H. Identification of motor symptoms related to Parkinson disease using motion-tracking sensors at home (KÄVELI): Protocol for an observational case-control study. *JMIR Res. Protoc.* **2019**, *8*, 12808. [CrossRef] [PubMed]
26. Phan, D.; Horne, M.; Pathirana, P.N.; Farzanehfar, P. Measurement of axial rigidity and postural instability using wearable sensors. *Sensors* **2018**, *18*, 495. [CrossRef] [PubMed]
27. Lonini, L.; Dai, A.; Shawen, N.; Simuni, T.; Poon, C.; Shimanovich, L.; Jayaraman, A. Wearable sensors for Parkinson's disease: Which data are worth collecting for training symptom detection models. *NPJ Digit. Med.* **2018**, *1*, 64. [CrossRef] [PubMed]
28. Luigi, B.; Marilena, V.; Gabriella, O.; Carlo, A.A.; Margherita, F.; Mario, G.R.; Alberto, R.; Maurizio, Z.; Leonardo, L. Home monitoring of motor fluctuations in Parkinson's disease patients. *J. Reliab. Intell. Environ.* **2019**, *5*, 145–162.
29. Rob, P.; Maryam, E.A.; Edith, M.A.; Sara, K.; Irida, M. Smartwatch inertial sensors continuously monitor real-world motor fluctuations in Parkinson's disease. *Sci. Transl. Med.* **2021**, *13*, 7865. [CrossRef]
30. Margot, H.; Jeroen, G.V.H.; Christian, H.; Jos, A.; An, S.; Mark, L.K.; Pieter, L.K. Monitoring Parkinson's disease symptoms during daily life: A feasibility study. *NPJ Parkinson's Dis.* **2019**, *5*, 21.
31. Francesco, M.; Alessandro, M.; Fiorella, P.; Mariagrazia, R.; Fioravante, C.; Keren, K.; Alit, S.I.; Ziv, Y.; Vincenzo, D.L.; Massimo, M. Parkinson's Disease Remote Patient Monitoring During the COVID-19 Lockdown. *Front. Neurol.* **2020**, *11*, 567413.
32. Satyabrata, A.; Jinyoung, Y.; Sabyasachi, C.; Pyari, M.P. A Supervised Machine Learning Approach to Detect the on/off State in Parkinson's Disease Using Wearable Based Gait Signals. *Diagnostics* **2020**, *10*, 421. [CrossRef]
33. Um, T.T.; Terry, T. Parkinson's disease assessment from a wrist-worn wearable sensor in free-living conditions: Deep ensemble learning and visualization. *arXiv* **2018**, arXiv:1808.02870.
34. Tobias, S.; Michele, M.; Ingrid, B.; Carlos, M.T. Analysis and Classification of Motor Dysfunctions in Arm Swing in Parkinson's Disease. *Electronics* **2019**, *8*, 1471. [CrossRef]
35. Pirker, W.; Katzenschlager, R. Gait disorders in adults and the elderly: A clinical guide. *Wien. Klin. Wochenschr.* **2017**, *129*, 81–95. [CrossRef] [PubMed]
36. Muthukrishnan, N.; Abbas, J.J.; Shill, H.A.; Krishnamurthi, N. Cueing Paradigms to Improve Gait and Posture in Parkinson's Disease: A Narrative Review. *Sensors* **2019**, *19*, 5468. [CrossRef] [PubMed]
37. Ebersbach, G.; Moreau, C.; Gandor, F.; Defebvre, L.; Devos, D. Clinical syndromes: Parkinsonian gait. *Mov. Disord.* **2013**, *28*, 1552–1559. [CrossRef]
38. Seong, B.C.; Kun, W.P.; Dae, H.L.; Se, J.K.; Joon, S.Y. Gait Analysis in Patients with Parkinson's Disease: Relationship to Clinical Features and Freezing. *J. Mov. Disord.* **2008**, *1*, 59–64.
39. Sofuwa, O.; Nieuwboer, A.; Desloovere, K.; Willems, A.M.; Chavret, F.; Jonkers, I. Quantitative gait analysis in Parkinson's disease: Comparison with a healthy control group. *Arch. Phys. Med. Rehabil.* **2005**, *86*, 1007–1020. [CrossRef]
40. Bovolenta, T.M.; de Azevedo Silva, S.; Saba, R.A.; Borges, V.; Ferraz, H.B.; Felicio, A.C. Average annual cost of Parkinson's disease in São Paulo, Brazil, with a focus on disease-related motor symptoms. *Clin. Interv. Aging* **2017**, *12*, 2095–2108. [CrossRef]
41. Hauser, R.A.; Lyons, K.E.; Pahwa, R. The UPDRS-8: A brief clinical assessment scale for Parkinson's disease. *Int. J. Neurosci.* **2012**, *122*, 333–340. [CrossRef]
42. Goetz, C.G.; Tilley, B.C.; Shaftman, S.R.; Stebbins, G.T.; Fahn, S.; Martinez-Martin, P.; Poewe, W.; Sampaio, C.; Stern, M.B.; Dodel, R. Movement Disorder Society-sponsored revision of the Unified Parkinson's Disease Rating Scale (MDS-UPDRS): Scale presentation and clinimetric testing results. *Mov. Disord.* **2008**, *23*, 2129–2170. [CrossRef]

43. Faragó, P.; Grama, L.; Farago, M.A.; Hintea, S. A Novel Wearable Foot and Ankle Monitoring System for the Assessment of Gait Biomechanics. *Appl. Sci.* **2021**, *11*, 268. [CrossRef]

44. Narwaria, R.P.; Verma, S.; Singhal, P.K. Removal of Baseline Wander and Power Line Interference from ECG Signal—A Survey Approach. *Int. J. Electron. Eng.* **2011**, *3*, 107–111.

45. Hung, C.C.; Halonen, I.; Ismail, M.; Porra, V. Micropower CMOS Gm-C Filters for Speech Signal Processing. In Proceedings of the 1997 IEEE International Symposium on Circuits and Systems, Hong Kong, China, 9–12 June 1997.

46. Mills, K.R. The basics of electromyography. *J. Neurol. Neurosurg. Psychiatry* **2005**, *76*, 32–35. [CrossRef] [PubMed]

47. Faragó, P.; Groza, R.; Hintea, S. High Precision Activity Tracker Based on the Correlation of Accelerometer and EMG Data. In Proceedings of the 42nd International Conference on Telecommunications and Signal Processing (TSP), Budapest, Hungary, 1–3 July 2019.

48. San-Segundo, R.; Zhang, A.; Cebulla, A. Parkinson's Disease Tremor Detection in the Wild Using Wearable Accelerometers. *Sensors* **2020**, *20*, 5817. [CrossRef] [PubMed]

49. Fridolfsson, J.; Börjesson, M.; Arvidsson, D. A Biomechanical Re-Examination of Physical Activity Measurement with Accelerometers. *Sensors* **2018**, *18*, 3399. [CrossRef]

50. Lee, H.J.; Lee, W.W.; Kim, S.K.; Park, H.; Jeon, H.S.; Kim, H.B.; Park, K.S. Tremor frequency characteristics in Parkinson's disease under resting-state and stress-state conditions. *J. Neurol. Sci.* **2016**, *362*, 272–277. [CrossRef]

51. Lin, W.Y.; Verma, V.K.; Lee, M.Y.; Lai, C.S. Activity Monitoring with a Wrist-Worn, Accelerometer-Based Device. *Micromachines* **2018**, *9*, 450. [CrossRef]

52. Karadsheh, M. Orthobullets—Gait Cycle. Available online: https://www.orthobullets.com/foot-and-ankle/7001/gait-cycle (accessed on 23 December 2021).

53. Alamdari, A.; Krovi, V.N. A Review of Computational Musculoskeletal Analysis of Human Lower Extremities. In *Human Modeling for Bio-Inspired Robotics: Mechanical Engineering in Assistive Technologies*; Ueda, J., Kurita, Y., Eds.; Academic Press: San Diego, CA, USA, 2017; pp. 37–73.

54. Hughes, J.R.; Bowes, S.G.; Leeman, A.L.; O'Neill, C.J.; Deshmukh, A.A.; Nicholson, P.W. Parkinsonian abnormality of foot strike: A phenomenon of ageing and/or one responsive to levodopa therapy? *Br. J. Clin. Pharmacol.* **1990**, *29*, 179–186. [CrossRef]

55. Murray, M.P.; Sepic, S.B.; Gardner, G.M.; Downs, W.J. Walking patterns of men with Parkinsonism. *Am. J. Phys. Med.* **1978**, *57*, 278–294.

56. O'Sullivan, S.O. Parkinson's Disease: Physical Therapy Intervention. In *Physical Rehabilitation*, 5th ed.; O'Sullivan, S.B., Schmitz, T.J., Eds.; EA Davis Company: Philadelphia, PA, USA, 2007; pp. 853–893.

57. Marple, S.L., Jr. *Digital Spectral Analysis*, 2nd ed.; Dover Publications: Mineola, NY, USA, 2019; ISBN 978-048-6780-52-8.

58. Park, Y.; Chung, W. A Novel EEG Correlation Coefficient Feature Extraction Approach Based on Demixing EEG Channel Pairs for Cognitive Task Classification. *IEEE Access* **2020**, *8*, 87422–87433. [CrossRef]

59. Diwaker, S.; Gupta, S.C.; Gupta, N.K. Classification of EEG Signal using Correlation Coefficient among Channels as Features Extraction Method. *Indian J. Sci. Technol.* **2016**, *9*, 1–7. [CrossRef]

60. Howard, A.G.; Zhu, M.; Chen, B.; Kalenichenko, D.; Wang, W.; Weyand, T.; Adam, H. Mobilenets: Efficient convolutional neural networks for mobile vision applications. *arXiv* **2017**, arXiv:1704.04861.

61. Tan, M.; Le, Q. EfficientNet: Rethinking Model Scaling for Convolutional Neural Networks. In Proceedings of the 36th International Conference on Machine Learning, Long Beach, CA, USA, 9–15 June 2019; pp. 6105–6114.

62. Moon, Y.; Wajda, D.A.; Motl, R.W.; Sosnoff, J.J. Stride-Time Variability and Fall Risk in Persons with Multiple Sclerosis. *Mult. Scler. Int.* **2015**, *2015*, 964790. [CrossRef] [PubMed]

63. Beauchet, O.; Annweiler, C.; Lecordroch, Y. Walking speed-related changes in stride time variability: Effects of decreased speed. *J. Neuroeng. Rehabil.* **2009**, *6*, 32. [CrossRef]

64. Beauchet, O.; Freiberger, E.; Annweiler, C.; Kressig, R.W.; Herrmann, F.R.; Allali, G. Test-retest reliability of stride time variability while dual tasking in healthy and demented adults with frontotemporal degeneration. *J. Neuroeng. Rehabil.* **2011**, *8*, 37. [CrossRef] [PubMed]

65. Jeon, H.-S.; Han, J.; Yi, W.-J.; Jeon, B.; Park, K.S. Classification of Parkinson gait and normal gait using Spatial-Temporal Image of Plantar pressure. In Proceedings of the 2008 30th Annual International Conference of the IEEE Engineering in Medicine and Biology Society, Vancouver, BC, Canada, 20–25 August 2008; pp. 4672–4675. [CrossRef]

66. Fahn, S.; Elton, R.L. *Recent Developments in Parkinson's Disease*; Fahn, S., Marsden, C.D, Goldstein, M., Calne, D.B., Eds.; Macmillan Health Care Information: Florham Park, NJ, USA, 1987; pp. 153–163.

67. Mirelman, A.; Bonato, P.; Camicioli, R.; Ellis, T.D.; Giladi, N.; Hamilton, J.L.; Hass, C.J.; Hausdorff, J.M.; Pelosin, E.; Almeida, Q.J. Gait impairments in Parkinson's disease. *Lancet Neurol.* **2019**, *18*, 697–708. [CrossRef]

68. Albani, G.; Cimolin, V.; Fasano, A.; Trotti, C.; Galli, M.; Mauro, A. "Masters and servants" in parkinsonian gait: A three-dimensional analysis of biomechanical changes sensitive to disease progression. *Funct. Neurol.* **2014**, *29*, 99–105. [PubMed]

69. Honeine, J.L.; Schiepati, M.; Gagey, O.; Do, M.C. The Functional Role of the Triceps Surae Muscle during Human Locomotion. *PLoS ONE* **2013**, *8*, 52943. [CrossRef]

70. Sterling, N.W.; Cusumano, J.P.; Shaham, N. Dopaminergic modulation of arm swing during gait among Parkinson's disease patients. *J. Parkinson's Dis.* **2015**, *5*, 141–191. [CrossRef]

71. Galna, B.; Lord, S.; Burn, D.J.; Rochester, L. Progression of gait dysfunction in incident Parkinson's disease: Impact of medication and phenotype. *Mov. Disord.* **2015**, *30*, 359–426. [CrossRef]
72. Mirelman, A.; Bernad-Elazari, H.; Thaler, A. Arm swing as a potential new prodromal marker of Parkinson's disease. *Mov. Disord.* **2016**, *31*, 1527–1561. [CrossRef]
73. Kwon, K.Y.; Kim, M.; Lee, S.M.; Kang, S.H.; Lee, H.M.; Koh, S.B. Is reduced arm and leg swing in Parkinson's disease associated with rigidity or bradykinesia? *J. Neurol. Sci.* **2014**, *341*, 32–35. [CrossRef] [PubMed]
74. Pistacchi, M.; Gioulis, M.; Sanson, F. Gait analysis and clinical correlations in early Parkinson's disease. *Funct. Neurol.* **2017**, *32*, 28–34. [CrossRef] [PubMed]
75. Rezvanian, S.; Lockhart, T.; Frames, C.; Soangra, R.; Lieberman, A. Motor subtypes of Parkinson's disease can be identified by frequency component of postural stability. *Sensors* **2018**, *18*, 1102. [CrossRef] [PubMed]
76. Okuma, Y.; Silva de Lima, A.L.; Fukae, J.; Bloem, B.R.; Snijders, A.H. A prospective study of falls in relation to freezing of gait and response fluctuations in Parkinson's disease. *Parkinsonism Relat. Disord.* **2018**, *46*, 30–35. [CrossRef]
77. Yusupov, E.; Chen, D.; Krishnamachari, B. Medication use and falls: Applying Beers criteria to medication review in Parkinson's disease. *SAGE Open Med.* **2017**, *5*, 2050312117743673. [CrossRef]
78. Barbosa, A.F.; Chen, J.; Freitag, F.; Valente, D.; Souza, C.O.; Voos, M.C.; Chien, H.F. Gait, posture and cognition in Parkinson's disease. *Dement. Neuropsychol.* **2016**, *10*, 280–286. [CrossRef]
79. Clark, D.J. Automaticity of walking: Functional significance, mechanisms, measurement and rehabilitation strategies. *Front. Hum. Neurosci.* **2015**, *9*, 246. [CrossRef]
80. Shane Anderson, A.; Loeser, R.F. Why is osteoarthritis an age-related disease? *Best Pract. Res. Clin. Rheumatol.* **2010**, *24*, 15–26. [CrossRef]
81. Zhong, L.L.; Song, Y.Q.; Cao, H.; Ju, K.J.; Yu, L. The non-motor symptoms of Parkinson's disease of different motor types in early stage. *Eur. Rev. Med. Pharmacol. Sci.* **2017**, *21*, 5745–5750. [CrossRef]
82. Limb, M. World will lack 18 million health workers by 2030 without adequate investment, warns un. *BMJ* **2016**, *354*, i5169. [CrossRef]
83. Turin, T.C.; Chowdhury, N.; Ekpekurede, M. Alternative career pathways for international medical graduates towards job market integration: A literature review. *Int. J. Med. Educ.* **2021**, *12*, 45–63. [CrossRef] [PubMed]

biosensors

Article

Explainable Artificial Intelligence and Wearable Sensor-Based Gait Analysis to Identify Patients with Osteopenia and Sarcopenia in Daily Life

Jeong-Kyun Kim [1,2], Myung-Nam Bae [2], Kangbok Lee [2], Jae-Chul Kim [2] and Sang Gi Hong [1,2,*]

[1] Department of Computer Software, University of Science and Technology, Daejeon 34113, Korea; kim.jk@etri.re.kr
[2] Intelligent Convergence Research Laboratory, Electronics and Telecommunications Research Institute, Daejeon 34129, Korea; mnbae@etri.re.kr (M.-N.B.); kblee@etri.re.kr (K.L.); kimjc@etri.re.kr (J.-C.K.)
* Correspondence: sghong@etri.re.kr; Tel.: +82-42-860-1795

Abstract: Osteopenia and sarcopenia can cause various senile diseases and are key factors related to the quality of life in old age. There is need for portable tools and methods that can analyze osteopenia and sarcopenia risks during daily life, rather than requiring a specialized hospital setting. Gait is a suitable indicator of musculoskeletal diseases; therefore, we analyzed the gait signal obtained from an inertial-sensor-based wearable gait device as a tool to manage bone loss and muscle loss in daily life. To analyze the inertial-sensor-based gait, the inertial signal was classified into seven gait phases, and descriptive statistical parameters were obtained for each gait phase. Subsequently, explainable artificial intelligence was utilized to analyze the contribution and importance of descriptive statistical parameters on osteopenia and sarcopenia. It was found that XGBoost yielded a high accuracy of 88.69% for osteopenia, whereas the random forest approach showed a high accuracy of 93.75% for sarcopenia. Transfer learning with a ResNet backbone exhibited appropriate performance but showed lower accuracy than the descriptive statistical parameter-based identification result. The proposed gait analysis method confirmed high classification accuracy and the statistical significance of gait factors that can be used for osteopenia and sarcopenia management.

Keywords: osteopenia; sarcopenia; XAI; SHAP; IMU; gait analysis

Citation: Kim, J.-K.; Bae, M.-N.; Lee, K.; Kim, J.-C.; Hong, S.G. Explainable Artificial Intelligence and Wearable Sensor-Based Gait Analysis to Identify Patients with Osteopenia and Sarcopenia in Daily Life. *Biosensors* **2022**, *12*, 167. https://doi.org/10.3390/bios12030167

Received: 28 December 2021
Accepted: 28 February 2022
Published: 7 March 2022

Publisher's Note: MDPI stays neutral with regard to jurisdictional claims in published maps and institutional affiliations.

1. Introduction

Osteopenia and sarcopenia can cause various senile disorders and are key factors related to the quality of life in old age [1–3]. Gait is a suitable indicator of musculo-skeletal diseases [4]. With the miniaturization of sensors and the development of intelligent monitoring technology, interest in wearable-sensor-based daily health management solutions is increasing [4,5]. Therefore, portable tools and methods that can analyze osteopenia and sarcopenia risks in our daily lives, rather than requiring a specialized hospital setting, can be considered.

Musculoskeletal disorders are increasingly being recognized as conditions that are associated with significant morbidity, mortality, and healthcare costs [1,2]. Osteopenia is a cause of fracture and increases the risk of complications, in addition to pain caused by fractures. Osteoporotic fractures generate costs that reach USD 25 billion, and sarcopenia generates costs of approximately USD 18 billion [2,6]. Patients with sarcopenia have a slow gait, reduced muscular endurance, face difficulty in daily living, and frequently need help from others. Osteoporosis, falls, and fractures can occur easily, whereas the blood and hormonal buffering action of the muscle are moderated, reducing the basal metabolic rate, making chronic diseases unmanageable, and increasing the likelihood of aggravating diabetes and cardiovascular disease [3].

Osteoporosis is defined by the World Health Organization (WHO) as a medical condition in which the bone mineral density (BMD) is less than -2.5 standard deviation (SD) below the mean level for young adults, and for osteopenia it is between -2.5 and -1.0 [7]. Sarcopenia is defined by the European Working Group on Sarcopenia in Older People (EWGSOP) as the presence of low muscle mass, reduced muscle strength, and physical performance [8]. BMD and muscle mass are diagnosed via dual-energy X-ray absorptiometry (DEXA) [1], although cannot be measured without expert assistance. Therefore, a system that can easily manage musculoskeletal diseases in daily life is required.

Human gait involves interactions between the musculoskeletal system and the nervous system. Thus, gait analysis is effective in identifying neuromusculoskeletal disorders such as Parkinson's diseases (PD) [9,10], fall risk [11,12], total hip arthroplasty (THA) [5,13], and sarcopenia [4]. Traditionally, cameras and force plates have been used as clinical gait assessment tools; however, these tools are used only in large institutions such as university hospitals and are difficult to apply in daily life or complex environments because of their high cost and large space requirements [14]. Given the recent miniaturization and increased accuracy of sensor technology, inertial measurement units (IMU) are increasingly being used for gait analysis [4].

Gait analysis methods include statistical comparisons of gait parameters obtained from control and target groups and a method of analysis of the classification results of the groups using machine learning. In the analysis of gait for osteopenia, osteoporosis, sarcopenia, and osteosarcopenia conducted by Intriago [2], the slowest walking speed was observed in osteosarcopenia: 0.9 m/s in osteopenia–osteoporosis, 0.893 in sarcopenia, and 0.7 in osteosarcopenia. Choi [15] investigated the correlation between kinetic gait parameters and femoral BMD of the femoral neck, trochanter, shaft, and total proximal femur. The highest correlation ($r = 0.153$, $p = 0.014$) was observed between the walking speed and femoral neck BMD among the older female participants. ElDeeb [16] aimed to investigate the gait characteristics of postmenopausal women with low BMD ($n = 17$) and to determine the predictive parameters of BMD. When the normal BMD group and women with low BMD were compared, the ankle joint showed less push-off ($p = 0.000$), which seemed to be used to obtain gait stability. Sung [17] divided 77 older participants ($n = 48$ female + 29 male) into normal BMD and low BMD groups using DEXA. The spatial–temporal gait parameters (speed, stride length, and support times) of both groups were subsequently investigated. The support times included those of the initial double support, single support, and terminal double support in the stance phase. The support time was confirmed to have a high ratio of the main foot (the foot mainly used), the stride length was found to be longer on the main foot side than on the other side, and the stride length was positively associated with the single support time on the dominant limb.

Although there are many studies on gait speed for osteopenia and sarcopenia, only a few studies have analyzed gait parameters such as PD, fall risk, and THA. Recently, explainable artificial intelligence (XAI) has received considerable attention as a method to analyze the importance and contribution of parameters. XAI presents predictive results for machine learning in a human-understandable form [18]. It is primarily used to enhance the reliability of machine learning results. Low machine learning accuracy results in the misinterpretation of XAI. The XAI technique detects feature importance and explains the influence of features on model decisions [19]. Therefore, for the management of osteopenia and sarcopenia in daily life, this study proposes an algorithm for detecting gait parameters and identifying patients based on inertia signals and interpreting the results using XAI.

2. Related Studies

This section describes research related to gait parameter detection techniques, wearable-sensor-based patient identification, and XAI. This information should help with the understanding of the proposed wearable-based gait analysis method for the management of osteopenia and sarcopenia in daily life. This study proposes an algorithm for detecting gait parameters based on inertial signals, identifying patients, and interpreting the results using XAI for gait analysis.

2.1. Gait Parameter

IMU-based gait analysis is used to identify PD, fall risk, THA, and sarcopenia. Gait parameters for analysis based on IMU include spatial–temporal parameters (e.g., step length, stance phase, swing phase, single support, double support, step time, cadence, and speed), kinematic parameters (the rotational angles of the sagittal, coronal, and transverse plane of the pelvis, hip, knee, and ankle), and descriptive statistical parameters (such as the maximum, mean, and standard deviation) of inertial signals for each gait phase. The results of gait analysis using spatial–temporal parameters can be compared with the results of other gait analysis tools such as cameras and force plates. However, the disadvantage is that the motion information acquired by the inertial sensor is reduced, resulting in a low classification result [4].

Gait events and phases are detected to extract the gait parameters. Taborri [20] classified gait into two to eight phases. Whittle [21] classified gait into seven phases, and this is the most widely used classification method. One stride is from a heel strike to the next heel strike. Gait is broadly classified into a stance phase from heel strike (HS) to toe off (TO) and a swing phase from TO to the next HS. The stance phase is classified into the loading response, mid stance, terminal stance, and pre swing phases, and the swing phase is classified into the initial swing, mid swing, and terminal swing phases. The spatial–temporal parameters can be obtained by extracting the HS, TO, opposite HS, opposite TO, and walking distance. HS and TO are detected by the time and frequency signal processing of the inertial signal. Kim [22] obtained HS and TO with high accuracy within 0.03 s through time–frequency analysis.

The inertial-sensor-based distance measurement algorithm is widely used as the basis for the distance measurement algorithm in indoor navigation research, and it is difficult to accurately measure the distance using only the inertial sensor [23,24]. Kinematic parameters are obtained by attaching inertial sensors to locations such as the pelvis, hip, knee, and ankle, but they are difficult to use in daily life due to the number of sensors required. Descriptive statistics and frequency analysis have been employed to analyze the signal obtained from the inertial sensor.

2.2. Identifying Patients Based on Inertial Signals

IMU-based gait analysis is used to identify PD, fall risk, THA, and sarcopenia. With gait-parameter-based disease identification, Caramia [9] classified PD using linear discriminant analysis (LDA), naïve Bayes (NB), k-nearest neighbor (k-NN), support vector machine (SVM), SVM radial basis function (RBF), decision tree (DT), and the majority of votes. The performance of the machine learning technique—SVM with a nonlinear kernel—was the best. Eskofier [10] analyzed the gait of PD using descriptive statistical parameters such as the energy maximum, minimum, mean, variance, skewness, and kurtosis of the signal measured by the inertial sensor and the fast Fourier transform, a frequency analysis method. Howcroft [11] predicted the risk of falls using accelerometer data and used temporal (cadence and stride time) and descriptive statistics (maximum, mean, and SD of acceleration). NB, SVM, and neural network (NN) were used as classification methods, and the best single-sensor model was NN. An advantage of deep learning is that it can detect features within the algorithm from a raw signal, although Tunca [12] achieved higher accuracy in long short-term memory (LSTM) when certain parameters (e.g., speed, stride length, cycle time, stance time, swing time, clearance, stance ratio, and cadence) were used as the inputs, compared with raw signals.

Teufl [5] classified THA patients using stride length, stride time, cadence, speed, hip, and pelvis range of motion (ROM) as features of the SVM, and obtained an accuracy of 97%. Dindorf [13] used local interpretable model-agnostic explanations (LIME) to understand the features for identifying THA, and found that the sagittal movement of the hip, knee, and pelvis, as well as transversal movement of the ankle, were particularly important for this specific classification task. Kim [4] obtained feature importance using the Shapley additive explanations (SHAP) approach for spatial–temporal parameters and descriptive statistical parameters detected from signals measured by inertial sensors on both feet of ten sarcopenia and control participants. Twenty descriptive statistical parameters of high importance were used as inputs to classification models such as SVM, RF, and multi –layer perceptron (MLP); the highest accuracy (95%) was achieved using the SVM model, as shown in Table 1.

Table 1. Existing studies on disease identification using gait parameters. Abbreviations are as shown in Table A1.

Reference	Parameter	Disease	Position	Classification	Accuracy
Caramia 2018 [9]	Step length, step time, stride time, speed, hip, knee, and ankle ROM	PD	R and L ankle, knee, hip, chest	LDA, NB, k-NN, SVM, SVM RBF, DT, majority of votes	96.88%
Eskofier 2016 [10]	Energy maximum, minimum, mean, variance, skewness, kurtosis, fast Fourier transform	PD	Upper limbs	AdaBoost, PART, k-NN, SVM, CNN	90.9%
Howcroft 2017 [11]	Cadence, stride time maximum, mean, and SD of acceleration	Faller	Head, pelvis, R and L shank	NB, SVM, NN	57%
Tunca 2019 [12]	Stride length, cycle time, stance time, swing time, clearance, stance ratio, cadence, speed	Faller	Both feet	SVM, RF, MLP, HMM, LSTM	94.30%
Teufl 2019 [5]	Stride length, stride time, cadence, speed, hip and pelvis ROM	THA	Hip, thigh, shank, foot	SVM	97%
Dindorf 2020 [13]	Various parameters	THA	Hip, knee, pelvis, ankle	RF, SVM, SVM RBF, MLP	100%
Kim 2021 [4]	Various parameters	Sarcopenia	Both feet	RF, SVM, MLP, CNN, BiLSTM	95%
Ours	Various parameters	Osteopenia Sarcopenia	Both feet	RF, SVM, XGBoost, CNN, BiLSTM, ResNet	88.69% 93.75%

2.3. Explainable Artificial Intelligence

XAI is a method that allows humans to understand the basis of decisions made by artificial intelligence models [25]. It is primarily used to enhance the reliability of machine learning results. Low machine learning accuracy results in a misinterpretation of XAI. The XAI technique detects feature importance and explains the influence of features on model decisions [19]. LIME and SHAP are often used to explain existing handcraft feature-based classification algorithms, and layer-wise relevance propagation (LRP) and class activation mapping (CAM) are used as algorithms to interpret deep learning.

LIME is effective for tabular data, text, and images. However, it is difficult to set the kernel width in tabular data, and different results are obtained during repeated execution because the sampling process is performed randomly. SHAP was proposed to consider the dependency between features. When the dependence between features is high, the SHAP feature importance is judged to be better than the permutation importance [19]. SHAP is based on Shaply values from game theory. The main advantages of the SHAP method are local explanations and consistency in the global model structure. SHAP is used in many machine learning models as a model-agnostic method.

LRP outputs a heatmap to the input image by tracing back the results of the deep learning model. Unlike LIME and SHAP, which interpret the model using the sensitivity analysis technique, LRP [25] is a mixture of relevance propagation and decomposition. Relevance propagation is a method for calculating the relevance of the contribution of the hidden layer to the output after the decomposition process. CAM visualizes model decisions by computing a weighted linear summation of the last convolutional feature map. CAM is limited to model architectures where the model must consist of one fully connected (FC) layer with global average pooling (GAP) [26]. Grad-CAM describes models without constraints on the model architecture. Gradient-based CAM methods share the problem of shattered gradients, causing noise saliency maps in the intermediate layer. An LRP-based Relevance-CAM has been proposed to solve the gradient problem [27].

3. Methods

To analyze the gait of the osteopenia and sarcopenia groups, the patients were identified using machine learning, and the machine learning model was interpreted using XAI. The inertial sensor signals and spatial–temporal and descriptive statistical parameters detected in the proposed algorithm were used as machine learning inputs. By analyzing the model that obtained high-accuracy identification results, the inertial signal and gait parameters of the osteopenia and sarcopenia groups were analyzed. The flowchart of patient identification and gait analysis for osteopenia and sarcopenia is shown in Figure 1.

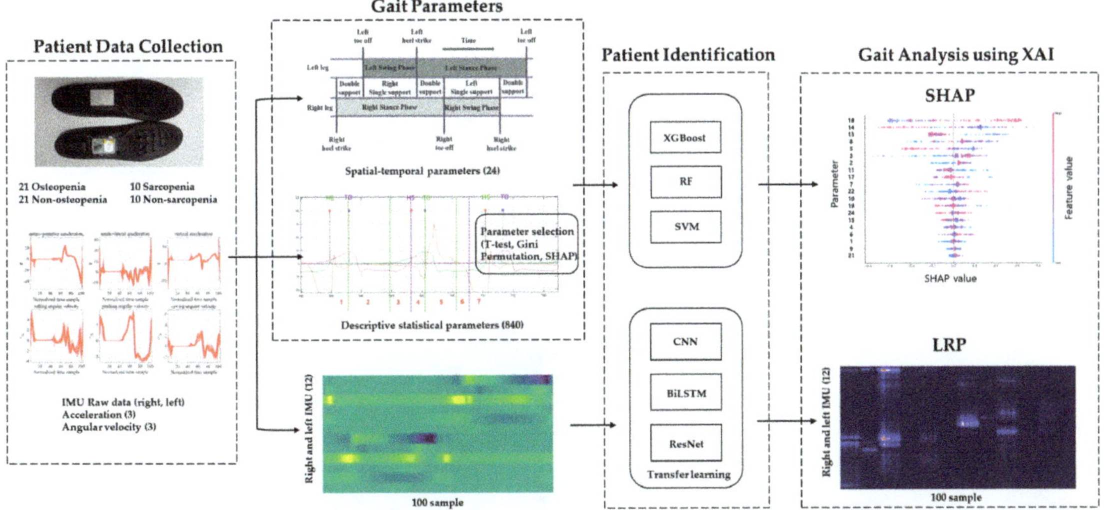

Figure 1. Gait analysis flowchart.

3.1. Patient Data Collection

Gait signals of 42 women over 65 years of age were obtained to analyze the gait characteristics for osteopenia and sarcopenia. Among the 42 subjects, there were 21 patients with osteopenia and 21 patients without osteopenia. The BMD obtained by measuring DEXA was compared with that of a healthy young person: when the T-score was -1 SD or higher, the data were assigned to the control group; when it was lower than -1, the data were assigned to the osteopenia group. Additionally, 10 sarcopenia and 10 non-sarcopenia patients were selected among the 42 subjects. Sarcopenia was diagnosed using the skeletal muscle mass index (SMI, appendicular skeletal muscle mass in kg/height in m^2) that was less than 5.4 kg/m^2 (as obtained through DEXA), whereas the grasp strength was less than 18 kg. The group without sarcopenia included participants with SMI of 5.5 or more and a grasp strength of 19 kg or more. Relevant statistics, including age, height, weight, foot size, Mini-Mental State Examination (MMSE) [28], the Mores Fall Scale (MFS) [29], SARC-F questionnaire [30], Berg Balance Scale (BBS) [31] and Timed Up and Go (TUG) scores [32], grasp power, T-score for DEXA, and SMI, are shown in Table 2.

Table 2. Group population statistics for osteopenia and sarcopenia groups.

Parameter	Osteopenia	Non-Osteopenia	Osteopenia *p*-Value	Sarcopenia	Non-Sarcopenia	Sarcopenia *p*-Value
Age (years)	70.48 ± 2.36	70.33 ± 2.56	0.852	71.10 ± 2.13	69.50 ± 3.14	0.199
Height (cm)	153.65 ± 4.83	152.80 ± 5.93	0.614	150.87 ± 4.66	153.10 ± 4.36	0.283
Weight (kg)	57.75 ± 6.12	59.57 ± 7.12	0.379	53.55 ± 5.62	61.20 ± 5.07	0.005
Feet_size (mm)	236.91 ± 7.66	238.57 ± 6.55	0.453	232.00 ± 5.87	239.50 ± 6.43	0.014
MMSE	27.62 ± 1.77	28.19 ± 1.78	0.303	27.80 ± 1.40	27.30 ± 2.16	0.547
SARC-F	3.19 ± 2.40	3.86 ± 2.15	0.349	2.90 ± 1.52	2.90 ± 2.85	1.000
MFS	23.10±17.92	26.43 ± 16.59	0.535	13.50 ± 12.92	23.50 ± 12.70	0.098
BBS	42.38 ± 8.48	42.19 ± 6.85	0.937	43.10 ± 6.26	41.90 ± 9.47	0.742
3m TUG	10.96 ± 1.64	11.50 ± 2.87	0.464	11.71 ± 1.62	9.85 ± 1.92	0.031
Grasp_right (kg)	17.29 ± 5.42	18.77 ± 4.71	0.351	14.42 ± 3.65	22.57 ± 2.73	0.000
Grasp_left (kg)	17.61 ± 4.67	18.04 ± 4.40	0.761	14.15 ± 3.97	22.17 ± 3.02	0.000
T_score (DEXA)	−1.85 ± 0.74	0.69 ± 1.49	0.000	−0.49 ± 2.08	−0.64 ± 2.03	0.872
SMI(ASM/height)	5.37 ± 0.55	5.38 ± 0.65	0.961	4.58 ± 0.32	5.93 ± 0.35	0.000

The limitations of this study were that the physiological and psychological variables of the participants could not be controlled, the age range of the participants could not be expanded, the study was conducted on women only, and the treadmill gait experiment with fall risk factors for older adults was excluded, and only the preferred speed through walking on flat ground was measured.

All participants wore the same sneaker model and walked the 27 m corridor four times in a straight line. The gait data were acquired from the right and left insoles using IMU, as shown in Figure 2. The IMU settings included an acceleration sensitivity of 8G, a gyro sensitivity of 1000°/s, and a sampling frequency of 100 Hz [4].

Figure 2. Sensor attachments to the insoles.

Additionally, 20 participants measured 9 m gait simultaneously with the clinical standard system and the proposed inertial system to verify the proposed device and algorithm. The clinical system consisted of ten cameras (Vicon, Oxford Metrics, Oxford, UK) and four force plates (Advanced Mechanical Technology, MA, USA). Data analysis was performed using the Vicon Polygon 3.5.2. Ethics approval was obtained from the Chungnam National University Hospital Institutional Review Board before conducting this study (File No: CNUH 2019-06-042).

3.2. Gait Signals and Parameters

Gait is a motion in which both feet alternately repeat the stance and swing phases, and the event points of gait that separate the stance and swing phases are called HS and TO. HS is at the start of the stance phase, and TO is at the start of the swing phase. The gait data obtained from the IMU sensor were 6-axis signals that included the *xyz*-axis acceleration and angular velocity signal. When the measured sensor data were separated based on HS and normalized to 100 samples, they exhibited periodic characteristics, as shown in Figure 3. The characteristics of the gait signal differ from person to person, and for IMU gait analysis, the spatial–temporal parameter was detected from the gait signal, and the gait signals were expressed as descriptive statistical parameters and analyzed. Additionally, patients were classified using raw data as inputs for deep learning without detecting the parameters; then, the gait signals were analyzed by interpreting the deep learning results.

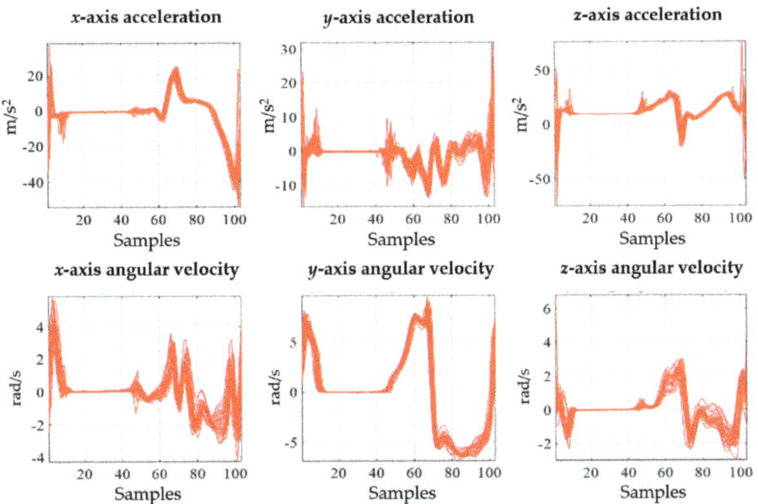

Figure 3. Acceleration and angular velocity signals.

The spatial–temporal parameters were extracted from the inertial signals using the proposed algorithm [4]. Twenty-four spatial–temporal parameters were detected: stance phase time right, stance phase time left, swing phase time right, swing phase time left, stance phase percent right, stance phase percent left, double support first phase time right, double support first phase time left, double support second phase time right, double support second phase time left, single support phase time right, single support phase time left, double support first phase percent right, double support first phase percent left, double support second phase percent right, double support second phase percent left, single support phase percent right, single support phase percent left, stride length right, stride length left, stance phase time SI, swing phase time SI, stance phase percent SI, and cadence. The definitions are summarized in Table 3.

After detecting HS and TO, the opposite HS, opposite TO, cadence, stance phase (time), swing phase (time), single support phase (time), and double support phase (time) could be obtained by arithmetic calculations. Secondary parameters, such as balance of difference between the right and left foot, were also collected through comparative analysis of both feet. Stride was detected through a distance estimation algorithm based on zero-velocity detection (zero-velocity update) using an extended Kalman filter [23,24].

To obtain descriptive statistical parameters, the six-axis gait signal was classified into seven phases, as proposed by Whittle. The detection of HS, TO, heel rise (HR), feet adjacent (FA), and tibia vertical (TV) is required to classify seven phases; it was detected using the method proposed in a previous study [4]. Ten descriptive statistical parameters were obtained from signals classified into seven phases, and the descriptive statistical parameters were max., min., SD, AbSum, root-mean-square (RMS), kurtosis, skewness, MMgr, DMM, and Mdif. A total of 840 descriptive statistical parameters (both feet (2) $\times$ sensor signal (6) $\times$ gait phase (7) $\times$ (10 parameters)) were detected.

Table 3. Definition of gait parameters.

Gait Parameters	Definition
	Spatial–temporal parameters
Cadence	Number of steps acquired per minute
Stance phase (time)	Percent (time) starting with HS and ending with TO of the same foot
Swing phase (time)	Percent (time) starting with TO and ending with HS of the same foot
Single support phase (time)	Percent (time) when only one foot is on the ground
Double support phase (time)	Percent (time) when both feet are on the ground
Stride length	Distance starting with HS and ending with next HS of the same foot
Symmetry indices (SI)	Absolute values of (right—left)/(0.5 × (right + left)
	Descriptive statistical parameters
Max	Greatest values
Min	Least or smallest values
SD	Standard deviation of values
AbSum	Absolute sum of values
Root-mean-square (RMS)	Arithmetic mean of the squares of a set of values
Kurtosis	Assesses whether the tails of a given distribution contain extreme values
Skewness	A measure of the asymmetry of the probability distribution of a real-valued random variable about its mean
MMgr	Gradient from maximum value to minimum value
DMM	Difference between maximum value and minimum value
Mdif	Maximum for the difference between two successive values

3.3. Patient Identification

To identify patients using the inertial gait signal and the proposed gait parameters, osteopenia and sarcopenia groups were classified through various models such as RF, XGBoost (Extreme Gradient Boosting), SVM, and deep learning models.

RF is a decision tree ensemble classifier that combines multiple single classifiers to obtain the result of each classification model either through majority vote or weighted average [33]. RF lowers the risk of overfitting by using some data and features from the training data. XGBoost is a decision tree ensemble model and improves the performance of the gradient boosting machine in terms of speed. Boosting models increase accuracy by iteratively updating the parameters of the previous classifier to reduce the slope of the loss function, thereby generating a robust classifier [33]. SVM is a binary classifier that aims to determine the optimal separation hyperplane that maximizes the margin between two classes. Kernel functions are used to map data to a higher-dimensional space; thus, an SVM can compute nonlinear decision boundaries [4].

The representative deep-learning-based models were convolutional neural network (CNN) and LSTM. A CNN is composed of one or more convolutional, pooling, FC, and dense layers. CNNs exhibit high performance in detecting and classifying features in images. Unlike LSTM, which only has forward hidden layers, BiLSTM has both forward and backward hidden layers. Therefore, it learns both before and after information and demonstrates high performance in time-series data. As a CNN backbone, ResNet exhibits excellent classification accuracy [34]. ResNet uses skip connections (or short connections) to pass the input from the previous layer to the next layer. This skip connection solves the gradient loss/burst problem, enabling deep neural networks. ResNet uses 18, 34, 50, 101, and 152 layers depending on the depth, and there are structural differences in approximately 50 layers. In particular, ResNet is a popular architecture despite the existence of other models that have improved performance in various fields. Moreover, it is a representative CNN architecture for which many supporting materials are available [34,35].

Transfer learning is applied as a solution to address the difficulty of training a model based on small datasets. In transfer learning, data similar to the target data are learned in advance and a specific layer is frozen, such that only the layer which is not frozen when learning the target data is learned [36]. Specific data characteristics can be overfitted, because patient identification is a binary classification. Therefore, person identification is pre-trained because high-resolution features can be detected by comparing and analyzing the gaits of multiple people.

3.4. Gait Analysis

The gait signals and parameters were analyzed using statistical methods and XAI techniques that interpret machine learning results. The independent t-test was used as a statistical method to compare the spatial–temporal parameters and descriptive statistical parameters. To improve the reliability of the machine-learning-based analysis method, a higher accuracy should first be obtained. Therefore, the osteopenia and sarcopenia groups were classified through various models, such as RF, XGBoost, SVM, and deep learning models. The CNN and LSTM models were used as the deep learning models.

Spatial–temporal parameters and 100 descriptive statistical parameters with low p-values of the t-test were used as inputs for RF, XGBoost, and SVM. The following RF parameters were used: number of trees = 50, max_depth = 30, and number of features = square root of the gait parameters. The XGBoost parameters were booster = gbtree, objective = binary:logistic, eta = 0.018, max_depth = 15, gamma = 0.009, subsample = 0.98, and colsample_bytree = 0.86. SVM explored the linear and RBF kernels, and the parameters were gamma = 1.0 and C = 5.0.

The 12 axes of acceleration and angular velocity signals obtained from both feet were applied to the deep learning models. We proposed a low-layer-based CNN and BiLSTM model and applied ResNet50. As the input of the deep learning model, a stride based on HS was detected and normalized to 100 samples using spline interpolation because the signal was collected at 100 Hz [4]. ResNet50 is reduced in size by pooling as the layers progress. The ResNet50 backbone cannot be used with an input of shape 12, and removing pooling lowers the accuracy. Therefore, the input shape (36,100) was generated by amplifying the signal of 12 axes threefold because the input size of ResNet50 must be 12 or more, and the kernel size was 3. The layers of each model are shown in Table 4. The parameters of the deep learning model were as follows: learning rate = 0.0005, training epoch = 100, batch size = 16, loss = CrossEntropyLoss, optimizer = Adam, and activation function = Rectified Linear Unit.

Table 4. Instantiation of deep learning model.

CNN		BiLSTM		ResNet50	
Input	None, 100, 36, 1	Input	None, 100, 36, 1	Input	None, 100, 36, 1
Conv1	$3 \times 3, 5$ 2×1 max pooling,	BiLSTM1	5	Conv1	$7 \times 7, 64$ stride 2 3×3 max pooling, stride 2
Conv2	$3 \times 3, 5$ 2×1 max pooling,	BiLSTM2	10	Conv2	$\begin{bmatrix} 1 \times 1,\ 64 \\ 3 \times 3,\ 64 \\ 1 \times 1,\ 256 \end{bmatrix} \times 3$
Conv3	$3 \times 3, 20$	Dropout	0.5	Conv3	$\begin{bmatrix} 1 \times 1,\ 128 \\ 3 \times 3,\ 128 \\ 1 \times 1,\ 512 \end{bmatrix} \times 4$
Dropout	0.5	FC, Dense		Conv4	$\begin{bmatrix} 1 \times 1,\ 256 \\ 3 \times 3,\ 256 \\ 1 \times 1,\ 1024 \end{bmatrix} \times 6$
FC, Dense				Conv5	$\begin{bmatrix} 1 \times 1,\ 512 \\ 3 \times 3,\ 512 \\ 1 \times 1,\ 2048 \end{bmatrix} \times 3$ GAP, FC

XGBoost can calculate the built-in importance (Gini importance) and permutation importance using the learned model. Permutation importance measures the increase or decrease in prediction error compared with the original data when the feature data are transformed [19]. Permutation importance does not consider the correlation between features; therefore, SHAP was proposed as a method to consider the dependency between features. In particular, the SHAP feature importance is considered to be better than the permutation importance because gait parameters have a high dependence on the features. The Gini, permutation, and SHAP importance of the spatial–temporal and descriptive statistical parameters were calculated to obtain important parameters of osteopenia and sarcopenia, and the results of deep learning were analyzed using LRP, Grad-CAM, and Relevance-CAM.

4. Results

4.1. Patient Identification

The identification results of 21 osteopenia and 21 non-osteopenia subjects showed the highest accuracy in SVM when 24 spatial–temporal parameters were used as inputs, but showed an accuracy of less than 65%. The descriptive statistics parameter obtained the highest accuracy of 68.45% in XGBoost by using 100 parameters with a low p-value as an input, as a result of an independent t-test. For training and testing, 21 cross-validations were performed on 21 subjects, and the average was obtained. Using an inertial sensor as an input for deep learning, ResNet showed the highest accuracy among CNN, BiLSTM, and ResNet. The results of applying transfer learning to the ResNet model showed lower accuracy than when no transfer learning was applied. However, when performing transfer learning, it was shown that the accuracy increased when features were extracted, including the test subject. This implies that ResNet was pre-trained for human identification using the data of 42 patients, and the patient identification was cross-validated for 21 patients. The osteopenia group obtained the highest recognition result in the transfer learning ResNet.

The identification results of 10 sarcopenia and 10 non-sarcopenia cases were over 70% accurate in terms of the spatial–temporal parameters, and the accuracy in case of sarcopenia was better than that in osteopenia. When the descriptive statistics parameter was used as the RF input, the highest accuracy was obtained, and the deep learning method of the inertial sensor input did not yield satisfactory identification results. Therefore, analysis of the results of XAI based on parameters is more reliable than the analysis of results based on deep learning. The patient identification results of machine learning are presented in Tables 5 and 6.

Table 5. Identification result of RF, XGBoost, and SVM (accuracy, precision, recall and F1-score).

Groups	Parameters	Models	Accuracy	Precision	Recall	F1-Score
Osteopenia	Spatial–temporal (24)	RF	0.494	0.476	0.370	0.393
		XGBoost	0.476	0.476	0.376	0.406
		SVM	0.637	0.619	0.511	0.544
	Descriptive statistical (100)	RF	0.649	0.655	0.612	0.607
		XGBoost	0.684	0.690	0.680	0.650
		SVM	0.607	0.678	0.590	0.604
Sarcopenia	Spatial–temporal (24)	RF	0.802	0.825	0.775	0.775
		XGBoost	0.752	0.725	0.667	0.677
		SVM	0.775	0.603	0.775	0.658
	Descriptive statistical (100)	RF	0.675	0.675	0.632	0.631
		XGBoost	0.603	0.675	0.557	0.591
		SVM	0.637	0.704	0.657	0.644

Table 6. Identification result of CNN, BiLSTM, and ResNet (accuracy, precision, recall and F1-score).

Groups	Models	Accuracy	Precision	Recall	F1-Score
Osteopenia	CNN	0.696	0.690	0.735	0.670
	BiLSTM	0.619	0.570	0.610	0.571
	ResNet	0.767	0.672	0.726	0.676
	ResNet(transfer)	0.786	0.869	0.747	0.787
Sarcopenia	CNN	0.600	0.437	0.525	0.447
	BiLSTM	0.425	0.300	0.350	0.299
	ResNet	0.612	0.337	0.500	0.394
	ResNet(transfer)	0.700	0.612	0.636	0.606

4.2. Importance of Descriptive Statistical Parameter

The order of Gini, permutation, and SHAP importance was obtained for the descriptive statistical parameters of osteopenia and sarcopenia. When using highly important parameters such as RF, XGBoost, and SVM inputs, SHAP obtained the highest identification rate; however, when using the inner 20 important parameters as inputs, more identification results than the 100 descriptive statistical parameters were obtained. Tables 7 and 8 showed the classification results as the number of parameters increased, and the average accuracy was obtained by performing 21 cross-validations for osteopenia and 10 cross-validations for sarcopenia. According to the result of each cross-validation, SHAP-based feature importance has different values. For example, osteopenia was trained with 40 datasets (20 osteopenia datasets and 20 non-osteopenia datasets) and tested with two datasets (1 osteopenia dataset and 1 non-osteopenia dataset) during 21 cross-validations. As a result of the training, the Shapley values were obtained based on the training data of 40 people, and the Shapley values were obtained in different orders. Table 9 shows the average results for 20 high-order Shapley values generated during 21 cross-validations. In osteopenia, the Shapley value is relatively high in the upper parameter and less than 0.1 from the 10th parameter. In sarcopenia, the difference in the Shapley value between the parameters is small. The parameter numbers of descriptive statistical parameters are shown in Table 10. The results of learning RF, XGBoost, and SVM with 20 parameters with high importance in Table 9 are shown in Table 11. Osteopenia obtained an accuracy of 88.69% in XGBoost using the top 4 parameters as inputs, and sarcopenia obtained an accuracy of 93.75% in RF using the top 18 parameters as inputs.

Table 7. Osteopenia identification results according to the number of important parameters (accuracy, %).

Class	ML	Number of Parameters										
		2	3	4	5	6	7	8	9	10	20	100
Gini	RF	70.83	70.23	64.88	72.02	68.45	63.69	61.30	60.11	60.71	61.30	64.88
	XGBoost	66.66	67.85	64.88	71.42	68.45	64.28	65.47	61.30	65.47	67.26	68.45
	SVM	64.28	64.88	64.88	64.28	61.30	61.30	59.52	55.35	57.73	58.33	60.71
Permutation	RF	73.21	70.83	69.64	67.26	64.28	68.45	70.23	69.04	67.26	67.26	64.88
	XGBoost	69.64	70.83	70.23	68.42	64.88	65.47	67.26	66.70	67.26	70.23	68.45
	SVM	65.47	68.45	66.07	64.28	66.66	66.66	64.28	64.88	64.88	60.71	60.71
SHAP	RF	73.80	76.19	70.23	63.69	63.09	63.69	63.09	63.69	57.73	60.11	64.88
	XGBoost	70.23	75	74.40	73.21	66.66	67.85	63.69	59.52	56.54	68.45	68.45
	SVM	71.42	71.42	67.26	61.30	58.33	58.33	57.14	55.95	57.14	62.5	60.71

Table 8. Sarcopenia identification results according to the number of important parameters (accuracy, %).

Class	ML	Number of Parameters										
		2	3	4	5	10	15	16	17	18	20	100
Gini	RF	50	58.75	62.5	65	68.75	67.5	68.75	71.25	71.25	62.5	67.5
	XGBoost	52.5	57.5	65	66.25	62.5	58.75	58.75	58.75	58.75	58.75	60
	SVM	52.5	58.75	66.25	65	72.5	57.5	56.25	56.25	58.75	60	63.75
Permutation	RF	62.5	60	56.25	53.75	57.5	67.5	55	60	70	62.5	67.5
	XGBoost	60	60	55	58.75	65	63.75	68.75	65	66.25	67.5	60
	SVM	61.25	60	60	55	65	66.25	66.25	68.75	63.75	60	63.75
SHAP	RF	56.25	60	57.5	65	67.5	62.5	72.5	73.75	68.75	67.5	67.5
	XGBoost	46.25	63.75	62.5	65	65	63.75	63.75	65	66.25	63.75	60
	SVM	58.75	67.5	60	61.25	675	66.25	68.75	62.5	60	58.75	63.75

Table 9. Feature importance and Shapley values of descriptive statistical parameters.

Class	Important Parameter	1	2	3	4	5	6	7	8	9	10
Osteopenia	Parameters	247	114	87	218	816	206	291	21	169	667
	Shapley value	0.97	0.28	0.27	0.2	0.18	0.17	0.16	0.13	0.1	0.09
Sarcopenia	Parameters	430	524	51	9	270	457	231	387	3	97
	Shapley value	0.66	0.28	0.25	0.22	0.17	0.16	0.15	0.13	0.13	0.13

Class	Important Parameter	11	12	13	14	15	16	17	18	19	20
Osteopenia	Parameters	774	117	45	802	312	23	542	242	554	422
	Shapley value	0.09	0.08	0.08	0.07	0.07	0.07	0.07	0.06	0.06	0.06
Sarcopenia	Parameters	5	67	521	690	607	704	380	469	8	257
	Shapley value	0.13	0.12	0.11	0.09	0.09	0.08	0.08	0.08	0.08	0.07

Table 10. Seven-phase descriptive statistical parameters.

Phase	Parameter	Right										Left									
		Max	Min	SD	AbSum	RMS	Ku	Ske	MMgr	DMM	Mdif	Max	Min	SD	AbSum	RMS	Ku	Ske	MMgr	DMM	Mdif
Loading response	AccX	1	2	3	4	5	6	7	8	9	10	421	422	423	424	425	426	427	428	429	430
	AccY	11	12	13	14	15	16	17	18	19	20	431	432	433	434	435	436	437	438	439	440
	AccZ	21	22	23	24	25	26	27	28	29	30	441	442	443	444	445	446	447	448	449	450
	GyroX	31	32	33	34	35	36	37	38	39	40	451	452	453	454	455	456	457	458	459	460
	GyroY	41	42	43	44	45	46	47	48	49	50	461	462	463	464	465	466	467	468	469	470
	GyroZ	51	52	53	54	55	56	57	58	59	60	471	472	473	474	475	476	477	478	479	480
Mid stance	AccX	61	62	63	64	65	66	67	68	69	70	481	482	483	484	485	486	487	488	489	490
	AccY	71	72	73	74	75	76	77	78	79	80	491	492	493	494	495	496	497	498	499	500
	AccZ	81	82	83	84	85	86	87	88	89	90	501	502	503	504	505	506	507	508	509	510
	GyroX	91	92	93	94	95	96	97	98	99	100	511	512	513	514	515	516	517	518	519	520
	GyroY	101	102	103	104	105	106	107	108	109	110	521	522	523	524	525	526	527	528	529	530
	GyroZ	111	112	113	114	115	116	117	118	119	120	531	532	533	534	535	536	537	538	539	540
Terminal stance	AccX	121	122	123	124	125	126	127	128	129	130	541	542	543	544	545	546	547	548	549	550
	AccY	131	132	133	134	135	136	137	138	139	140	551	552	553	554	555	556	557	558	559	560
	AccZ	141	142	143	144	145	146	147	148	149	150	561	562	563	564	565	566	567	568	569	570
	GyroX	151	152	153	154	155	156	157	158	159	160	571	572	573	574	575	576	577	578	579	580
	GyroY	161	162	163	164	165	166	167	168	169	170	581	582	583	584	585	586	587	588	589	590
	GyroZ	171	172	173	174	175	176	177	178	179	180	591	592	593	594	595	596	597	598	599	600
Pre swing	AccX	181	182	183	184	185	186	187	188	189	190	601	602	603	604	605	606	607	608	609	610
	AccY	191	192	193	194	195	196	197	198	199	200	611	612	613	614	615	616	617	618	619	620
	AccZ	201	202	203	204	205	206	207	208	209	210	621	622	623	624	625	626	627	628	629	630
	GyroX	211	212	213	214	215	216	217	218	219	220	631	632	633	634	635	636	637	638	639	640
	GyroY	221	222	223	224	225	226	227	228	229	230	641	642	643	644	645	646	647	648	649	650
	GyroZ	231	232	233	234	235	236	237	238	239	240	651	652	653	654	655	656	657	658	659	660
Initial swing	AccX	241	242	243	244	245	246	247	248	249	250	661	662	663	664	665	666	667	668	669	670
	AccY	251	252	253	254	255	256	257	258	259	260	671	672	673	674	675	676	677	678	679	680
	AccZ	261	262	263	264	265	266	267	268	269	270	681	682	683	684	685	686	687	688	689	690
	GyroX	271	272	273	274	275	276	277	278	279	280	691	692	693	694	695	696	697	698	699	700
	GyroY	281	282	283	284	285	286	287	288	289	290	701	702	703	704	705	706	707	708	709	710
	GyroZ	291	292	293	294	295	296	297	298	299	300	711	712	713	714	715	716	717	718	719	720

Table 10. *Cont.*

	Parameter	Right										Left									
		Max	Min	SD	AbSum	RMS	Ku	Ske	MMgr	DMM	Mdif	Max	Min	SD	AbSum	RMS	Ku	Ske	MMgr	DMM	Mdif
Mid swing	AccX	301	30	303	304	305	306	307	308	309	310	721	722	723	724	725	726	727	728	729	730
	AccY	311	312	313	314	315	316	317	318	319	320	731	732	733	734	735	736	737	738	739	740
	AccZ	321	322	323	324	325	326	327	328	329	330	741	742	743	744	745	746	747	748	749	750
	GyroX	331	332	333	334	335	336	337	338	339	340	751	752	753	754	755	756	757	758	759	760
	GyroY	341	342	343	344	345	346	347	348	349	350	761	762	763	764	765	766	767	768	769	770
	GyroZ	351	352	353	354	355	356	357	358	359	360	771	772	773	774	775	776	777	778	779	780
Terminal swing	AccX	361	362	363	364	365	366	367	368	369	370	781	782	783	784	785	786	787	788	789	790
	AccY	371	372	373	374	375	376	377	378	379	380	791	792	793	794	795	796	797	798	799	800
	AccZ	381	382	383	384	385	386	387	388	389	390	801	802	803	804	805	806	807	808	809	810
	GyroX	391	392	393	394	395	396	397	398	399	400	811	812	813	814	815	816	817	818	819	820
	GyroY	401	402	403	404	405	406	407	408	409	410	821	822	823	824	825	826	827	828	829	830
	GyroZ	411	412	413	414	415	416	417	418	419	420	831	832	833	834	835	836	837	838	839	840

Table 11. Osteopenia and sarcopenia identification results with the 20 parameters from Table 9 (accuracy, %).

Class	Important Parameter	1	2	3	4	5	6	7	8	9	10
	RF	x	75	85.11	85.71	78.57	82.14	80.95	81.54	77.97	76.78
Osteopenia	XGBoost	x	72.02	80.95	88.69	87.69	87.5	85.11	82.73	81.54	83.33
	SVM	x	74.40	75	75.59	83.92	82.73	80.95	81.54	80.35	78.57
	RF	x	85	82.5	83.75	85	85	86.25	82.5	8	82.5
Sarcopenia	XGBoost	x	80	72.5	78.75	76.25	73.75	75	71.25	73.75	71.25
	SVM	x	81.25	80	82.5	81.25	82.5	86.25	86.25	87.5	81.25

Class	Important Parameter	11	12	13	14	15	16	17	18	19	20
	RF	77.38	72.61	78.57	74.40	79.16	82.14	79.76	73.80	82.14	82.73
Osteopenia	XGBoost	76.78	76.78	76.19	77.97	80.95	81.54	77.38	74.40	73.80	74.40
	SVM	76.19	77.97	74.40	72.61	75.59	76.78	76.19	77.97	79.16	74.40
	RF	81.25	83.75	86.25	88.75	86.25	87.5	91.25	93.75	86.25	92.5
Sarcopenia	XGBoost	71.52	75	71.25	70	71.25	75.	72.5	72.5	71.25	72.5
	SVM	80	83.75	86.25	83.75	86.25	81.25	83.75	78.75	78.75	78.75

4.3. Gait Analysis

In the spatial–temporal parameters of the osteopenia group, the stance phase percentage decreased, double support percentage (time) decreased, and single support percentage increased. The sarcopenia group showed an increase in the value of the SI parameter compared with the non-sarcopenia group, implying that the difference between both feet was large. Except for the SI parameter, the *p*-value did not have a statistical significance of less than 0.001. Table 12 shows the mean and Shapley values of the spatial–temporal parameters; * indicates that the *p*-value is less than 0.025, and ** indicates that the *p*-value is less than 0.001.

From the result of the SHAP plot of osteopenia, as the value of the single support phase percent left (parameter 18) increased, the risk of osteopenia increased, as indicated by the positive SHAP value. As the value decreased, the risk also decreased, with the SHAP value being negative. When the single support phase percent left value increased, the risk increased linearly, and the osteopenia risk was low, at 39 or lower, and the risk increased at 42 or higher. A low double support first phase percent (parameter 13 and 14) increased the risk of osteopenia, with a decreased risk above 9 and an increased risk below 9. A low value of double support first phase time left (parameter 8) increased the risk, a high value decreased the risk of osteopenia, and a double support first phase time left lower than 0.075 led to an increase in risk. A low value of the stance phase percent right (parameter 5) increased the risk.

As a result of the SHAP plot of sarcopenia, the risk of sarcopenia increased when the double support first phase time left (parameter 8) had a very low value (less than 0.07). Stance phase percent left (parameter 6) increased the risk above 60 and decreased below 60, but did not show linearity. The risk increased when the value of the stance phase time SI (parameter 21) increased, and the risk was high at 0.35 or higher, although it was low at less than 0.35. SHAP plots of the spatial–temporal parameters of osteopenia and sarcopenia are shown in Figure 4.

Table 12. Spatial–temporal parameters of osteopenia and sarcopenia.

	Parameter	Osteopenia	Non-Osteopenia	Shapley Value	Sarcopenia	Non-Sarcopenia	Shapley Value
1	Stance phase time right (s)	0.61	0.645	0.034 **	0.614	0.608	0.014
2	Stance phase time left (s)	0.612	0.641	0.084 *	0.617	0.604	0.18
3	Swing phase time right (s)	0.427	0.419	0.156	0.416	0.414	0.143
4	Swing phase time left (s)	0.424	0.422	0.04	0.412	0.417	0.039
5	Stance phase percent right (%)	58.77	60.442	0.196 **	59.468	59.445	0.235
6	Stance phase percent left (%)	59.05	60.124	0.035 **	59.853	59.114	0.345
7	Double support first phase time right (s)	0.1	0.115	0.074 **	0.112	0.099	0.005
8	Double support first phase time left (s)	0.085	0.106	0.197 **	0.09	0.090	0.551
9	Double support second phase time right (s)	0.085	0.106	0.031 **	0.09	0.090	0.097
10	Double support second phase time left (s)	0.1	0.115	0.072 **	0.111	0.099	0.007
11	Single support phase time right (s)	0.424	0.422	0.078	0.412	0.418	0.007
12	Single support phase time left (s)	0.427	0.419	0.017	0.416	0.414	0.018
13	Double support first phase percent right (%)	9.66	10.711	0.224	10.802	9.692	0
14	Double support first phase percent left (%)	8.18	9.857	0.311 **	8.563	8.858	0.248
15	Double support second phase percent right (%)	8.17	9.846	0.046 **	8.556	8.855	0.072
16	Double support second phase percent left (%)	9.606	10.686	0.017 **	10.727	9.677	0.001
17	Single support phase percent right (%)	40.939	39.884	0.077 **	40.11	40.897	0.035
18	Single support phase percent left (%)	41.262	39.58	0.416 **	40.562	40.578	0.02
19	Stride length right (m)	0.95	0.93	0.065	0.94	0.979	0.022
20	Stride length left (m)	0.918	0.892	0.015	0.896	0.942	0.011
21	Stance phase time SI	0.031	0.032	0.018	0.036	0.025	0.250 **
22	Swing phase time SI	0.041	0.046	0.073	0.053	0.034	0.049 **
23	Stance phase percent SI	0.026	0.028	0.013	0.0325	0.021	0.007 **
24	Cadence (steps/min)	115.781	113.859	0.047	116.21	117.469	0

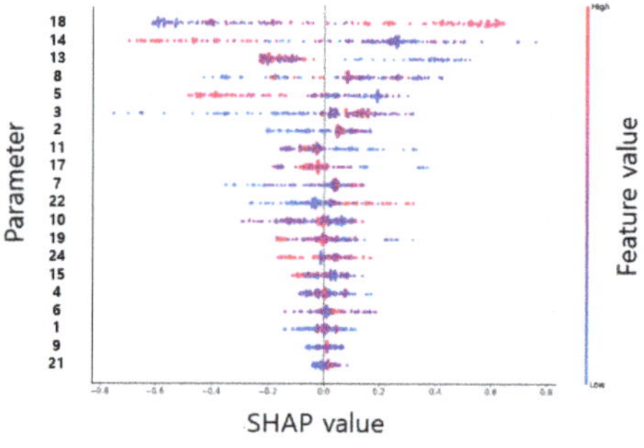

(a) Osteopenia.

Figure 4. *Cont.*

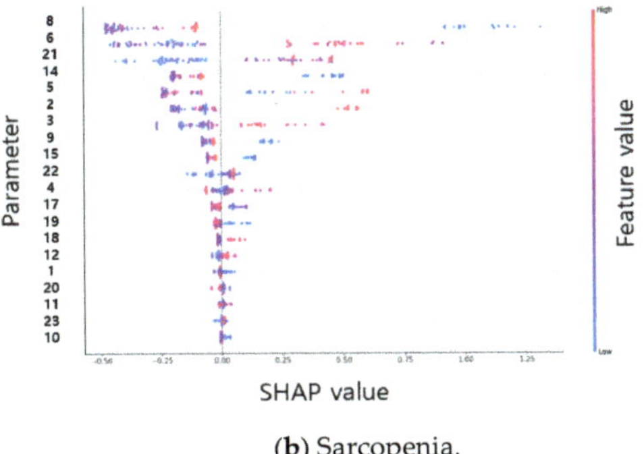

(b) Sarcopenia.

Figure 4. SHAP plots of the spatial–temporal parameters of osteopenia (**a**) and sarcopenia (**b**).

The parameter with the highest SHAP value within the descriptive statistical parameters of osteopenia is the skewness of the x-axis of the accelerometer in the initial swing phase (parameter number 247). Initial swing refers to the FA after TO. When the skewness is negative, the probability density function has a long tail on the left side, and the data, including the median, are more distributed on the right side. When the skewness is positive, there is a long tail on the right side of the probability density function, indicating that the data are more distributed on the left side. Skewness has a positive value when the mean is smaller than the median, negative when the mean is larger, and has a larger value as the difference between the median and the mean becomes larger. When there is a negative value, the right part of Figure 5a has a long tail, but the average decreases and the skewness value becomes small or negative.

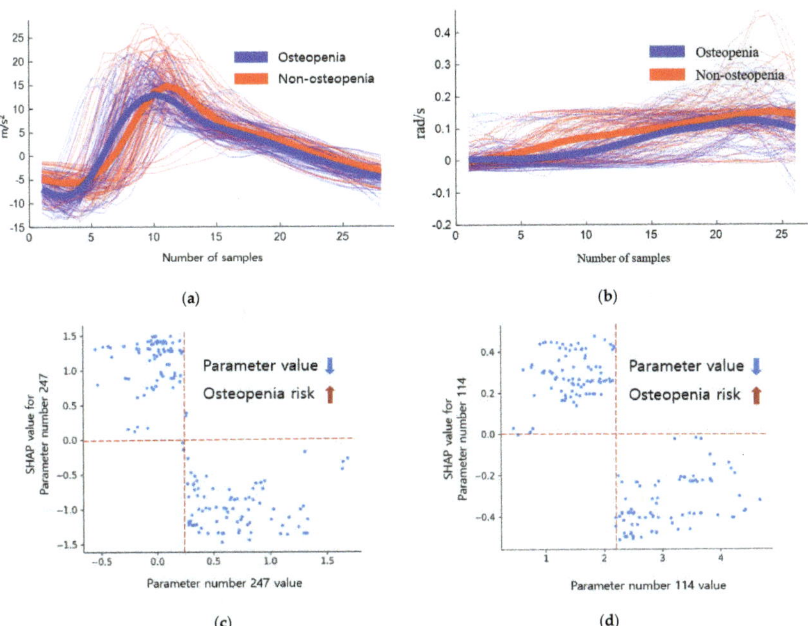

Figure 5. Inertial signals and SHAP dependence plots of descriptive statistical parameters 247 and 114 of osteopenia. (**a**) Inertial signal 247. (**b**) Inertial signal 114. (**c**) SHAP dependence plot 247. (**d**) SHAP dependence plot 114.

The inertial signals and SHAP dependence plots of the descriptive statistical parameters 247 and 114 of osteopenia cases are shown in Figure 5. The blue signal represents osteopenia, and the red signal represents non-osteopenia. A skewness of 0.5 or higher shows a low risk of osteopenia, whereas a negative value shows an increased risk. The absolute sum of gyro z values in the mid stance (parameter number 114) decreased in osteopenia. This implies that there is no rotation of the z-axis in the mid stance. When the absolute sum of values was 2.6 or higher, the risk decreased, and when the absolute sum of values was 1.89 or lower, the risk increased, as shown in Table 13.

Table 13. Descriptive statistical parameters of osteopenia and sarcopenia. * indicates that the p-value is less than 0.025, and ** indicates that the p-value is less than 0.001.

	Osteopenia				Sarcopenia			
	Parameter	Osteopenia	Non-Osteopenia	Shapley Value	Parameter	Sarcopenia	Non-Sarcopenia	Shapley Value
1	247	0.126	0.548	1.033 **	430	2.748	3.797	0.921 **
2	114	1.892	2.613	0.312 **	524	4.925	2.403	0.113 **
3	87	0.357	1.201	0.247 **	51	0.813	0.463	0.189 **
4	218	5.671	7.065	0.200 **	9	8.121	11.813	0.142 **
5	816	3.091	2.502	0.055 **	270	16.417	13.079	0.304 **
6	206	1.926	2.089	0.119 *	457	−0.352	0.047	0.003 **
7	291	3.774	3.129	0.020 **	231	1.532	0.891	0.002 **
8	21	35.175	29.313	0.023 **	387	−0.17	0.042	0.002 **
9	169	3.563	2.823	0.032 **	3	2.267	3.44	0.129 **
10	667	0.135	0.481	0.153 **	97	−0.425	0.274	0.021 **

The inertial signals and SHAP dependence plots of the descriptive statistical parameters 430 and 524 of sarcopenia are shown in Figure 6. Here, the blue signal represents sarcopenia, whereas the red signal represents non-sarcopenia. The maximum difference between two successive values of accelerometer x in the loading response (parameter number 430) was lower in the sarcopenia group than in the non-sarcopenia group. When the maximum value was less than 2.74, the risk of sarcopenia increased; however, when the maximum value was 3.79 or more, the risk of sarcopenia decreased. In the sarcopenia group, the change in the acceleration was smooth. The absolute sum of gyro y values in the mid stance (parameter number 524) increased in the sarcopenia group. As the absolute sum value increased, the risk of sarcopenia increased.

The output for layer2, when the deep-learning-based XAI technique, LRP, Grad-CAM, and Relevance-CAM were applied to ResNet50, is shown in Figure 7. In ResNet50, the CAM technique shows low resolution in small-sized images because the feature map is reduced in layer2. It is difficult to interpret the CAM results for ResNet50 with input sizes of 100 horizontal and 36 vertical. Therefore, it is desirable to interpret ResNet results as LRP. Figure 8 shows the analysis results of LRP for the ResNet of osteopenia and sarcopenia. The LRP attention map of the osteopenia group had high values in 64~67 samples of the acceleration x-axis of the right foot. Its position is the section where the acceleration value rises after TO, and it is the same as the position in Figure 5a, the parameter-based SHAP result. Osteopenia pays attention to changes in acceleration after TO in SHAP and LRP. The LRP attention map of sarcopenia group has a high value at 99~100 positions of the right acceleration x. The position is the section where HS occurs, and it is the same section as Figure 6a. The result of paying attention to the various sections of the acceleration left is similar to having a high identification result when used as various parameter inputs in SHAP. The sarcopenia group pays attention to the HS section and the various sections of the signal.

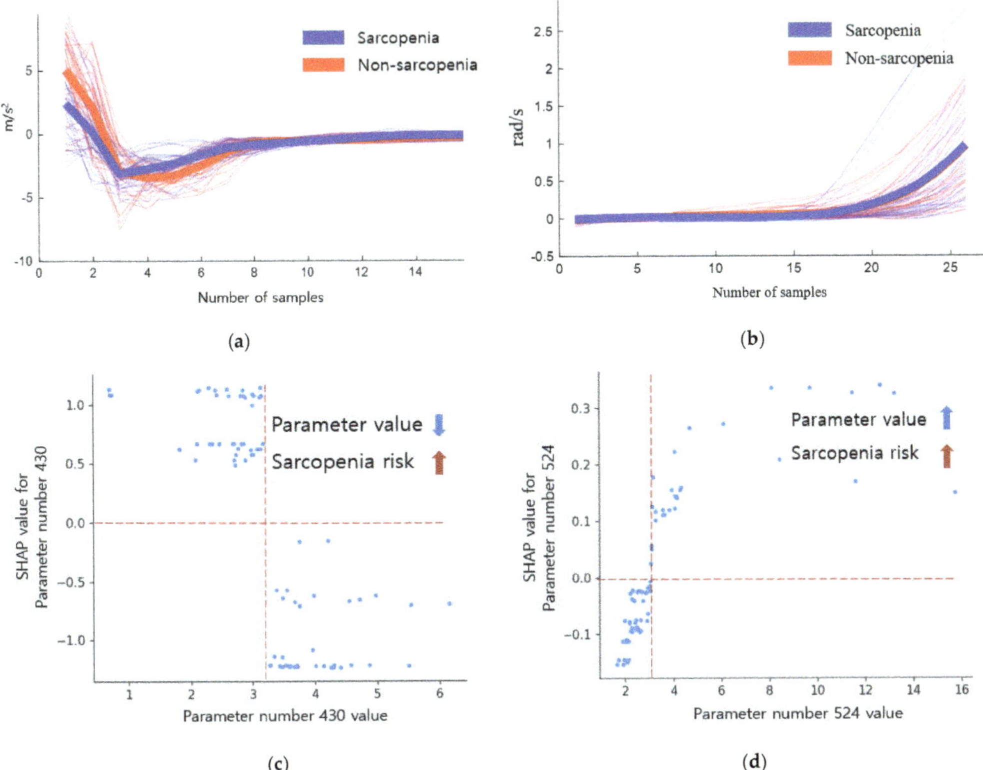

Figure 6. Inertial signals and SHAP dependence plots of descriptive statistical parameters 430 and 524 of sarcopenia. (**a**) Inertial signal 430. (**b**) Inertial signal 524. (**c**) SHAP dependence plot 430. (**d**) SHAP dependence plot 524.

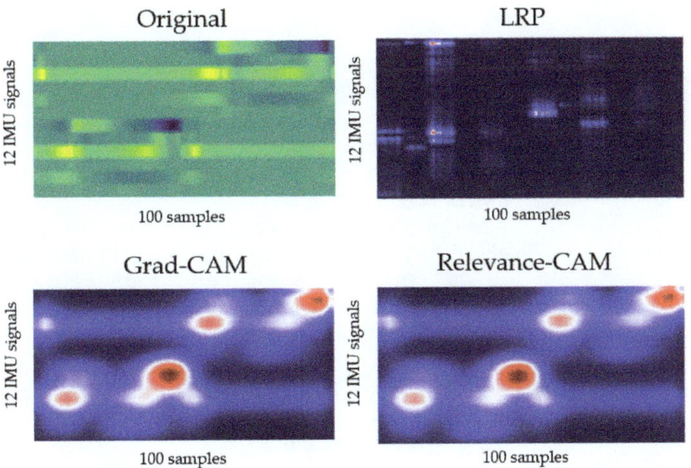

Figure 7. Layer2 result of applying LRP, Grad-CAM, and Relevance-CAM to ResNet50.

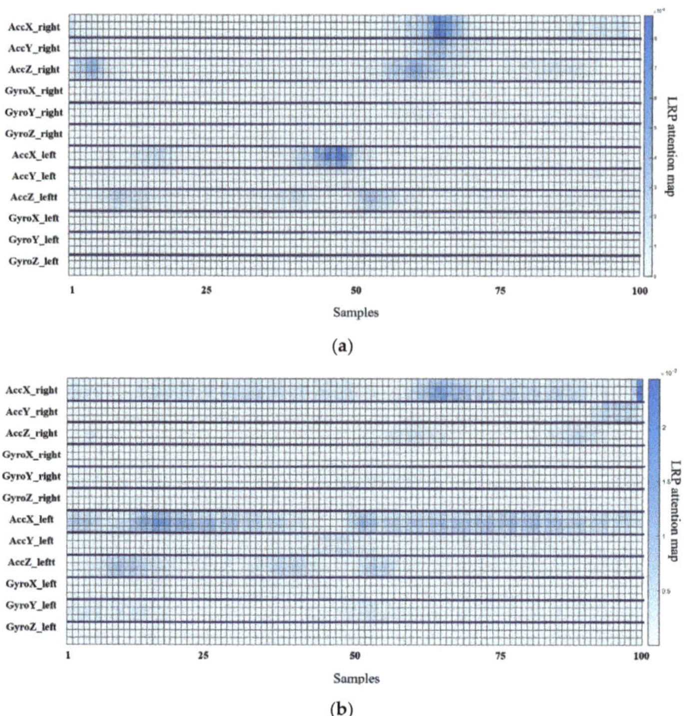

Figure 8. Osteopenia and sarcopenia result of applying LRP to ResNet50. (**a**) LRP result of osteopenia. (**b**) LRP result of sarcopenia.

5. Discussion

The objective of this study was to propose and evaluate a method that can utilize the gait parameters obtained from a wearable device with an inertial sensor in the health management of patients suffering from sarcopenia and osteopenia in daily life.

In the proposed method, the patient was identified using gait phase description-based descriptive statistical parameters as the handcrafted feature-based machine learning input and the original signal of the inertial sensor as the input for the deep learning algorithm. For gait analysis, the identification results were analyzed using XAI tools, such as SHAP and LRP. To verify the proposed gait analysis method, the results of functional tests and questionnaires obtained at the hospital for participants, the results using the existing gait parameters, and the results of the proposed method are discussed.

To identify osteopenia and sarcopenia, a decrease in walking speed and poor body balance has been reported in previous studies. It has been reported that patients with sarcopenia have a slower walking speed than those with osteopenia. The result of the 3 m TUG was 11.71 s in the sarcopenia group and 10.96 s in the osteopenia group, indicating that the walking speed was slower in the patients with sarcopenia. Except for TUG, statistical significance was not obtained for the MMSE, MFS, SARC-F questionnaire, or BBS.

In gait analysis using inertial sensors, spatial–temporal parameters have traditionally been used as tools to conveniently identify diseases such as Faller, PD, and THA in everyday life. In this study, to identify patients with osteopenia and sarcopenia, 24 spatial–temporal parameters used for conventional disease identification were detected, and descriptive statistical parameters were detected to analyze the inertial sensor signals according to the gait phase. Statistical significance was obtained for the stance phase, double support phase, and single support phase percent in osteopenia, and SI in the stance and swing phase in sarcopenia.

With gait analysis using XAI, SHAP demonstrates the importance of parameters and the positive/negative contribution of parameters to the classification results of machine learning. To apply SHAP to machine learning classifiers, it is necessary to obtain a high machine learning accuracy. Good classification results were obtained for osteopenia in XGBoost and sarcopenia in RF. It has been reported that XGBoost has the advantage of being the most accurate among tree-based classifiers, and RF has a strong advantage in terms of overfitting. Comparing various machine learning results, osteopenia showed an accuracy of lower than 70% and sarcopenia showed overfitting. Overfitting in sarcopenia was inferred from the results of the deep learning model.

From the SHAP results of spatial–temporal parameters, single support phase percent left, double support first phase percent left, double support first phase percent right, double support first phase time left, and stance phase percent right were highly important for the osteopenia group. In the sarcopenia group, double support first phase time left, stance phase percent left, stance phase time SI, double support first phase percent left, and stance phase percent right were found to be of high importance. The important parameters obtained similar results to the statistical analysis; in osteopenia, the phase had a high contribution, whereas in sarcopenia, SI had a high contribution. Double support first phase time left and stance phase percent right showed a high contribution in both groups. The double support first phase time decreased in the osteopenia group compared with that in the sarcopenia group, and the stance phase percentage decreased. An increase in the double support first phase time and an increase in the stance phase percentage indicate a decrease in the walking speed. The time-related parameters were lower in the osteopenia group than in the non-osteopenia group; therefore, it is difficult to identify the osteopenia group based on walking speed. Sarcopenia significantly contributed to the reductions in walking speed and balance parameters, as in the previous study results [1–3].

The accuracy of identification of osteopenia patients with spatial–temporal parameters, which is the existing gait analysis parameter, was lower than 70%; thus, it was difficult to analyze the results in SHAP. The inertial sensor had a high temporal resolution; therefore, it was possible to obtain differences between groups by segmenting and analyzing the gait. As a result of SHAP for 840 descriptive statistical parameters, a high contribution from the skewness of the x-axis of the accelerometer in the initial swing phase (247) and the absolute sum of values of gyro z in the mid stance (114) were observed for osteopenia. For sarcopenia, the maximum difference between two successive values of accelerometer x in loading response (430) and the absolute sum of values of gyro y in the mid stance (524) showed a high contribution.

The descriptive statistical parameter 247 for osteopenia was smaller than that for sarcopenia, and the low skewness of osteopenia is due to the rapid occurrence of the maximum value of the acceleration x-axis after TO and a large negative value. This result is related to the increase in the swing phase time. The descriptive statistical parameter 430 represents the change in acceleration, and its value of osteopenia is larger than that of the sarcopenia group, implying that the gait speed of the sarcopenia group is slow because the acceleration x is the walking direction. Parameters 114 and 524 are the absolute sum of the gyro values in the mid stance. The absolute sum decreases in osteopenia indicating less foot movement in the mid stance, as shown in Table 14.

The interpretation of results for LRP-based deep learning was similar to the results of descriptive statistical parameter analysis based on SHAP. However, the reliability was low due to the change in the attention map of the inertial sensor according to the learning results of deep learning as a result of repeated experiments and low identification accuracy. If high-accuracy identification results are obtained, it is expected that the inertial signal characteristics of osteopenia and sarcopenia can be obtained using deep learning.

Functional tests and questionnaires conducted in the hospital were not statistically significant, except for TUG in the sarcopenia group. Spatial–temporal parameters, which have previously been used as gait parameters, showed statistical significance in the sarcopenia group and the osteopenia group, but showed a low identification accuracy of 63% in the osteopenia group. The proposed descriptive statistical parameters obtained an accuracy of 76% or more, and the descriptive statistical parameters attributed similar meanings to the results of the spatial–temporal parameters, had high statistical significance, and can be used as a new clinical tool because the difference in parameter values between the osteopenia and sarcopenia groups is remarkable. Descriptive statistical parameters can be used as useful tools for patient identification and risk detection.

Table 14. Top 2 descriptive statistical parameters of osteopenia and sarcopenia.

Parameter	Osteopenia	Non-Osteopenia	Sarcopenia	Non-Sarcopenia
247	0.126 + 0.425	0.548 + 0.382	0.364 + 0.483	0.327 + 0.534
114	1.892 + 0.86	2.613 + 0.938	2.078 + 1.088	2.217 + 0.591
430	3.292 + 1.05	3.285 + 0.818	2.748 + 0.833	3.797 + 0.813
524	3.317 + 2.098	4.297 + 4.873	4.925 + 3.479	2.403 + 0.473

6. Conclusions

The inertial-sensor-based gait signal was acquired and analyzed for patients with osteopenia and sarcopenia. Spatial–temporal parameters used in conventional clinical evaluation and diagnosis are effective tools for understanding gait. However, they have poor temporal resolution and do not include the function of kinematic signals during the gait cycle. Therefore, the inertial sensor data can obtain descriptive statistical parameters for each gait phase.

For analyzing the patients and control groups, parameters can be statistically analyzed or analyzed through machine-learning-based XAI. To apply XAI, high-accuracy machine learning is required; thus, useful parameters obtained from parameter analysis are used to increase the accuracy of machine learning. Therefore, parameter interpretation is important for patient identification and risk estimation. As a machine learning algorithm, XGBoost for osteopenia and RF for sarcopenia showed high performance, whereas for deep learning, ResNet50, which transfer-learned a human identification model, achieved high accuracy. For the analysis of gait parameters, SHAP was applied to the machine learning model to detect the importance and contribution of the parameters. Unlike Gini and permutation importance, SHAP has advantages of lowering the importance of a parameter when there are similar characteristics between the high-importance parameters. When deep learning identifies patient, the attention map of the inertial sensor signal was analyzed using LRP.

Analyzing the signal of the inertial sensor through XAI, we can diagnose and manage osteopenia and sarcopenia in daily life using a smart insole rather than an expensive clinical tool because the inertial sensor signal contains abundant information on gait. Although the number of participants in this study was extremely small to enable fully understanding the gait characteristics of osteopenia and sarcopenia, the proposed method is effective in analyzing osteopenia and sarcopenia. Therefore, in future studies, additional clinical evaluations will be performed to obtain and analyze many patients and segment data according to sex, age, and dominant leg.

Author Contributions: Conceptualization, J.-K.K., M.-N.B., K.L., J.-C.K. and S.G.H.; Data curation, J.-K.K. and M.-N.B.; Formal analysis, J.-K.K.; Funding acquisition, K.L. and J.-C.K.; Investigation, J.-K.K.; Methodology, J.-K.K. and S.G.H.; Project administration, M.-N.B., K.L. and J.-C.K.; Software, J.-K.K.; Supervision, K.L. and J.-C.K.; Validation, J.-K.K. and S.G.H.; Visualization, J.-K.K.; Writing—original draft, J.-K.K.; Writing—review and editing, J.-K.K., M.-N.B. and S.G.H. All authors have read and agreed to the published version of the manuscript.

Funding: This study was supported by a National Research Foundation of Korea (NRF) grant funded by the Korea government (MSIT) (No. 2021M3I2A1077405) and by a grant (22DRMS-B146826-05) from the Development of Customized Contents Provision Technology for Realistic Disaster Management Based on Spatial Information Program funded by the Ministry of the Interior and Safety of the Korean government.

Institutional Review Board Statement: The study was conducted according to the guidelines of the Declaration of Helsinki, and approved by the Institutional Review Board of Chungnam National University Hospital Institutional (File No: CNUH 2019-06-042).

Informed Consent Statement: Informed consent was obtained from all subjects involved in the study.

Data Availability Statement: The data are not publicly available due to company security policy and personal protection of subjects. Data are available from the authors upon reasonable request.

Conflicts of Interest: The authors declare no conflict of interest.

Appendix A

Table A1. Abbreviations.

Abbreviations	Raw	Abbreviations	Raw
XAI	eXplainable Artificial Intelligence	LDA	Linear Discriminant Analysis
BMD	Bone Mineral Density	NB	Naïve Bayes
SD	Standard Deviation	k-NN	k-Nearest Neighbor
DEXA	Dual-Energy X-ray Absorptiometry	SVM	Support Vector Machines
PD	Parkinson's Diseases	RBF	Radial Basis Function
THA	Total Hip Arthroplasty	DT	Decision Tree
IMU	Inertial Measurement Unit	XGBoost	Extreme Gradient Boosting
HS	Heel Strike	HMM	Hidden Markov Model
TO	Toe Off	RF	Random Forest
LIME	Local Interpretable Model-agnostic Explanations	ANN	Artificial Neural Network
SHAP	SHapley Additive exPlanations	CNN	Convolutional Neural Network
SMI	Skeletal Muscle mass Index	LSTM	Long Short-Term Memory
MMSE	Mini-Mental State Examination	ResNet	Residual neural Network
MFS	Mores Fall Scale	GAP	Global Average Pooling
TUG	Timed Up and Go	FC	Fully Connected
BBS	Berg Balance Scale	LRP	Layer-wise Relevance Propagation
ROM	Range of Motion	CAM	Class Activation Mapping

References

1. Coll, P.P.; Phu, S.; Hajjar, S.H.; Kirk, B.; Duque, G.; Taxel, P. The prevention of osteoporosis and sarcopenia in older adults. *J. Am. Geriatr. Soc.* **2021**, *69*, 1388–1398. [CrossRef] [PubMed]
2. Intriago, M.; Maldonado, G.; Guerrero, R.; Messina, O.D.; Rios, C. Bone Mass Loss and Sarcopenia in Ecuadorian Patients. *J. Aging Res.* **2020**, *2020*, 1072675. [CrossRef] [PubMed]
3. Stephen, W.C.; Janssen, I. Sarcopenic-obesity and cardiovascular disease risk in the elderly. *J. Nutr. Health Aging* **2009**, *13*, 460–466. [CrossRef] [PubMed]
4. Kim, J.-K.; Bae, M.-N.; Lee, K.; Hong, S. Identification of Patients with Sarcopenia Using Gait Parameters Based on Inertial Sensors. *Sensors* **2021**, *21*, 1786. [CrossRef]
5. Teufl, W.; Taetz, B.; Miezal, M.; Lorenz, M.; Pietschmann, J.; Jöllenbeck, T.; Fröhlich, M.; Bleser, G. Towards an Inertial Sensor-Based Wearable Feedback System for Patients after Total Hip Arthroplasty: Validity and Applicability for Gait Classification with Gait Kinematics-Based Features. *Sensors* **2019**, *19*, 5006. [CrossRef]
6. Tatangelo, G.; Watts, J.; Lim, K.; Connaughton, C.; Abimanyi-Ochom, J.; Borgström, F.; Nicholson, G.C.; Shore-Lorenti, C.; Stuart, A.L.; Iuliano-Burns, S.; et al. The cost of osteoporosis, osteopenia, and associated fractures in Australia in 2017. *J. Bone Miner. Res.* **2019**, *34*, 616–625. [CrossRef]
7. Kanis, J.A. Assessment of fracture risk and its application to screening for postmenopausal osteoporosis: Synopsis of a WHO report. *Osteoporos. Int.* **1994**, *4*, 368–381. [CrossRef]
8. Cruz-Jentoft, A.J.; Baeyens, J.P.; Bauer, J.M.; Boirie, Y.; Cederholm, T.; Landi, F.; Martin, F.C.; Michel, J.-P.; Rolland, Y.; Schneider, S.M. Sarcopenia: European consensus on definition and diagnosisReport of the European Working Group on Sarcopenia in Older People. *Age Ageing* **2010**, *39*, 412–423. [CrossRef]

9. Caramia, C.; Torricelli, D.; Schmid, M.; Muñoz-Gonzalez, A.; Gonzalez-Vargas, J.; Grandas, F.; Pons, J.L. IMU-based classification of Parkinson's disease from gait: A sensitivity analysis on sensor location and feature selection. *IEEE J. Biomed. Health Inform.* **2018**, *22*, 1765–1774. [CrossRef]

10. Eskofier, B.M.; Lee, S.I.; Daneault, J.F.; Golabchi, F.N.; Ferreira-Carvalho, G.; Vergara-Diaz, G.; Sapienza, S.; Costante, G.; Klucken, J.; Kautz, T.; et al. Recent machine learning advancements in sensor-based mobility analysis: Deep learning for Parkinson's disease assessment. In Proceedings of the 2016 38th Annual International Conference of the IEEE Engineering in Medicine and Biology Society (EMBC), Orlando, FL, USA, 16–20 August 2016; pp. 655–658.

11. Howcroft, J.; Kofman, J.; Lemaire, E. Prospective Fall-Risk Prediction Models for Older Adults Based on Wearable Sensors. *IEEE Trans. Neural Syst. Rehabil. Eng.* **2017**, *25*, 1812–1820. [CrossRef]

12. Tunca, C.; Salur, G.; Ersoy, C. Deep Learning for Fall Risk Assessment with Inertial Sensors: Utilizing Domain Knowledge in Spatio-Temporal Gait Parameters. *IEEE J. Biomed. Health Inform.* **2020**, *24*, 1994–2005. [CrossRef] [PubMed]

13. Dindorf, C.; Teufl, W.; Taetz, B.; Bleser, G.; Fröhlich, M. Interpretability of Input Representations for Gait Classification in Patients after Total Hip Arthroplasty. *Sensors* **2020**, *20*, 4385. [CrossRef] [PubMed]

14. Ren, L.; Peng, Y. Research of Fall Detection and Fall Prevention Technologies: A Systematic Review. *IEEE Access* **2019**, *7*, 77702–77722. [CrossRef]

15. Choi, W.; Choi, J.H.; Chung, C.Y.; Sung, K.H.; Lee, K.M. Can gait kinetic data predict femoral bone mineral density in elderly men and women aged 50 years and older? *J. Biomech.* **2021**, *123*, 110520. [CrossRef] [PubMed]

16. ElDeeb, A.M.; Khodair, A.S. Three-dimensional analysis of gait in postmenopausal women with low bone mineral density. *J. Neuroeng. Rehabil.* **2014**, *11*, 55. [CrossRef]

17. Sung, P.S. Increased double limb support times during walking in right limb dominant healthy older adults with low bone density. *Gait Posture* **2018**, *63*, 145–149. [CrossRef]

18. Gunning, D.; Aha, D. DARPA's Explainable Artificial Intelligence (XAI) Program. *AI Mag.* **2019**, *40*, 44–58. [CrossRef]

19. Parsa, A.B.; Movahedi, A.; Taghipour, H.; Derrible, S.; Mohammadian, A. (Kouros) toward safer highways, application of XGBoost and SHAP for real-time accident detection and feature analysis. *Accid. Anal. Prev.* **2020**, *136*, 105405. [CrossRef]

20. Taborri, J.; Palermo, E.; Rossi, S.; Cappa, P. Gait Partitioning Methods: A Systematic Review. *Sensors* **2016**, *16*, 66. [CrossRef]

21. Whittle, M.W. *Gait Analysis: An Introduction*; Butterworth-Heinemann: Oxford, UK, 2014.

22. Kim, J.; Bae, M.-N.; Lee, K.B.; Hong, S.G. Gait event detection algorithm based on smart insoles. *ETRI J.* **2019**, *42*, 46–53. [CrossRef]

23. Nilsson, J.-O.; Gupta, A.K.; Handel, P. Foot-mounted inertial navigation made easy. In Proceedings of the 2014 International Conference on Indoor Positioning and Indoor Navigation (IPIN), Busan, Korea, 27–30 October 2014; pp. 24–29.

24. Skog, I.; Nilsson, J.-O.; Handel, P. Pedestrian tracking using an IMU array. In Proceedings of the 2014 IEEE International Conference on Electronics, Computing and Communication Technologies (CONECCT), Bangalore, India, 6–7 January 2014; pp. 1–4.

25. Bach, S.; Binder, A.; Montavon, G.; Klauschen, F.; Müller, K.-R.; Samek, W. On Pixel-Wise Explanations for Non-Linear Classifier Decisions by Layer-Wise Relevance Propagation. *PLoS ONE* **2015**, *10*, e0130140. [CrossRef] [PubMed]

26. Zhou, B.; Khosla, A.; Lapedriza, A.; Oliva, A.; Torralba, A. Learning deep features for discriminative localization. In Proceedings of the IEEE Conference on Computer Vision and Pattern Recognition, Las Vegas, NV, USA, 27–30 June 2016; pp. 2921–2929.

27. Lee, J.R.; Kim, S.; Park, I.; Eo, T.; Hwang, D. Relevance-CAM: Your Model Already Knows Where to Look. In Proceedings of the 2021 IEEE/CVF Conference on Computer Vision and Pattern Recognition (CVPR), Nashville, TN, USA, 20–25 June 2021; pp. 14939–14948.

28. Kang, Y.; Na, D.L.; Hahn, S. A validity study on the Korean Mini-Mental State Examination (K-MMSE) in dementia patients. *J. Korean Neurol. Assoc.* **1997**, *15*, 300–308.

29. Baek, S.; Piao, J.; Jin, Y.J.; Lee, S.-M. Validity of the Morse Fall Scale implemented in an electronic medical record system. *J. Clin. Nurs.* **2014**, *23*, 2434–2441. [CrossRef] [PubMed]

30. Kim, S.; Kim, M.; Won, C.W. Validation of the Korean version of the SARC-F questionnaire to assess sarcopenia: Korean frailty and aging cohort study. *J. Am. Med. Dir. Assoc.* **2018**, *19*, 40–45. [CrossRef]

31. Berg, K.; Wood-Dauphine, S.; Williams, J.I.; Gayton, D. Measuring balance in the elderly: Preliminary development of an instrument. *Physiother. Can.* **1989**, *41*, 304–311. [CrossRef]

32. Podsiadlo, D.; Richardson, S. The timed "Up & Go": A test of basic functional mobility for frail elderly persons. *J. Am. Geriatr. Soc.* **1991**, *39*, 142–148.

33. Zhang, D.; Qian, L.; Mao, B.; Huang, C.; Huang, B.; Si, Y. A Data-Driven Design for Fault Detection of Wind Turbines Using Random Forests and XGboost. *IEEE Access* **2018**, *6*, 21020–21031. [CrossRef]

34. Duan, H.; Zhao, Y.; Chen, K.; Shao, D.; Lin, D.; Dai, B. Revisiting Skeleton-based Action Recognition. *arXiv* **2021**, arXiv:2104.13586.

35. Wu, Z.; Shen, C.; Van Den Hengel, A. Wider or deeper: Revisiting the resnet model for visual recognition. *Pattern Recog.* **2019**, *90*, 119–133. [CrossRef]

36. Wan, Z.; Yang, R.; Huang, M.; Zeng, N.; Liu, X. A review on transfer learning in EEG signal analysis. *Neurocomputing* **2021**, *421*, 1–14. [CrossRef]

Article

A Study on Dictionary Selection in Compressive Sensing for ECG Signals Compression and Classification

Monica Fira [1], Hariton-Nicolae Costin [1,*] and Liviu Goraş [1,2]

[1] Institute of Computer Science, Romanian Academy, 700481 Iasi, Romania; monica.fira@iit.academiaromana-is.ro (M.F.); lgoras@etti.tuiasi.ro (L.G.)

[2] Faculty of Electronics, Telecomunications & Information Technology, Gheorghe Asachi Technical University of Iasi, 700050 Iasi, Romania

[*] Correspondence: hariton.costin@iit.academiaromana-is.ro

Abstract: The paper proposes a comparative analysis of the projection matrices and dictionaries used for compressive sensing (CS) of electrocardiographic signals (ECG), highlighting the compromises between the complexity of preprocessing and the accuracy of reconstruction. Starting from the basic notions of CS theory, this paper proposes the construction of dictionaries (constructed directly by cardiac patterns with R-waves, centered or not-centered) specific to the application and the results of their testing. Several types of projection matrices are also analyzed and discussed. The reconstructed signals are analyzed quantitatively and qualitatively by standard distortion measures and by the classification of the reconstructed signals. We used a k-nearest neighbors (KNN) classifier to evaluate the reconstructed models. The KNN module was trained with the models from the mega-dictionary used in the classification block and tested with the models reconstructed with class-specific dictionaries. In addition to the KNN classifier, a neural network was used to test the reconstructed signals. The neural network was a multilayer perceptron (MLP). Moreover, the results are compared with those obtained with other compression methods, and ours proved to be superior.

Keywords: compressed sensing; ECG signal; reconstruction dictionaries; projection matrices; signal classifications

Citation: Fira, M.; Costin, H.-N.; Goraş, L. A Study on Dictionary Selection in Compressive Sensing for ECG Signals Compression and Classification. *Biosensors* **2022**, *12*, 146. https://doi.org/10.3390/bios12030146

Received: 28 December 2021
Accepted: 24 February 2022
Published: 27 February 2022

Publisher's Note: MDPI stays neutral with regard to jurisdictional claims in published maps and institutional affiliations.

1. Introduction

Compressed sensing (CS) is a method of signals acquisition and processing based on the fact that sparse or rare signals can be reconstructed from a relatively small number of projections on a set of random signals [1]. This technique is relatively new compared to classical techniques, so in recent years, a large number of papers on implementation, applicability, advantages and the pertinence to dedicated types of signals have been published [2–12].

Many of the papers that address CS focus on how to build specific dictionaries for signal reconstruction [13–26]. In the case of the ECG signal, due to its particularities, namely, the quasi-periodicity of the P, Q, R and S waves and the preservation of their shapes, many of the methods proposed in the literature focus on the advantages offered by these features specific to the ECG signal [27–37]. Thus, a large part of the methods proposed regarding CS of ECG signals aim at building dictionaries specific to these signals. In many cases, building these dictionaries involves a preprocessing step with or without signal segmentation, with or without QRS wave alignment. Another aspect regarding CS applied to ECG signals is the optimization of the compression matrix.

In the following lines, we will briefly present some specific ECG methods proposed in the literature over the past years, which contain results similar to the methods we presented in this paper, except for the fact of using patient-specific dictionaries or involving updating the dictionary when there are changes in the ECG signal. In general, there is a big inconvenience in the situation of using such a system in practice, because it involves resubmitting the dictionary and necessary calculations in real time to see if the dictionary is good or needs to be updated. All these calculations imply additional hardware needs, which can make the method less practical in real-time acquisition situations. On the other hand, our approach is based on the use of non-modified patient-specific dictionaries or pathology-specific dictionaries; these are established once and updating can be done less frequently than in other techniques and does not require real-time decisions.

In one paper [33], the presented method uses an over complete wavelet dictionary, a dictionary that is later reduced due to a training phase. In addition, it is proposed to align the beats according to the position of the R-peak. This alignment aims to exploit the different scaling characteristics of ECG waves in the wavelet dictionary optimization process. Three different methods are tested for dictionary optimization. It should be mentioned that this optimized dictionary is specific to the patient and for its construction, the first 5 minutes of registration are taken. For acquisition, the authors use a matrix optimized for the ECG signal to be acquired through CS. The use of an optimized compression matrix leads to improved results, but has the disadvantage that once this matrix is changed it must be sent together with the compressed ECG signal. That means both the compressed ECG signal and the compression matrix must be sent to restore the ECG signal.

Another approach is presented in [34], where the quasi-periodic character of the ECG signal is used to detect similarities between ECG pulses and to transmit segments that show dissimilarities normally, without compression. This approach is proposed because abnormal frames, which could be signs of heart disease, are not similar to normal frames. Thus, only the ECG segments considered normal are transmitted by CS, the rest being transmitted normally. Once it is determined, whether the heartbeat is acquired normally or by CS, a quantization step follows and then a Huffman compression. These two steps lead to improved compression results. A critical point in the method is the correct detection of normal vs. abnormal beats, because this automated detection is debatable in the light of the fact that normality or abnormality is determined by a cardiologist and the accuracy of the acquisition should not be influenced by this decision.

In paper [35], the authors also used CS associated to dictionaries built specifically for the ECG signal, thus using the dictionary learning technique to construct a better sparsifying basis to improve the compression ratio. Moreover, the authors consider the change of ECG signal characteristics and propose a physiological variation detection technique and a low-complexity dictionary refreshing algorithm to update the dictionary from time to time when the current dictionary is no longer suitable for the patient.

Many papers in the CS field focus on optimizing the measurement matrix, i.e., the matrix is used in the acquisition stage or on optimizing the necessary calculations in this stage by arranging this matrix in a way that allows easy hardware implementation of the necessary calculations. In practical implementations, the simple random or Bernoulli matrix may have the inconvenience of the required number of operations. Thus, in paper [36], the authors propose an optimized algorithm for collecting the compressed ECG signal, based on the proposed optimization of a deterministic binary block diagonal matrix. The blocks, which make up the diagonal of the matrix, are identical and contain $m = N/M$ elements each, where M and N, respectively, represent the number of rows and columns.

In paper [37], a new method of compressive sampling of ECG signals is presented, which is based on the idea of building the compression matrix adapted to the frame of the ECG signal to be compressed. Thus, a circulating matrix is proposed, containing zeros and ones, obtained by quantizing (with 1-bit resolution) the size of the ECG signal. The detection matrix adapted in this way guarantees that the significant portions of the waveform of the compressed ECG signal are in fact contained in the compressed version. In this way, a more precise reconstruction is guaranteed in relation to the methods already available in the literature. For the reconstruction stage, the acquisition matrix is then used in combination with a modified wavelet dictionary, which also allows the reconstruction of the signal deviation for each processed frame. The big disadvantage of the method is that whenever the acquisition matrix has to be updated, it has to be sent to the receivers and for reconstruction we have to know each frame with which matrix was collected.

In this paper, we propose a detailed comparative study of two different approaches regarding the possibility of compressed sensing specific ECG signals. This study considers several acquisition techniques/projection matrices used in the acquisition stage and several dictionaries used in the ECG signal reconstruction stage. We will also analyze the effect of preprocessing on the results.

Broadly speaking, we analyze and discuss two CS approaches dedicated to ECG signals, namely:

1. An approach that is based on the direct CS obtaining of the signal, without preprocessing it prior acquiring the projections. This "genuine" CS we call patient-specific classical compressed sensing (PSCCS), since the dictionary is constructed from a patient's initial signals.
2. A variant that involves a module of pre-processing and segmentation of the ECG signal. This stage aims at improving the scatter and recoverability of the ECG signal. In this additional stage of preprocessing, the ECG signal results in the rhythmicization of the ECG signal and divides it into cardiac cycles—hereinafter referred to as cardiac patterns compressed sensing (CPCS). Now the acquired signals and atoms of dictionary are segmented heartbeats pre-processed without or with the R-waves centered.

For both approaches from above, we will analyze several projection matrices, namely, matrices with random independent and identically distributed (i.i.d.) elements taken from the Gaussian or Bernoulli distribution and project matrices optimized for the particular dictionary used in the reconstruction. To optimize the projection matrix, the method presented in [7] will be used.

Furthermore, we will pay special attention to the way the dictionary is built. We will also present the advantages and disadvantages of each and the choice of the method that depends on the available hardware and software resources.

The paper is organized as follows: Section 2 is dedicated to the types of sampling vectors, projection matrices and dictionary construction methods. Section 3 presents the CS methods dedicated to ECG signals. Section 4 shows the results obtained. In Section 5 the results from the previous section are compared and in Section 6 conclusions are drawn.

2. Compressed Sensed Overview

Traditionally, signals are acquired according to the sampling theorem [8] that states that an f_0-bandlimited signal can be recovered from its samples if the sampling frequency is at least 2 f_0, i.e., twice the highest frequency of the signal spectrum. Thus, in a time window W, an f_0-bandlimited analog signal can be represented by N = 2 f_0 W samples equally spaced at T = 1/2 f_0, i.e., as a vector belonging to the space R^N. Such a signal can be alternatively defined by using any complete set of orthogonal functions in R^N. In fact, sampling is nothing else than taking projections (scalar products) on the elements of the canonical basis. In the general case the signal can be reconstructed from its projections on N orthogonal (or only linear independent) elements in RN the canonical basis being the most frequently used. However, in practice, there are cases in which a signal can be reconstructed from fewer samples or projections on an appropriate set of signals, compared to the number prescribed by the sampling theorem. This is possible since the samples contain unnecessary information, and thus, these signals can be compressed and recovered using projections and previous known information. An example would be the class of sparse or rare signals [9–11] that allow a representation based on a small number of elements/atoms in RN. In signal processing literature, the name "k-sparse" denotes signals that can be reconstructed by means of k of elements of RN, the most significant situation being that in which $k \ll N$. A discrete signal or vector $x \in R^N$ is k-sparse if there exists a base $\Psi = \{\Psi_i, i = 1, \ldots, N\}$ in R^N so that most of the elements $\alpha = \{\alpha_i, i = 1, \ldots, N\}$ of its representation in that basis, $x = \Psi\alpha$, are zero. Alternatively, they can be approximately zero, so that the signal can be represented accurate enough with the k's largest terms α_i from its expansion with respect to that basis. The CS concept is based on theory that a k-sparse signal, i.e., a signal that can be compressed into a base (or, more general, dictionary) Ψ can be recuperated with very good quality from a number m of the order of scale $m = O(k \log(N/k))$ of non-adaptive linear projections on a set of vectors Φ, which are not comprehensible with the first, i.e., their elements cannot be used for a compressed representation of any $\Psi_i, i = 1, \ldots, N$. Therefore, for obtaining the measurement signal instead of measuring the N components of the signal in the canonic base, a number of m $(k < m << N)$ linear projections on the elements of the matrix Φ^{N*m} are acquired:

$$y = \Phi x = \Phi \Psi \alpha = \Theta \alpha \tag{1}$$

where the measurement noise was not taken into account. If we use as a projection matrix (noted with Φ) a matrix with dimensions mxN, with $m < N$, then it means that we will make a number of m measurements, each measurement of size N. That is, the vectors on which x is projected represent the rows of the projection matrix.

The main idea regarding (1) is that, because $m < N$, the rebuilding of the original signal cannot be realized, but only under the compressibility hypothesis. It has been shown that if Φ and Ψ satisfy certain conditions, the original vector α can be obtained as the unique result to the optimization problem:

$$\hat{\alpha} = \text{argmin} \|\alpha\|_{l_0} \text{ subject to } y = \Phi \Psi \alpha, \tag{2}$$

where l_0 is the (pseudo)norm consisting of the number of nonzero entries of α.

The reconstructed signal has the form:

$$\hat{x} = \Psi \hat{\alpha} \tag{3}$$

corresponding to the sparsest representation of y in terms of the dictionary $\Phi \Psi$. To circumvent the problems of combinatorial nature and noise effect in the case of almost sparse signals, two directions evolved:

(i)   seeking for a suboptimal solution of problem (2) and
(ii) using the Basis Pursuit (BP) procedure [1] that consists of replacing l_0 with l_1 minimization, by resolving problem (4) instead of the initial one:

$$\hat{\alpha} = \text{arg}_\alpha \min \|\alpha\|_{l1} \text{ subject to } y = \Phi \Psi \alpha \tag{4}$$

Let us stress the fact that although pure sparse signals (built of exactly k<<N atoms from a specified dictionary) are difficult to find, conventional results are valid for signals that are "almost sparse" (which can be built of k<<N non-negligible atoms) with respect to dictionaries that can be overcomplete (contain more atoms than their intrinsic dimension), as in the case of some classes of biomedical signals. Taking into consideration this fact, it has been found useful to adapt the theory of CS to the field of processing ECG and electroencephalographic (EEG) signals [2–4] as well as for applications [5] such as compression, transmission, reconstruction of ECG signals, ECG filtering and monitoring [6,27,30–32].

For a better understanding of the algorithm, in the following we present a pseudocode summary.

INPUTS: ORIGINAL SIGNAL = x

Acquisition Stage:

Step 1: Compute random measurements

$y = \Phi x$, where Φ is a MxN matrix of random independent and identically distributed (i.i.d.) entries.

Reconstruction Stage:

Step 2: Compute α coefficients using L1 minimization

$\hat{\alpha} = \text{arg}_\alpha \min \|\alpha\|_{l1} \text{ subject to } y = \Phi \Psi \alpha$

Step 3: Reconstruct original signal x'
$x' = \alpha \Psi$
OUTPUTS: RECONSTRUCTED SIGNAL = x'

3. Sample Vectors, Projection Matrices and Dictionaries

Here, we briefly show several ways of segmentation, a couple of projection matrices, as well as some several ways of building various types of dictionaries specific to the ECG signal. Depending on the chosen CS method, the way of building the dictionary which is used to reconstruct the ECG signal is different.

3.1. Sample Vectors

First of all, let us mention that we will refer to ECG signals with a sampling frequency according to the Nyquist–Shannon sampling theorem of 360 Hz and 300 (or 301 for case with R-wave centered) samples/vector, respectively. Each vector is projected on a number of random vectors with identic size and the obtained values are utilized for recovering through a dictionary.

In the simplest way, the first 300 samples of the ECG signal set up the first vector; then, the succeeding 300 samples form the second vector, etc. The place of the R-wave can be anywhere in a vector or it may be missing sometimes, which is, obviously, not desirable.

In order to take advantage of the cyclicity of the ECG signal and of the changes produced on the ECG signal in case of some diseases, we proposed some modified acquisition techniques that requires preprocessing [13–16]. Thus, samples of the ECG signal are stored in a buffer zone and a series of preprocessing can be performed on these stored signals. The R-waves can be detected, and based on them, the ECG signal can be segmented into cardiac patterns. A cardiac pattern is delimited by the halves of the RR intervals of two adjacent intervals and re-sampled by interpolation so that the pattern has a fixed number of 301 samples. The above segmentation and preprocessing technique contain simple calculations and, as will be presented, notably increases results for the compression and reconstruction processes.

Starting from the method described above, an improvement of the cardiac model can be obtained by centering the R-wave on sample 151. Thus, a resampling to the left of the R-wave will be performed and another resampling to the right of the R-wave and the final cardiac model will have 301 samples with the R-wave centered. This alignment of the R-wave can be a reversible process, provided that the reduction/stretch ratio from left to right is known. To make a much clearer picture of the re-sampling and alignment effect, we provide in the following examples of unfocused (misaligned) heartbeats and the same cycles prepared to be aligned. These segments constitute atoms in the dictionary or preprocessed sample vectors.

Figure 1 shows examples of cardiac models with and without a centered R-wave.

In conclusion, the sampling vectors and the atoms of the dictionary can be: (i) unprocessed or pre-processed through segmentation and resampling or (ii) segmented and resampled with a centered R-wave.

3.2. Projection Matrices

A key element in the CS method is the projection matrix for the acquisition of the ECG signal. The reconstruction quality of the ECG signal is considerably decided by the kind of the matrix used in the compression stage [7,9,10,13].

Moreover, the number of random vectors (and respectively the number of calculated scalar products) considered is based on the tolerated tradeoff between the compression ratio and the reconstruction error: thus, the compression ratio is directly related to the reconstruction error.

In Section 3, we will analyze and determine which is a stop ratio and we will determine in the case of our ECG signals how many projections we need for a good ECG compression.

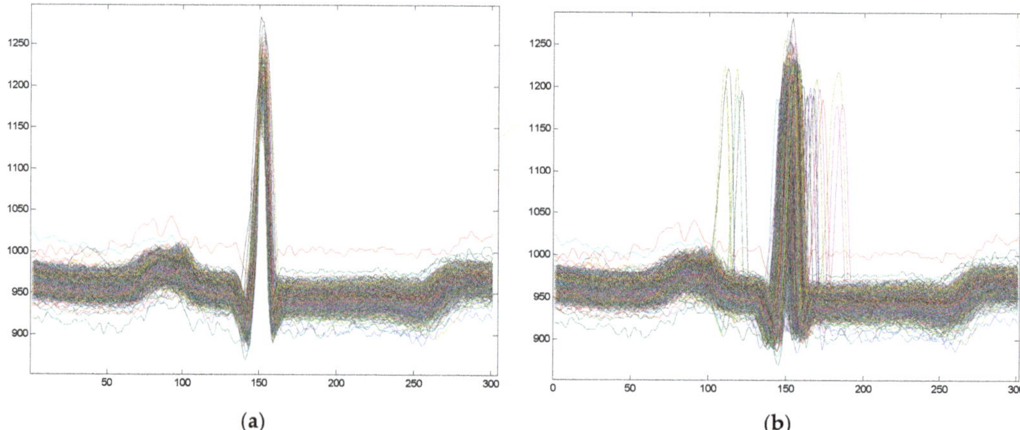

Figure 1. Examples of cardiac patterns obtained by centered or non-centered R-wave: (**a**) Cardiac patterns with a centered R-wave; (**b**) Cardiac patterns without a centered R-wave.

In the following, we use and discuss three types of projection matrices.

- As so far shown in Introduction and CS theory, projecting on a matrix Φ results in a system. A simple approach is to use as Φ a random matrix with i.i.d. normal elements. Nevertheless, this matrix has a higher Restricted Isometry Property (RIP) constant and, thus, it is inappropriate for reconstruction [7].
- Another possibility is to build a projection matrix specific to the dictionary used in the reconstruction phase. Thus, we can define such a matrix as a product of the random matrix and the transposition of a square matrix containing an arbitrary selection of N dictionary atoms [7]. In this way, the reconstruction errors will be smaller. In the tables with results, we denote this matrix with "Random * Dict †".
- A third possibility of projection matrix analyzed in this paper is the Bernoulli type matrix built only of elements of 0s and 1s, with symmetric distribution (half of the inputs of a row are created with the Bernoulli distribution and the other half reversing the first half) [14]. The advantage of this matrix is the low computational complexity, and thus, saving of IT resources.

In this paper, we examine the consequence of these three types of projection matrices on various dictionaries.

3.3. Dictionaries

Using standard Discrete Cosinus/Sinus Transform (DCT/DST), Wavelet or other typical dictionaries is not always the best choice if we are referring to ECG signal reconstruction errors [15]. Thus, we will analyze the use of dictionaries dedicated to ECG signals, dictionaries that can be specific to the patient, specific to the pathology or universal. The way dictionaries are built is closely related to the segmentation methods of the ECG signals presented above. Thus, concerning the preprocessing stage, we used dictionaries with three types of atoms: (1) Unprocessed (patient-specific only) and processed atoms; (2) Segmented atoms; (3) Segmented plus R-wave centered. The last two types contain either patient-specific beats, or normal beats and/or seven types of pathological beats.

3.3.1. Patient-Specific Dictionaries

In order to build patient-specific dictionaries, we used the first minutes of each patient's record and then the rest of the ECG signal was used for testing. Thus, the atoms represent ECG segments of size 300, successive segments of vectors, without any processing. In our studies, such dictionaries were constructed (only) from the first few minutes of the patient's records (patient-specific dictionary), the atoms being further used for CS with various projection matrices.

In order to maintain uniformity in the size of the dictionaries, we chose to build patient-specific dictionaries of 700 atoms, each atom having a size of 300. A size of 300 for atoms was determined considering the sampling frequency (360 Hz) and the average beat frequency heart rate (~70 beats/min for normal patients). In this way, the dictionary is actually a matrix with a size of 300 × 700. We highlight that the atoms of the dictionary were aleatory sequences of the ECG recording, and therefore, the R-wave can appear anywhere in the 300 samples or even be missing (not a happy case).

We note that besides the simplicity of the ECG signal segmentation method, another advantage is the capture of the specificity of the patient's ECG particularities in the moment the recording has started.

An improved version of the method is to preprocess the ECG segments to build the dictionary. Thus, segmentation can be performed by detecting heartbeats (i.e., R-waves) and then the R-wave centers. Therefore, patient-specific dictionaries can be constructed without or with preprocessing for R-wave alignment. However, in all cases, the first portion of an ECG recording is used to construct the dictionary, while the rest of the signal (the unused part in the dictionary) was used in the testing techniques.

The next two types of dictionaries contain only atoms obtained through segmentation, normalized to 301 elements with or without a centered R-wave.

3.3.2. Universal Mega-Dictionaries

The mega-dictionary used consists of 1472 atoms (i.e., 184 beats from each of the 8 classes discussed, 7 pathological and the normal beat class). Depending on the preprocessing tested, the atoms of the dictionary may or may not have a centered R-wave.

3.3.3. Pathology-Specific Dictionaries

When the reconstruction stage considers the pathologic class that the cardiac beat belongs to, a particular or specific dictionary has been constructed for each pathological class. Because the ECG recordings include heartbeats from several pathological classes, we tested the variant in which, for each pathological class, we made a specific dictionary. Thus, analyzing 7 pathological classes and the normal class, we built 8 dictionaries, each with 700 atoms specific to each class. Atoms may or may not have a centered R-wave. Thus, we note that the number of atoms in each of the dictionaries is higher than the number of atoms related to a certain pathology contained in the mega-dictionary.

4. Proposed Methods for Dictionary-Based ECG Compression

In the Introduction, we talked about the presentation of two totally different methods of CS specific to ECG signals, but both have in common the need of building specific dictionaries. However, the use of ECG signal characteristics and how to build dictionaries differ remarkably.

Thus, the PSCCS method is based on ECG signal specific features of each patient, while CPCS on the cyclical patterns of the heartbeat.

In the next subsection, we present two methods for CS of ECG signals with some dissimilarity associated to the projection matrices.

4.1. Patient-Specific Classical Compressed Sensing—PSCCS

A first variant of compressed acquisition of the ECG signal is presented in Figure 2. It can be implemented even on hardware system and involves the compressed collection of the ECG signal using the CS technique and a patient-specific dictionary together with the Basis Pursuit technique [14].

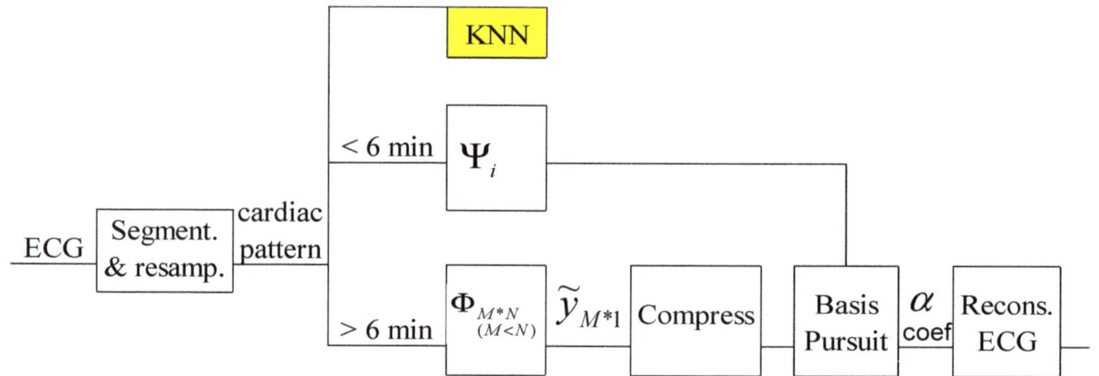

Figure 2. Principle of the PSCCS method.

In this method, the compression of the ECG signal involves the classic use of the CS technique, without any additional signal processing. The advantage of the method is that it speculates on the specific features of the patient. Another advantage is the reduced complexity equal to that of the traditional CS algorithm. The particularity of this procedure is the need for a classic 6-minute ECG acquisition to build the dictionary. In order to obtain improved results, the dictionary can be upgraded in case of long recordings or in case the patient has undergone changes on the ECG signal from one recording to another.

4.2. Cardiac Patterns Compressed Sensing—CPCS

Below, we present a different approach from the classic CS, which involves a preprocessing stage used both for segmentation of the ECG signal for compressed acquisition and for building useful dictionaries in the signal reconstruction stage.

Figure 3 shows the block diagram of the method. As we can see, at the level of the reconstruction stage there are two approaches, namely, a way of reconstruction using a mega-dictionary or another variant in which dictionaries specific to pathologies are used. The first two operations are common to both approaches and are colored in yellow in the block diagram.

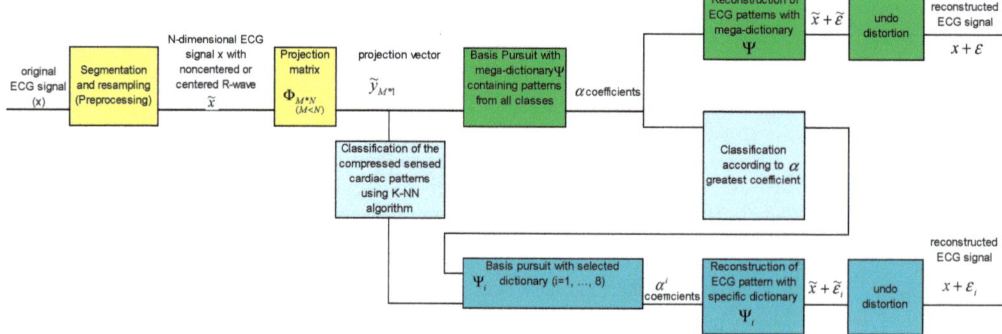

Figure 3. Block diagram of the CSCP method, using the mega-dictionary and/or a pathology-specific dictionary.

The upper branch of the block scheme, colored in green, is for the version with the universal mega-dictionary and the lower part of the figure, colored in blue, is for the version with dictionaries specific to pathologies.

In the case of reconstruction with dictionaries specific to pathologies, it is necessary to know the pathological class to which each cardiac pattern belongs. Therefore, it is necessary to classify the heartbeats. One option is to use a KNN classifier or any other classifier trained with various compressed beats [15,17]. Another option for classifying the heartbeats is a first reconstruction with the mega-dictionary on the upper branch of Figure 3 and the analysis of alpha coefficients corresponding to the mega-dictionary, i.e., the pathological class associated with the heartbeats is the same as the class in which the atom in the mega-dictionary with the highest coefficient belongs at reconstruction with the BP algorithm. Once the pathological class is established, the final reconstruction will be performed with the dictionary specific to that pathology [16].

For the classification of the ECG pattern and the establishment of the dictionary with which the signal will be reconstructed, the KNN classifier trained with the compressed version of the heartbeat from the universal mega-dictionary can be used.

Thus, a first step is to establish the class of the pattern. For this, we will use the KNN classifier based on the highest coefficient corresponding to the mega-dictionary, shown in light blue in Figure 3. Once the membership class is established, the Basis Pursuit algorithm together with the calculation of α coefficients necessary for the reconstruction of the ECG pattern are used. In addition, the almost insignificant distortions due to the centering of the R-wave can be improved by means of the knowledge about the original location of the R-wave.

4.3. Acceptance of the Compression Methods

To evaluate the compression and reconstruction performances, we assess the distortion between the original and the reconstructed signals by standard *PRD* and *PRDN* measures. Most ECG compression algorithms in the literature evaluate the errors using the percentage root-mean-square difference (*PRD*) measure and its normalized version, *PRDN*, defined as:

$$PRD\% = 100 \sqrt{\frac{\sum\limits_{n=1}^{N} (x(n) - \widetilde{x}(n))^2}{\sum\limits_{n=1}^{N} x^2(n)}}$$

and:

$$PRDN\% = 100 \sqrt{\frac{\sum\limits_{n=1}^{N} (x(n) - \widetilde{x}(n))^2}{\sum\limits_{n=1}^{N} (x(n) - \overline{x})^2}}$$

where $x(n)$ and $\widetilde{x}(n)$ are the samples of the original and the reconstructed signals, respectively, $\overline{x}$ is the mean value of the original signal and N is the length of the window over which the *PRD* is calculated.

For the evaluation of the compression, we used the compression rate (*CR*) defined as the ratio between the number of bits needed to represent the original and the compressed signal:

$$CR = \frac{b_{orig}}{b_{comp}}$$

where b_{orig} and b_{comp} represent the number of the bits required for the original and compressed signals, respectively.

We also used an alternative measure defined in [19], the Quality Score (*QS*), which is the ratio between the *CR* and the *PRD*:

$$QS = \frac{CR}{PRD}.$$

In addition to the quantitative measure related to the reconstruction of ECG signals, we also used a qualitative evaluation of the signals by classifying them. For classification, we used the KNN classifier. Thus, in the CPCS method version with a pathology-specific dictionary, in order to estimate the signal classification ratio in one of the eight possible classes, we used a KNN classifier to evaluate the reconstructed models. We mention that the KNN was trained with the models from the mega-dictionary used in the classification block (models that were not subjected to compression with the known class for each atom) and tested with the models reconstructed with class-specific dictionaries.

In addition to the KNN classifier, a neural network was used to test the reconstructed signals. The neural network was a multilayer perceptron (MLP) with 10 neurons in the hidden layer with backpropagation gradient descent for training.

However, the final verdict on the fidelity and clinical acceptability of the reconstructed signal should be validated by visual inspection by the cardiologist.

5. Experimental Results

In this study, we used 24 ECG recordings from the MIT-BIH Arrhythmia database acquired at a sampling frequency of 360 Hz, with 11 bits/sample [18]. Besides the ECG signals, the database also includes annotation files containing the index of the R-wave and the class to which each ECG pattern belongs.

In the CPCS method, we used the annotation databases in the preprocessing step (segmentation of cardiac cycles and forming of dictionaries) and in the reconstructed signal validation phase (KNN classifier-training stage).

The PSCCS technique used only the ECG signals from the MIT-BIH database, without requiring additional knowledge (ECG annotated files).

5.1. Results for the Patient-Specific Classical Compressed Sensing (PSCCS) Method

To test the PSCCS procedure, we used several compression ratios, namely, 4:1, 10:1 and 15:1. We also used several types of projection matrices (Bernoulli, Gaussian distribution random and dictionary specifics). The data used are 24 records from the MIT-BIH Database. In Table 1, we present the average results for 24 ECG records.

Table 1. Average results for 24 ECG records processed with the PSCCS method.

Projection Matrix and Its Size	CR	AVG. PRD	AVG. PRDN	QS
Gaussian distribution Random * Dict [†]	4:1	0.31	6.47	12.9
Bernoulli with 0 and 1 (75 × 300)	4:1	0.41	7.96	9.75
Gaussian distribution Random (75 × 300)	4:1	0.43	8.56	9.30
Gaussian distribution Random * Dict [†]	10:1	0.67	13.42	14.92
Bernoulli with 0 and 1 (30 × 300)	10:1	0.81	15.49	12.34
Gaussian distribution Random (30 × 300)	10:1	0.82	16.48	12.19
Gaussian distribution Random * Dict [†]	15:1	0.97	21.31	15.46
Bernoulli with 0 and 1 (20 × 300)	15:1	1.31	23.28	11.45
Gaussian distribution Random (20 × 300)	15:1	1.13	25.37	13.27

In addition to the average results reported for the MIT-BIH database, a number of authors reported the results for record no. 117 (in Table 2), which is why we will report these results as well.

Table 2. Results for the 117 records processed with the PSCCS method.

Projection Matrix and Its Size	CR	AVG. PRD	AVG. PRDN	QS
Gaussian distribution Random * Dict [†]	4:1	0.19	4.69	21.05
Bernoulli with 0 and 1 (75 × 300)	4:1	0.40	7.20	10
Gaussian distribution Random (75 × 300)	4:1	0.45	8.12	8.88
Gaussian distribution Random * Dict [†]	10:1	0.45	11.19	22.22
Bernoulli with 0 and 1 (30 × 300)	10:1	0.70	12.67	14.28
Gaussian distribution Random (30 × 300)	10:1	0.73	13.21	13.69
Gaussian distribution Random * Dict [†]	15:1	0.63	15.61	23.80
Bernoulli with 0 and 1 (20 × 300)	15:1	0.96	17.28	15.62
Gaussian distribution Random (20 × 300)	15:1	1.01	18.24	14.85

In Figure 4a, we present a part of the registration no. 117 in the initial version and its version reconstructed following the compression of 4:1, 10:1 and 15:1 for the application of a Bernoulli type projection matrix. It is observed that for CR = 15:1, especially in the noisy region (sample from 2000 to 2200), there are some visible reconstruction differences due to this noise. There are no significant differences in the rest of the signal.

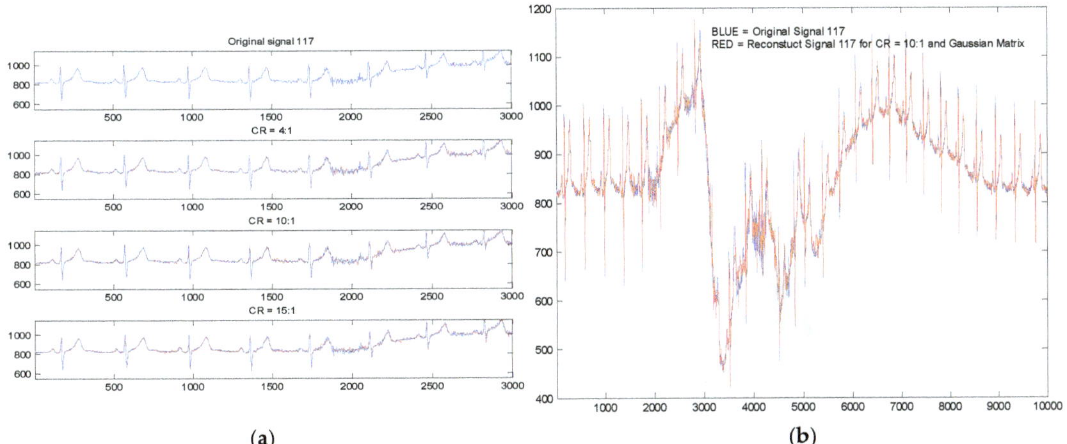

Figure 4. Original (blue) and reconstruct (red) ECG signal with PSCCS method (registration no. 117): (**a**) for CR 4:1, 10:1 and 15:1 with a Bernoulli projection matrix; (**b**) for CR 10:1 with random projection (Gaussian distribution).

In Figure 4b, we also present from the recording 117 an original ECG signal segment and its variant reconstructed subject to a CR = 10:1 (for random projection matrix with Gaussian distribution). The segment shown is the segment with the highest noise in the entire recording. In this way, we wanted to highlight the robustness of the method to noise and artifacts due to the patient's movement and breathing.

The results obtained on 14 ECG signals, for a compression ratio of 15:1, for centered and non-centered R-wave are shown in Table 3. We used the KNN and MLP algorithm for the evaluation by classification.

Table 3. Average results for 14 ECG records with the PSCCS method.

Projection Matrix and Its Size	CR	AVG. PRD	AVG. PRDN	Classif. Rate with KNN	Classif. Rate with MLP
Patient-specific dictionary with a non-centered R-wave					
Gaussian distribution Random * Dict [†] (20 × 301)	15:1	0.78	11.98	92.24%	93.7%
Bernoulli with 0 and 1 (20 × 301)	15:1	0.94	16.06	84.71%	86.2%
Gaussian distribution Random (20 × 301)	15:1	0.82	13.82	91.14%	93.4%
Patient-specific dictionary with a centered R-wave					
Gaussian distribution Random * Dict [†] (20 × 301)	15:1	0.51	9	93.41%	95.2%
Bernoulli with 0 and 1 (20 × 301)	15:1	0.71	12.4	88.06%	90.3%
Gaussian distribution Random (20 × 301)	15:1	0.72	12.51	89.70%	91.6%

The KNN and MLP classifiers were trained with normal and abnormal heart beats evenly distributed on both classes. The beats used to train the classifier were extracted from the dictionary constructed for the compressed acquisition. In this case, the classification was on two classes, normal or abnormal, and it did not follow the seven pathological classes.

The advantage of the KNN classifier is the simplicity of the calculations, this classifier assuming only the calculation of some Euclidean distances. In the case of MLP networks, the calculations are more complex, but the results are better compared to the KNN classifier.

5.2. Results for the Cardiac Patterns Compressed Sensing (CPCS) Method

5.2.1. Universal Mega-Dictionary

For the construction of a mega-dictionary, from all the 24 ECG recordings, we randomly chose 184 patterns from the 8 cardiac classes, thus obtaining a dictionary with 1472 patterns with the size 1472 × 301.

The testing was performed on 200 patterns from each class, chosen at random from the 24 records, with the mention that special attention was paid to random choice, namely, the models used to build the dictionary could no longer be used for testing.

Table 4 shows the average results obtained on all 24 records, with R-wave alignment and centering and without R-wave centering, for all the projection matrices presented.

Table 4. Average results for 24 ECG records processed with the CSCP method with the mega dictionary.

Projection matrix and Its Size	CR	AVG. PRD	AVG. PRDN	QS
Mega-dictionary with a non-centered R-waves				
Gaussian distribution Random * Dict [†]	15:1	0.88	13.67	17.04
Bernoulli with 0 and 1 (20 × 301)	15:1	1.44	21.43	10.41
Gaussian distribution Random (20 × 301)	15:1	1.62	24.33	9.25
Mega-dictionary with a centered R-waves				
Gaussian distribution Random * Dict [†]	15:1	0.67	9.99	22.38
Bernoulli with 0 and 1 (20 × 301)	15:1	1.08	15.47	13.88
Gaussian distribution Random (20 × 301)	15:1	1.19	17.18	12.60

5.2.2. Pathology-Specific Dictionaries

Each of the eight pathology-specific dictionaries is made up of 700 atoms that actually represent patterns with or without a centered R-waves. Dictionaries are matrices of size 700 × 301.

For testing, we used a number of 2000 cardiac patterns chosen at random from the 24 records with the mention that the patterns used for testing are different from those used for training (see Table 5).

Table 5. Average results for 24 ECG records for CSCP method with a specific dictionary and classification based on the largest coefficient of the sparsest decomposition for the mega-dictionary.

Projection Matrix and Its Size	CR	AVG. PRD	AVG. PRDN	QS
Pathological specific dictionaries with a non-centered R-wave				
Gaussian distribution Random * Dict [†]	15:1	0.77	11.76	19.48
Bernoulli with 0 and 1 (20×301)	15:1	1.23	17.90	12.19
Gaussian distribution Random (20×301)	15:1	1.37	20.25	10.94
Pathological specific dictionaries with a centered R-wave				
Gaussian distribution Random * Dict [†]	15:1	0.62	6.14	24.19
Bernoulli with 0 and 1 (20×301)	15:1	0.97	13.93	15.46
Gaussian distribution Random (20×301)	15:1	1.04	14.80	14.42

In this variant, with dictionaries specific to the pathological class, in the reconstruction stage, it is necessary to identify the class to which the pattern belongs. The reconstruction results are strongly influenced by the correctness of establishing the pathological class to which the model belongs. Thus, for patterns classification, a KNN type classifier will be used or it will be made based on the highest alpha coefficient. Once the pathological class is established, the Basis Pursuit algorithm, the dictionary specific to that pathology and the projection matrix will be used for reconstruction.

Thus, using the classification of patterns based on the highest alpha coefficient in the mega-dictionary version, a pattern classification rate of 88.75% is obtained [16]. Using the KNN classifier with training on 1472 compressed cardiac patterns (uniformly distributed in the eight classes), a classification rate of 93.77% is obtained [15].

In Figure 5, we present examples of reconstructed cardiac beats for every pathology class.

Qualitative estimation of reconstructed signals based on classification. In addition to the quantitative measures of the distortions between the original and reconstructed ECG signals, for a further verification of the quality of the proposed compression scheme, we performed a classification of reconstituted models with the KNN algorithm. The classifier was trained with the atoms from the mega-dictionary. A first check of the method is to test the performance of the KNN classifier, and for this, we initially tested the original models (i.e., the uncompressed models that we used to test the compression scheme). For these patterns, we obtained a classification rate of 93.75%. The results presented below are obtained on the reconstructed patterns [28].

- Classifying the patterns reconstructed with the mega-dictionary (with patterns out of all classes) yielded an accuracy of 92.5%.
- Classifying the patterns reconstructed with the class-specific dictionaries provided an accuracy of 95.5%.

In addition to KNN, an MLP classifier was also tested. This second classification aims to strengthen the correctness of the idea of testing the reconstructed patterns from a qualitative point of view. This test is based on a classifier and is needed to compare the results obtained with these two different classifiers. Thus, there is a slight and almost insignificant improvement of the classification rate in the case of MLP compared to KNN. However, in practical implementations, the MLP classifier should be chosen according to the available hardware resources. Table 6 shows obtained results for dictionaries with a centered R-wave.

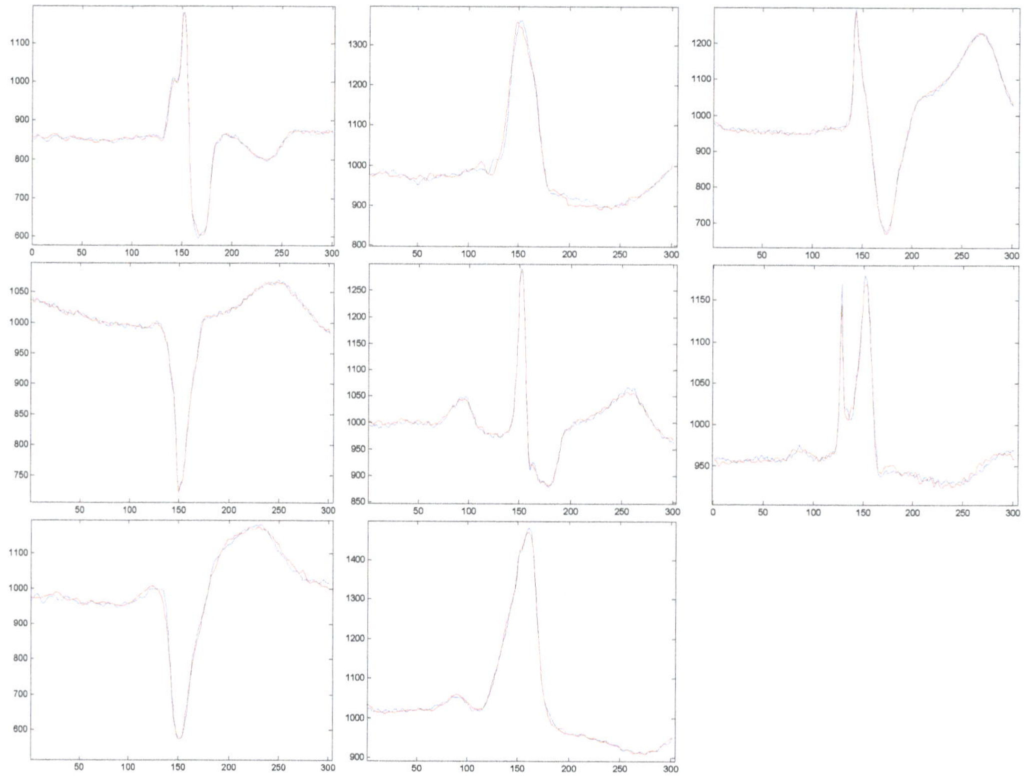

Figure 5. Original and reconstructed signals with pathology-specific dictionaries.

Table 6. Results summary for dictionaries with a centered R-wave.

Dictionary with a Centered R-Wave	Compression Rate	AVG. PRD	AVG. PRDN	KNN Classif. Rate	MLP Classif. Rate
mega-dictionary	10:1	0.47	6.24	93.2%	93.8%
mega-dictionary	15:1	0.67	9.99	92.5%	93.1%
specific dictionaries	10:1	0.43	6.02	95.2%	96%
specific dictionaries	15:1	0.62	6.14	95.5%	96.2%
KNN classification results with original patterns				95.5%	96%
PRDN and KNN classification rate for the case with correct identification (100%) of the specific dictionary	0.55	8.53	93%	93.7%	

It is known that in a classification process, especially when it applies to several classes, special attention must be assigned to the confusion matrix, to see if the classification is uniform on all classes or only certain classes are detected. For this we have exemplified in Table 7 a confusion matrix for the classification variant with a mega-dictionary. It can be seen that the classification rate is evenly distributed over all eight classes.

Table 7. Confusion matrix for KNN classification of the reconstructed patterns with a mega-dictionary.

	Class1	Class2	Class3	Class4	Class5	Class6	Class7	Class8
class1	**90**	10	0	0	0	0	0	0
class2	20	**70**	0	0	0	10	0	0
class3	0	0	**100**	0	0	0	0	0
class4	0	0	0	**100**	0	0	0	0
class5	0	0	0	0	**100**	0	0	0
class6	0	0	0	0	0	**100**	0	0
class7	0	0	0	0	0	0	**100**	0
class8	0	10	0	0	0	0	10	**80**

5.2.3. Patient-Specific Dictionaries

The patient-specific dictionaries were constructed from the patient's first 700 heartbeats, and preprocessed as previously described (i.e., with or without R-wave alignment). Thus, the dictionary is made up of 700 atoms, each of size 300, i.e., it is actually a matrix of size 301 × 700. This method has the advantage of speculating quasi-periodicity and the particular characteristics of the ECG signal of a particular patient. Table 8 shows average results for 24 ECG recordings for the CSCP method and it can be seen that the best results are obtained if we refer to QS.

Table 8. Average results for 24 ECG Records for the CSCP method with a patient-specific dictionary built from the first 700 cardiac cycles.

Projection Matrix and Its Size	CR	AVG. PRD	AVG. PRDN	QS
Patient-specific dictionary with a non-centered R-wave				
Gaussian distribution Random * Dict [†]	15:1	0.78	11.98	19.23
Bernoulli with 0 and 1 (20 × 301)	15:1	0.94	16.06	15.87
Gaussian distribution Random (20 × 301)	15:1	0.82	13.82	18.29
Patient-specific dictionary with a centered R-wave				
Gaussian distribution Random * Dict [†]	15:1	**0.51**	**9**	29.13
Bernoulli with 0 and 1 (20 × 301)	15:1	0.71	12.4	20.98
Gaussian distribution Random (20 × 301)	15:1	0.72	12.51	20.59

Because our results are generally obtained by mediating the results obtained by processing 24 records from MIT-BIH Arrhythmia database, we present in Figure 6 the histograms of PRD and PRDN, respectively, for the method of CS with patient-specific dictionaries with a centered R-wave and projection matrix by type Gaussian distribution Random * Dict †. For this case, PRD_average = 0.51 and PRDN_average = 9 (see Table 8).

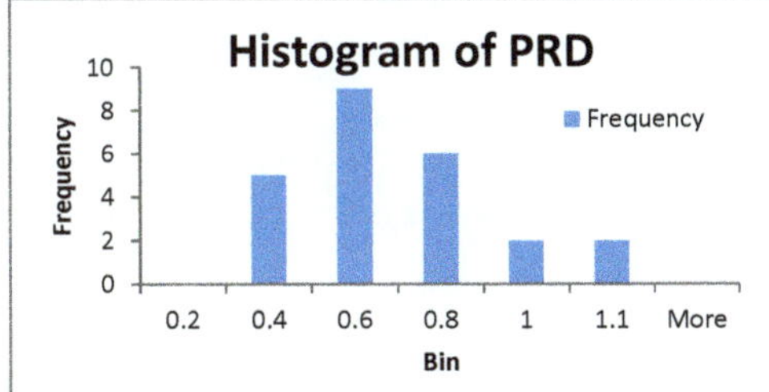

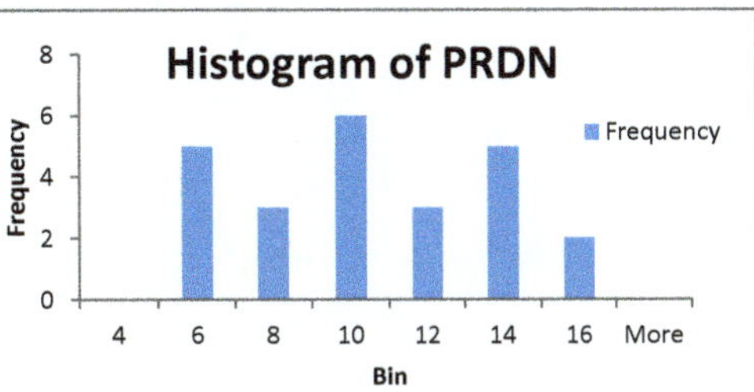

Figure 6. Histogram of PRD and PRDN for 24 ECG records for the CSCP method with a patient-specific dictionary with projection matrix by type of Gaussian distribution Random * Dict †.

6. Discussions

In Table 9, we resume the results previously presented for the two analyzed methods, for a CR = 15:1 with all investigated projection matrices and with all discussed reconstruction and preprocessing dictionaries. We marked in bold the best results obtained on QS (Quality Score) for each method.

Table 9. Results summary for CR = 15:1.

Projection Matrix and Its Size	CR	AVG. PRD	AVG. PRDN	QS
PSCCS METHOD				
Gaussian distribution Random * Dict [†]	15:1	0.97	21.31	**15.46**
Bernoulli with 0 and 1 (20 × 300)	15:1	1.31	23.28	11.45
Gaussian distribution Random (20 × 300)	15:1	1.13	25.37	13.27
CPCS METHOD				
Universal mega-dictionary without a centered R-wave				
Gaussian distribution Random * Dict [†]	15:1	0.88	13.67	**17.04**
Bernoulli with 0 and 1 (20 × 301)	15:1	1.44	21.43	10.41
Gaussian distribution Random (20 × 301)	15:1	1.62	24.33	9.25
Universal mega-dictionary with a centered R-wave				
Gaussian distribution Random * Dict [†]	15:1	0.67	9.99	**22.38**
Bernoulli with 0 and 1 (20 × 301)	15:1	1.08	15.47	13.88
Gaussian distribution Random (20 × 301)	15:1	1.19	17.18	12.60
Pathological specific dictionaries without a centered R-wave				
Gaussian distribution Random * Dict [†]	15:1	0.77	11.76	**19.48**
Bernoulli with 0 and 1 (20 × 301)	15:1	1.23	17.90	12.19
Gaussian distribution Random (20 × 301)	15:1	1.37	20.25	10.94
Pathological specific dictionaries with a centered R-wave				
Gaussian distribution Random * Dict [†]	15:1	0.62	6.14	**24.19**
Bernoulli with 0 and 1 (20 × 301)	15:1	0.97	13.93	15.46
Gaussian distribution Random (20 × 301)	15:1	1.04	14.80	14.42
Patient-specific dictionaries without a centered R-wave				
Gaussian distribution Random * Dict [†]	15:1	0.78	11.98	**19.23**
Bernoulli with 0 and 1 (20 × 301)	15:1	0.94	16.06	15.87
Gaussian distribution Random (20 × 301)	15:1	0.82	13.82	18.29
Patient-specific dictionaries with a centered R-wave				
Gaussian distribution Random * Dict [†]	15:1	0.51	9	**29.13**
Bernoulli with 0 and 1 (20 × 301)	15:1	0.71	12.4	20.98
Gaussian distribution Random (20 × 301)	15:1	0.72	12.51	20.59

It can be seen that the best QS result is obtained for dictionary specific to the patient in which the R-wave is centered and a projection matrix is optimized to the dictionary. In addition, it has been also found that in all cases optimization of dictionaries improves the results. Moreover, it has also been observed that preprocessing improves the results, namely, for PSCCS (i.e., without preprocessing) for CR = 15:1 the best QS equals 15.46, i.e., almost half of the value obtained with CPCS with a patient-specific centered R-wave dictionary when QS = 29.13.

It should be noted that any preprocessing means hardware resources and choosing a method with preprocessing means additional hardware resources. However, we must mention that the detection of the QRS complex and the R-wave is a problem that can be implemented in real time in the Matlab® environment, an example of implementation being even available in Help Matlab® [29].

Table 10 shows the average results on the 24 records and for the 117 records obtained by other authors.

Table 10. Average values for 24 records and 117 record for other compression algorithms.

	Record/Ave.	CR	AVG. PRD	AVG. PRDN
Other Compression Algorithms				
Polania [20,21]	117	8:1	2.18	Notspec.
Polania [20,21]	117	10:1	2.5	Notspec.
Mamaghanian [22] for before and after inter-packet redundancy removal and Huffman coding	Ave. for 24 records	4:1 (75)	Before Huffman 35 After Huffman 15
		10:1 (90)	Before Huffman >45 After Huffman >45
		15:1 (93)	Before Huffman >45 After Huffman >45

We note that Mamaghanian in [22] presents a classical CS compression method followed by Huffman coding, the final CR being higher due to the additional Huffman compression. For a more accurate comparison, we must compare our results with those obtained by Mamaghanian before Huffman compression. Additionally, the same author uses in [22] the compression ratio defined as:

$$CR = \frac{b_{orig} - b_{comp}}{b_{orig}} * 100,$$

which is not the same as ours and gives a very different gamut of values compared with ours.

The results we obtained with the proposed method are compared in Table 11 with the results of other compression methods in the literature.

Table 11. Quality score for compression algorithms for average values for 24 records.

Algorithm	Average of Errors (PRD or RMS)	Average of CR	QS
Wavelet [23]	18.2 RMS	21.4:1	
	3.57 PRD	12:1	3.39
SPHIT [24]	4.85 PRD	16:1	3.29
	6.49 PRD	20:1	3.08
	2.19 PRD	12:1	5.47
JPEG2000 [25]	2.74 PRD	16:1	5.8
	3.26 PRD	20:1	6.1
QLV–Skeleton–Huffman * [26]	0.641 PRD *	16.9:1 *	29.36 *
Skeleton [10]	1.17 PRD 11.35 RMS	18.27:1	15.61
PSCCS method	**0.97 PRD**	**15:1**	**15.46**
CS with patient-specific dictionaries with a centered R-wave	**0.51 PRD**	**15:1**	**29.13**

NOTE: The results reported in [26] marked with * in Table 11 were obtained using a combined ECG compression method consisting of a preprocessing stage with quad level vector (QLV) for the extraction of the ECG skeleton achieving an 8.4:1 compression and a coding block (consisting of delta and Huffman Coding). The results referenced in Table 3 are the final one, improved by the Huffman coding stage.

7. Conclusions

The results presented in this paper reveal several interesting aspects, as follows.

It has been revealed that the first stage of the CS method, i.e., the signal acquisition part, based on the projection matrices, has only a relatively small influence on the decompression or classification results.

On the other hand, for the second stage, namely signal reconstruction, the dictionary used for reconstruction of the compressed sensed ECG signals has an essential role in obtaining good results. Therefore, depending on the application targeted with the used CS technique, namely, Holter monitoring or recorded ECG signal classification, a dictionary that leads to optimal final results can be selected.

Thus, in a Holter monitoring application, where the ECG signal is recorded for 24 h from the same patient, one can choose the Patient-Specific Classical Compressed Sensing (PSCCS) method. By analyzing the first minutes of the recording, a dictionary specific to the patient will be built, and then it will be used to reconstruct the ECG segments of interest to the specialist.

Otherwise, if the CS-based application aims at classifying heartbeats for ECG monitoring or abnormality identification applications, the Cardiac Patterns Compressed Sensing (CPCS) method will be chosen, where each pathological heart beat class will be associated to a specific dictionary.

The above discussed methods are primarily based on waveform segmentation (cardiac beats) with no preprocessing. Yet, depending on the available hardware resources and the time constraints in which the application should run, the results can be significantly improved by centering the R-wave using ECG preprocessing i.e., segmented cardiac patterns with a centered R-wave.

This choice is related to the idea that any ECG signal preprocessing leads to higher hardware requirements and slowdowns in the acquisition and reconstruction processes over time. However, these aspects can be easily dealt with, aiming at better results. However, we must mention that the detection of the QRS complex and the R-wave is a problem that can be implemented in real time in the Matlab® environment, with an example of implementation even being available in Help Matlab®.

Author Contributions: Conceptualization, M.F., H.-N.C. and L.G.; methodology, M.F.; software, M.F.; validation, M.F., H.-N.C. and L.G.; formal analysis, M.F.; investigation, M.F.; resources, M.F.; data curation, M.F.; writing—original draft preparation, M.F., H.-N.C. and L.G.; writing—review and editing, M.F., H.-N.C. and L.G. All authors have read and agreed to the published version of the manuscript.

Funding: This research received no external funding.

Institutional Review Board Statement: Not applicable.

Informed Consent Statement: Not applicable.

Data Availability Statement: The data presented in this study are openly available in [physionet] at [10.1109/51.932724 and 10.1161/01.cir.101.23.e215], reference number [18]. The webpage of the MIT-BIH Arrhythmia Database is "https://www.physionet.org/content/mitdb/1.0.0/" (accessed on 27 December 2021).

Conflicts of Interest: The authors declare no conflict of interest.

References

1. Chen, S.; Donoho, D.; Saunders, M. Atomic decomposition by basis pursuit. *SIAM Rev.* **2001**, *43*, 129–159. [CrossRef]
2. Djelouat, H.; Amira, A.; Bensaali, F.; Boukhennoufa, I. Secure compressive sensing for ECG monitoring. *Comput. Secur.* **2020**, *88*, 101649. [CrossRef]
3. Daponte, P.; De Vito, L.; Iadarola, G.; Picariello, F. ECG Monitoring Based on Dynamic Compressed Sensing of multi-lead signals. *Sensors* **2021**, *21*, 7003. [CrossRef] [PubMed]
4. Wang, C.; Liu, J.; Sun, J. Compression algorithm for electrocardiograms based on sparse decomposition. *Front. Electr. Electron. Eng. China* **2009**, *4*, 10–14. [CrossRef]
5. Mitra, D.; Zanddizari, H.; Rajan, S. Investigation of Kronecker-based recovery of compressed ECG signal. *IEEE Trans. Instrum. Meas.* **2020**, *69*, 3642–3653. [CrossRef]
6. Craven, D.; McGinley, B.; Kilmartin, L.; Glavin, M.; Jones, E. Compressed sensing for bioelectric signals: A review. *IEEE J. Biomed. Health Inform.* **2015**, *19*, 529–540. [CrossRef] [PubMed]
7. Cleju, N.; Fira, M.; Barabasa, C.; Goras, L. Robust reconstruction of compressively sensed ECG patterns. In Proceedings of the ISSCS 2011 (The 10-th International Symposium on Signals, Circuits and Systems), Iasi, Romania, 30 June–1 July 2011; pp. 507–510.

8. Shannon, C.E. Communication in the presence of noise. *Proc. Inst. Radio Eng.* **1949**, *37*, 10–21. [CrossRef]

9. Elad, M. Optimized projections for compressed sensing. *IEEE Trans. Signal Process.* **2007**, *55*, 5695–5702. [CrossRef]

10. Duarte, M.; Davenport, M.; Wakin, M.; Laska, J.; Takhar, D.; Kelly, K.; Baraniuk, R. Multiscale random projections for compressive classification. In Proceedings of the 2007 IEEE International Conference on Image Processing (ICIP), San Antonio, TX, USA, 16–19 September 2007.

11. Donoho, D. Compressed sensing. *IEEE Trans. Inf. Theory* **2006**, *52*, 1289–1306. [CrossRef]

12. Candès, E. Compressive sampling. In Proceedings of the international Congress of Mathematicians, Madrid, Spain, 22–30 August 2006; Volume 3, pp. 1433–1452.

13. Fira, M.C.; Goras, L.; Barabasa, C.; Cleju, N. On ECG compressed sensing using specific overcomplete dictionaries. *Adv. Electr. Comput. Eng.* **2010**, *10*, 23–28. [CrossRef]

14. Fira, M.C.; Goras, L.; Barabasa, C. Reconstruction of compressed sensed ECG signals using patient specific dictionaries. In Proceedings of the International Symposium on Signals, Circuits and Systems (ISSCS 2013), Iasi, Romania, 11–12 July 2013.

15. Fira, M.C.; Goras, L.; Barabasa, C.; Cleju, N. ECG compressed sensing based on classification in compressed space and specified dictionaries. In Proceedings of the 2011 European Signal Processing Conference (EUSIPCO 2011), Barcelona, Spain, 29 August–2 September 2011; pp. 1573–1577.

16. Fira, M.C.; Goras, L.; Cleju, N.; Barabasa, C. Results on ECG compressed sensing using specific dictionaries and its validation. In Proceedings of the International Conference on Information Technology Interfaces–ITI 2012, Dubrovnik, Croatia, 25–28 June 2012; pp. 423–428.

17. Fira, M.C.; Goras, L.; Cleju, N.; Barabasa, C. On the projection matrices influence in the classification of compressed sensed ECG signals. *Int. J. Adv. Comput. Sci. Appl. IJACSA* **2012**, *3*, 141–145. [CrossRef]

18. MIT-BIH. Arrhythmia Database. Available online: http://www.physionet.org/physiobank/database/mitdb/ (accessed on 27 December 2021).

19. Fira, M.C.; Goras, L. An ECG signals compression method and its validation using NNs. *IEEE Trans. Biomed. Eng.* **2008**, *55*, 1319–1326. [CrossRef]

20. Polania, L.F.; Carrillo, R.E.; Blanco-Velasco, M.; Barner, K.E. ECG compression via matrix completion. In Proceedings of the 19th European Signal Processing Conference (EUSIPCO), Barcelona, Spain, 29 August–22 September 2011.

21. Polania, L.F.; Carrillo, R.E.; Blanco-Velasco, M.; Barner, K.E. Compressed sensing based method for ECG compression. In Proceedings of the 2011 IEEE International Conference on Acoustics, Speech and Signal Processing (ICASSP), Prague, Czech Republic, 22–27 May 2011.

22. Mamaghanian, H.; Khaled, N.; Atienza, D.; Vandergheynst, P. Compressed sensing for real-time energy-efficient ECG compression on wireless body sensor nodes. *IEEE Trans. Biomed. Eng.* **2011**, *58*, 2456–2466. [CrossRef]

23. Al-Shrouf, A.; Abo-Zahhad, M.; Ahmed, S.M. A novel compression algorithm for electrocardiogram signal based on the linear prediction of the wavelet coefficients. *Digit. Signal Process.* **2003**, *13*, 604–622. [CrossRef]

24. Lu, Z.; Kim, D.Y.; Pearlman, W.A. Wavelet Compression of ECG signals by the set partitioning in hierarchical trees (SPIHT) algorithm. *IEEE Trans. Biomed. Eng.* **2000**, *47*, 849–856. [PubMed]

25. Bilgin, A.; Marcellin, M.W.; Altbach, M.I. Wavelet compression of ECG signals by JPEG2000. In Proceedings of the Data Compression Conference DDC2004, Snowbird, UT, USA, 23–25 March 2004; p. 527.

26. Kim, H.; Yazicioglu, R.F.; Merken, P.; Van Hoof, C.; Yoo, H.-J. ECG signal compression and classification algorithm with quad level vector for ECG Holter system. *IEEE Trans. Inf. Technol. Biomed.* **2010**, *14*, 93–100. [PubMed]

27. Chou, C.-Y.; Chang, E.-J.; Li, H.-T.; Wu, A.-Y. Low-complexity privacy-preserving compressive analysis using subspace-based dictionary for ECG telemonitoring system. *IEEE Trans. Biomed. Circuits Syst.* **2018**, *12*, 801–811. [PubMed]

28. Fira, M.; Goras, L. On projection matrices and dictionaries in ECG compressive sensing—A comparative study. In Proceedings of the 12th Symposium on Neural Network Applications in Electrical Engineering (NEUREL), Belgrade, Serbia, 25–27 November 2014; pp. 3–8. [CrossRef]

29. Mathworks. Available online: https://www.mathworks.com/help/dsp/ug/real-time-ecg-qrs-detection.html (accessed on 27 December 2021).

30. Turnip, A.; Turnip, M.; Dharma, A.; Paninsari, D.; Nababan, T.; Ginting, C.N. An application of modified filter algorithm fetal electrocardiogram signals with various subjects. *Int. J. Artif. Intell.* **2020**, *18*, 207–217.

31. Ciucu, R.I.; Dragomir, D.A.; Adochiei, I.R.; Seritan, G.C.; Cepisca, C.; Adochiei, F.C. A non-contact heart-rate monitoring system for long-term assessments of HRV. In Proceedings of the 2017 10th International Symposium on Advanced Topics in Electrical Engineering (ATEE), Bucharest, Romania, 23–25 March 2017.

32. Seritan, G.; Bujanovschi, M.; Adochiei, F.C.; Voiculescu, D.; Argatu, F.C.; Grigorescu, S.D. Alert and surveillance system for newborns. In Proceedings of the 2019 E-Health and Bioengineering Conference (EHB), Iasi, Romania, 21–23 November 2019. [CrossRef]

33. Lopes, A.R.; Nihei, O.K. Depression, anxiety and stress symptoms in Brazilian university students during the COVID-19 pandemic: Predictors and association with life satisfaction, psychological well-being and coping strategies. *PLoS ONE* **2021**, *16*, e0258493. [CrossRef]

34. Nasimi, F.; Khayyambashi, M.R.; Movahhedinia, N.; Law, Y.W. Exploiting similar prior knowledge for compressing ECG signals. *Biomed. Signal Process. Control* **2020**, *60*, 101960. [CrossRef]

35. Lin, Y.-M.; Chen, Y.; Kuo, H.-C.; Wu, A.-Y.A. Compressive sensing based ECG telemonitoring with personalized dictionary basis. In Proceedings of the 2015 IEEE Biomedical Circuits and Systems Conference (BioCAS), Atlanta, GA, USA, 22–24 October 2015; pp. 1–4. [CrossRef]
36. Ravelomanantsoa, A.; Rabah, H.; Rouane, A. Compressed sensing: A simple deterministic measurement matrix and a fast recovery algorithm. *IEEE Trans. Instrum. Meas.* **2015**, *64*, 3405–3413. [CrossRef]
37. Picariello, F.; Iadarola, G.; Balestrieri, E.; Tudosa, I.; De Vito, L. A novel compressive sampling method for ECG wearable measurement systems. *Measurement* **2020**, *167*, 108259. [CrossRef]

Article

Machine Learning Based Lens-Free Shadow Imaging Technique for Field-Portable Cytometry

Rajkumar Vaghashiya [1,†], Sanghoon Shin [2,†], Varun Chauhan [1], Kaushal Kapadiya [1], Smit Sanghavi [1], Sungkyu Seo [2,*] and Mohendra Roy [3,*]

1   Department of Computer Engineering, Pandit Deendayal Energy University, Gandhinagar 382007, India; rajkumar.vce16@sot.pdpu.ac.in (R.V.); varun.cce16@sot.pdpu.ac.in (V.C.); kaushal.kce16@sot.pdpu.ac.in (K.K.); smit.sce16@sot.pdpu.ac.in (S.S.)
2   Department of Electronics and Information Engineering, Korea University, Sejong 30019, Korea; ghost10s@korea.ac.kr
3   Department of Information and Communication Technology, Pandit Deendayal Energy University, Gandhinagar 38207, India
*   Correspondence: sseo@korea.ac.kr (S.S.); mohendra.roy@ieee.org (M.R.); Tel.: +82-44-860-1427 (S.S.); +91-79-2327-5483 (M.R.); Fax: +82-44-860-1585 (S.S.); +91-79-2327-5030 (M.R.)
†   These authors contributed equally to this study.

Abstract: The lens-free shadow imaging technique (LSIT) is a well-established technique for the characterization of microparticles and biological cells. Due to its simplicity and cost-effectiveness, various low-cost solutions have been developed, such as automatic analysis of complete blood count (CBC), cell viability, 2D cell morphology, 3D cell tomography, etc. The developed auto characterization algorithm so far for this custom-developed LSIT cytometer was based on the handcrafted features of the cell diffraction patterns from the LSIT cytometer, that were determined from our empirical findings on thousands of samples of individual cell types, which limit the system in terms of induction of a new cell type for auto classification or characterization. Further, its performance suffers from poor image (cell diffraction pattern) signatures due to their small signal or background noise. In this work, we address these issues by leveraging the artificial intelligence-powered auto signal enhancing scheme such as denoising autoencoder and adaptive cell characterization technique based on the transfer of learning in deep neural networks. The performance of our proposed method shows an increase in accuracy >98% along with the signal enhancement of >5 dB for most of the cell types, such as red blood cell (RBC) and white blood cell (WBC). Furthermore, the model is adaptive to learn new type of samples within a few learning iterations and able to successfully classify the newly introduced sample along with the existing other sample types.

Keywords: artificial intelligence; lens-free shadow imaging technique; cell-line analysis; cell signal enhancement; deep learning

Citation: Vaghashiya, R.; Shin, S.; Chauhan, V.; Kapadiya, K.; Sanghavi, S.; Seo, S.; Roy, M. Machine Learning Based Lens-Free Shadow Imaging Technique for Field-Portable Cytometry. *Biosensors* **2022**, *12*, 144. https://doi.org/10.3390/bios12030144

Received: 30 January 2022
Accepted: 25 February 2022
Published: 27 February 2022

1. Introduction

The lens-free shadow imaging technique (LSIT) is a well-established technique for the characterization of microparticles and biological cells [1]. This technique is widely popular for its simple imaging structure and cost-effectiveness. It comprises a lens-less detector, such as a complementary metal-oxide semiconductor (CMOS) image sensor, a semi-coherent light source, such as light-emitting diode (LED), and a disposable cell chip (C-Chip). The absence of a lens or other optical arrangements allows it to fit into a very small space, thereby reducing the size of the overall system (as described in Figure 1a in the LSIT platform (Cellytics) built within a dimension of $100 \times 120 \times 80$ mm^3). Since this arrangement consists of a few components, most of which are easily available at a low price, it therefore reduces the overall cost of the system [2]. This simple and cost-effective nature facilitates the feasibility of the LSIT for the applications in the fields of point-of-care systems or telemedicine systems [3–5].

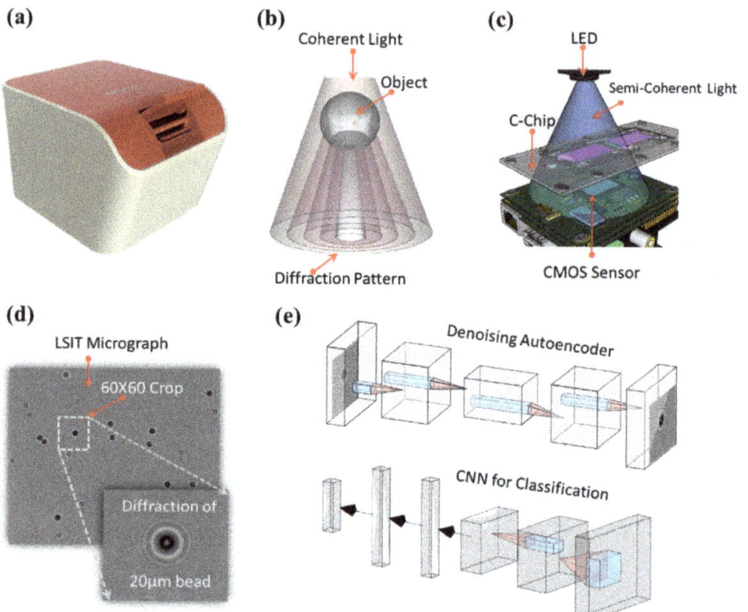

Figure 1. Schematics of the LIST setup and the proposed neural network architecture for the auto characterization of LSIT micrographs. (**a**) LSIT platform (Cellytics) (**b**) schematic of the principle of diffraction, i.e., shadow, pattern generation of a micro-object, (**c**) schematic of the LSIT imaging setup showing the simplicity of the setup, (**d**) schematic of the dataset creation process by automatic cropping individual cell diffraction pattern from the whole LSIT micrograph, and (**e**) the schematic of the proposed denoising and classification architecture. Here, the denoising autoencoder enhances the signal of the individual cells which is then fed to the CNN module for classification.

Recent advancements in machine learning, especially deep learning, have facilitated many applications concerning medical diagnostics [6–12], and have been widely adopted in the field of microscopy [13–15]. In particular, deep learning has been incorporated with the LSIT [14], where it is has been used to enhance the resolution of the LSIT micrographs [16] and enabled polarization-based holographic microscopy [17].

In our previous work, we have successfully developed the LSIT imaging system for the complete blood count using an analytical model based on handcrafted features [3] that can automatically segment out the individual cells from a whole frame LSIT micrograph and subsequently analyze them based on the handcrafted parameters. However, the performance of the system is dependent on the uniform illumination as well as the strong signatures of the microparticle samples. Since the diffraction signature of a microparticle depends on the size as well as the signal-to-noise ratio of the particle, therefore any background noise can affect the overall performance of the auto characterization system. Further, the handcrafted approach of finding the features for every additional cell line is time-consuming and prone to subjective errors. To address these limitations, in this work, we have developed an artificial intelligence (AI) powered signal enhancement scheme for the LSIT micrographs that can enhance the signal quality (signal to noise ratio (SNR)) for various cell lines in a heterogeneous cell sample. For this, we employed the autoencoder-based denoising scheme [18]. Further, we have developed an auto characterization method based on a convolutional neural network [19,20] (CNN) architecture to classify the various cell lines from the LSIT micrograph. Here, we have first introduced the transfer of learning scheme in a neural network, which can leverage the feasibility to introduce new cell types to the algorithm and thus learn their characteristics within a few iterations. Thus, the LSIT platform saves time and computation resources required to learn to classify the additional cell types along with the existing ones.

In this article, we have described the detailed methods adopted for the designing as well as optimization of various parameters to design a suitable model with better accuracy. These optimized models are simple and light-weight, and require a smaller number of samples for effectively learning the cell signatures. The details are as given in the following sections.

2. Methods

2.1. LSIT Imaging Setup

The schematic of our proposed setup (Figure 1a) is as shown in Figure 1. When light from the coherent or semi-coherent source passes through a micro-object, it produces characteristic diffraction, i.e., shadow, pattern of the object [21,22] as shown in Figure 1b. These diffraction patterns are prominent just beneath the sample, typically a few hundred micrometers away from the sample plane, from where they are captured using a high-density image sensor such as CCD or CMOS [21] (Figure 1c). As these signatures are significant enough to be captured by the bared image sensor, it does not require any kind of lens arrangement [23]. In our proposed setup, we used a pinhole conjugated semi-coherent LED light source with a peak wavelength of 470 ± 5 nm (HT-P318FCHU-ZZZZ, Harvatek, Hsinchu, Taiwan). The diffraction patterns were captured using a 5-megapixel CMOS image sensor (EO-5012M, Edmund Optics, Barrington, NJ, USA), and a custom-developed C-Chip (Infino, Seoul, Korea) was used to hold the cell samples [2,5]. All of these components can fit in a compact dimension of 100 mm $\times$ 120 mm $\times$ 80 mm. Due to the absence of a lens-based setup, the field-of-view of this system is about 20 times that of a conventional optical microscope at 100$\times$. This high-throughput nature provides an extra advantage to characterize several thousand cells within a single digital frame.

2.2. Preparation of Various Cell Lines

In this work we used various cell lines, starting from red blood cell (RBC), white blood cell (WBC), cancer cell lines HepG2 (human liver cell-line) and MCF7 (human breast cancer cell-line), and polystyrene microbeads of 10 μm and 20 μm. The preparations of these cell lines are as follows [2,3]. The use of human whole blood in the experiment was approved by the Institutional Review Board (Approval No. # 2021AN0040 of Korea University Anam Hospital (Seoul, Korea).

RBC: The RBC samples were prepared from the whole blood samples that were collected from the Korea University Anam Hospital under IRB approval. The samples were diluted about 16,000 times by using RPMI solution (Thermo Scientific, Waltham, MA, USA) [2,3].

WBC: First, Ficoll solution (Ficoll-Paque™ Plus, GE Healthcare, Chicago, IL, USA) was used to isolate mononuclear cells from the whole blood. The samples of peripheral blood mononuclear cells (PBMCs) obtained using the Ficoll solution are mixtures of lymphocytes and monocytes. To separate these two cell types, the MACS (Magnetic-activated cell sorting) device and antibodies (Miltenyi Biotec, Bergisch Gladbach, Germany) were utilized. The helper-T cells in the lymphocytes were separated using the CD4 antibody (#130-090-877), and the cytotoxic-T cells with the CD8 antibody (# 130-090-878). Finally, 10 μL of this solution was then loaded into the unruled C-Chip cell counting chamber [2,3].

HepG2: The HepG2 cell lines were prepared from the American Type Culture Collection (ATCC HB-8065) and incubated in a high-glucose medium (DMEM, Merck, Darmstadt, Germany) with 10% heat-inactivated fetal bovine serum, 0.1% gentamycin, and a 1 penicillin/streptomycin solution under 95% relative humidity and 5% CO_2 at 370 °C. The developed cells were then trypsinized and separated from 24 well pate and incubated from 2–5 min at 370 °C. These cells were then diluted with DMEM solution [2,3].

MCF7: The MCF7 cell samples were prepared from the American Type Culture Collection (ATCC HTB-22). The cells were preserved in a solution of DMEM containing 1% penicillin/streptomycin solution, 0.1% gentamycin, and 10% calf serum at 95% relative humidity and 5% CO_2 at 370 °C. These cells were then trypsinized and separated from the 24 well pate. These separated cells were then incubated for 2–5 min at 370 °C. The cells were then washed with DMEM solution. 10 μL of this solution was then loaded in the C-Chip [2,3].

Polystyrene microbead: The 10 μm and 20 μm bead samples were prepared by diluting the respective polystyrene microbeads (Thermo Scientific, Waltham, MA, USA) with de-ionized water [2,3].

2.3. Dataset Creation

A whole frame LSIT image (of cell diffraction patterns) contains an average of ~500 diffraction patterns of microparticles. Deep learning-based architectures utilize the features of each class, and typically require a minimum of a few hundred diffraction patterns of each cell type for optimal learning. Therefore, we cropped individual diffraction patterns of each cell type (that were verified using a traditional microscope) with a window of 66×66 pixels as shown in Figure 1d. This window size included the complete sample signature along with a minimal background that would provide complete cell-line information during the auto-feature selection process in learning algorithms. We further augment this base sample set by rotating the individual diffraction patterns with an increasing angle of 10 degrees clockwise. Finally, a dataset of 1980 samples for each of the six cell lines and microparticles was created, totaling 11,880 samples for all of the classes under study. The typical architecture of a CNN is illustrated in Figure 1e. As many learning algorithms are black-box models, it is difficult to ascertain the optimal cell-signal size that covers the majority of the information and minimal background. Naturally, a smaller cell-size would need lesser computation and have lesser noise. Hence, we created the dataset for $60 \times 60, 56 \times 56, 50 \times 50,$ $46 \times 46, 40 \times 40,$ and 36×36 cell sizes as input sets, with each set further divided into training and test folds. As the data augmentation used sample rotation, the splitting of the dataset into the train, validation, and test folds needs to be carried out while keeping a check on data leakage. Augmented samples distributed across the train and test sets may bias the model and may give a wrong estimate of its performance as the test data may not be of entirely "unseen" samples. Accounting for this, the 1980 samples of each class were carefully split into 1490 training samples, 166 validation samples, and 324 testing samples.

Though the cell-lines may seem visually similar, there are significant differences in the statistical distributions of the pixel illumination intensity in the cell diffraction pattern as revealed in our exploratory data analysis. The 2D contour plots (in Figure 2) show the observed variances. Hence, it is possible for intelligent algorithms to automatically identify and utilize the descriptive features in signal enhancement as well as classification.

2.4. Denoising Modality

For denoising of the LSIT micrographs, we adopted the concept of autoencoder [24]. An autoencoder is an unsupervised scheme that scuffles to recreate the input at its output. It consists of an input layer (x), an output layer (r), and a hidden layer (h). The hidden layer h termed as a code layer stands for the input in a revised dimension. The whole network structure can be labelled into two parts. The first part is an encoder, which tries to code the input as h = f(x), and the second part is a decoder which tries to recreate the input from the reduced code layer as r = g(h), where r is the recreated assortment of input x. Basically, it tries to attain r = g(f(x)). However, this is not a linear transformation since the model is enforced to learn the significant features of the input to encode it into the code layer.

In this work, we specifically used the denoising version of the autoencoder. Traditionally, the autoencoders try to reduce the loss as L(x, g(f(x))). However, the denoising autoencoder attempts to reduce the cost as L(x, g(f(x'))) where x' is the noisy form of the input x. We tried two different methods to design the denoising architectures, namely, extreme learning machine (ELM) and convolutional neural network (CNN).

2.4.1. ELM

This is a single hidden layer fully connected architecture [25]. In this method, the input weights are initiated randomly and kept intact. Only the output weights take part in the learning process through a straightforward learning method [25–27]. For N arbitrary input samples $x_i \in R^n$ and their counterpart targets $t_i \in R^m$, the ELM achieves this mapping using the following relation as shown in Equation (1).

$$H\beta = T \tag{1}$$

Here, H is the hidden layer output matrix, β is the output weight matrix (i.e., between the hidden layer and the output layer) and T is the target matrix or matrix of desired output [26]. From Equation (1), we can obtain the β using Moore–Penrose pseudoinverse [25] as shown in Equation (2).

$$\beta = \left(H^T H\right)^{-1} H^T T \tag{2}$$

In the extended sequential learning form of ELM, the β can update sequentially. This provides an added advantage of updating the learning whenever a new type of sample is available, thus providing the flexibility of transfer of learning. The β update mechanism [28,29] is as shown in Equation (3).

$$\beta_n = \beta_{n-1} + P_n^{-1}\left(T_n - H_n\beta_{n-1}\right)H_n^T \tag{3}$$

Here

$$P_n = P_{n-1} + H_n H_n^T \tag{4}$$

For $n = 1$,

$$P_{n-1} = P_0 = \left(\frac{1}{C} + H_0 H_0^T\right) \tag{5}$$

Here H_0 is the hidden layer output with the first sample or first batch of samples [30].

2.4.2. CNN

Convolutional neural networks (CNN) [6,19,20] are a type of neural network widely used in the analysis of spatial data such as image classification and object segmentation. In this network, two-dimensional kernels are used to extract the spatial features from the input patterns, using a convolution operation between the kernel and the input. The typical architecture of a CNN is as shown in Figure 1e. Here, the kernel is shared spatially by the input or by the feature map. The feature at the location (i, j) in the kth feature map of the lth layer can be evaluated as shown in Equation (6).

$$Z^l_{i,j,k} = \left(W^l_K\right)^T X^l_{i,j} + b^l_k \tag{6}$$

Here, W^l_K and b^l_k are the weights and the bias vector of the kth filter in the lth layer. Here the weight layer is shared spatially which reduces the complexity. $X^l_{i,j}$ is the value of the input at location (i, j) of the lth layer. The nonlinearity in this network can be obtained by introducing the activation function, denoted here as $g(.)$. The activated output can be represented as shown in Equation (7).

$$a^l_{i,j,k} = g\left(Z^l_{i,j,k}\right) \tag{7}$$

Additionally, there are pooling layers that introduce shift-invariance by reducing the resolution of the activated feature maps. Each pooling layer connects the feature map to the preceding convolutional layer. The expression for pooling is as shown in Equation (8).

$$y^l_{i,j,k} = P\left(a^l_{n,m,k}\right), \forall (m,n)\epsilon R_{ij} \tag{8}$$

Here $P(.)$ is a pooling operation for the local neighborhood R_{ij} around the location (i, j). In this work, we used CNN for both denoising as well as classification. The details of their architectures and their impacts are discussed in the Section 3.

3. Results and Discussion

3.1. Performance of Denoising Algorithms

For efficient and adaptive denoising, we analyzed various autoencoder schemes, starting with the fully-connected autoencoder. In our first iteration, we experimented with the fully connected network having three hidden layers with 512, 256, and 512 neurons, respectively. The input layer is the 1D vectorized array of the input cell diffraction pattern, e.g., of 66×66 pixels. The input to the model is the noisy version of the input cell diffraction pattern and the expected target output is the original cell diffraction pattern. The noisy cell diffraction patterns were created using a Gaussian distribution with variance ranging from 100 to 600 with zero mean (refer to the supplementary section for detail). Further, we experimented with an increased network size having five hidden layers with 256, 128, 64, 128, and 256 neurons, respectively. In all of these networks, rectified linear unit (ReLU) was used as the activation function while mean squared error (MSE) [31–34] was used to calculate the loss. The Adam optimizer [35,36] was found to deliver better convergence and hence used to perfect the weight and biases. The denoising performance was quantified in terms of the improvement in SNR, measured in dB, denoted here by SNR_{imp}, as given by Equation (9) [37].

$$SNR_{imp} = SNR_{out} - SNR_{in} \tag{9}$$

where $SNR_{out} = 10log_{10}\left(\frac{\sum_{n=1}^{N} x_i^2}{\sum_{n=1}^{N}(\hat{x}_i - x_i)^2}\right)$, and $SNR_{in} = 10log_{10}\left(\frac{\sum_{n=1}^{N} x_i^2}{\sum_{n=1}^{N}(\tilde{x}_i - x_i)^2}\right)$. Here x_i is the value of sampling point i in the original LSIT signal, $\tilde{x}_i$ is the value of sampling point i in the noisy LSIT, and $\hat{x}_i$ is the value of sampling point i in the denoised version of the same cell diffraction pattern. N is the total number of sample points in that LSIT image (cell diffraction patterns).

The fully connected network for both the above configuration shows no significant improvement in SNR_{imp} after reaching saturation at around -10.08 dB. For further improvement, we experimented with CNN architecture using various models with a different number of convolution layers and distinct kernel sizes. The configuration of the model which accomplished the best outcomes is $3 \times 3, 3 \times 3, 5 \times 5, 5 \times 5, 7 \times 7, 7 \times 7, 1 \times 1$ with 32 filters in each layer except the last layer. The last layer consists of a single pixel filter (1×1 filter) that is used to condense the output across all the 32 filters. Here, the input and output size are the same. Padding was used to maintain the original size after the output of each convolutional layer. The Adam optimizer was used to optimize the network to reduce the mean squared error loss. The CNN results show a better reconstruction as shown in Figure 2.

The CNN network has been optimized for various parameters. First, the optimization of the network for various design parameters, such as varying the convolution layers and the kernel sizes, was carried out. The results in Figure 3a show that the architecture with kernel sizes $3 \times 3, 3 \times 3, 5 \times 5, 5 \times 5, 7 \times 7, 7 \times 7, 1 \times 1$ has a better performance in terms of SNR_{imp}. The performance of the optimized network for various noise parameters, as shown in Figure 3b, indicates the network performs better reconstruction with increasing noise variance in the image (cell diffraction pattern). An increase in the variance results in a noisier image (cell diffraction pattern), which warrants a detailed reconstruction to reverse it to the original form, and hence larger the value of SNR_{imp}. Therefore, a higher improvement in SNR_{imp} implies the network has learned the optimal representational features for the cell types which enables it to perform a better qualitative reconstruction. Figure 3c compares the reconstruction performance of the model on different sizes of the input image (cell diffraction pattern). Due to the black-box nature of deep learning methods, we had to create datasets with multiple cell-signature dimensions, such that the smallest size just covered the central signature of the cell and increased the window size till it covered a significant background portion as well. The models were evaluated across varying cell sizes to determine the optimal signal to background ratio, the spatial extent up to which the models covered the features, and to study its effects on the model performance. This analysis is critical in understanding the model explainability and interpretability since having a size larger than the optimum increases the inclusion of background artifacts that affect denoising as well as overpower the cell signal while having a smaller one could exclude the important deterministic features of the cell signature. The convergence in the training phase of the network is as shown in Figure 3d. The results depict that the loss across the first epoch, with a high variation in the initial phase, gets smoother towards the end of the first iteration. The advantage of this system is that it generalizes well for all of the types of cell lines using the same model.

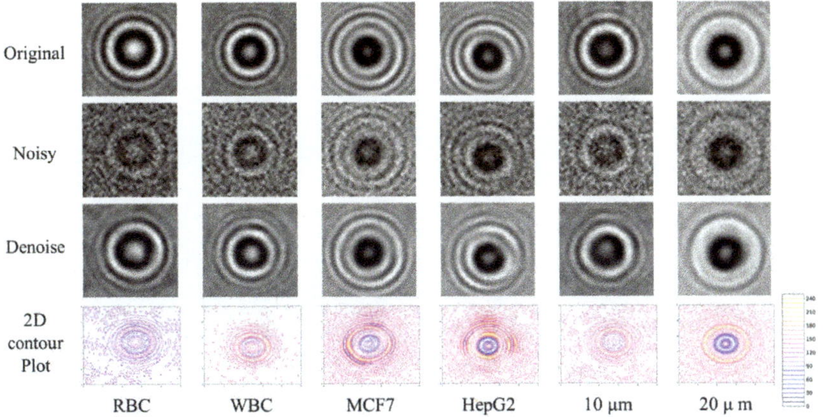

Figure 2. Reconstructed results from the optimized CNN. The top row is the original LSIT image (cell diffraction pattern) of a single RBC, WBC, MCF7, HepG2, 10 μm beam, and 20 μm bead. The second row is the noisy version (with variance 100) of the corresponding original images (cell diffraction pattern). The third row is the denoised version of the corresponding original images (cell diffraction pattern) from the noisy image (cell diffraction pattern). The fourth row is the 2D intensity contour plot of the original image (cell diffraction pattern) to show the unique signature of each of these cell lines.

Further, we tried the ELM architecture which is well known for its fast convergence [25]. The results in Figure 3e–h show the performance of the ELM architecture with varying number of neurons in the hidden layer. As it can be concluded from Figure 3e, the model with 2000 neurons provides better performance in terms of SNR_{imp}. Further, the optimized model has been used to test the performance across various noise levels as shown in Figure 3f. It is observed that the model maintains the SNR_{imp} value on increasing the noise in the input image (cell diffraction pattern), i.e., the image (cell diffraction pattern) quality of the output relative to the input remains the same. The results of the model performance across different sizes of the input image (cell diffraction pattern), as shown in Figure 3g, indicate that the 40×40 is having a higher value of SNR. However, the variation is of 2 as compared to the variation for the size 50×50, which is of 1.5, representing the lowest compared to all of the other sizes. Since CNN shows a substantial performance with lower variance for the 50×50 input size, therefore we fixed it as optimal for all of the further studies and comparisons. Figure 3h shows the loss across the first epoch for ELM which is remarkably high initially but converges faster, as compared to CNN, after training with only a few thousand samples. This faster convergence may help save time and resources during incremental training phases for newer cell types. The performances of these optimized models have been compared with the traditional denoising methods as shown in Table 1 (The comparative visual reconstructions are provided in the supplementary document). It is concluded from the data that CNN shows better performance compared to the other modalities (The details of the traditional methods are provided in the supplementary document). Therefore, we prefer to use CNN for denoising.

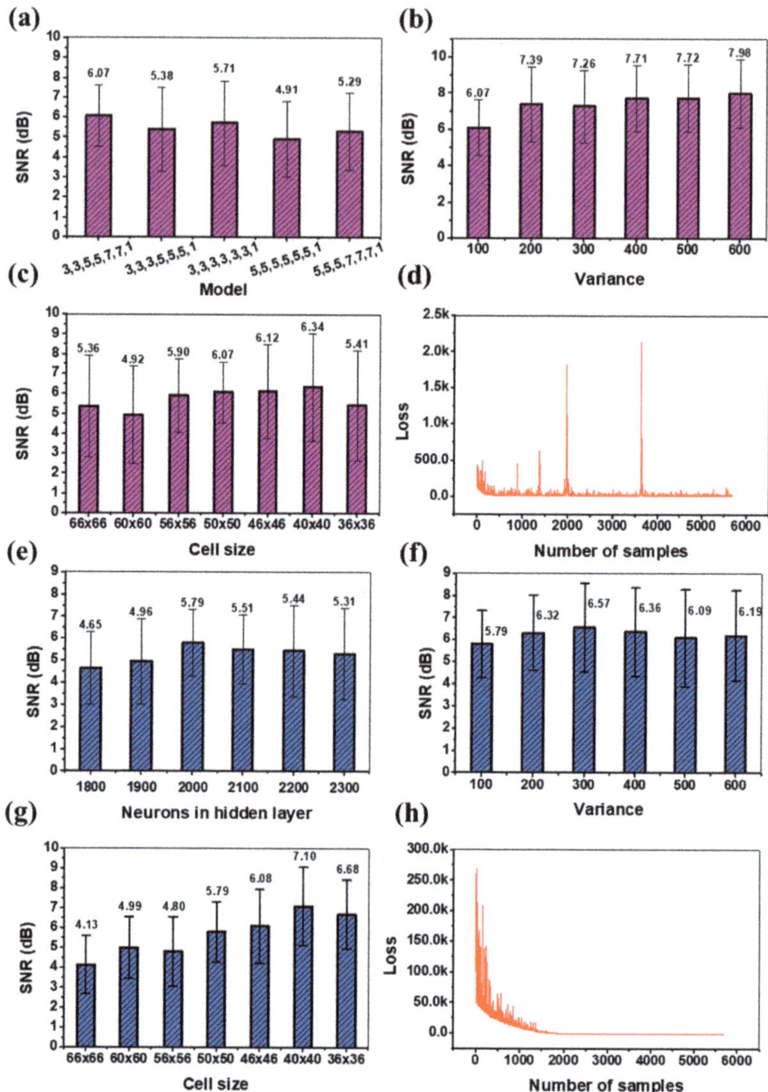

Figure 3. Results from the CNN and ELM autoencoder. (**a**) The performance of the CNN autoencoder for improved *SNR* (average of all of the classes) across various layers and kernel sizes. (**b**) The performance of the CNN autoencoder across varying noise levels. Here, the variance of the Gaussian noise ranges from 100 to 600, and is evaluated on the optimal network architecture, i.e., 3,3,5,5,7,7,1. (**c**) The performance of the CNN autoencoder across various input sizes (cropping size). Here, the sizes vary from 66 × 66 to 36 × 36. (**d**) The convergence of the optimal CNN network with the number of samples for the first epoch. (**e**) The number of hidden layer neurons in ELM autoencoder vs. improvement in *SNR*, (**f**) variance vs. improved SNR for ELM autoencoder, (**g**) Input size vs. improved SNR for ELM autoencoder, and (**h**) The convergence of ELM autoencoder within the first epoch.

Table 1. Comparison of improved *SNR* with respect to the variance for various denoising modalities (here we keep input and output size as 50×50).

Variance	Gaussian	Average	Median	Bilateral	BM3D	CNN	ELM
100	4.525580	4.237872	3.199475	-0.02635	5.684597	6.06905	5.79161
200	4.613044	4.198564	3.068866	-0.00640	5.713797	7.38544	6.31896
300	4.476552	4.259405	3.078091	-0.10943	5.119646	7.26487	6.57256
400	4.674922	4.703331	3.405470	-0.05448	3.792226	7.71265	6.36135
500	4.128021	4.201562	3.198964	-0.14542	2.206876	7.72364	6.09254
600	4.023621	4.183404	2.908946	-0.16924	1.683290	7.97877	6.19359

3.2. Performance of Classification Algorithm

Since the diffraction patterns of cells and microparticles in a LSIT micrograph depend upon their physical and optical properties, therefore, the diffraction patterns carry the unique signatures of each of the cell types as shown in the 2D contour plot in Figure 2. These unique signatures can be utilized for the classification of these cell types. Since our previous inference concludes that CNN works better for denoising, therefore we experimented with the same modality for the classification as well. In this work, in order to determine the optimal architecture of CNN for cell-line recognition, we first proceeded to find the optimal depth of the network by studying the classification performance of the model on increasing the depth, by adding convolutional and pooling layers, as well as by varying number of kernels and kernel size, till we reached performance saturation. We have experimented with and evaluated various shallow and deep CNN models to classify cell lines. The details of the model architecture are as described in Figure 4.

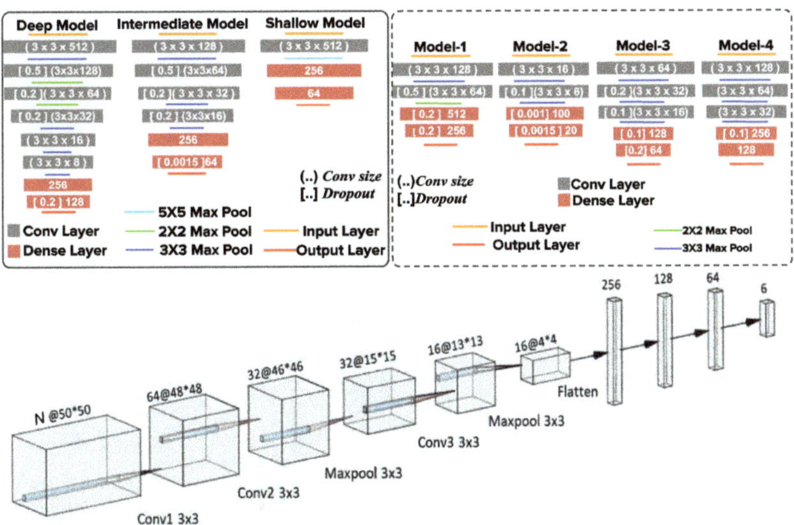

Figure 4. The CNN Architectures with varying depth. The models in the solid box are for the optimization of the model with varying depth. The models inside the dotted box are for the optimization of the parameters. Here, the orange line is the input layer and the rectangle represents the 2D convolution layer. The kernel size and number of kernels are indicated inside the round brackets, the dropout rate in square brackets, and the number of neurons in fully connected layers are placed directly within the rectangle. The aqua blue, green, and purple lines are the max-pool layers. The red line is the output layer and uses a SoftMax activation function. The final optimized architecture is as represented by the last 3D figure.

The Deep Model starts with a convolutional (Conv2D) layer having 512 kernels of size 3×3, followed by a max-pooling layer of the same kernel size. The output from this is further convoluted with 128 kernels of 3×3 size with a dropout rate of 0.5, and then a max pool with 2×2 kernel. This output goes to a Conv2D layer with 64 of 3×3 sized kernels and a dropout rate of 0.2. We further reduce the dimension using a 2×2 max pool kernel. The output of this layer further convolves with 32 of 3×3 kernels, then a dropout of 0.2. The output dimension from this convoluted layer is further reduced by using the max pool with a 3×3 kernel. This again convolves with 16 of 3×3 kernels, and a max pool layer with 3×3 kernel. The output of which is again convoluted with 8 of 3×3 kernels, followed by a 3×3 max pool. This output is then vectorized and input to a fully connected (FC) layer having 256 nodes, and then to another FC with 128 nodes and having a dropout of 0.2. The final layer is a SoftMax function, with six output nodes. The model architectures used for studying the impact of the network depth and breadth on performance are well described in Figure 4. Once the approximate optimal depth and breadth had been determined, we proceeded to fine-tune the hyper-parameters such as the number of kernels, kernel size, and dropouts, to reach the best performance of the models across varying cell sizes (i.e., input dimension of cells). In all of the models, Adam [38] provided better convergence as compared to other optimizers and has been used as the model optimizer, with categorical cross-entropy [39,40] as a loss estimator. The result for depth and breadth optimization indicates that the average accuracy of the intermediate model is >0.85 (including all of the classes), whereas it is <0.85 for shallow and deep networks. Therefore, we proceed with this intermediate model as an optimum model for further study. Considering the intermediate model has the optimum breadth and depth, the further optimization of the parameters and the results are as shown in Figure 5a.

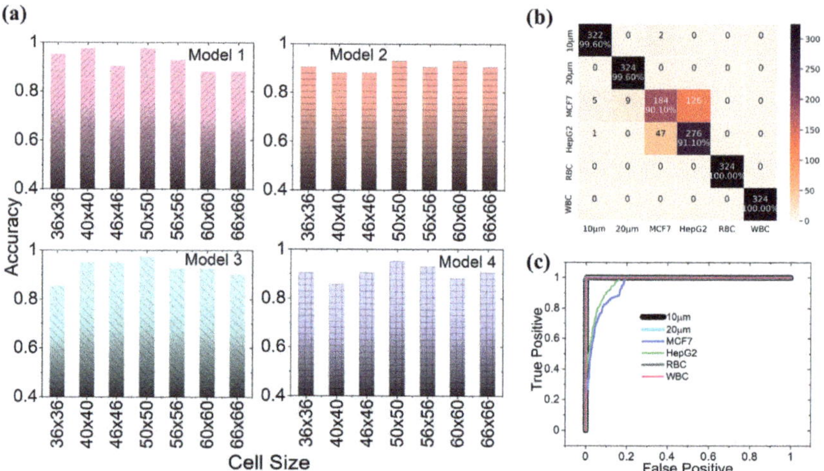

Figure 5. Results for the optimization of the CNN model. (**a**) Performance of the fine-tuned intermediate architecture. (**b**) The confusion matrix showing classification accuracy on the test dataset across all cell lines using the fine-tuned and optimized model. (**c**) The receiver operating characteristic (ROC) curve for each of the cell lines for the optimized model.

From Figure 5a, it is inferred that Model 3 shows better classification performance, on the validation fold, of all of the models. The results depict that there is consistency in performance for the input sizes 40×40 to 66×66, with a very small variance in the accuracy. The performance of this optimized model is further evaluated over the test dataset containing 324 samples of each cell type. The per-class performance of this model is shown in the confusion matrix of Figure 5b. The results depict that the model can classify

RBC, WBC, 10 μm, and 20 μm bead with over 99% accuracy. However, the comparatively poor performance of about 90% for the cancer cells, HepG2 and MCF7, can be attributed to the non-homogeneity in their signature characteristics as well as the lack of sufficient original samples which further complicates the issue. This is well depicted in our previous work [4] (see Figure 2 of the reference). From the receiver operating characteristic (ROC) curve for all of the cell lines shown in Figure 5c, the area under the curve (AUC) for all of the cell lines is >0.99, except MCF7 (~0.95) and HepG2 (~0.96). From these results, it can be inferred that the classifier is working well, especially for RBC, WBC, 10 μm, and 20 μm beads. The visualization of the internal activation maps, as shown in the Supplementary Information, implies that the network is learning core descriptive features in the diffraction signatures rather than using some random features.

The performance evaluation of the proposed AI model with various matrices such as true positive (TP), true negative (TN), false positive (FP), false negative (FN), accuracy, recall, specificity, sensitivity, F1 score, positive predictive value (PPV) and negative predictive value (NPV) is summarized in Table 2.

Table 2. Performance evaluation scores of the proposed AI model using various metrics for different cell types.

Sample Type	TP	TN	FP	FN	Accuracy	Precision	Recall	Specificity	Sensitivity	F1	PPV	NPV
10 μm bead	322	1614	2	6	0.9959	0.9938	0.9817	0.9988	0.9817	0.9877	0.9938	0.9963
20 μm bead	324	1611	0	9	0.9954	1.0000	0.9730	1.0000	0.9730	0.9863	1.0000	0.9944
MCF7	184	1571	140	49	0.9028	0.5679	0.7897	0.9182	0.7897	0.6607	0.5679	0.9698
HepG2	276	1494	48	126	0.9105	0.8519	0.6866	0.9689	0.6866	0.7603	0.8519	0.9222
RBC	324	1620	0	0	1.0000	1.0000	1.0000	1.0000	1.0000	1.0000	1.0000	1.0000
WBC	324	1620	0	0	1.0000	1.0000	1.0000	1.0000	1.0000	1.0000	1.0000	1.0000

Additionally, we also investigate the transfer of learning to gauge the ability of the trained network to adapt to newer cell types (Figure 6). For this scenario, the CNN was initially trained with all of the cell lines except RBC. From the epoch vs. accuracy graph in Figure 6a, the transfer training achieved higher accuracy with the same number of epochs compared to the initial training. This is also validated by the epoch vs. loss graph in Figure 6b. From these results, it can be inferred that the network can be effectively used to adapt to newer cell lines with very less amount of training. From the per-class test accuracy shown in Figure 6c, it is observed that the model misclassified all of the RBC samples as WBC. In the transfer of the learning phase, the initially trained network is frozen except for the last layer, which is modified to accommodate the newer class and kept trainable. The network is then re-trained with a mix of RBC samples. The per-class test accuracy of the re-trained model is shown in the confusion matrix of Figure 6d, where it is inferred that the re-trained network can classify RBC correctly with substantial accuracy.

The comparison between the proposed AI method and the manual method for counting various cell types from a heterogeneous sample is shown in Figure 7. This comparison shows the robustness of the model.

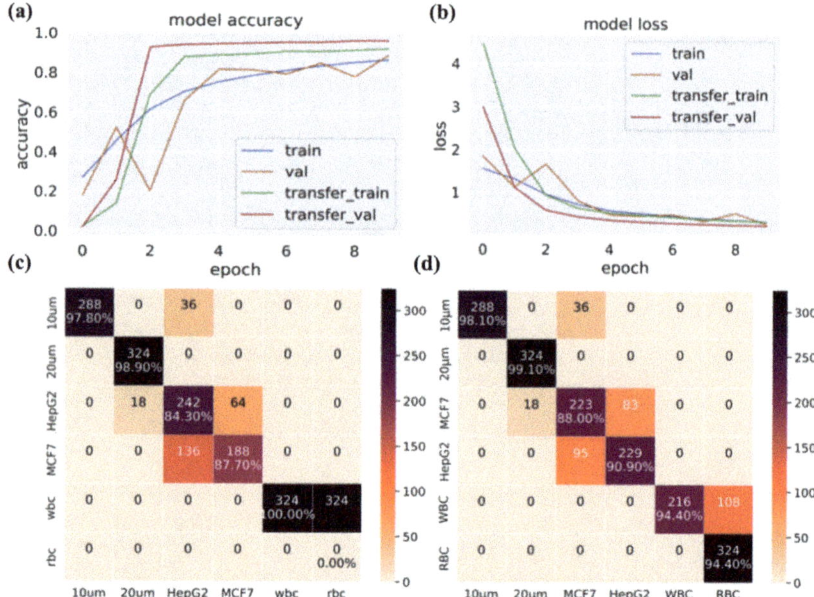

Figure 6. The results from the transfer of learning study. (**a**) Epoch vs. accuracy graph for the initial epochs in the initial training phase without RBC and then the transfer of training with RBC. The blue and saffron colors represent the initial training accuracy and validation curves. The green and red lines represent the transfer learning accuracy and validation curves, (**b**) epoch vs. loss, (**c**) the confusion matrix from the pre-trained model, and (**d**) the confusion matrix with the re-trained model.

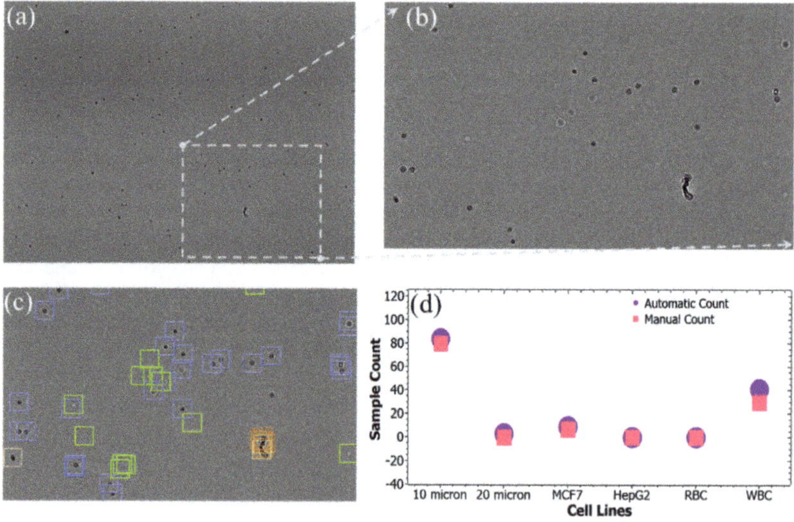

Figure 7. Comparison between the proposed AI method and the manual method for counting. (**a**) A whole frame LSIT micrograph, (**b**) magnified region-of-interest (ROI) of the whole frame LSIT micrograph, (**c**) automatically classified cell diffraction patterns of the ROI, and (**d**) comparison of the cell counts between those two different modalities.

4. Conclusions

In conclusion, we have explored the advantages of using neural networks in the characterization of LSIT micrographs. Here, we have perfected neural networks that can automatically improve the signal quality and classify the cell types. We find that this neural network can classify the RBC and WBC with great accuracy (i.e., over 98%), and the cancer cell with an accuracy of about 90%. This network is also flexible for adapting to newer cell lines by retraining the trained network with very few samples. Together with this algorithm, the lightweight and cost-effective LSIT setup can be utilized as a point of care system for the diagnosis of pathological disorders in the resource-limited setup of our world. In our future work, we aim to combine the denoising and classification modalities due to the significant overlap in their operation. This will remove the dual training times as well as minimize computation costs. Also, we aim to work on improving their performance and deploying it in real scenarios.

Supplementary Materials: The following supporting information can be downloaded at: https://www.mdpi.com/article/10.3390/bios12030144/s1, 1. Convolutional Neural Network workflow; 2. Gaussian Noise; 3. Traditional denoising methods; 4. Comparison of the denoised outputs from various modalities; 5. SNR of individual samples of various cell types; 6. Grad-CAM and Saliency maps of the individual cell-lines.

Author Contributions: Conceptualization, S.S. (Sungkyu Seo) and M.R.; methodology, R.V., S.S. (Sanghoon Shin), V.C., K.K. and S.S. (Smit Sanghavi); validation, R.V., V.C., K.K., S.S. (Smit Sanghavi) and M.R.; formal analysis, R.V., K.K., S.S. (Sungkyu Seo) and M.R.; investigation, R.V., V.C., K.K. and S.S. (Smit Sanghavi); resources, S.S. (Sungkyu Seo) and M.R.; data curation, R.V., S.S. (Sanghoon Shin), V.C., K.K. and S.S. (Smit Sanghavi); writing—original draft preparation, R.V., V.C., K.K., S.S. (Smit Sanghavi) and M.R.; writing—review and editing, R.V., S.S. (Sanghoon Shin), S.S. (Sungkyu Seo) and M.R.; visualization, R.V., S.S. (Sanghoon Shin), V.C., S.S. (Sungkyu Seo) and M.R.; supervision, M.R.; project administration, S.S. (Sungkyu Seo) and M.R.; funding acquisition, S.S. (Sungkyu Seo) and M.R. All authors have read and agreed to the published version of the manuscript.

Funding: M.R. acknowledges the seed grant No. ORSP/R&D/PDPU/2019/MR/RO051 of PDPU (the AI software development part) and the core research grant No. CRG/2020/000869 of the Science and Engineering Research Board (SERB), India. S.S. acknowledges the support of the Basic Science Research Program (Grant#: 2014R1A6A1030732, Grant#: 2020R1A2C1012109) through the National Research Foundation (NRF) of Korea, the Korea Medical Device Development Fund (Project#: 202012E04) granted by the Korean government (the Ministry of Science and ICT, the Ministry of Trade, Industry and Energy, the Ministry of Health & Welfare, the Ministry of Food and Drug Safety), and the project titled 'Development of Management Technology for HNS Accident' and 'Development of Technology for Impact Assessment and Management of HNS discharged from Marine Industrial Facilities', funded by the Ministry of Oceans and Fisheries of Korea.

Institutional Review Board Statement: All human blood samples were approved by the Institutional Review Board of Anam Hospital, Korea University (# 2021AN0040).

Informed Consent Statement: Not applicable.

Data Availability Statement: The data underlying the results presented in this paper are not publicly available at this time but may be obtained from the authors upon reasonable request.

Conflicts of Interest: The authors declare that they have no known competing financial interests or personal relationships that could have appeared to influence the work reported in this paper. The funders had no role in the design of the study; in the collection, analyses, or interpretation of data; in the writing of the manuscript, or in the decision to publish the results.

References

1. Mudanyali, O.; Erlinger, A.; Seo, S.; Su, T.W.; Tseng, D.; Ozcan, A. Lensless On-Chip Imaging of Cells Provides a New Tool for High-Throughput Cell-Biology and Medical Diagnostics. *J. Vis. Exp.* **2009**, e1650. [CrossRef] [PubMed]
2. Roy, M.; Seo, D.; Oh, C.H.; Nam, M.H.; Kim, Y.J.; Seo, S. Low-Cost Telemedicine Device Performing Cell and Particle Size Measurement Based on Lens-Free Shadow Imaging Technology. *Biosens. Bioelectron.* **2015**, *67*, 715–723. [CrossRef] [PubMed]

3. Roy, M.; Jin, G.; Seo, D.; Nam, M.H.; Seo, S. A Simple and Low-Cost Device Performing Blood Cell Counting Based on Lens-Free Shadow Imaging Technique. *Sens. Actuators B Chem.* **2014**, *201*, 321–328. [CrossRef]
4. Roy, M.; Seo, D.; Oh, S.; Chae, Y.; Nam, M.H.; Seo, S. Automated Micro-Object Detection for Mobile Diagnostics Using Lens-Free Imaging Technology. *Diagnostics* **2016**, *6*, 17. [CrossRef]
5. Roy, M.; Jin, G.; Pan, J.H.; Seo, D.; Hwang, Y.; Oh, S.; Lee, M.; Kim, Y.J.; Seo, S. Staining-Free Cell Viability Measurement Technique Using Lens-Free Shadow Imaging Platform. *Sens. Actuators B Chem.* **2016**, *224*, 577–583. [CrossRef]
6. Liu, Y.; Gadepalli, K.; Norouzi, M.; Dahl, G.E.; Kohlberger, T.; Boyko, A.; Venugopalan, S.; Timofeev, A.; Nelson, P.Q.; Corrado, G.S.; et al. Detecting Cancer Metastases on Gigapixel Pathology Images. Available online: http://arxiv.org/abs/1703.02442 (accessed on 23 February 2022).
7. Yu, K.H.; Beam, A.L.; Kohane, I.S. Artificial Intelligence in Healthcare. *Nat. Biomed. Eng.* **2018**, *2*, 719–731. [CrossRef]
8. Liu, Y.; Kohlberger, T.; Norouzi, M.; Dahl, G.E.; Smith, J.L.; Mohtashamian, A.; Olson, N.; Peng, L.H.; Hipp, J.D.; Stumpe, M.C. Artificial Intelligence–Based Breast Cancer Nodal Metastasis Detection Insights into the Black Box for Pathologists. *Arch. Pathol. Lab. Med.* **2019**, *143*, 859–868. [CrossRef]
9. Ardila, D.; Kiraly, A.P.; Bharadwaj, S.; Choi, B.; Reicher, J.J.; Peng, L.; Tse, D.; Etemadi, M.; Ye, W.; Corrado, G.; et al. End-to-End Lung Cancer Screening with Three-Dimensional Deep Learning on Low-Dose Chest Computed Tomography. *Nat. Med.* **2019**, *25*, 954–961. [CrossRef]
10. Im, H.; Pathania, D.; McFarland, P.J.; Sohani, A.R.; Degani, I.; Allen, M.; Coble, B.; Kilcoyne, A.; Hong, S.; Rohrer, L.; et al. Design and Clinical Validation of a Point-of-Care Device for the Diagnosis of Lymphoma via Contrast-Enhanced Microholography and Machine Learning. *Nat. Biomed. Eng.* **2018**, *2*, 666–674. [CrossRef]
11. Wang, P.; Xiao, X.; Glissen Brown, J.R.; Berzin, T.M.; Tu, M.; Xiong, F.; Hu, X.; Liu, P.; Song, Y.; Zhang, D.; et al. Development and Validation of a Deep-Learning Algorithm for the Detection of Polyps during Colonoscopy. *Nat. Biomed. Eng.* **2018**, *2*, 741–748. [CrossRef]
12. Ballard, Z.S.; Joung, H.A.; Goncharov, A.; Liang, J.; Nugroho, K.; di Carlo, D.; Garner, O.B.; Ozcan, A. Deep Learning-Enabled Point-of-Care Sensing Using Multiplexed Paper-Based Sensors. *Npj Digit. Med.* **2020**, *3*, 66. [CrossRef] [PubMed]
13. Rivenson, Y.; Göröcs, Z.; Günaydin, H.; Zhang, Y.; Wang, H.; Ozcan, A. Deep Learning Microscopy. *Optica* **2017**, *4*, 1437–1443. [CrossRef]
14. Ozcan, A. Deep Learning-Enabled Holography (Conference Presentation). In Proceedings of the Label-Free Biomedical Imaging and Sensing (LBIS) 2020, San Francisco, CA, USA, 10 March 2020; Shaked, N.T., Hayden, O., Eds.; SPIE: Bellingham, DC, USA, 2020; p. 60.
15. Wang, H.; Koydemir, H.C.; Qiu, Y.; Bai, B.; Zhang, Y.; Jin, Y.; Tok, S.; Yilmaz, E.C.; Gumustekin, E.; Luo, Y.; et al. Deep Learning Enables High-Throughput Early Detection and Classification of Bacterial Colonies Using Time-Lapse Coherent Imaging (Conference Presentation). In Proceedings of the Optics and Biophotonics in Low-Resource Settings VI, San Francisco, CA, USA, 9 March 2020; Levitz, D., Ozcan, A., Eds.; SPIE: Bellingham, DC, USA, 2020; p. 13.
16. Liu, T.; de Haan, K.; Rivenson, Y.; Wei, Z.; Zeng, X.; Zhang, Y.; Ozcan, A. Enhancing Resolution in Coherent Microscopy Using Deep Learning (Conference Presentation). In Proceedings of the Quantitative Phase Imaging VI, San Jose, CA, USA, 11 March 2020; p. 3.
17. Liu, T.; de Haan, K.; Bai, B.; Rivenson, Y.; Luo, Y.; Wang, H.; Karalli, D.; Fu, H.; Zhang, Y.; Fitzgerald, J.; et al. Deep Learning-Based Holographic Polarization Microscopy. *ACS Photonics* **2020**, *7*, 3023–3034. [CrossRef]
18. Kunapuli, S.S.; Bh, P.C.; Singh, U. Enhanced Medical Image De-Noising Using Auto Encoders and MLP. In Proceedings of the International Conference on Intelligent Technologies and Applications, Bahawalpur, Pakistan, 6–8 November 2019; pp. 3–15.
19. Indolia, S.; Goswami, A.K.; Mishra, S.P.; Asopa, P. Conceptual Understanding of Convolutional Neural Network—A Deep Learning Approach. *Procedia Comput. Sci.* **2018**, *132*, 679–688. [CrossRef]
20. Yamashita, R.; Nishio, M.; Do, R.K.G.; Togashi, K. Convolutional Neural Networks: An Overview and Application in Radiology. *Insights Imaging* **2018**, *9*, 611–629. [CrossRef]
21. Seo, S.; Su, T.W.; Tseng, D.K.; Erlinger, A.; Ozcan, A. Lensfree Holographic Imaging for On-Chip Cytometry and Diagnostics. *Lab Chip* **2009**, *9*, 777–787. [CrossRef]
22. Mudanyali, O.; Tseng, D.; Oh, C.; Isikman, S.O.; Sencan, I.; Bishara, W.; Oztoprak, C.; Seo, S.; Khademhosseini, B.; Ozcan, A. Compact, Light-Weight and Cost-Effective Microscope Based on Lensless Incoherent Holography for Telemedicine Applications. *Lab Chip* **2010**, *10*, 1417–1428. [CrossRef]
23. Jin, G.; Yoo, I.H.; Pack, S.P.; Yang, J.W.; Ha, U.H.; Paek, S.H.; Seo, S. Lens-Free Shadow Image Based High-Throughput Continuous Cell Monitoring Technique. *Biosens. Bioelectron.* **2012**, *38*, 126–131. [CrossRef]
24. Vincent, P.; Larochelle, H.; Lajoie, I.; Bengio, Y.; Manzagol, P.-A. Stacked Denoising Autoencoders: Learning Useful Representations in a Deep Network with a Local Denoising Criterion. *J. Mach. Learn. Res.* **2010**, *11*, 3371–3408.
25. Cambria, E.; Huang, G.-B.; Kasun, L.L.C.; Zhou, H.; Vong, C.M.; Lin, J.; Yin, J.; Cai, Z.; Liu, Q.; Li, K.; et al. Extreme Learning Machines. *IEEE Intell. Syst.* **2013**, *28*, 30–59. [CrossRef]
26. Bucurica, M.; Dogaru, R.; Dogaru, I. A Comparison of Extreme Learning Machine and Support Vector Machine Classifiers. In Proceedings of the 2015 IEEE 11th International Conference on Intelligent Computer Communication and Processing (ICCP), Cluj-Napoca, Romania, 3–5 October 2015; Institute of Electrical and Electronics Engineers Inc.: Piscataway, NJ, USA; pp. 471–474.

27. Huang, G.; Huang, G.-B.; Song, S.; You, K. Trends in Extreme Learning Machines: A Review. *Neural Netw.* **2015**, *61*, 32–48. [CrossRef]
28. Liang, N.Y.; Huang, G.-B.; Saratchandran, P.; Sundararajan, N. A Fast and Accurate Online Sequential Learning Algorithm for Feedforward Networks. *IEEE Trans. Neural Netw.* **2006**, *17*, 1411–1423. [CrossRef] [PubMed]
29. van Schaik, A.; Tapson, J. Online and Adaptive Pseudoinverse Solutions for ELM Weights. *Neurocomputing* **2015**, *149*, 233–238. [CrossRef]
30. Roy, M.; Bose, S.K.; Kar, B.; Gopalakrishnan, P.K.; Basu, A. A Stacked Autoencoder Neural Network Based Automated Feature Extraction Method for Anomaly Detection in On-Line Condition Monitoring. In Proceedings of the 2018 IEEE Symposium Series on Computational Intelligence (SSCI), Bangalore, India, 18–21 November 2018; pp. 1501–1507.
31. Ahmed, F.E. Artificial Neural Networks for Diagnosis and Survival Prediction in Colon Cancer. *Mol. Cancer* **2005**, *4*, 29. [CrossRef]
32. Zhang, Y.D.; Pan, C.; Chen, X.; Wang, F. Abnormal Breast Identification by Nine-Layer Convolutional Neural Network with Parametric Rectified Linear Unit and Rank-Based Stochastic Pooling. *J. Comput. Sci.* **2018**, *27*, 57–68. [CrossRef]
33. Munir, K.; Elahi, H.; Ayub, A.; Frezza, F.; Rizzi, A. Cancer Diagnosis Using Deep Learning: A Bibliographic Review. *Cancers* **2019**, *11*, 1235. [CrossRef]
34. Shanthi, P.B.; Faruqi, F.; Hareesha, K.S.; Kudva, R. Deep Convolution Neural Network for Malignancy Detection and Classification in Microscopic Uterine Cervix Cell Images. *Asian Pac. J. Cancer Prev.* **2019**, *20*, 3447–3456. [CrossRef]
35. Shen, L.; Margolies, L.R.; Rothstein, J.H.; Fluder, E.; McBride, R.; Sieh, W. Deep Learning to Improve Breast Cancer Detection on Screening Mammography. *Sci. Rep.* **2019**, *9*, 12495. [CrossRef]
36. Kingma, D.P.; Ba, J. Adam: A Method for Stochastic Optimization. *arXiv* **2014**, arXiv:1412.6980.
37. Chiang, H.T.; Hsieh, Y.Y.; Fu, S.W.; Hung, K.H.; Tsao, Y.; Chien, S.Y. Noise Reduction in ECG Signals Using Fully Convolutional Denoising Autoencoders. *IEEE Access* **2019**, *7*, 60806–60813. [CrossRef]
38. Wang, Y.; Liu, J.; Misic, J.; Misic, V.B.; Lv, S.; Chang, X. Assessing Optimizer Impact on DNN Model Sensitivity to Adversarial Examples. *IEEE Access* **2019**, *7*, 152766–152776. [CrossRef]
39. Zhang, L.; Gao, H.J.; Zhang, J.; Badami, B. Optimization of the Convolutional Neural Networks for Automatic Detection of Skin Cancer. *Open Med.* **2020**, *15*, 27–37. [CrossRef] [PubMed]
40. Rączkowski, Ł.; Możejko, M.; Zambonelli, J.; Szczurek, E. ARA: Accurate, Reliable and Active Histopathological Image Classification Framework with Bayesian Deep Learning. *Sci. Rep.* **2019**, *9*, 14347. [CrossRef] [PubMed]

Article

Decoding Vagus-Nerve Activity with Carbon Nanotube Sensors in Freely Moving Rodents

Joseph T. Marmerstein, Grant A. McCallum and Dominique M. Durand *

Neural Engineering Center, Biomedical Engineering, Case Western Reserve University, Cleveland, OH 44106, USA; jtm124@case.edu (J.T.M.); gam19@case.edu (G.A.M.)
* Correspondence: dxd6@case.edu

Abstract: The vagus nerve is the largest autonomic nerve and a major target of stimulation therapies for a wide variety of chronic diseases. However, chronic recording from the vagus nerve has been limited, leading to significant gaps in our understanding of vagus nerve function and therapeutic mechanisms. In this study, we use a carbon nanotube yarn (CNTY) biosensor to chronically record from the vagus nerves of freely moving rats for over 40 continuous hours. Vagal activity was analyzed using a variety of techniques, such as spike sorting, spike-firing rates, and interspike intervals. Many spike-cluster-firing rates were found to correlate with food intake, and the neural-firing rates were used to classify eating and other behaviors. To our knowledge, this is the first chronic recording and decoding of activity in the vagus nerve of freely moving animals enabled by the axon-like properties of the CNTY biosensor in both size and flexibility and provides an important step forward in our ability to understand spontaneous vagus-nerve function.

Keywords: vagus nerve; intraneural; decoding; intrafascicular; recording; carbon nanotube

1. Introduction

The vagus nerve innervates nearly every internal organ, providing sensory input to the brain and parasympathetic-control inputs to the viscera. Therefore, abnormal vagus-nerve activity has been linked to many chronic diseases, such as epilepsy, diabetes, hypertension, and cancer [1–5]. Vagus-nerve stimulation has been used to treat a wide variety of diseases [6], most successfully implemented for the treatment of epilepsy [7], even while the mechanisms are not well understood and direct recordings of vagal activity associated with disease are not available [8]. The majority of vagal afferent fibers come from the gut [9,10], and abnormal vagal activity has been clearly implicated in eating and metabolic disorders [11–15]. In this study, we analyze the first chronic recordings of vagal spikes and the correlation of signals to several behaviors in healthy rats.

The chronic recording of vagal signals has been limited, partially due to the difficulty in chronically recording high-quality signals in small autonomic nerves. Extraneural cuff electrodes have proven to be very effective peripheral nerve interfaces, allowing for selective stimulation [16] and some selectivity in recording [17–20]. However, the insulating perineurium layer between the active nerve fibers and the recording electrodes results in a low signal-to-noise ratio (SNR) or requires desheathing of the nerve. Thus, intraneural or intrafascicular electrodes placed much closer to active fibers may be necessary for certain recording applications, providing a higher SNR and higher selectivity for multifascicular nerves. However, intraneural electrodes are more invasive and thus, have issues related to long-term stability [21–25]. In particular, small autonomic nerves necessitate smaller and more flexible neural sensors. We have previously shown that carbon nanotube yarn (CNTY) electrodes have favorable properties for nerve interfacing: specifically, their small size, high flexibility, and low impedance. Thus, they provide a stable, high-SNR interface for chronic recording in small autonomic nerves in rats, with high-quality signals continuing up to four months after implantation [26]. Furthermore, we have developed techniques

Citation: Marmerstein, J.T.; McCallum, G.A.; Durand, D.M. Decoding Vagus-Nerve Activity with Carbon Nanotube Sensors in Freely Moving Rodents. *Biosensors* **2022**, *12*, 114. https://doi.org/10.3390/bios12020114

Received: 8 January 2022
Accepted: 5 February 2022
Published: 11 February 2022

Publisher's Note: MDPI stays neutral with regard to jurisdictional claims in published maps and institutional affiliations.

for recording activity in the cervical vagus nerves of rats without anesthesia, allowing for the first chronic recordings of truly spontaneous vagal activity. This technology has been successfully applied to make the first direct measurements of vagal tone in freely moving animals [27].

Due to the prevalence of gastric afferents in the vagus nerve, we expect signaling from the gut to be the dominant activity present in vagal recordings. Previous studies have shown that vagal afferents are sensitive to mechanical stimulation of the gut [28] and to gastric hormones which regulate food intake and gastric motility [29–31]. There are also reflex pathways which modulate efferent vagal activity in response to gastric distention and contractions [32]. However, such signals have not been reported from chronic, unanesthetized animals, and this CNTY-electrode biosensor demonstrates the ability to decode vagal activity related to various animal behaviors, such as eating.

In this study, we continuously record spontaneous vagal-spiking activity from awake, freely moving rats for >48 h up to two weeks after implantation. To our knowledge, this is the first time this has been successfully demonstrated. The neural-recording data was synchronized with continuous video recording of the subjects. Spike sorting is used to separate semi-distinct spike clusters, which are then correlated to animal behavior identified from the video recordings. Interspike interval distributions are also found to change in response to food intake, presenting another neural feature that can be used to decode spontaneous vagal activity. We report several spike clusters that show tuning to animal eating, and the firing dynamics of multiple decoded spike clusters can be used to classify eating compared to drinking, grooming, and resting behaviors.

2. Materials and Methods

2.1. CNTY Electrode Manufacture

CNT yarns were manufactured at Case Western Reserve University, as described previously [26]. CNTYs were then connected to 35NLT®-DFT® wire (Fort Wayne Metals, Fort Wayne, IN, USA) with silver conductive epoxy (H20E, EPO-TEK), creating a CNTY-DFT® junction. Dacron mesh and silicone elastomer (MED-4211/MED-4011, NuSil Silicone Technology, Carpinteria, CA, USA) were added to seal the junction, confirmed by measuring the impedance of the junction at 1kHz in a saline bath. The free end of the CNTY was tied to the end of an 11-0 nylon suture (S&T 5V33) using a fisherman's knot, as shown in Figure 1A. The entire CNTY was coated with parylene-C (5 μm thick vapor deposition coating, SMART Microsystems, Elyria, OH, USA) on a custom rack which masks the suture needle from coating. Then, a small section (~200 μm long) of parylene-C was removed approximately 500 μm behind the CNTY-suture knot using a laser spot welder (KelanC Laser, set to 1A current, 0.3 ms pulse width, and 300 μm diameter), as shown in Figure 1B. Figure 1C shows the CNTY-suture knot outside of the nerve after implantation. Electrode viability was confirmed by measuring the impedance of the recording site before and after using the laser.

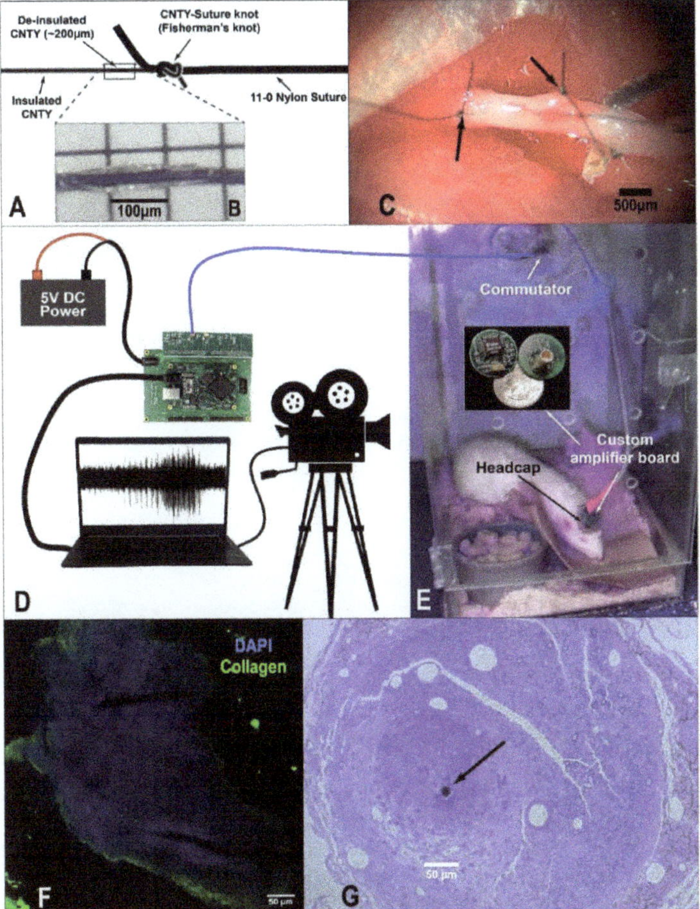

Figure 1. Electrode implantation, histology, and recording methods. (**A**) Diagram of CNTY electrode mated with an 11-0 nylon suture with a fisherman's knot. (**B**) Section of CNTY electrode deinsulated by laser. (**C**) Vagus nerve with two implanted CNTY electrodes. CNTY-suture knots are shown with arrows. (**D/E**) Diagram showing the setup for continuous recording of vagal activity and video for behavior identification. Signals travel from the implants to the headcap connector mounted on the animal's skull, where they are digitized and amplified by the custom amplifier board shown. These signals are then routed through a commutator, which can rotate and allows the animal to move freely without twisting or pulling on the cable. From the commutator, the signals are sent to an Intan USB interface board, which is powered by an external DC-power source and finally sends the signals to a computer, where they are saved and can be viewed in real time. A video camera is manually synced to the vagal recordings. (**F**) Fluorescent images showing collagen + cellular encapsulation of CNTY electrodes implanted in the vagus nerve for seven days. (**G**) Toluidine blue-stained nerve section showing encapsulation of a CNTY electrode implanted for two weeks.

2.2. Surgery

All surgical and experimental procedures were done with the approval and oversight of the Case Western Reserve University Institutional Animal Care and Use Committee to ensure compliance with all federal, state, and local animal welfare laws and regulations. Electrodes were implanted in male Sprague Dawley rats between 7–12 weeks of age.

To expose the left cervical vagus nerve, a midline incision was made along the neck. The muscles and salivary glands were separated and held in place, revealing the carotid sheath which contains the carotid artery and vagus nerve. The vagus nerve was carefully separated from the carotid artery using blunt dissection and held in slight tension using a glass hook. CNTY electrodes were implanted by sewing the suture through the nerve for ~2 mm, then pulling the suture until the CNTY-suture knot was pulled through. Then, the electrode was pulled back so that the knot sat against the epineurium, ensuring the recording site remained inside the nerve, as shown in Figure 1C. Two electrodes were implanted with ~2 mm separation; the extra suture and needles were cut off after implantation, and the nerve, electrodes, and junctions were covered with ~1 mL of fibrin glue (Tisseel, Baxter International Inc., Deerfield, IL, USA) to help secure the area for recovery. Next, the DFT wires were tunneled from the neck to the back of the skull and soldered to a 5-pin Omnectics connector (Omnetics Connector Corporation MCP-5-SS). The skin on top of the skull was opened, and the connector was fixed on top of the skull with dental cement. The amplifier ground was connected to a screw placed in the skull, which also helps keep the headcap in place. Electrodes were implanted for chronic recording in two animals, and animals were given one week for recovery before recording.

2.3. Recording

Recordings were carried out continuously in awake, behaving animals for 56 and 40 h (Rat 1 and 2, respectively). A custom-built PCB with an Intan RHD2216 recording chip was attached to the headcap connector, which was secured to the animal with a 3D-printed locking mechanism and attached to a PlasticsOne® (Roanoke, VA, USA) commutator, allowing the rat to move around the cage without tangling or pulling on the connector cable [27]. Input signals were routed to eight amplifier channels, using 8-channel hardware averaging to decrease amplifier noise. Output from the amplifier board was run through the commutator into an Intan RHD USB Interface board (Intan part #C3100), which is powered by an external battery supplying 5V DC power. Signals are then routed to a computer where they are saved for offline analysis and can be viewed in real time.

Neural recordings were sampled at 20 kHz with a 5 kHz low-pass filter. Recordings were started around 10 AM (approximately four hours after the start of the light cycle). During ENG recording, a video camera was used for simultaneous video recording. The camera was equipped with an infrared light and infrared sensor, allowing for filming even during the dark cycle. The camera was connected to the recording computer and manually synced to the recording. A diagram of the recording setup can be seen in Figure 1D,E.

2.4. Signal Processing

ENG data were imported into MATLAB, where they were further processed. ENG was band-pass filtered from 500–5000 Hz to minimize interference from EMG, ECG, or other possible sources. The filter bandwidth was kept relatively wide to minimize distortion of spike waveforms. Spikes were detected and sorted into clusters using the UltraMegaSort2000 software in MATLAB, using a threshold of eight times the RMS of baseline. Spike waveforms (3 ms long) were transformed into the principal component space, and principal components accounting for 95% of the total waveform variance were used for spike clustering. Spike clustering was done using k-means clustering of spike waveform principal components, with a maximum of k = 256 clusters. Using the UMS2000 software, clusters were further analyzed for better separation and exclusion of artifacts. First, outliers were removed if they had a z-score greater than 500 on the χ^2 distribution of distance to the cluster center. Clusters were removed from analysis if the spike waveform contained a second, larger threshold crossing (i.e., removing of spikes which were detected twice due to threshold-crossing of the spike tail). Clusters were also removed if spike width was less than 0.3 ms and amplitude was greater than 1mV (presumed recording artifacts) or if spike width was greater than 2 ms. Spike waveform values were used to calculate spike amplitudes (difference between the maximum and minimum voltage values) and the spike RMS. Spike-cluster-firing timings were also used to calculate cluster-firing rates and interspike intervals (ISI). Average spike amplitudes over time are shown in Supplementary Figure S1, and spike RMS was used to calculate average SNR, shown in Supplementary Figure S2. Animal behaviors (eating, drinking, grooming, and resting) were identified via video recording. The overall data processing and analysis workflow is diagrammed in Figure 2.

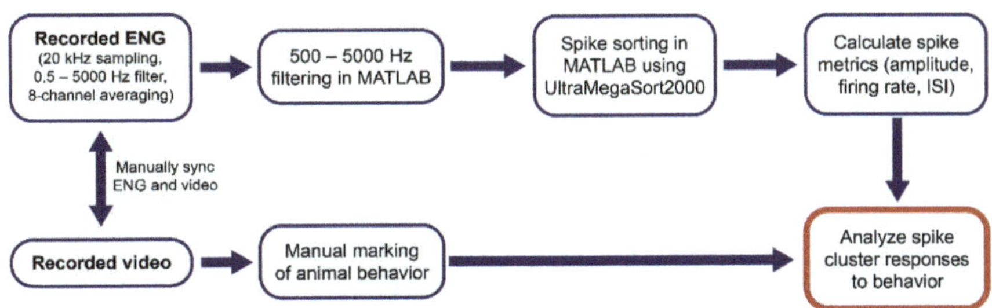

Figure 2. Diagram of data processing and analysis workflow. Vagal ENG and video are recorded simultaneously from freely moving rats. Spike sorting is used to decode spike metrics, which are analyzed with respect to animal behaviors identified from the video.

2.5. Histology

Toluidine blue staining: the image shown in Figure 1G was obtained from an implanted nerve which was fixed, sectioned, and stained with toluidine blue. Two weeks after implantation, animals were perfused with 1.25% glutaraldehyde, 1% formalin, and 0.1 M phosphate buffer. This fixative solution is approximately 640 mOsM/kg. Animals were injected with 0.2–0.5 mL of 1% procaine at 37 °C through the left ventricle. Followed by 200 mL of the fixative solution perfused at 37 °C using a variable speed peristaltic pump. After completing the perfusion process, the vagus nerve was dissected at the implant location. The complete nerve section was transferred into a postfixative solution (1% osmium tetroxide in 100-mM phosphate buffer) for two hours at room temperature before being transferred to 4 °C. Following postfixation, the nerve tissue was dissected in 1-mm-long pieces and embedded in an epoxy resin. Sections (0.7 m) were cut from the epoxy blocks using a diamond knife (DiATOM) microtome. Toluidine blue (1% toluidine blue and 2% borate) was used to stain the nerve axons.

Fluorescent staining: the image shown in Figure 1F was obtained from an implanted nerve which was optically cleared using the CLARITY protocol [33,34]. Seven days after implantation, the vagus nerve was extracted and immediately placed into hydrogel monomer solution. The sample was passively cleared and stained with a collagen antibody, as described by our group previously [26]. DAPI staining was done by placing the sample in VectaShield with DAPI (Vector Laboratories) on a glass-bottom petri dish (Ted Pella, Inc., Redding, CA, USA). Samples were imaged on a Leica SP8 gSTED Super-Resolution Confocal microscope (Leica Microsystems, Wetzlar, Germany).

2.6. Statistical Methods

Where relevant, results are reported as mean ± standard deviation. Average spike waveforms in Figure 2 are shown with shaded areas representing the 95% confidence interval. Overall spike-firing rate, median spike amplitudes, and average spike SNR over time were fitted with a linear regression to determine if the slope was different from zero, with slopes and p-values shown in Figure 2 and Supplemental Figures S1 and S2. Spike clusters were grouped based on their response to eating, and firing rate changes before, during, or after eating for each group were compared to baseline group firing rates using a one-sample t-test, with a significance level of 0.01 and a Bonferroni correction ($\alpha = 5.6 \times 10^{-4}$), as shown in Table 1. ISI distributions of the before, during, and after eating periods were compared to noneating periods using a two-sample Kolmogorov–Smirnov test, with a significance level of 0.01 and a Bonferroni correction for the number of tested distributions ($\alpha = 2.2 \times 10^{-5}$, Supplementary Table S1). All tests performed were two tailed.

Table 1. Firing rates of cluster groups relative to eating. Sorted clusters are separated into five cluster groups based on their response to eating. Table shows the number of clusters of each group recorded in both animals and the behavior of those cluster groups before, during, and after eating: up arrow/green color means an increased firing rate, dash/yellow color means no change in firing rate, and down arrow/red color means a decreased firing rate for the cluster group.

Cluster Group	Rat 1	Rat 2	Before Eating		During Eating		After Eating	
Group I	19	0	↑	$p \ll 0.0001$	↑	$p \ll 0.0001$	↑	$p \ll 0.0001$
Group II	13	13	↑	$p \ll 0.0001$	–	$p = 0.024$	↑	$p \ll 0.0001$
Group III	24	0	↑	$p \ll 0.0001$	↓	$p \leq 0.001$	↑	$p \ll 0.0001$
Group IV	0	59	↑	$p \ll 0.0001$	↑	$p = \ll 0.0001$	–	$p = 0.95$
Group V	0	1	↑	$p \ll 0.0001$	–	$p = 0.0093$	↓	$p = \ll 0.0001$

3. Results

3.1. CNTY Electrodes Record Stable Spikes from Freely Moving Animals

We have previously shown that CNTY electrodes can record spikes from the glossopharyngeal and vagus nerves in anesthetized rats and can be used to measure vagal tone in freely moving animals [26,27]. Here, we demonstrate a novel continuous chronic-recording setup (shown in Figure 1D,E) to record unanesthetized spiking activity which can be sorted into semi-distinct clusters. A total of four electrodes were implanted, two each in the left cervical vagus nerves of two rats, with an average impedance of 11.7 ± 6.5 kΩ at the time of implantation (measured at 1 kHz). Further measurements of CNTY electrode impedances for long-term implants have been published previously [26,27]. Figure 3A shows an example of filtered ENG with several recorded spikes, and Figure 3B–E show several example spike clusters from two animals. A total of 132 spike clusters were identified (56 in Rat 1, and 76 in Rat 2). Clusters are referred to as RatNumber.ClusterNumber (e.g., Cluster 1.21 is Cluster 21 from Rat 1). Average peak-to-peak amplitude of recorded spikes was 152 ± 97 μV for Rat 1 and 180 ± 162 μV for Rat 2. Spike SNR, defined as the average RMS of the spike waveforms compared to the RMS of the baseline, was 7.0 ± 4.9 for Rat 1, and 9.1 ± 5.3 for Rat 2. This is significantly larger than published SNR for acute recording with either the TIME or the LIFE electrodes [33,34]. Furthermore, median spike amplitude for all recorded spikes was stable over the recording time for Rat 2 and slightly increased over time for Rat 1, as shown in Supplementary Figure S1. Overall spike-firing rates were also consistent over the recording periods for both animals: Figure 3F,G show the average firing rates for each hour of recording, with least-squares regression lines showing no significant change in firing rate over time. Similarly, average spike SNR was stable over the recording time for both animals, as shown in Supplemental Figure S2. Thus, we are able to continuously record vagal spikes which have stable amplitude, SNR, and firing rates over time.

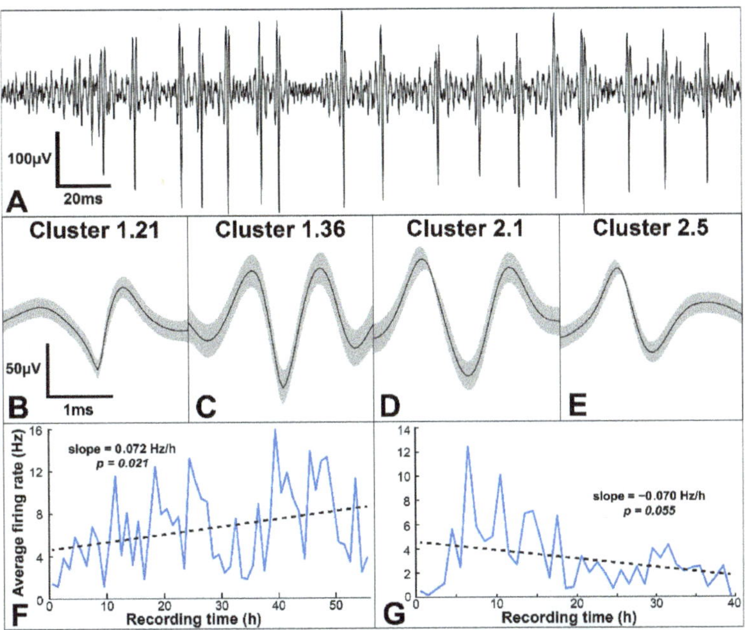

Figure 3. Spontaneous spikes recorded in freely moving animals. (**A**) Filtered ENG showing example recording spikes. (**B–E**) Example clusters sorted from recorded spikes in two animals. (**F,G**) Spike firing rate over recording time for two animals. Neither animal had a significant change in firing rate over time.

3.2. Spike Clusters' Activity Is Correlated with Eating

Identifying the function of spontaneous spikes in freely moving animals is important to understanding how vagal fibers modulate their activity during normal animal behavior. Given the high ratio of gastric afferents in the vagus, most vagal spiking is involved with gastric signaling.

After animal-eating times were identified from video recordings, they were compared to the firing rates of individual spike clusters. In both animals, several clusters show a significant increase in firing rate that occurs <25 min before eating. Some clusters also had increased or decreased firing that occurred during eating, while others had increased firing that occurred <10 min after eating. Figure 4A,B show raster plots for one such spike cluster from each animal, with each row representing one eating event (shown by the shaded grey area). Figure 4C,D show the average firing rate of these clusters relative to the eating events, along with the overall average firing rate for each cluster. Cluster 1.36 (Figure 4A,C) had higher-than-average activity in the 25 min before eating, and higher-than-average activity in the 10 min following eating, with no change occurring during food consumption. Similarly, the firing rate of Cluster 2.1 is increased before and during eating, and unchanged after eating.

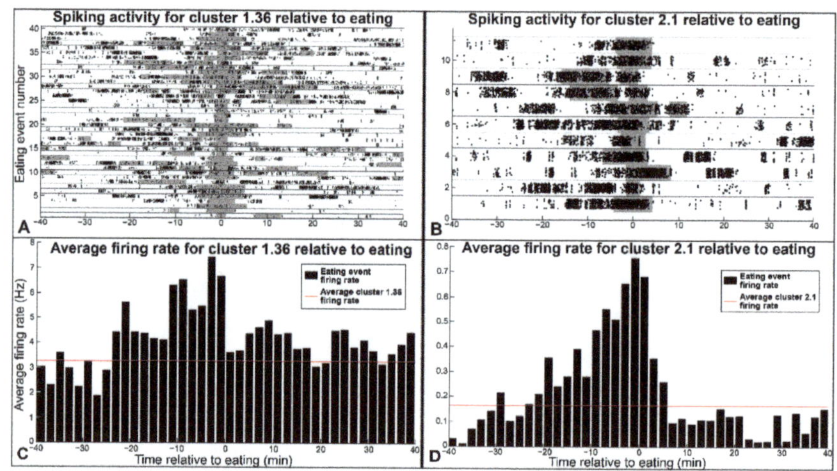

Figure 4. Example spiking activity related to eating. (**A**) Raster plot for Cluster 1.36. Grey-shaded boxes represent eating events, with dots representing spikes. (**B**) Raster plot for Cluster 2.1. (**C**) Firing rate of Cluster 1.36 relative to eating, averaged for all eating events. Red line represents the overall average firing rate of Cluster 1.36. (**D**) Firing rate of Cluster 2.1 relative to eating.

Many clusters exhibited a mix of behaviors, showing firing rates before, during, or after eating that were significantly different from baseline activity ($p < 0.01$ with Bonferroni correction). To analyze cluster behavior related to eating, clusters were sorted into groups based on their firing rate response before eating (from 25 min before, until the start of eating), during eating, and after eating (end of the eating event, until 10 min after eating). These data are summarized in Table 1 for both rats, which show how the cluster-firing rates changed for each group and the number of clusters from each animal which make up each group. The table shows the direction of change and associated p-value for the changes in firing rate of each group in the different eating-related periods (sum of the spiking activity in all clusters within a group compared to the baseline firing activity for the clusters in that group). Only 3 of the 132 recorded clusters did not showing any significant tuning to eating behavior. While specific spiking correlations are unique to each subject, they are consistent within each animal, and Figure 3 and Table 1 show that we can identify spike clusters that exhibit firing rate changes before, during, and after eating in both subjects.

3.3. Spike Cluster Interspike Intervals Show Changes in Bursting Related to Eating

Spikes are often observed exhibiting bursting behavior, where fibers tend to fire at specific frequencies. Bursting behavior can be seen in Figure 4A,B, where spikes appear in clumps. To quantify bursting, spike cluster interspike intervals (ISIs) were calculated for noneating, pre-eating, during eating, and posteating time windows. Eating-related distributions were compared to noneating distributions using a two-sample Kolmogorov–Smirnov test and were plotted in a histogram. Figure 5 shows ISI distributions for noneating, pre-eating, during eating, and posteating periods for one example cluster (Cluster 1.8, which is part of Group II and has increased activity before and after eating). In Figure 5A, we can see that the peak ISI of this cluster during noneating times is around 21 ms or a 48 Hz firing rate. However, in the 25 min before eating, this distribution shifts to the left, peaking instead at 7 ms or 143 Hz, signifying an increase in the bursting firing rate before eating. In the 10 min following eating, the bursting rate returns to the noneating value, though the ISI peak is more pronounced, meaning that bursting is a more prevalent spike behavior after eating. After eating, we also observe a secondary peak around 47 ms (21 Hz). During eating, the ISI distribution is not significantly different from noneating; thus, the bursting activity of Cluster 1.8 is changed before and after, but not during, eating behavior. In total,

10 clusters in Rat 1 and 18 clusters in Rat 2 demonstrated changes in ISI distribution related to eating.

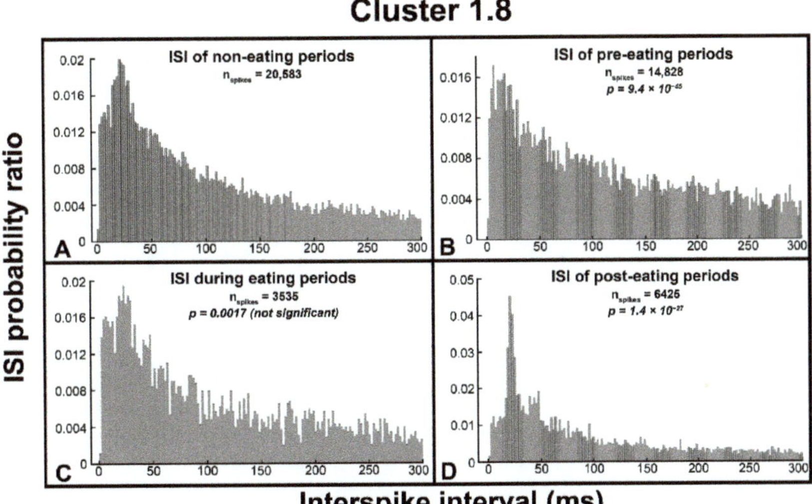

Figure 5. Interspike interval histograms for Cluster 1.8. (**A**) ISI histogram for noneating periods, which has a peak around 21 ms. (**B**) ISI histogram for pre-eating periods, which has a peak around 7ms, and a significantly different ISI distribution compared to noneating periods. (**C**) ISI histogram for eating periods, which has a peak around 23 ms and is not significantly different from noneating periods. (**D**) ISI histogram for post-eating periods, which has a peak around 21 ms and a secondary peak around 47 ms, and a significantly different ISI distribution compared to noneating periods.

These data are summarized in Supplemental Table S1, which shows *p*-values comparing noneating and eating-related ISI distributions for any cluster which showed a significant change. The 18 clusters in Rat 2 only showed a change in ISI distribution during eating, with no changes either before or after. The 10 clusters in Rat 1 each showed changes before eating, while some also had a significantly different ISI distribution during or after eating as well. Figure 5 and Supplementary Table S1 show that some of the spike clusters which are tuned to eating are observed to change bursting activity related to eating, though not all the clusters which show changes in overall activity have altered ISI/bursting behavior.

3.4. Spike-Cluster-Firing Rates Can Be Used to Classify Eating Compared to Other Behavior

In addition to showing that individual spike clusters are correlated with food intake, we also examined whether spike-firing rates are sufficient to classify the times during which the animal is eating, compared to other behaviors, such as drinking, grooming, and resting. A multinomial logistic regression model was constructed, with behaviors and spike-cluster-firing rates averaged over 30 s. The model uses firing rates from each of the recorded clusters, as well as firing rates during peak delayed or preceding correlations with eating. The models were trained on the first 2/3 of recording data and tested on the final 1/3 of recording data. Figure 6 shows the confusion matrices for both animals, which show the performance of the model for classifying behavior with a probability threshold of $\pi = 0.5$ for classification. Percentages on the y-axis show the amount of time spent doing each behavior as a percentage of total recording time. In Rat 1, the model was able to classify eating most accurately, with a 73.1% accuracy. In Rat 2, the model performed best at classifying resting, with a 93.8% accuracy. Additionally, we can see by plotting the receiver operating characteristic (ROC) curves and the associated areas under the curve (AUC) in Supplementary Figure S3 that both models performed better than random chance for almost all behaviors (the only exception being classifying other activity in Rat 1). Overall, these results show that the firing rates of spontaneous vagal spikes sorted into clusters are sufficient to classify eating behavior in freely moving animals.

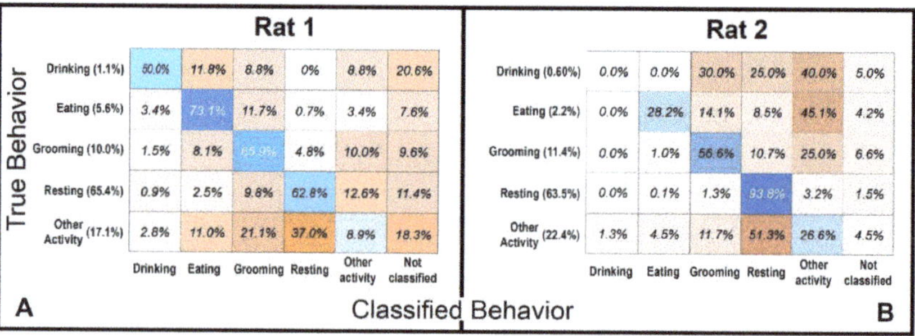

Rat 1

True Behavior	Drinking	Eating	Grooming	Resting	Other activity	Not classified
Drinking (1.1%)	50.0%	11.8%	8.8%	0%	8.8%	20.6%
Eating (5.6%)	3.4%	73.1%	11.7%	0.7%	3.4%	7.6%
Grooming (10.0%)	1.5%	8.1%	65.9%	4.8%	10.0%	9.6%
Resting (65.4%)	0.9%	2.5%	9.8%	62.8%	12.6%	11.4%
Other Activity (17.1%)	2.8%	11.0%	21.1%	37.0%	8.9%	18.3%

Rat 2

True Behavior	Drinking	Eating	Grooming	Resting	Other activity	Not classified
Drinking (0.60%)	0.0%	0.0%	30.0%	25.0%	40.0%	5.0%
Eating (2.2%)	0.0%	28.2%	14.1%	8.5%	45.1%	4.2%
Grooming (11.4%)	0.0%	1.0%	56.6%	10.7%	25.0%	6.6%
Resting (63.5%)	0.0%	0.1%	1.3%	93.8%	3.2%	1.5%
Other Activity (22.4%)	1.3%	4.5%	11.7%	51.3%	26.6%	4.5%

Classified Behavior

Figure 6. Confusion matrix for classifying animal behavior based on spike firing rates. Blue-colored cells show rates of correct classification, and orange-colored cells show rates of incorrect classification, such that each row sums to 100%. *Y*-axis labels show the percentage of recording time spent doing each behavior. (**A**) Confusion matrix for the classification of behavior in Rat 1. (**B**) Confusion matrix for the classification of behavior in Rat 2.

4. Discussion

CNTY electrodes are a promising neural interface: a small, low-impedance, and highly flexible biosensor ideal for interfacing with small peripheral nerves. Flexural rigidity, measured with an atomic-force microscope, shows that the CNTY electrodes are >10 times more flexible than PtIr electrodes of the same diameter ($3.3 \pm 1.5 \times 10^{-12}$ N$\times$ m^2 for the CNTY compared to $2.0 \pm 0.57 \times 10^{-10}$ N$\times$ m^2 for the PtIr) [35]. Partly due to its small size (10 µm diameter) and flexibility, this axon-like biosensor has demonstrated stable, low impedance with chronic implantation, stable high SNR, and minimal evidence of chronic inflammation or nerve damage [26]. Furthermore, we have shown that CNTYs can be used for chronic recording in small autonomic nerves, such as the vagus and glossopharyngeal nerves, and for stimulation in larger somatic nerves and fascicles, such as the rat sciatic nerve [26]. Compared to previous intrafascicular interfaces, CNTYs provide higher SNR and improved stability and should be further investigated as a component of neural sensor devices.

Vagus-nerve-stimulation therapy is a rapidly growing field, with a wide variety of companies and studies investigating its use for treatment of a wide variety of diseases, including epilepsy, obesity, and heart failure [7,36–38]. However, many VNS studies have reported ambiguous results, pointing to the need for an improved understanding of natural vagal function, VNS mechanisms, and closed loop control of stimulation. Neural interfaces which allow for stable, high-SNR recordings are necessary for high-fidelity closed-loop control, and chronic recording in animal models may be used to better understand vagal function and response to therapy. In this study, we utilize the CNTY neural interface to show that eating-related spikes can be decoded from continuous chronic recordings in the vagus nerve, providing the first demonstration that spontaneous, physiologically specific signals can be recorded. We also show that spike-firing rate and interspike interval distributions show differing responses to physiological changes, which may be used as neural features for long-term recording and closed-loop systems, though further histological analysis may also be necessary to show the impact of chronic implantation on nerve health.

The vagus nerve contains afferent and efferent fibers that sense and control nearly every internal organ, playing a vital role in homeostasis, reflex pathways, and responses to physiological changes. While individual vagal-spiking activity has been recorded from isolated fibers and from acute intraneural recordings, to our knowledge, this is the first time spikes have been recorded in the vagus nerve in a chronic model. Combined with previous studies on recording average vagal RMS [27], we have demonstrated that various spontaneous and physiologically relevant signals can be recorded from the rat vagus nerve from freely moving animals using CNTY electrodes. Spike-firing rates stayed consistent for up to 56 recording hours (7–10 days after implantation), suggesting that the electrode interface is relatively stable during that time. Though spike clusters detected in the peripheral nervous system likely represent multiunit activity, Figure 2 shows that CNTY-recorded spikes can be sorted into clusters, allowing for more specialized decoding of relevant signals. Individual spike clusters recorded in freely moving rats show changes in firing rate before, during, and after eating; clusters were further sorted into different groups based on their firing behavior, as shown in Figure 3 and Table 1. Almost all recorded spike clusters showed increased activity up to 25 min before the start of eating. Firing rate also increased during and after eating for several cluster groups. Additionally, Figure 4 and Supplemental Table S1 show that some spike clusters exhibit changes in their ISI distribution during different eating phases. ISI distribution and peak ISI values are an important metric for describing spike-bursting behavior, and bursting frequencies may change independently of overall spiking activity in the cluster (due to clusters containing recordings from multiple individual axons). Thus, analysis of overall cluster activity and cluster ISI distributions are important metrics of vagal activity. Finally, Figure 5 and Supplementary Figure S3 shows that spike-cluster-firing rates can be used to accurately classify eating, drinking, grooming, and resting behavior, an important proof of concept for a closed-loop VNS system. Future studies could analyze how overall vagal-spiking activity and bursting rates respond to models of chronic disease, physiological stimuli (such as fasting), or to VNS treatment.

There are several important questions regarding the behavior of these fibers and how they might respond to physiological changes. One possibility is that these fibers may be related to the secretion or sensing of ghrelin and cholecystokinin (CCK) in the gut, peptide hormones that are known to regulate hunger and satiety via the vagus nerve and can even be found in small concentrations in the brain [30,39]. Ghrelin increases food intake and weight gain in rats and suppresses vagal activity of some gastric afferent fibers [30]. CCK, on the other hand, suppresses appetite while stimulating gastric afferent discharge [40,41]. However, the full picture relating the secretion of these hormones and their effects to vagus-nerve activity is not known. Future studies could investigate how administration of these peptides alters vagal activity in freely moving animals or how changes in vagal-spiking behavior correlate with changes in the concentration of gastric hormones in the gastrointestinal tract, in the blood, or in the CNS. Furthermore, changes in diet, such as high-fat or high-carbohydrate models of diet-induced obesity, have been shown to alter vagal satiety signaling and may have an effect on hunger signaling as well [14,42]. Utilizing the CNTY chronic-recording model described here would allow for insights into the effects of diet on vagal gastric signaling, and a more detailed measurement of food intake would allow for analysis of how differences in vagal spiking relate to differences in dietary behavior.

One important application for chronic recording of vagal activity is measuring the acute and chronic response of vagal signaling to various therapy approaches. High frequency VNS (vBLoc®) is thought to suppress afferent hunger signaling to reduce overeating in obese patients [43]. vBLoc® stimulation is performed with a fixed on/off cycle; when stimulation is on, vagal activity is suppressed, which would decrease hunger signaling (e.g., the activity observed in cluster groups 1–5), and satiety signaling (e.g., cluster groups 1–3). However, studies have shown that vBLoc® reduces hunger and *increases* satiety, though it is not clear how significant this effect is compared to a placebo [44]. More thorough investigation of the effects of vBLoc® on hunger, satiety, and vagal activity are necessary to better understand therapy performance. The abiliti® gastric stimulation system, on the other hand, is closed-loop, producing satiety during preprogrammed periods and in response to eating, thus reducing food intake and helping create a more stable meal schedule [45]. Patients with this system report improved self-control while eating, decreased binge eating, and reduced sensitivity to hunger [46], likely as a result of increased afferent activity during and after eating.

Combining animal models of these stimulation systems with chronic spike recording using CNTY electrodes would allow for direct investigation of the mechanisms of these therapies [43]. Furthermore, chronic recording could be paired with other VNS paradigms, such as VNS for epilepsy or depression, to investigate possible off-target side effects.

Overall, our results show that it is possible to record chronic signals in the rat vagus nerve continuously, opening the door for studies that were previously not possible. Furthermore, the ability to detect and decode spontaneous spiking activity from chronic vagal recordings could allow for a more detailed analysis of vagus-nerve response to changes in diet, therapy, and behavior. This technology could be used to develop closed-loop VNS for metabolic disorders, which adapt stimulation based on recorded vagal activity. Chronic recording in animal models can also be used to further study the vagal pathways that control food intake and how they respond to VNS and other treatments.

Supplementary Materials: The following supporting information can be downloaded at: https://www.mdpi.com/article/10.3390/bios12020114/s1, Supplementary Figure S1. Median spike amplitude over time. (A). Median spike amplitude for spikes recorded in Rat 1 slightly increase over the recording time, with a statistically significant slope of 0.39µV per hour. (B). Median spike amplitude of spikes recorded in Rat 2 did not change over the recording time. Supplementary Figure S2. Average spike SNR (spike RMS divided by noise RMS) over time. (A). SNR of spikes recorded in Rat 1 did not change over the recording time. (B). SNR of spikes recorded in Rat 2 did not change over the recording time. Supplementary Table S1. Differences in ISI distributions for before, during, and after eating periods, compared to non-eating periods, for all clusters which had at least one group with a significant change. Cluster groups are shown for each cluster (see Table 1), and non-significant p-values are not shown. Supplementary Figure S3. Receiver operating characteristic (ROC) curves and area-under-the-curve (AUC) values to assess performance of a multinomial logistic regression model to classify animal behaviors based on spike cluster firing rates. Dotted lines show the expected ROC curve for a random classifier. (A). ROC curve for classifying drinking in Rat 1, with AUC = 0.86. (B). ROC curve for classifying drinking in Rat 2, with AUC = 0.62. (C). ROC curve for classifying eating in Rat 1, with AUC = 0.94. (D). ROC curve for classifying eating in Rat 2, with AUC = 0.82. (E). ROC curve for classifying grooming in Rat 1, with AUC = 0.88. (F). ROC curve for classifying grooming in Rat 2, with AUC = 0.90. (G). ROC curve for classifying resting in Rat 1, with AUC = 0.82. (H). ROC curve for classifying resting in Rat 2, with AUC = 0.86. I: ROC curve for classifying other activity in Rat 1, with AUC = 0.47. (J). ROC curve for classifying other activity in Rat 2.

Author Contributions: Conceptualization, J.T.M., G.A.M. and D.M.D.; methodology, J.T.M., G.A.M. and D.M.D.; software, J.T.M.; validation, J.T.M. and G.A.M.; formal analysis, J.T.M.; investigation, J.T.M.; resources, D.M.D.; data curation, J.T.M. and G.A.M.; writing—original draft preparation, J.T.M.; writing—review and editing, J.T.M., G.A.M. and D.M.D.; visualization, J.T.M.; supervision, G.A.M. and D.M.D.; project administration, D.M.D.; funding acquisition, D.M.D. All authors have read and agreed to the published version of the manuscript.

Funding: This work was funded by the Congressionally Directed Medical Research Programs (Autonomic nervous system activity and the implications on breast cancer metastasis, W81XWH-18-1-0581), and the National Institutes of Health, (Nerve reshaping for improved selectivity, 5R01NS032845-22). Confocal microscopy was performed at the Light Microscopy Imaging Facility at Case Western Reserve University, made available through the Office of Research Infrastructure (NIH-ORIP) Shared Instrumentation Grant (S10OD016164).

Institutional Review Board Statement: All surgical and experimental procedures were done with the approval and oversight of the Case Western Reserve University Institutional Animal Care and Use Committee to ensure compliance with all federal, state, and local animal welfare laws and regulations.

Data Availability Statement: The data that support the findings of this study are available from the corresponding author upon reasonable request. Some custom-analysis code was created to analyze spiking activity and is available from the corresponding author upon reasonable request.

Acknowledgments: Assistance for tissue fixation, sectioning, staining, and imaging for Figure 1G was provided by Hisashi Fujioka of the Case Western Reserve University Cryo-Electron Microscopy Core. Special thanks to William Marcus, who helped monitor animals during surgery and recovery and assisted with recording.

Conflicts of Interest: The authors declare no conflict of interest.

References

1. Liao, D.; Cai, J.; Brancati, F.L.; Folsom, A.; Barnes, R.W.; Tyroler, H.A.; Heiss, G. Association of vagal tone with serum insulin, glucose, and diabetes mellitus—The ARIC Study. *Diabetes Res. Clin. Pract.* **1995**, *30*, 211–221. [CrossRef]
2. Ronkainen, E.; Korpelainen, J.T.; Heikkinen, E.; Myllyla, V.V.; Huikuri, H.V.; Isojarvi, J.I.T. Cardiac autonomic control in patients with refractory epilepsy before and during vagus nerve stimulation treatment: A one-year follow-up study. *Epilepsia* **2006**, *47*, 556–562. [CrossRef] [PubMed]
3. Sivakumar, S.; Namath, A.G.; Tuxhorn, I.E.; Lewis, S.J.; Galán, R.F. Decreased heart rate and enhanced sinus arrhythmia during interictal sleep demonstrate autonomic imbalance in generalized epilepsy. *J. Neurophysiol.* **2016**, *115*, 1988–1999. [CrossRef] [PubMed]
4. Grassi, G. Assessment of sympathetic cardiovascular drive in human hypertension: Achievements and perspectives. *Hypertension* **2009**, *54*, 690–697. [CrossRef]
5. Porges, W. Cardiac Vagal Tone: A Physiological Index of Stress. *Neurosci. Biobehav. Rev.* **1995**, *19*, 225–233. [CrossRef]
6. Gidron, Y.; Deschepper, R.; De Couck, M.; Thayer, J.F.; Velkeniers, B. The Vagus Nerve Can Predict and Possibly Modulate Non-Communicable Chronic Diseases: Introducing a Neuroimmunological Paradigm to Public Health. *J. Clin. Med.* **2018**, *7*, 371. [CrossRef]
7. Fan, J.J.; Shan, W.; Wu, J.P.; Wang, Q. Research progress of vagus nerve stimulation in the treatment of epilepsy. *CNS Neurosci. Ther.* **2019**, *5*, 1222–1228. [CrossRef]
8. Cracchiolo, M.; Ottaviani, M.M.; Panarese, A.; Strauss, I.; Vallone, F.; Mazzoni, A.; Micera, S. Bioelectronic medicine for the autonomic nervous system: Clinical applications and perspectives. *J. Neural Eng.* **2021**, *18*, 041002. [CrossRef]
9. Mei, N.; Condamin, M.; Boyer, A. The composition of the vagus nerve of the cat. *Cell Tissue Res.* **1980**, *209*, 423–431. [CrossRef]
10. Prechtl, J.C.; Powley, T.L. The fiber composition of the abdominal vagus of the rat. *Anat. Embryol.* **1990**, *181*, 101–115. [CrossRef]
11. Kral, J.G.; Görtz, L.; Hermansson, G.; Wallin, G.S. Gastroplasty for obesity: Long-term weight loss improved by vagotomy. *World J. Surg.* **1993**, *17*, 75–78. [CrossRef] [PubMed]
12. Smith, D.; Sarfeh, J.; Howard, L. Truncal vagotomy in hypothalamic obesity. *Lancet* **1983**, *321*, 1330–1331. [CrossRef]
13. Prologo, J.D.; Lin, E.; Bergquist, S.H.; Knight, J.; Matta, H.; Brummer, M.; Singh, A.; Patel, Y.; Corn, D. Percutaneous CT-Guided Cryovagotomy in Patients with Class I or Class II Obesity: A Pilot Trial. *Obesity* **2019**, *27*, 1255–1265. [CrossRef] [PubMed]
14. Loper, H.; Leinen, M.; Bassoff, L.; Sample, J.; Romero-Ortega, M.; Gustafson, K.J.; Taylor, D.M.; Schiefer, M.A. Both high fat and high carbohydrate diets impair vagus nerve signaling of satiety. *Sci. Rep.* **2021**, *11*, 10394. [CrossRef] [PubMed]
15. Kral, J.G.; Wencesley, A.E.; Ae, P.; Wolfe, B.M. Vagal Nerve Function in Obesity: Therapeutic Implications. *World J. Surg.* **2009**, *33*, 1995–2006. [CrossRef]
16. Tan, D.W.; Schiefer, M.A.; Keith, M.W.; Anderson, J.R.; Tyler, J.; Tyler, D.J. A neural interface provides long-term stable natural touch perception. *Sci. Transl. Med.* **2014**, *6*, 257ra138. [CrossRef]
17. Dweiri, Y.M.; Eggers, T.E.; Gonzalez-Reyes, L.E.; Drain, J.; McCallum, G.A.; Durand, D.M. Stable Detection of Movement Intent from Peripheral Nerves: Chronic Study in Dogs. *Proc. IEEE* **2016**, *105*, 50–65. [CrossRef]
18. Eggers, T.E.; Dweiri, Y.M.; McCallum, G.A.; Durand, D.M. Recovering Motor Activation with Chronic Peripheral Nerve Computer Interface. *Sci. Rep.* **2018**, *8*, 14149. [CrossRef]
19. Sabetian, P.; Sadat-Nejad, Y.; Yoo, P.B. Classification of directionally specific vagus nerve activity using an upper airway obstruction model in anesthetized rodents. *Sci. Rep.* **2021**, *11*, 10682. [CrossRef]
20. Koh, R.G.L.; Nachman, A.I.; Zariffa, J. Classification of naturally evoked compound action potentials in peripheral nerve spatiotemporal recordings. *Sci. Rep.* **2019**, *9*, 11145. [CrossRef]
21. Lago, N.; Yoshida, K.; Koch, K.P.; Navarro, X. Assessment of Biocompatibility of chronically implanted polyimide and platinum intrafascicular electrodes. *IEEE Trans. Biomed. Eng.* **2007**, *54*, 281–290. [CrossRef] [PubMed]
22. Badia, J.; Boretius, T.; Pascual-Font, A.; Udina, E.; Stieglitz, T.; Navarro, X. Biocompatibility of chronically implanted transverse intrafascicular multichannel electrode (TIME) in the rat sciatic nerve. *IEEE Trans. Biomed. Eng.* **2011**, *58*, 2324–2332. [CrossRef] [PubMed]
23. Rijnbeek, E.H.; Eleveld, N.; Olthuis, W. Update on peripheral nerve electrodes for closed-loop neuroprosthetics. *Front. Neurosci.* **2018**, *12*, 350. [CrossRef] [PubMed]
24. Jiman, A.A.; Ratze, D.C.; Welle, E.J.; Patel, P.R.; Richie, J.M.; Bottorff, E.C.; Seymour, J.P.; Chestek, C.A.; Bruns, T.M. Multi-channel intraneural vagus nerve recordings with a novel high-density carbon fiber microelectrode array. *Sci. Rep.* **2020**, *10*, 15501. [CrossRef]

25. Patel, P.R.; Popov, P.; Caldwell, C.M.; Welle, E.J.; Egert, D.; Pettibone, J.R.; Roossien, D.H.; Becker, J.B.; Berke, J.D.; Chestek, C.A.; et al. High density carbon fiber arrays for chronic electrophysiology, fast scan cyclic voltammetry, and correlative anatomy. *J. Neural Eng.* **2020**, *17*, 056029. [CrossRef]

26. McCallum, G.A.; Sui, X.; Qiu, C.; Marmerstein, J.; Zheng, Y.; Eggers, T.E.; Hu, C.; Dai, L.; Durand, D.M. Chronic interfacing with the autonomic nervous system using carbon nanotube (CNT) yarn electrodes. *Sci. Rep.* **2017**, *7*, 11644. [CrossRef]

27. Marmerstein, J.T.; McCallum, G.A.; Durand, D.M. Direct measurement of vagal tone in rats does not show correlation to HRV. *Sci. Rep.* **2021**, *11*, 1210. [CrossRef]

28. Ozaki, N.; Sengupta, J.N.; Gebhart, G.F. Mechanosensitive Properties of Gastric Vagal Afferent Fibers in the Rat. *J. Neurophysiol.* **1999**, *82*, 2210–2220. [CrossRef]

29. Schwartz, G.J.; McHugh, P.R.; Moran, T.H. Integration of vagal afferent responses to gastric loads and cholecystokinin in rats. *Am. J. Physiol. Integr. Comp. Physiol.* **1991**, *261*, R64–R69. [CrossRef]

30. Date, Y.; Murakami, N.; Toshinai, K.; Matsukura, S.; Niijima, A.; Matsuo, H.; Kangawa, K.; Nakazato, M. The role of the gastric afferent vagal nerve in ghrelin-induced feeding and growth hormone secretion in rats. *Gastroenterology* **2002**, *123*, 1120–1128. [CrossRef]

31. Strader, A.D.; Woods, S.C. Gastrointestinal hormones and food intake. *Gastroenterology* **2005**, *128*, 175–191. [CrossRef] [PubMed]

32. Davison, J.S.; Grundy, D. Modulation of single vagal efferent fibre discharge by gastrointestinal afferents in the rat. *J. Physiol.* **1978**, *284*, 69–82. [CrossRef] [PubMed]

33. Deisseroth, K.; Clarity Resource Center. Available online: http://clarityresourcecenter.org/ (accessed on 21 January 2017).

34. Chung, K.; Deisseroth, K. CLARITY for mapping the nervous system. *Nat. Methods* **2013**, *10*, 508–513. [CrossRef] [PubMed]

35. Zheng, Y. Design and Fabrication of a Highly Flexible Neural Interface. Ph.D. Dissertation, Case Western Reserve University, Cleveland, OH, USA, 2017.

36. Groves, D.A.; Brown, V.J. Vagal nerve stimulation: A review of its applications and potential mechanisms that mediate its clinical effects. *Neurosci. Biobehav. Rev.* **2005**, *29*, 493–500. [CrossRef]

37. Pardo, J.V.; Sheikh, S.A.; Kuskowski, M.A.; Surerus-Johnson, C.; Hagen, M.C.; Lee, J.T.; Rittberg, B.R.; Adson, D.E. Weight loss during chronic, cervical vagus nerve stimulation in depressed patients with obesity: An observation. *Int. J. Obes.* **2007**, *31*, 1756–1759. [CrossRef]

38. Johnson, R.L.; Wilson, C.G. A review of vagus nerve stimulation as a therapeutic intervention. *J. Inflamm. Res.* **2018**, *11*, 203–213. [CrossRef]

39. Murakami, N.; Hayashida, T.; Kuroiwa, T.; Nakahara, K.; Ida, T.; Mondal, M.S.; Nakazato, M.; Kojima, M.; Kangawa, K. Role for central ghrelin in food intake and secretion profile of stomach ghrelin in rats. *J. Endocrinol.* **2002**, *174*, 283–288. [CrossRef]

40. Schwartz, G.J.; Moran, T.H.; White, W.O.; Ladenheim, E.E. Relationships between gastric motility and gastric vagal afferent responses to CCK and GRP in rats differ. *Am. J. Physiol. Content* **1997**, *272*, R1726–R1733. [CrossRef]

41. Schwartz, G.J.; McHugh, P.R.; Moran, T.H. Gastric loads and cholecystokinin synergistically stimulate rat gastric vagal afferents. *Am. J. Physiol. Integr. Comp. Physiol.* **1993**, *265*, R872–R876. [CrossRef]

42. Marques, C.; Meireles, M.; Norberto, S.; Leite, J.; Freitas, J.; Pestana, D.; Faria, A.; Calhau, C. High-fat diet-induced obesity Rat model: A comparison between Wistar and Sprague-Dawley Rat. *Adipocyte* **2015**, *5*, 11–21. [CrossRef]

43. Johannessen, H.; Révész, D.F.; Kodama, Y.; Cassie, N.; Skibicka, K.P.; Barrett, P.; Dickson, S.; Holst, J.J.; Rehfeld, J.F.; van der Plasse, G.; et al. Vagal blocking for obesity control: A possible mechanism-of-action. *Obes. Surg.* **2017**, *27*, 177–185. [CrossRef] [PubMed]

44. Sarr, M.G.; The EMPOWER Study Group; Billington, C.J.; Brancatisano, R.; Brancatisano, A.; Toouli, J.; Kow, L.; Nguyen, N.T.; Blackstone, R.; Maher, J.W.; et al. The EMPOWER Study: Randomized, Prospective, Double-Blind, Multicenter Trial of Vagal Blockade to Induce Weight Loss in Morbid Obesity. *Obes. Surg.* **2012**, *22*, 1771–1782. [CrossRef] [PubMed]

45. Horbach, T.; Thalheimer, A.; Seyfried, F.; Eschenbacher, F.; Schuhmann, P.; Meyer, G. Abiliti® Closed-Loop Gastric Electrical Stimulation System for Treatment of Obesity: Clinical Results with a 27-Month Follow-Up. *Obes. Surg.* **2015**, *25*, 1779–1787. [CrossRef] [PubMed]

46. Horbach, T.; Meyer, G.; Morales-Conde, S.; Alarcón, I.; Favretti, F.; Anselmino, M.; Rovera, G.M.; Dargent, J.; Stroh, C.; Susewind, M.; et al. Closed-loop gastric electrical stimulation versus laparoscopic adjustable gastric band for the treatment of obesity: A randomized 12-month multicenter study. *Int. J. Obes.* **2016**, *40*, 1891–1898. [CrossRef]

Article

Detection of Liver Dysfunction Using a Wearable Electronic Nose System Based on Semiconductor Metal Oxide Sensors

Andreas Voss [1,2,*], Rico Schroeder [1,3], Steffen Schulz [1], Jens Haueisen [2], Stefanie Vogler [4], Paul Horn [4], Andreas Stallmach [4] and Philipp Reuken [4]

[1] Institute of Innovative Health Technologies IGHT, Ernst-Abbe-Hochschule Jena, 07745 Jena, Germany; schroederrico@gmx.de (R.S.); steffen.schulz@charite-berlin.de.com (S.S.)

[2] Institute of Biomedical Engineering and Informatics (BMTI), Technische Universität Ilmenau, 98693 Ilmenau, Germany; jens.haueisen@tu-ilmenau.de

[3] UST Umweltsensortechnik GmbH, 99331 Geratal, Germany

[4] Clinic for Internal Medicine IV, University Hospital Jena, 07747 Jena, Germany; Gastro@med.uni-jena.de (S.V.); P.Horn@bham.ac.uk (P.H.); andreas.stallmach@med.uni-jena.de (A.S.); philipp.reuken@med.uni-jena.de (P.R.)

* Correspondence: Andreas.Voss@tu-ilmenau.de; Tel.: +49-3677-69-2861

Abstract: The purpose of this exploratory study was to determine whether liver dysfunction can be generally classified using a wearable electronic nose based on semiconductor metal oxide (MOx) gas sensors, and whether the extent of this dysfunction can be quantified. MOx gas sensors are attractive because of their simplicity, high sensitivity, low cost, and stability. A total of 30 participants were enrolled, 10 of them being healthy controls, 10 with compensated cirrhosis, and 10 with decompensated cirrhosis. We used three sensor modules with a total of nine different MOx layers to detect reducible, easily oxidizable, and highly oxidizable gases. The complex data analysis in the time and non-linear dynamics domains is based on the extraction of 10 features from the sensor time series of the extracted breathing gas measurement cycles. The sensitivity, specificity, and accuracy for distinguishing compensated and decompensated cirrhosis patients from healthy controls was 1.00. Patients with compensated and decompensated cirrhosis could be separated with a sensitivity of 0.90 (correctly classified decompensated cirrhosis), a specificity of 1.00 (correctly classified compensated cirrhosis), and an accuracy of 0.95. Our wearable, non-invasive system provides a promising tool to detect liver dysfunctions on a functional basis. Therefore, it could provide valuable support in preoperative examinations or for initial diagnosis by the general practitioner, as it provides non-invasive, rapid, and cost-effective analysis results.

Keywords: electronic nose; liver dysfunction; cirrhosis; semiconductor metal oxide gas sensor

Citation: Voss, A.; Schroeder, R.; Schulz, S.; Haueisen, J.; Vogler, S.; Horn, P.; Stallmach, A.; Reuken, P. Detection of Liver Dysfunction Using a Wearable Electronic Nose System Based on Semiconductor Metal Oxide Sensors. *Biosensors* **2022**, *12*, 70. https://doi.org/10.3390/bios12020070

Received: 29 December 2021
Accepted: 24 January 2022
Published: 26 January 2022

Publisher's Note: MDPI stays neutral with regard to jurisdictional claims in published maps and institutional affiliations.

1. Introduction

Metabolic disorders are sometimes connected with typical odors which can be measured on breath, sweat, or other excreta from humans. Examples are ammonia odor, which is related to renal diseases, and acetone odor, which is related to diabetes.

The beginnings of the use of electronic noses (e-noses) date back to pioneering work by a few research groups, such as Hartman, Wilkens, Dodd, and Moncrieff [1–4]. Here, the foundation was laid for specific odors to be detectable and, thus, evaluable with suitable electronics and analysis technology. The concept of sampling breath for health monitoring was initially conceived in the 20th century. In 1952, Henderson [5] reported on the increased acetone content of breath samples from young diabetics, promoting an interest in the content of breath [6].

In recent decades, improvements in materials, sensors, electronics, and signal processing technologies have led to a rapid increase in the development and application of e-noses [7–9]. E-noses are used, among other things, to analyze, detect, discriminate, classify, and monitor gas components or odors in many fields of science and industry, and are of interest for numerous applications. For example, e-noses are used in the food and beverage industry to monitor processing and determine the quality of the final product [10,11], in pharmaceutical science for formulation development and quality assurance [8], and for air quality monitoring [12]. In addition, e-noses are also used in agriculture, water management, medicine, security systems, and many other fields [13].

In the following, we will only deal with the e-noses that meet Gardner's definition [14]. He stated that an e-nose is an instrument, which comprises an array of electronic chemical sensors with partial specificity and an appropriate pattern-recognition system, capable of recognizing simple or complex odors. However, unlike other analytical methods, an e-nose does not detect directly specific volatile organic components (VOCs); rather, it builds chemical patterns to form an identity. The sensor array produces output patterns that represent VOCs in the breath (or different substances), and the data processing extracts a set of mathematical descriptors that represent the signature of the breath sample as a pattern [15]. The detection of the input signal occurs depending on the operating principle implemented in the sensor arrays. There are a variety of sensor types used in e-nose technology. These include the following in particular [13,16]:

- Metal oxide,
- Conducting polymer,
- Quartz crystal microbalance,
- Acoustic wave,
- Electro-chemical,
- Catalytic bead,
- Optical.

Among these available gas sensing methods, semiconducting metal oxide gas (MOx) sensor devices have several unique advantages, such as low cost, small size, easy measurement, durability, ease of fabrication, and low detection limits (low ppm level). In addition, most MOx-based sensors are relatively resistant to poisoning. For these reasons, they have quickly gained popularity and have become the most widely used gas sensors today [13].

E-noses have also been developed for medical applications. Here, e-noses can distinguish between different types of diseases and their severity by analyzing body odor. This includes disease-related metabolic changes especially [17], but any kind of drug consumption [18] can also be detected on the skin surface and/or exhaled breath.

One can show that such e-noses can be successfully used to improve the diagnosis of various diseases, ranging from kidney disease [19] and diabetes [20], to various types of respiratory diseases [21] and carcinoma [22,23], up to heart diseases [24]. These and other studies provide evidence that, after a necessary validation, a cost-effective, portable, and fast working e-nose system could be useful for improved diagnostics and health protection.

The diagnosis of chronic liver disease is usually based on a combination of clinical signs, laboratory parameters, and imaging results [25]. However, this approach has several important weaknesses. First, the prognosis of cirrhosis depends on the structure and function of the liver, but even more important is the occurrence of complications, such as variceal bleeding or infections [26,27]. Second, laboratory values can be influenced by other conditions that present in the same way as cirrhosis, which may lead to misinterpretation. Third, some imaging techniques, such as transient elastography, are influenced by "non-liver" factors, such as central venous pressure [28]. Fourth, there is a great need for an exact measurement of the current patient's situation to choose the optimal treatment, e.g., if transplantation is needed or a non-hepatic or hepatic surgery must be performed. In current scoring systems, such as the Model for End-stage Liver Disease (MELD) [29], patients with portal hypertension as a decompensating event (ascites, variceal bleeding) are poorly represented due to the nature of the score.

Based on a proof-of-concept study, De Vicentis et al. [30] showed that an e-nose based on piezoelectric gas sensors could be a valid non-invasive instrument for characterizing chronic liver disease and monitoring hepatic function over time.

The advantages of piezoelectric gas sensors are high sensitivity, small size, fast response time, low power consumption, and robustness [31]. However, these piezoelectric sensors have a poor signal-to-noise ratio, as they operate at very high frequencies and require complex electronic circuits to delineate the signal response, making it difficult for them to act as a supportive element for an efficient e-nose system.

Therefore, the objective of this study was to determine, within the framework of an explorative study, whether liver dysfunction is generally recognizable and whether the level of this dysfunction can be classified utilizing a wearable semiconducting MOx gas sensor-based e-nose.

2. Materials and Methods

2.1. Electronic Nose and Signal Processing

In this study, a system called "LiverTracer" was developed. It is based on an e-nose system that detects changes in the VOCs from exhaled breath caused by liver dysfunctions and their severity. This system consists of a measuring head, which contains the sensor array, and a base unit for measurement control and data analysis (Figure 1).

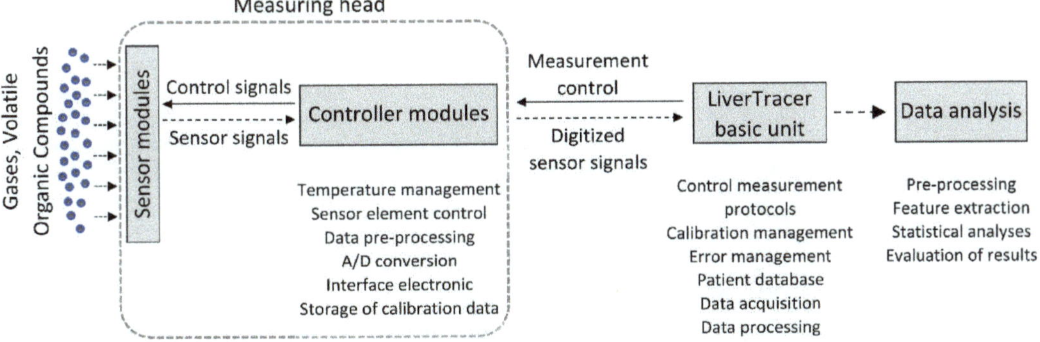

Figure 1. Setup of the electronic nose system "LiverTracer".

The sensor head contains three active MOx semiconductor gas sensor modules (TripleSensor®, UST Umweltsensortechnik GmbH, Geschwenda, Germany). Each sensor module consists of three different gas-sensitive MOx layers that can detect reducible, easily oxidizable, and highly oxidizable gases. The selectivity and sensitivity of the sensor layers for different gas molecules depend mainly on the MOx semiconductor materials and specific catalyst additives used, and can additionally be varied by temperature changes. The latter are controlled by a platinum (PD) heater integrated into each sensor module. Depending on the type of gas, the gas molecules interact specifically with the surface of the different sensor layers, resulting in changes to their electrical conductivity. This conductivity (here measured as resistance) is registered and evaluated. According to the type of gas and sensor layer, concentration ranges from a few ppb up to the percentage range can be detected.

The used gas sensor elements (Triplesensor®) S1, S2, and S3 are realized through hybrid technology: they include a ceramic carrier substrate (aluminum oxide (Al2O3)) with a micro-structured PD thin-film layer, covered with a passivation layer, specific layers for contacts, as well as a gas-sensitive metal-oxide semiconductor layer (or layers) [32–34].

S1 is a ceramic MOx semiconductor gas sensor element with PD multi-electrodes, with a length × width × height (L × W × H) of 2.1 mm × 2.3 mm × 0.63 mm, respectively, with one sensitive MOx layer 2000C2+ (tin oxide (SnO2) thick film layer with a specific catalyst, for the detection of easily oxidizable gases, mainly carbon monoxide (CO), as well as hydrogen (H2) and ethanol (C2H5OH)). The processing of multi-electrode structure signals will be used for the detection of non-desorbing components/contaminations.

S2 is a ceramic MOx semiconductor gas sensor element UST Triplesensor® (type 3A4P10), with an L × W × H of 2.1 mm × 2.3 mm × 0.63 mm, with three sensitive MOx layers: 2000C2+ (specific SnO2 compound with a specific catalyst and a thick film, for the detection of easily oxidizable gases, mainly CO, as well as H2 and C2H5OH), 3000C2+ (a specific SnO2 compound with a specific Pd catalyst and a thick film, for the detection of heavily oxidizable gases, mainly hydrocarbons (CxHy), and which is optimal especially for a number of carbon atoms (C1 to C8)), and 5000C2+ (a specific tungsten trioxide (WO3) compound with a thick film, for the detection of reducible gases, e.g., nitrogen dioxide (NO2)).

S3 is a ceramic MOx semiconductor gas sensor element UST Triplesensor® (type 3A4P10), with a L × W × H of 2.1 mm × 2.4 mm × 0.63 mm, with three sensitive metal-oxide layers: 1000C2+ (a specific SnO2 compound with a catalyst and a thick film), 2000C2+ (a specific SnO2 compound with a specific catalyst and a thick film, for the detection of easily oxidizable gases, mainly CO, as well as H2 and C2H5OH), and 9000C2+ (a specific SnO2 compound with a catalyst and a thick-film, for the detection of long chain hydrocarbons).

The electronic microcontroller modules installed in the measuring head, with an analog-to-digital converter for each sensor element, control the heating temperature, the preprocessing of the sensor signals, the storage of the calibration data and the communication with the basic unit.

A spirometer "SPIROSTIK COMPLETE" (Geratherm Respiratory GmbH, Bad Kissingen, Germany) was used as the basic unit. It contains a Windows 10 computer system. This device was modified according to our requirements. In particular, the sensor control, data storage, operator guidance (semi-automatic patient measurement), and data analysis were developed and integrated on the software side, as were the pump system for flushing and calibrating the measuring head on the hardware side. The principle of the measurement regime is shown in Figure 2.

After starting the system, it is checked whether a scheduled calibration of the e-nose is necessary to verify the correctness of the reference resistance values of the sensor layers to avoid measurement errors. For this purpose, a commercially available test gas (consisting of the components carbon monoxide, oxygen, and nitrogen) is used. Strong deviations of the measurement results from the resistance pattern typical for the applied calibration gas indicate the contamination or aging of the sensor layers. In this case, suitable countermeasures (cleaning, sensor replacement . . .) must be carried out.

If calibration is not required, or after successful calibration, preparation for the actual patient measurement begins. For this purpose, the operator selects an existing patient from the patient database (in case of repetition) or enters the required data for a new patient into the patient database and starts the patient data acquisition.

The processing of the patient measurement protocol (based on a predefined temperature control of the sensor heater optimized in preliminary studies [35]; see Figure 3a) is started with a cyclic thermal cleaning of the sensors until the sensor layers reach their original reference resistances (time-variable process). This is followed by the recording of the room air composition and the actual two patient measurements. By controlling the sensor heating temperature, it is possible to influence the sensitivity of the sensors for different VOCs (extension of the detection range). Burn-off cleaning phases serve to burn or evaporate impurities that may have adhered to the sensor surface. The measurement protocol has a duration of about 16 min. The temperature profile and the associated resistance data curves of all sensor layers are stored for subsequent analysis.

Data analyses were performed using MATLAB R2019a (The MathWorks, Inc., Natick, MA, USA). The 9 raw resistance waveforms were evaluated for outliers, technical problems, artifacts, and measurement errors. No measurement had to be discarded. For the analysis of the respective breathing air segments, the relevant 30 s segments were extracted from the measurement (Figure 3b, marked by vertical dashed lines). This was performed automatically based on the specified temperature measurement protocol, which clearly defines where the breath measurement starts and ends (Figure 3a, "bc1" and "bc2").

The data analysis is based on the extraction of 10 features (time domain and non-linear dynamics domain) from the resistance time series of the extracted breathing gas measurement cycles for each sensor layer.

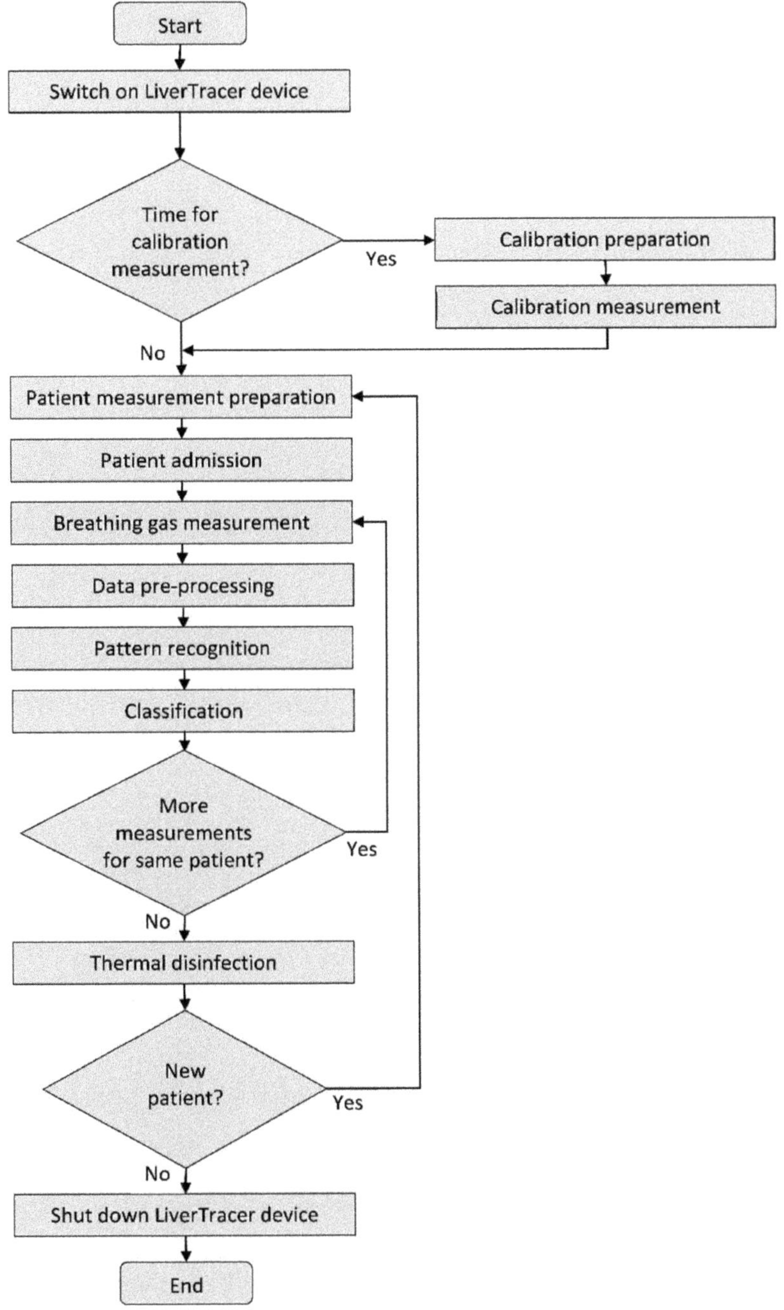

Figure 2. Flow chart of the LiverTracer measurement regime.

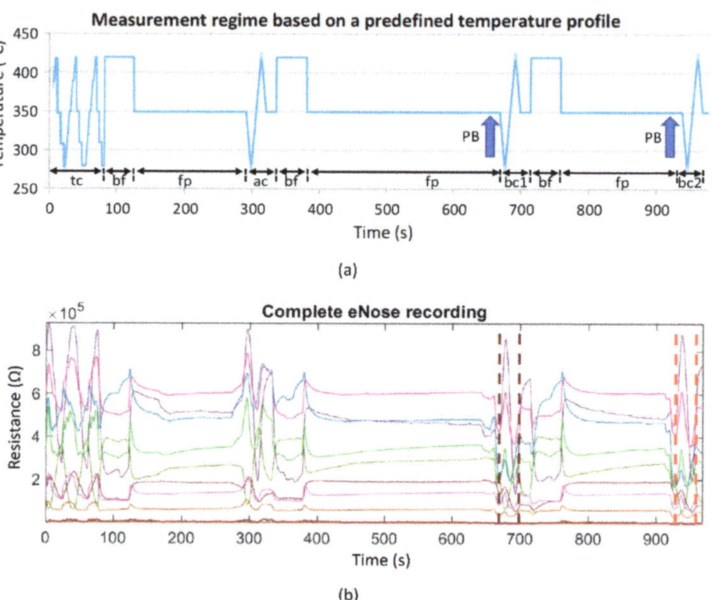

Figure 3. (**a**) Schematic representation of the measurement protocol based on a predefined temperature control of the sensor heater. It contains time-variable cyclic thermal cleaning cycles "tc", burn-off cleaning phases "bf" (rectangle functions), subsequent flushing phases "fp" (horizontal lines), one ambient air measurement cycle "ac", and two breathing gas measurement cycles "bc1" and "bc2". The arrows mark the exhalation cycles (patient breathing: PB); (**b**) example of a recording of 9 sensor layer resistance curves. Vertical dashed lines mark the two breathing gas measurements.

In the time domain, the following features (Figure 4) were calculated:

- slope_startmax (Ω/s): slope from cycle start (Start) to absolute maximum (Max);
- s_slope_startmax (Ω/s): steepest slope of 1s duration from cycle start to Max;
- s_slope_startmax_pos (s): corresponding position of s_slope_startmax;
- s_slope_maxmin (Ω/s): steepest slope of 1s duration from Max to minimum (Min);
- area1 (Ω·s): area under the curve from cycle start to Max;
- area2 (Ω·s): area under the curve from cycle Max to midpoint;
- area3 (Ω·s): area under the curve from cycle midpoint to Min;
- area4 (Ω·s): area under the curve from cycle Min to cycle end;
- area3sec_9 (Ω·s): ninth subarea (24 s to 27 s); area under the curve is evaluated incrementally in 3 s subareas beginning from cycle start.

From the nonlinear dynamics domain, a feature of classical symbolic dynamics and one entropy measure were used. By employing symbolic dynamics [36,37], the original time series is transformed into a symbolic sequence and, thus, presented in a coarser form. Detailed information is lost, which allows the quantification of the dynamics contained in the time series. In the present study, for the quantification of symbolic dynamics of the resistance time series R of the breathing gas measuring cycles, the symbols 0 and 1 were assigned according to the following transformation rules:

$$0 : R_{n+1} - R_n \leq 0,$$
$$1 : R_{n+1} - R_n > 0. \tag{1}$$

Here, R_n and R_{n+1} are the resistance values at the time points n and $n + 1$. While symbol 0 indicates decreasing resistance values, symbol 1 reflects increasing resistance values. Based on the transformed symbol string, words were formed consisting of two successive symbols. The frequency distribution of the word type 00 was determined (this was less dependent on minor fluctuations):

- $p00$—probability for the occurrence of the word type 00 within the resistance value time series.

The entropy measure, Renyi entropy, was calculated [37]. The density distribution (histogram) of resistance values in the resistance time series required for entropy calculations was determined using six classes. The optimal number of classes k was calculated using Sturges' criterion [38]:

$$k = 1 + 3.32 * \log(N), \ N \dots number\ of\ resistance\ values. \tag{2}$$

Based on the density distributions, the individual class probabilities p_i were calculated (with i = 1 to k), followed by the estimation of the following Renyi entropy measures:

$$\text{Renyi}\_\alpha\,[bit] = \frac{1}{1-\alpha} * \log_2 \sum_{i=1}^{k} p_i^{\alpha} \tag{3}$$

Renyi entropy was estimated considering the coefficient value $\alpha = 4$, which influences the weighting of the probabilities p_i (weights larger fluctuations stronger than smaller ones).

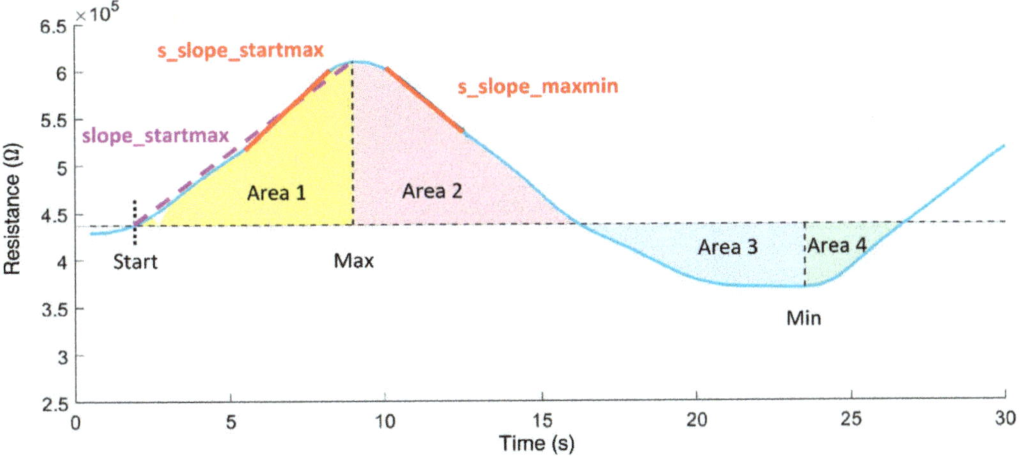

Figure 4. Time domain features extracted from the resistance curve of an exhalation cycle.

2.2. Patients

A total of 30 participants were enrolled, 10 of them being healthy controls, 10 with compensated cirrhosis, and 10 with decompensated cirrhosis, between October 2019 and March 2020. Participating patients were randomly recruited consecutively according to availability in the normal care unit. Patients with ongoing acute-on-chronic liver failure, mechanical cholestasis, acute renal failure, malignant disease, severe cardiopulmonary disease (New York Heart Association classification severity level of heart failure NYHA III/IV (severe heart failure) [39] and/or chronic obstructive pulmonary disease (according to Global Initiative for Chronic Obstructive Lung Disease (GOLD) categories C (high risk/less symptoms) and D (high risk/more symptoms)) [40], and uncontrolled diabetes mellitus were excluded from the study. Control patients were either admitted to the hospital for elective hospitalization for non-liver disease (n = 8) or healthy medical staff (n = 2). Controls were matched for age, sex, and bodyweight. Decompensation was classified according to the Child–Pugh classification score (CPS) [41]. Patients that were classified as CPS B (significant functional compromise) or C (decompensated disease) were allocated to the decompensated group. In addition, patients with variceal hemorrhage were classified as decompensated.

The patients with compensated cirrhosis were male in 7 cases, had a median body weight of 94 kg and a median age of 57 years. Four of them were smokers. The etiology of cirrhosis was ethanol in 6 of these patients and four had other reasons for cirrhosis (2 viral hepatitis, 2 cholestatic liver disease). Patients with decompensated cirrhosis were male in 8 cases, had a median bodyweight of 80 kg and a median age of 62 years. Three of them were smokers and, again, the main etiology of cirrhosis was ethanol consumption in 8 of the patients (the others were 1 autoimmune hepatitis and 1 nonalcoholic steatohepatitis).

Control participants were male in 5 cases and had a median bodyweight of 81 kg. They had a median age of 58 years and one of them was a smoker. They had no history of known liver disease. None of the demographic parameters showed significant differences between the three groups. Vital parameters at inclusion between these groups did not differ as well (Table 1).

Table 1. Patient data (values in parentheses represent the respective minimum and maximum values or describe percentages).

	Control (n = 10)	Compensated Cirrhosis (n = 10)	Decompensated Cirrhosis (n = 10)	p-Value
Sex (f/m)	5/5	3/7	2/8	0.500
Age (years)	58 (51; 65)	57 (52; 64)	62 (56; 67)	0.543
Bodyweight (kg)	81 (68; 96)	94 (79; 101)	80 (68; 97)	0.136
Height (cm)	175 (167; 178)	176 (167; 178)	176 (169; 181)	0.712
Smoker (n,%)	1 (10%)	4 (40%)	3 (30%)	0.450
Vital signs				
RR systolic (mmHg)	135 (118; 161)	126 (107; 155)	122 (103; 136)	0.266
RR diastolic (mmHg)	81 (76; 104)	76 (61; 92)	72 (63; 79)	0.146
Heart rate (pbm)	78 (67; 102)	85 (71; 88)	92 (81; 104)	0.212
Temperature (°C)	36.8 (3.4; 37.0)	36.6 (36.1; 37.0)	36.7 (36.4; 37.1)	0.523
Etiology of cirrhosis (n,%)				
Ethanol	N/A	6 (60%)	8 (80%)	0.628
Other	N/A	4 (40%)	2 (20%)	
Co-medication (n,%)				
Lactulose	1 (10%)	3 (30%)	8 (80%)	0.009
Proton pump inhibitors	5 (50%)	7 (70%)	9 (90%)	0.262
B-Blocker	5 (50%)	4 (40%)	5 (50%)	0.897
Antibiotics	1 (10%)	3 (30%)	7 (70%)	0.016
Rifaximin	0	1 (10%)	6 (60%)	
other	1 (10%)	2 (20%)	1 (10%)	

f—females; m—males; n—number of patients; p—significance.

Relevant co-medication with known influence on intestinal flora and, therefore, on the results of the LiverTracer was analyzed. Lactulose was taken by 1 control patient, 3 patients with compensated cirrhosis, and 8 patients with decompensated cirrhosis (p = 0.009). Antibiotics were taken by 1 control patient, 3 patients with compensated cirrhosis, and 8 patients with decompensated cirrhosis (p = 0.016); however, the difference in antibiotics were caused by rifaxmin, which was taken by 1 patient with compensated and 6 with decompensated cirrhosis. Protone pump inhibitors (p = 0.262) and betablockers (p = 0.897) did not show differences between both groups (Table 1).

All procedures performed in the study involving human participants were approved by the Institutional Ethics Commission of the University Hospital Jena (5359-11/17), and were performed in accordance with the 1964 Helsinki declaration and its later amendments. Written informed consent was obtained from all individual participants prior to inclusion in the study.

2.3. Statistics

Statistical analyses were performed using IBM SPSS 21.0 (IBM Corp. Released 2012. IBM SPSS Statistics for Windows, version 21.0. Armonk, NY, USA: IBM Corp). Descriptive statistics were used to calculate means, standard deviations, medians, and interquartile ranges for all features calculated from the resistance time series separately for all nine sensor layers for respiratory gas measurement. The Kolmogorov–Smirnov test was applied to check the normal distribution of the features. The presence of statistical differences between the respiratory gas analysis characteristics of the control group (CON) and the two groups of patients with compensated (COMP) and decompensated (DECOMP) cirrhosis was tested with Welch's t-test for normally distributed characteristics and with the nonparametric exact two-sided Mann–Whitney U test for non-normally distributed characteristics. A significance level of $p < 0.05$ was considered to be the criterion for statistical differences. Consistent with most of the published work on this topic, this paper presents only means and standard deviations for the identified features, regardless of the distribution or significance test applied, which improves the comparability of study results. Forward stepwise linear discriminant analyses combined with the leave-one-out cross-validation procedure were performed, and receiver operator characteristic (ROC) curves were calculated to assess the classification strength of the feature sets. Sensitivity (SENS), specificity (SPEC), area under the ROC curve (AUC), and accuracy (ACC) were determined for significant features and feature sets, each consisting of 2 or 3 uncorrelated (Pearson correlation coefficient) significant features. The resulting discriminant function analysis was then determined to be the classifier for automatic classification.

3. Results

We report below only the results of the first breathing gas cycle, as we did not find significant differences between the first and second breathing gas cycles. Let us first consider the classification results of the LiverTracer e-nose (Tables 2 and 3). The separation of the patient groups (Table 2) from the controls was 100% successful in each case. Between the patient groups, a correct classification of 95% was achieved, where 90% of the patients from the DECOMP group and 100% of patients from the COMP group were correctly classified. Interestingly, these remarkable classification results were reached using only the features of sensors 1 and 3. Sensor 3 mainly contributed to the result. Sensor 2 did not make any significant contribution. Table 3 shows the descriptive statistics of those features that were automatically selected by the discriminant analysis to obtain the optimal separation results.

Table 2. Percentage classification rate of e-nose features. The optimal parameter set (consisting of either double or triple sets) is shown for each group comparison.

Group	Features	SENS	SPEC	ACC	AUC
CON—COMP	RS11_s_slope_maxmin (Ohm/s) RS32_area3sec_9 (Ohm·s) RS32_p00	1.00	1.00	1.00	1.00
CON—DECOMP	RS31_slope_startmax (Ohm/s) RS32_s_slope_startmax_pos (s) RS33_p00	1.00	1.00	1.00	1.00
COMP—DECOMP	RS32_Renyi4_entropy (bit) RS33_area2 (Ohm·s)	0.90	1.00	0.95	0.97

CON—control group; COMP—patients with compensated cirrhosis; DECOMP—patients with decompensated cirrhosis; RSxy—R denotes resistance measurement values of sensor layer y of sensor Sx (e.g., RS12 describes the resistance readings of sensor layer 2 of sensor S1); SENS—sensitivity; SPEC—specificity; ACC—Accuracy; AUC—area under the receiver operator characteristic curve.

Table 3. Classification results of features automatically selected by discriminant analysis (mv—mean value, sd—standard deviation).

Group				CON	COMP	DECOMP
Test	Features	p		mv ± sd	mv ± sd	mv ± sd
CON vs. COMP	RS11_s_slope_maxmin (Ohm/s)	0.046		−86,258 ± 5225	−81,023 ± 5676	
	RS32_area3sec_9 (Ohm·s)	0.038		1,807,616 ± 207,540	2,071,884 ± 309,151	
	RS32_p00	0.017		0.336 ± 0.050	0.276 ± 0.045	
CON vs. DECOMP	RS31_slope_startmax (Ohm/s)	0.029		8901 ± 3207		6956 ± 1845
	RS32_s_slope_startmax_pos (s)	0.019		6.250 ± 1.161		6.900 ± 0.211
	RS33_p00	0.041		0.369 ± 0.045		0.319 ± 0.056
COMP vs. DECOMP	RS32_Renyi4_entropy (bit)	0.028			1.843 ± 0.386	2.179 ± 0.185
	RS33_area2 (Ohm·s)	0.131			48,252 ± 23,296	34,507 ± 14,547

CON—control group; COMP—patients with compensated cirrhosis; DECOMP—patients with decompensated cirrhosis; RSxy—R denotes the resistance measurement values of sensor layer y of sensor Sx (e.g., RS12 describes the resistance readings of sensor layer 2 of sensor S1); p—significance value; mv ± sd—mean value ± standard deviation.

In Table 4, we included four clinical parameters for the stratification, which are based on the Child–Pugh score and represent different aspects of liver disease, including two laboratory values and two clinical aspects. Bilirubin, the end product of hemoglobin degradation, is cleared from circulation via hepatic elimination and, therefore, elevated in patients with cirrhosis and disturbed liver function. The international normalized ratio (INR), a marker of coagulation, includes proteins synthetized in the liver, which are therefore lowered in cirrhosis. Ascites is frequently present in advanced cirrhosis and is a consequence of cirrhosis-associated portal hypertension, while the occurrence of a hepatic encephalopathy is a typical complication of disturbed detoxification. We decided to skip the fifth parameter, albumin, as this also represents liver synthesis. Except for hepatic encephalopathy, no parameter was convincingly successful. While the controls could still be separated successfully, the detection of liver dysfunction severity was not convincing. The successful classification by hepatic encephalopathy is not surprising, since it was a component of clinical diagnostics.

Table 4. Classification rate (in %) of the clinical parameters that achieved an overall accuracy for discriminating the groups greater than 50%.

	Categorized Bilirubin	Categorized INR	Ascites	Hepatic Encephalopathy
CON	100	86	100	100
COMP	10	40	70	100
DECOMP	90	60	50	50
ACC	63	59	73	83

CON—control group; COMP—patients with compensated cirrhosis; DECOMP—patients with decompensated cirrhosis; INR—international normalized ratio of blood clotting test; ACC—Accuracy.

4. Discussion

This exploratory pilot study extracted and analyzed unique VOC fingerprints in the breath of patients and provides initial evidence that breath VOC analysis using MOx sensors is a potential diagnostic tool for detecting liver dysfunction of different severities.

The sensitivity, specificity, and accuracy for distinguishing compensated and decompensated cirrhosis from healthy controls was 1.00 in all cases. Compensated and decompensated cirrhosis patients could be distinguished with a sensitivity of 0.90, a specificity of 1.00, and an accuracy of 0.95. Sensor 3 (with its three layers) showed the highest discriminatory power, and sensor 1, layer 1 could improve the result of sensor 3 by up to 5%. It was quite sufficient to evaluate only the first exhalation cycle of the patient. The inclusion of the second exhalation cycle did not bring any improvement.

In this study, we included patients with different stages of liver cirrhosis. Differentiation between patients with and without early stages of cirrhosis is challenging, but of great clinical importance. It is usually based on a combination of clinical, imaging, and laboratory parameters, but all of these can be influenced by non-liver related factors as well. Despite these weaknesses, the differentiation between cirrhosis and non-cirrhosis is of great clinical relevance, as the rate of postoperative complications and the mortality are higher in patients with cirrhosis [42]. However, the main predictor of these complications is the hepatic portal venous pressure gradient [43], which is not routinely measured. Using single laboratory parameters or clinical features does not result in the satisfying identification of patients with especially compensated cirrhosis in our study.

A study by Germanese et al. [44] that attempted to discriminate the severity of liver disease, particularly based on detected breath ammonia with MOx sensors, showed that the accuracy of discriminating between non-cirrhotic patients with chronic liver disease and cirrhotic ones was only 0.63, while that of discriminating between liver diseases and healthy controls was 0.81.

The generation of specific VOCs within the body can be the result of metabolic derangement, toxin or teratogen exposure, and finally microbiological processes [45]. Breath tests, which provide an indirect, non-invasive, and relatively low-input evaluation of various diseases, are used as diagnostic tools for quantifying the presence of one or more metabolites of a particular substrate in exhaled breath. Qin et al. [46] analyzed breath samples in hepatocellular carcinoma patients and controls by means of gas chromatography–mass spectrometry (GC/MS) combined with solid phase microextraction. Three potential VOCs, 3-hydroxy-2-butanone, styrene, and decane, were selected as promising biomarkers. A survey of other potential biomarkers in various liver diseases can be found in the publication by De Vincentis et al. [47]. Interestingly, alkanes (decanes) are precisely the group of markers that are particularly favorably detected by the sensors used in our study.

Even though these preliminary results are very promising, several limitations of this explorative pilot study are worth noting. First, it must be noted that the number of patients included is relatively small. However, it should be noted that this is a proof-of-concept study with a new sensor technology compared to previous studies [48]. It should also be noted that, in general, e-noses allow only indirect gas compound detection. In future studies, we intend to combine them with classical laboratory methods (e.g., GC/MS) to enable a direct assignment of biomarkers to the sensors. This would also have the advantage whereby the sensors could be optimally adapted to the pathology via the appropriate doping of the sensor layers. Another limitation of this study is that only a single measurement was performed per patient. Therefore, the system should be validated in a long-term and repeatability study. Additionally, we must mention that the influence of acute events, such as infections, was not studied in detail. This should be addressed in a subsequent study. Finally, MOx sensors also have drawbacks that are mainly related to the lack of sensor stability and the production of sensors with nearly identical sensor characteristics [13,49,50]. Among other factors, contamination and aging of the sensors may lead to short- and long-term drift of the sensors, causing differences in the measured sensor values compared to the originally measured values of new sensors, and reducing the accuracy of pattern recognition based on a trained pattern. Time-consuming recalibrations are often required to compensate for the drift [51]. Replacing a nearly identical sensor is usually difficult. Our approach to significantly reduce the drift and aging problems of the MOx sensors we use is to automatically assess the quality of the sensors before each breath measurement based on the resistance values of the individual sensor layers. In doing so, we compare these with stored threshold values of the original resistances (values of the sensor when it was newly installed). If there are deviations from the original resistance thresholds beyond a certain threshold, cyclic thermal cleaning is automatically performed until the sensor reaches the stored thresholds. If the thresholds are not reached within a specified cycle frequency, the sensor is recalibrated with a test gas. If all these measures fail, the sensor should be replaced with an adequate sensor with as close to identical resistance values as possible, and the e-nose may need to be recalibrated. However, throughout the study period, the values of our applied sensor array remained within the approved quality level. We therefore assume that the drift problem can be largely compensated for by sensor monitoring and calibration, but there remains some residual risk (especially in the case of sensor failure), the impact of which is currently being investigated in a validation study.

The results from this pilot study are very promising and suggest the principal suitability (especially by using the complex feature extraction method) of the MOx multisensory signals for the analysis of breath changes and, thus, for the identification of liver dysfunctions. Among the sensors used in e-noses for medical diagnostics, MOx semiconductor sensors are by far the most popular. They have high sensitivity, are durable, and, probably most importantly, are relatively inexpensive. Price is an important factor when considering large-scale commercial deployment, especially in developing countries. In addition, because they can operate in a wide range of relative humidity, they are particularly suitable for outdoor use [13,31].

In medicine and biology, e-noses are intelligent biosensor-based systems for the rapid detection, analysis, and classification of complex gaseous odors (usually as VOC mixtures of compound metabolite profiles). These instruments are innovative diagnostic tools with great potential for the non-invasive early detection of many types of diseases based on the analysis of VOC metabolites in the form of gaseous clinical samples [52]. They are inexpensive, have low operating and maintenance costs, and provide real-time analysis. Due to the growing demand for improved healthcare devices and procedures, the need for simpler and wearable e-nose systems that can provide fast and accurate diagnostic results and replace traditional, complex, often expensive and time-consuming clinical and laboratory methods has permanently increased. Such systems should non-invasively detect VOCs and accelerate on-site testing, allowing earlier diagnosis, faster treatment of disease, better prognosis, shorter hospital stays, faster recovery, and ultimately lower healthcare costs. Further development and point-of-care testing of new e-nose technologies and the development of standardized diagnostic methods will help bring these e-noses into routine clinical practice.

In summary, the multisensory analyses performed in this study based on a wearable MOx sensor array showed high separation accuracies of 95% to 100% between the studied groups. It was not only possible to distinguish liver dysfunctions of different severity from controls at 100%, but also to discriminate between the severities of liver dysfunction at 95% with a correct identification of 100% of all COMP cirrhosis and of 90% of all DECOMP cirrhosis).

Based on a semiconductor MOx sensor array, the wearable e-nose system for detecting disease—in this case liver dysfunction—offers significant advantages over conventional laboratory analysis and the use of other sensor systems when combined with the nonlinear processing of sensor signals. Our system thus represents a promising tool for distinguishing between patients with compensated and decompensated cirrhosis on a functional basis, and can thus make an important contribution, e.g., in preoperative workup or at the level of the general practitioner for the initial diagnosis and, thus, early detection of liver dysfunction.

Author Contributions: Conceptualization, A.V. and A.S.; methodology, A.V., A.S., P.R., P.H., R.S. and S.S.; software, R.S. and S.S.; validation, A.V., A.S. and P.R.; formal analysis, A.V., R.S. and S.S.; investigation, S.V., P.H., A.S. and P.R.; resources, A.V. and A.S.; data curation, S.V.; writing—original draft preparation, A.V., P.R. and R.S.; writing—review and editing, A.V., P.R., J.H., A.S. and R.S.; visualization, R.S. and S.S.; supervision, A.V.; project administration, A.V. and A.S.; funding acquisition, A.V. and A.S. All authors have read and agreed to the published version of the manuscript.

Funding: These studies were supported by grants from the Thuringian Ministry of Economy, Labor and Technology (TMBWAT/TAB 20018), the Federal Ministry for Economic Affairs and Energy (ZF4485201SB7), and the ERDF—European Regional Development Fund.

Institutional Review Board Statement: All procedures performed in the study involving human participants were approved by the Institutional Ethics Commission of the University Hospital Jena (5359-11/17) and were performed in accordance with the 1964 Helsinki declaration and its later amendments.

Informed Consent Statement: Written informed consent was obtained from all individual participants prior to inclusion in the study.

Data Availability Statement: The data of this study are not publicly available due to the fact that the study has not yet been completed, and further evaluations are currently in progress. However, the data are available on request from the corresponding author.

Acknowledgments: We would like to thank the companies UST Umweltsensortechnik GmbH, Geschwenda, Germany and Geratherm Respiratory GmbH, Bad Kissingen, Germany for their support in the use of their technical components. We acknowledge support for the publication costs by the Open Access Publication Fund of the Technische Universität Ilmenau.

Conflicts of Interest: The authors declare no conflict of interest.

References

1. Hartman, J. A possible objective method for the rapid estimation of flavors in vegetables. *Proc. Amer. Soc. Hort. Sci* **1954**, *64*, 335–342.
2. Wilkens, W.F.; Hartman, J.D. An Electronic Analog for the Olfactory Processes. *Ann. N. Y. Acad. Sci.* **1964**, *116*, 608–612. [CrossRef]
3. Dodd, G.H.; Bartlett, P.N.; Gardner, J.W. Odours—The stimulus for an electronic nose. In Proceedings of the NATO Advanced Research Workshop on Sensors and Sensory Systems for an Electronic Nose, Reykjavik, Iceland, 5–8 August 1991; pp. 1–11.
4. Moncrieff, R.W. An instrument for measuring and classifying odors. *J. Appl. Physiol.* **1961**, *16*, 742–749. [CrossRef]
5. Henderson, M.J.; Karger, B.A.; Wren Shall, G.A. Acetone in the breath: A study of acetone exhalation in diabetic and nondiabetic human subjects. *Diabetes* **1952**, *1*, 188–193. [CrossRef]
6. Gardner, J.W.; Vincent, T.A. Electronic Noses for Well-Being: Breath Analysis and Energy Expenditure. *Sensors* **2016**, *16*, 947. [CrossRef]
7. Farraia, M.V.; Cavaleiro Rufo, J.; Paciencia, I.; Mendes, F.; Delgado, L.; Moreira, A. The electronic nose technology in clinical diagnosis: A systematic review. *Porto Biomed. J.* **2019**, *4*, e42. [CrossRef]
8. Wasilewski, T.; Migon, D.; Gebicki, J.; Kamysz, W. Critical review of electronic nose and tongue instruments prospects in pharmaceutical analysis. *Anal. Chim. Acta* **2019**, *1077*, 14–29. [CrossRef]
9. Wilson, A.D.; Baietto, M. Advances in electronic-nose technologies developed for biomedical applications. *Sensors* **2011**, *11*, 1105–1176. [CrossRef]
10. Majchrzak, T.; Wojnowski, W.; Dymerski, T.; Gębicki, J.; Namieśnik, J. Electronic noses in classification and quality control of edible oils: A review. *Food Chem.* **2018**, *246*, 192–201. [CrossRef]
11. Tan, J.; Xu, J. Applications of electronic nose (e-nose) and electronic tongue (e-tongue) in food quality-related properties determination: A review. *Artif. Intell. Agric.* **2020**, *4*, 104–115. [CrossRef]
12. He, J.; Xu, L.; Wang, P.; Wang, Q. A high precise E-nose for daily indoor air quality monitoring in living environment. *Integration* **2017**, *58*, 286–294. [CrossRef]
13. Karakaya, D.; Ulucan, O.; Turkan, M. Electronic Nose and Its Applications: A Survey. *Int. J. Autom. Comput.* **2020**, *17*, 179–209. [CrossRef]
14. Gardner, J.W.; Bartlett, P.N. A brief history of electronic noses. *Sens. Actuators B* **1994**, *18*, 211–220. [CrossRef]
15. Gupta, A.; Singh, T.S.; Yadava, R.D.S. MEMS sensor array-based electronic nose for breath analysis-a simulation study. *J. Breath Res.* **2018**, *13*, 016003. [CrossRef]
16. Wojnowski, W.; Dymerski, T.; Gebicki, J.; Namiesnik, J. Electronic Noses in Medical Diagnostics. *Curr. Med. Chem.* **2019**, *26*, 197–215. [CrossRef]
17. Wilson, A.D. Advances in electronic-nose technologies for the detection of volatile biomarker metabolites in the human breath. *Metabolites* **2015**, *5*, 140–163. [CrossRef]
18. Voss, A.; Witt, K.; Kaschowitz, T.; Poitz, W.; Ebert, A.; Roser, P.; Bar, K.J. Detecting cannabis use on the human skin surface via an electronic nose system. *Sensors* **2014**, *14*, 13256–13272. [CrossRef]
19. Voss, A.; Baier, V.; Reisch, R.; von Roda, K.; Elsner, P.; Ahlers, H.; Stein, G. Smelling renal dysfunction via electronic nose. *Ann. Biomed. Eng.* **2005**, *33*, 656–660. [CrossRef]
20. Wang, P.; Tan, Y.; Xie, H.; Shen, F. A novel method for diabetes diagnosis based on electronic nose. *Biosens. Bioelectron.* **1997**, *12*, 1031–1036. [CrossRef]
21. Dragonieri, S.; Pennazza, G.; Carratu, P.; Resta, O. Electronic Nose Technology in Respiratory Diseases. *Lung* **2017**, *195*, 157–165. [CrossRef]
22. Baldini, C.; Billeci, L.; Sansone, F.; Conte, R.; Domenici, C.; Tonacci, A. Electronic Nose as a Novel Method for Diagnosing Cancer: A Systematic Review. *Biosensors* **2020**, *10*, 84. [CrossRef] [PubMed]

23. Capelli, L.; Taverna, G.; Bellini, A.; Eusebio, L.; Buffi, N.; Lazzeri, M.; Guazzoni, G.; Bozzini, G.; Seveso, M.; Mandressi, A.; et al. Application and Uses of Electronic Noses for Clinical Diagnosis on Urine Samples: A Review. *Sensors* **2016**, *16*, 1708. [CrossRef] [PubMed]

24. Voss, A.; Witt, K.; Fischer, C.; Reulecke, S.; Poitz, W.; Kechagias, V.; Surber, R.; Figulla, H.R. Smelling heart failure from human skin odor with an electronic nose. In Proceedings of the 2012 Annual International Conference of the IEEE Engineering in Medicine and Biology Society, San Diego, CA, USA , 28 August–1 September 2012; pp. 4034–4037. [CrossRef] [PubMed]

25. European Association for the Study of the Liver. EASL Clinical Practice Guidelines for the management of patients with decompensated cirrhosis. *J. Hepatol.* **2018**, *69*, 406–460. [CrossRef]

26. Moreau, R.; Jalan, R.; Gines, P.; Pavesi, M.; Angeli, P.; Cordoba, J.; Durand, F.; Gustot, T.; Saliba, F.; Domenicali, M.; et al. Acute-on-chronic liver failure is a distinct syndrome that develops in patients with acute decompensation of cirrhosis. *Gastroenterology* **2013**, *144*, 1426–1437. [CrossRef]

27. Trebicka, J.; Fernandez, J.; Papp, M.; Caraceni, P.; Laleman, W.; Gambino, C.; Giovo, I.; Uschner, F.E.; Jimenez, C.; Mookerjee, R.; et al. The PREDICT study uncovers three clinical courses of acutely decompensated cirrhosis that have distinct pathophysiology. *J. Hepatol.* **2020**, *73*, 842–854. [CrossRef]

28. Millonig, G.; Friedrich, S.; Adolf, S.; Fonouni, H.; Golriz, M.; Mehrabi, A.; Stiefel, P.; Pöschl, G.; Büchler, M.W.; Seitz, H.K.; et al. Liver stiffness is directly influenced by central venous pressure. *J. Hepatol.* **2010**, *52*, 206–210. [CrossRef]

29. Kamath, P.S.; Wiesner, R.H.; Malinchoc, M.; Kremers, W.; Therneau, T.M.; Kosberg, C.L.; D'Amico, G.; Dickson, E.R.; Kim, W.R. A model to predict survival in patients with end-stage liver disease. *Hepatology* **2001**, *33*, 464–470. [CrossRef]

30. De Vincentis, A.; Pennazza, G.; Santonico, M.; Vespasiani-Gentilucci, U.; Galati, G.; Gallo, P.; Zompanti, A.; Pedone, C.; Antonelli Incalzi, R.; Picardi, A. Breath-print analysis by e-nose may refine risk stratification for adverse outcomes in cirrhotic patients. *Liver Int.* **2017**, *37*, 242–250. [CrossRef]

31. Behera, B.; Joshi, R.; Anil Vishnu, G.K.; Bhalerao, S.; Pandya, H.J. Electronic nose: A non-invasive technology for breath analysis of diabetes and lung cancer patients. *J. Breath Res.* **2019**, *13*, 024001. [CrossRef]

32. Kiesewetter, O.; Ewert, A.; Melchert, V.; Kittelmann, S. Arrangement for the Detection of Air Constituents and Method for Operating the Arrangement. DE102004060101 B4, 13 December 2004.

33. Kiesewetter, O.; Ewert, A.; Melchert, V.; Kittelmann, S. Assembly for Detecting Air Components. EP 1602924 B1, 2 June 2005.

34. Kittelmann, S.; Ewert, A.; Kiesewetter, O. Air Content Detection Arrangement for Use in e.g., Building, HAS EVALUating Unit with Evaluating Resistance That Is Connected in Parallel with Individual or Combinations of Effective Resistances of Set of Operating Layers Connected with Unit. DE 102006033528 B3, 20 July 2006.

35. Witt, K.; Fischer, C.; Reulecke, S.; Kechagias, V.; Surber, R.; Figulla, H.R.; Voss, A. Electronic nose detects heart failure from exhaled breath. *Biomed. Tech.* **2013**, *58* (Suppl. S1). [CrossRef] [PubMed]

36. Kurths, J.; Voss, A.; Saparin, P.; Witt, A.; Kleiner, H.J.; Wessel, N. Quantitative analysis of heart rate variability. *Chaos* **1995**, *5*, 88–94. [CrossRef] [PubMed]

37. Voss, A.; Kurths, J.; Kleiner, H.J.; Witt, A.; Wessel, N.; Saparin, P.; Osterziel, K.J.; Schurath, R.; Dietz, R. The application of methods of non-linear dynamics for the improved and predictive recognition of patients threatened by sudden cardiac death. *Cardiovasc. Res.* **1996**, *31*, 419–433. [CrossRef]

38. Sturges, H.A. The Choice of a Class Interval. *J. Am. Stat. Assoc.* **1926**, *21*, 65–66. [CrossRef]

39. Hunt, S.A. ACC/AHA 2005 guideline update for the diagnosis and management of chronic heart failure in the adult: A report of the American College of Cardiology/American Heart Association Task Force on Practice Guidelines (Writing Committee to Update the 2001 Guidelines for the Evaluation and Management of Heart Failure). *J. Am. Coll. Cardiol.* **2005**, *46*, e1–e82. [CrossRef]

40. Patel, A.R.; Patel, A.R.; Singh, S.; Singh, S.; Khawaja, I. Global Initiative for Chronic Obstructive Lung Disease: The Changes Made. *Cureus* **2019**, *11*, e4985. [CrossRef]

41. Child, C.G.; Turcotte, J.G. Surgery and portal hypertension. *Major Probl. Clin. Surg.* **1964**, *1*, 1–85.

42. Parikh, N.D.; Chang, Y.H.; Tapper, E.B.; Mathur, A.K. Outcomes of Patients With Cirrhosis Undergoing Orthopedic Procedures: An Analysis of the Nationwide Inpatient Sample. *J. Clin. Gastroenterol.* **2019**, *53*, e356–e361. [CrossRef]

43. Wong, M.; Busuttil, R.W. Surgery in Patients with Portal Hypertension. *Clin. Liver Dis.* **2019**, *23*, 755–780. [CrossRef]

44. Germanese, D.; Colantonio, S.; D'Acunto, M.; Romagnoli, V.; Salvati, A.; Brunetto, M. An E-Nose for the Monitoring of Severe Liver Impairment: A Preliminary Study. *Sensors* **2019**, *19*, 3656. [CrossRef]

45. Arasaradnam, R.P.; McFarlane, M.; Ling, K.; Wurie, S.; O'Connell, N.; Nwokolo, C.U.; Bardhan, K.D.; Skinner, J.; Savage, R.S.; Covington, J.A. Breathomics–exhaled volatile organic compound analysis to detect hepatic encephalopathy: A pilot study. *J. Breath Res.* **2016**, *10*, 016012. [CrossRef] [PubMed]

46. Qin, T.; Liu, H.; Song, Q.; Song, G.; Wang, H.Z.; Pan, Y.Y.; Xiong, F.X.; Gu, K.S.; Sun, G.P.; Chen, Z.D. The screening of volatile markers for hepatocellular carcinoma. *Cancer Epidemiol. Biomark. Prev.* **2010**, *19*, 2247–2253. [CrossRef] [PubMed]

47. De Vincentis, A.; Vespasiani-Gentilucci, U.; Sabatini, A.; Antonelli-Incalzi, R.; Picardi, A. Exhaled breath analysis in hepatology: State-of-the-art and perspectives. *World J. Gastroenterol.* **2019**, *25*, 4043–4050. [CrossRef] [PubMed]

48. De Vincentis, A.; Pennazza, G.; Santonico, M.; Vespasiani-Gentilucci, U.; Galati, G.; Gallo, P.; Vernile, C.; Pedone, C.; Antonelli Incalzi, R.; Picardi, A. Breath-print analysis by e-nose for classifying and monitoring chronic liver disease: A proof-of-concept study. *Sci. Rep.* **2016**, *6*, 25337. [CrossRef]

49. Esfahani, S.; Tiele, A.; Agbroko, S.O.; Covington, J.A. Development of a Tuneable NDIR Optical Electronic Nose. *Sensors* **2020**, *20*, 6875. [CrossRef]
50. Ruffer, D.; Hoehne, F.; Buhler, J. New Digital Metal-Oxide (MOx) Sensor Platform. *Sensors* **2018**, *18*, 1052. [CrossRef]
51. Padilla, M.; Perera, A.; Montoliu, I.; Chaudry, A.; Persaud, K.; Marco, S. Drift compensation of gas sensor array data by Orthogonal Signal Correction. *Chemom. Intell. Lab. Syst.* **2010**, *100*, 28–35. [CrossRef]
52. Wilson, A.D. Application of Electronic-Nose Technologies and VOC-Biomarkers for the Noninvasive Early Diagnosis of Gastrointestinal Diseases (dagger). *Sensors* **2018**, *18*, 2613. [CrossRef]

Article

Phase-Based Grasp Classification for Prosthetic Hand Control Using sEMG

Shuo Wang [1], Jingjing Zheng [1], Bin Zheng [2] and Xianta Jiang [1,*]

[1] Department of Computer Science, Memorial University of Newfoundland, St. John's, NL A1C 5S7, Canada; shuow@mun.ca (S.W.); jzheng20@mun.ca (J.Z.)

[2] Department of Surgery, University of Alberta, Edmonton, AB T6G 2R3, Canada; bzheng1@ualberta.ca

* Correspondence: xiantaj@mun.ca; Tel.: +1-709-8069-4096

Abstract: Pattern recognition using surface Electromyography (sEMG) applied on prosthesis control has attracted much attention in these years. In most of the existing methods, the sEMG signal during the firmly grasped period is used for grasp classification because good performance can be achieved due to its relatively stable signal. However, using the only the firmly grasped period may cause a delay to control the prosthetic hand gestures. Regarding this issue, we explored how grasp classification accuracy changes during the reaching and grasping process, and identified the period that can leverage the grasp classification accuracy and the earlier grasp detection. We found that the grasp classification accuracy increased along the hand gradually grasping the object till firmly grasped, and there is a *sweet period* before firmly grasped period, which could be suitable for early grasp classification with reduced delay. On top of this, we also explored corresponding training strategies for better grasp classification in real-time applications.

Keywords: myoelectric prosthesis; sEMG; grasp phases analysis; grasp classification; machine learning

1. Introduction

Losing a hand is a tremendously physical trauma to any individual. Amputated individuals face a huge difficulty in performing daily activities independently [1], which can also lead to unemployment and social isolation [2]. According to statistics, only 66% of them can resume work after that [3].

To restore the functionality of hands in daily life and in work place, wearing prostheses is one of the necessary options for amputees. There are three types of prosthetic hands: cosmetic hand, body-power hand and Myoelectric hand [4,5]. Among them, Myoelectric prosthesis hand is the most promising one, which allows an amputee to controls the robotic hand by reading his/her muscle actives using Surface Electromyography (sEMG) sensors on the residual forearm. The computer chip will read in muscle signals and convert signals into executable commands.

Interpretation on muscle signals is essential for the control of electric powered prosthetic hands, which requires machine learning algorithms to classify muscular electric signals into corresponding hand movement patterns. In most of the published papers, scientists use myoelectric signals recorded during firmly grasped periods for grasp classification, which yielded satisfactory classification outcomes [6–11]. For instance, the research done by Jiang et al. [7] using 3 s firm grasp sEMG signals achieved approximately 85% accuracy for classifying 16 grasp gestures. However, the firmly grasped periods occur at the end of reaching and grasping, giving no time to control arm movement in a real-life environment [12].

Citation: Wang, S.; Zheng, J.; Zheng, B.; Jiang, X. Phase-Based Grasp Classification for Prosthetic Hand Control Using sEMG. *Biosensors* **2022**, *12*, 57. https://doi.org/10.3390/bios12020057

Received: 15 December 2021

Accepted: 18 January 2022

Published: 21 January 2022

Publisher's Note: MDPI stays neutral with regard to jurisdictional claims in published maps and institutional affiliations.

When including muscle activities recorded from entire grasp period, the classification accuracy decreased. In Cognolato et al.'s report [13], the accuracy of the classification for 10 grasp gestures was approximately 63% to 82% by using the sEMG signals during the whole grasp period.

To solve this problem, developing a method to classify grasp pattern using sEMG data recorded in the earlier grasp period with a high accuracy is necessary. In this study, we investigate how grasp classification accuracy changes over the entire reaching and grasping process, and identify a period in early grasp phase that can achieve the best grasp classification outcome. We call this period as *sweet period*. Once the sweet period is identified, we can develop a better classification strategy that can be used in the real-time environment.

Specifically, we first apply and compare several processing methods for the feature extraction of the sEMG signals. Then, we conduct an experiment to find the sweet period that is suitable for early grasp classification with the best classification outcome. Finally, we will conduct another experiment to compare several common training and testing strategies to identify an effective strategy for better real-time grasp classification. We hypothesize that the muscle activities recorded in the early period of hand grasping can provide sufficient information to achieve the same or higher accuracy of grasp classification than other time periods with a reduced delay for prosthetic hand control.

2. Materials and Methods

2.1. Data Collection

The data used in this study were from an open-source dataset collected by Cognolato et al. [13], where the sEMG data were recorded from 30 healthy subjects (27 male and 3 female), with an average age of 46.63 ± 15.11 years.

Twelve sEMG sensors were placed on the forearm of each subject, producing twelve columns of sEMG data, respectively. Due to the hardware problem, no myoelectric data were received from electrode number eight during the acquisition of subject S024. Therefore, the sEMG data for this subject were recorded from eleven electrodes instead of twelve [13].

Ten grasp gestures were performed in this data collection which were selected based on the hand taxonomies [14–17] and grasp frequency in Activities of Daily Living [18]. The participant performed each gesture for four repetitions, and in each repetition, the same gesture was performed three times using three different objects, respectively. A designated experimenter vocally guided the participant to perform which gestures and grasp which objects. The data were labelled according to the vocal instruction. The list of gestures and objects are shown in Table 1.

In the data post-processing part, the abnormal samples were replaced with the precedent valid samples when filtering outliers [13]. As there might be a delay between the participants' response to the vocal instructions [13], the sEMG activation time might not be matched perfectly with the stimulus time. Therefore, relabeling was performed to calibrate this difference using the method described by Kuzborskij et al. [19].

2.2. Electromyography Feature Extraction and Selection

In the feature extraction process, we first determined the suitable window size for deriving features [20]. As shown in Table 2, several sizes of the overlapped window were tested, which are 50 ms, 100 ms, 200 ms, 500 ms, and 1000 ms. As the increase of the window size, the accuracy keeps increasing, which means that the more data we used to derive features, the better performance we could get. However, considering the capability of Myoelectric prosthesis in the real-life condition, a large window would delay the grasp action from the prosthetic hand. On the other hand, it can be seen that, when increasing the window size over 200 ms, the increase of the accuracy is less than 1%, which is a very small increase. Therefore, to keep the balance between accuracy and implementation speed, we chose 200 ms as the window length with the step of 50 ms, which is a 75% overlap between successive windows.

Table 1. The columns indicate the ID and name of the grasp gestures, the name of the object, and the name of the part of the object involved in the grasping. Adapted from ref. [13].

ID	Grasp Gesture	Object	Grasp Location
1	medium wrap	bottle can door handle	bottle body can body door handle stick
2	lateral	mug key pencil case	mug handle key body case zip
3	parallel extension	plate book drawer	plate edge book body drawer edge
4	tripod grasp	bottle mug drawer	bottle cap mug body drawer knob
5	power sphere	ball bulb key	ball body bulb body key chain
6	precision disk	jar bulb ball	jar lid bulb body ball body
7	prismatic pinch	clothespin key can	clothespin body key ring can pull tab
8	index finger extension	remote knife fork	remote button knife body fork body
9	adducted thumb	screwdriver remote wrench	screwdriver body remote body wrench body
10	prismatic four finger	knife fork wrench	knife handle fork handle wrench handle

Table 2. Window Length Analysis. Both training and test data used the whole grasp period. The classifier used was lightGBM. The features used were STD, RMS, IEMG, MAV, WL, SSI, AAC, and DASDV mentioned in Figure 1. The cross-validation method used was leave-one-repetition-out cross-validation which used one repetition data for testing and the rest three repetitions for training the model, and repeated this process four times to cover all repetitions for testing.

Window Length	Accuracy
50 ms	77.02%
100 ms	78.79%
200 ms	79.98%
500 ms	80.04%
1000 ms	80.33%

To assure the recognition accuracy by using proper features, we tested eleven commonly used features, which were Standard Deviation (STD), Root Mean Square (RMS), Integrated EMG (IEMG), Mean Absolute Value (MAV), Waveform Length (WL), Log Detector (LOG), Simple Square Integral (SSI), Skewness (SKW), Kurtosis (KURT), Average Amplitude Change (AAC) and Difference Absolute Standard Deviation Value (DASDV) [21]. We dropped three lowest performance features, whichwere LOG, SKW and KURT and chose the rest eight with the highest accuracy as the final features for the following research. The performance of these features are shown in Figure 1. After applying the eight features to the sEMG signals, the data set was converted from 12 columns to 96 columns. Due to the sensor hardware issue mentioned in the first subsection, the sEMG data of subject S024 was changed from 11 columns to 88 columns.

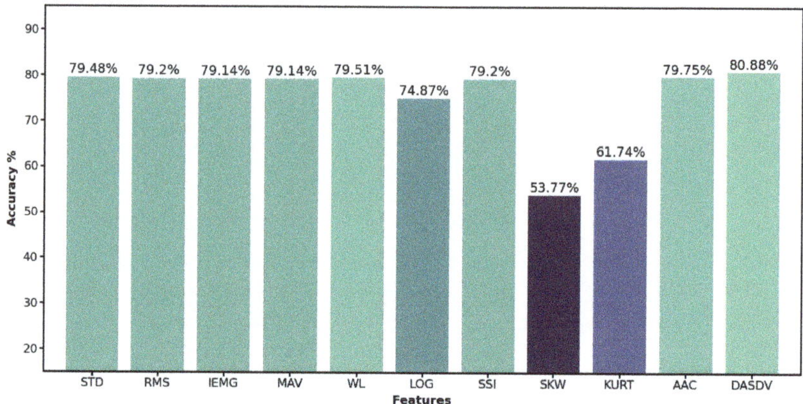

Figure 1. Single feature performance with window size 200 ms. The eleven features are Standard Deviation (STD), Root Mean Square (RMS), Integrated EMG (IEMG), Mean Absolute Value (MAV), Waveform Length (WL), Log Detector (LOG), Simple Square Integral (SSI), Skewness (SKW), Kurtosis (KURT), Average Amplitude Change (AAC) and Difference Absolute Standard Deviation Value (DASDV). The classifier used was lightGBM. The cross-validation method used was leave-one-repetition-out cross-validation which used one repetition data for testing and the rest three repetitions for training the model, and repeated this process four times to cover all repetitions for testing.

2.3. Classification Models

Gradient boosting decision tree, such as XGBoost [22] and Light Gradient Boosting Machine (LightGBM) [23], is a popular machine learning algorithm used by a large amount of data scientists recently, which can achieve a high performance by using decision trees as weak learners and assembling them to come up with one strong learner. Considering the high feature dimensions and large data size, we chose LightGBM as the classifier which runs faster while maintaining a high level of accuracy by utilizing two novel techniques called Gradient-Based One-Side Sampling (GOSS) and Exclusive Feature Bunding (EFB) [23]. In the experiment of Ke et al. (2017), LightGBM can accelerate the training process up to twenty times than XGBoost.

We tuned the hyperparameters by using the training set of all the subjects and obtained the best results as follows: the learning rate is 0.1; no limit was set for the maximum depth; the number of estimators is 100; the number of leaves is 31; the remaining parameters are set to the default values.

2.4. Phase-Based Grasp Analysis

Normally, a typical reaching and grasping process can be divided into three phases [24,25]:

1. The Reaching Phase: starts from the hand lifting off, and ends by touching the object. During this phase, the hand is accelerated to a peak velocity and then is decelerated and brought to touch the target object. The hand usually opens to be configured to the target grasp gesture (pre-shape) [26].

2. The Early Grasping Phase: begins at the moment when the hand initially contacts the object, and gradually closes the fingers until the hand starts to firmly grasp the object.
3. The Firm Grasping Phase: the target object is firmly grasped and hand shape is maintained relatively steady.

We segmented the Reaching, Early grasping, and Firm Grasping phases of each grasp gesture from each subject by observing corresponding videos frame by frame and calculated the average duration of each phase from all the observations. The judgment criteria for entering an Early Grasping Phase was the moment that the hand started to touch the target object, the judgement criteria for entering a Firm Grasping Phase was the moment that the target grasp gesture was completely formed and the hand started to keep relatively steady. According to the segmentation, Early Grasping Phase and Firm Grasping Phase started averagely 1020 ms and 1604 ms from the beginning of Reaching Phase, respectively. An example of grasp phases overlaid with sEMG signals during a full grasp trail is shown in Figure 2.

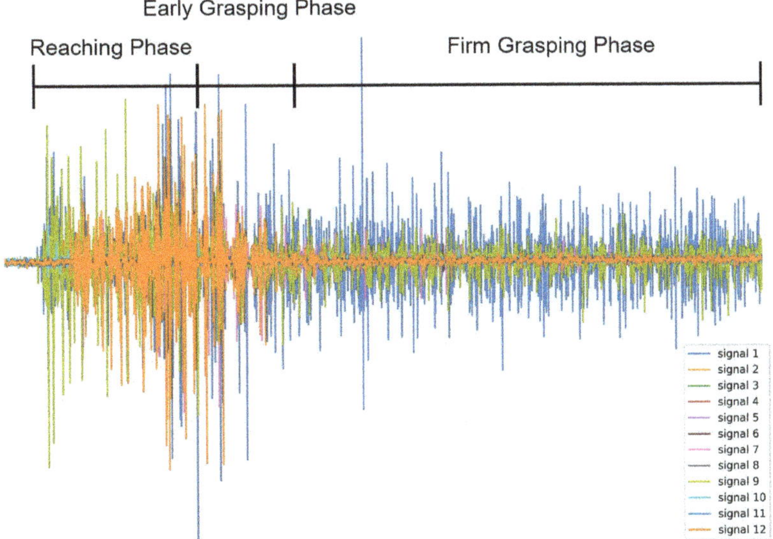

Figure 2. An example of grasp phases overlaid with sEMG signals during a full grasp trial. The start and end positions of these three phases were determined by observing corresponding videos frame by frame.

3. Experiments and Results

We conducted two experiments, the first one aimed to analyse the grasp classification accuracy during the three grasping phases and find out the best position and length of sweet period, another was to find out the best training strategy.

3.1. Data Processing

The grasp trials performed by the participants lasted approximately 4.5–5 s [13]. We removed the data after 4.5 s to align all the trials the same length. Because the overlapped window step is 50 ms and the grasp period length is 4.5 s, 90 pieces of data were reminded for each trial.

In this study, each participant performed one grasp gesture four times (repetitions) which allowed us to split the sEMG data by repetitions to validate testing results. For all the cases in this study, we used three repetitions (75%) for training and one repetition (25%) for testing with leave-one-repetition-out cross-validation, which used one repetition data for testing and the rest three repetitions for training the model, and repeated this process four times to cover all repetitions for testing. To increase the reliability of the sEMG data set, there were three objects being grasped in each repetition with the same gesture as mentioned in the data collection section. In other words, there were 324,000 data samples (90 samples/grasp × 10 grasp gestures × 4 repetitions × 3 objects × 30 subjects) in the data set.

3.2. Phases and Sweet Period Analysis

Figure 3 shows the mean changes of testing accuracy of grasp classification during all the three phases. Each data point is averaged across all 900 trials from 30 participants.

From Figure 3 we can see that the accuracy increases from 42% to 84% during the Reaching phase and then becomes stable at the start of Early Grasping phase at around the time of 1000 ms, fluctuating between 84% and 87% during the rest of the grasp period. The mean accuracy further increases to relatively stable at around the time of 1250 ms, where we then define the location of the sweet period.

To find the optimal length of the sweet period, we designed different sliding windows with sizes of 300 ms, 400 ms, 500 ms, 600 ms, 700 ms, 800 ms, 900 ms and 1000 ms. The sliding window moved along with the time with step 50 ms, and in each move, it calculated and recorded the mean accuracy. We analyzed the records from the sliding window, and the results are given in Figure 4.

From Figure 4, we can see that the mean accuracy increases with the increase of window length significantly during the Reaching phase and beginning of the Early Grasping phase (at about 1100 ms) but not significantly afterward. For instance, although the window length of 1000 ms can reach the highest accuracy of 86.3%, it takes a much longer time than the length of 300 ms with an accuracy of 85.5%. Therefore, the length of the sweet period is set as 300 ms, and the position is set from 1100 ms to 1400 ms, which makes it entirely located in the Early Grasping phase as the blue region shown in Figure 3.

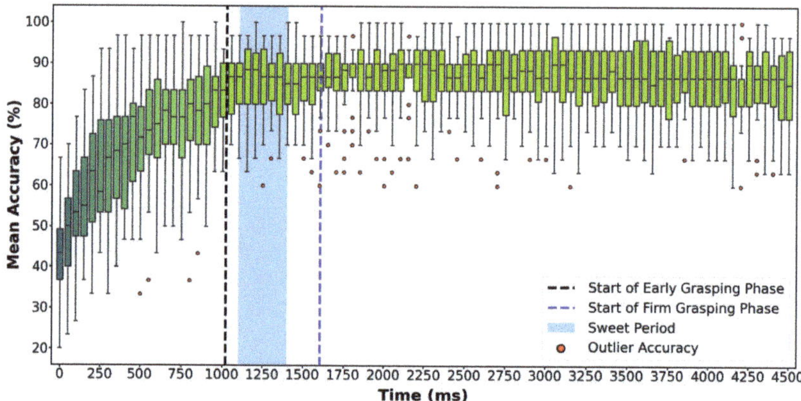

Figure 3. Mean accuracy at each time point during the entire grasp period. This result is from the model which was trained using all three phases data using leave-one-repetition-out cross-validation, and the mean accuracy represents the average accuracy of 30 subjects. The blue region, starts from 1100 ms and ends from 1400 ms, is the sweet period which was confirmed from the first experiment. The vertical dashed lines are averaged starting times of Early Grasping and Firm Grasping phases, which locates at 1020 ms and 1604 ms, respectively. The red dots are outliers.

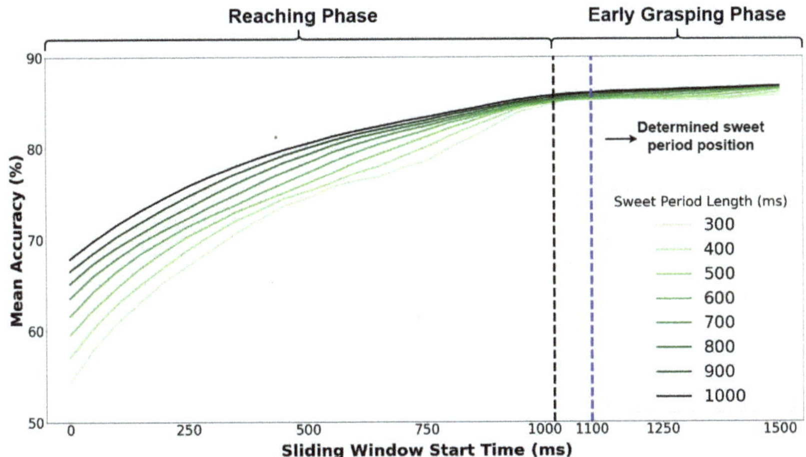

Figure 4. Mean accuracy with different sweet period lengths at different start time.

3.3. Comparison Experiment

In the comparison experiment, we tested six strategies using different training and testing data, as shown in Table 3.

Table 3. Analysis Results for Six Cases. All Three Phases include signal from the time of 0 ms to 4500 ms, Firm Grasping Phase is from the time of 2000 ms to 4500 ms, sweet period is from the time of 1100 ms to 1400 ms. Leave-one-repetition-out cross-validation was employed for all cases, such that all testing data was excluded from training the model.

Case Number	Training Data	Testing Data	Accuracy
1	All Three Phases	All Three Phases	79.98%
2	All Three Phases	Firm Grasping Phase	81.68%
3	All Three Phases	Sweet Period	85.50%
4	Firm Grasping Phase	Firm Grasping Phase	80.39%
5	Firm Grasping Phase	Sweet Period	60.80%
6	Reaching Phase and Early Grasping Phase	Sweet Period	81.01%
7	Early and Firm Grasping Phase	Sweet Period	82.51%
8	Sweet Period	Sweet Period	74.99%

In cases 1–3, we used all the three grasp phases as training data and reduced the testing data size, from all three phases to only the firm grasping phase, then to the sweet period. The purpose of performing these three comparisons was to study which phase/period was the better choice for testing data when using all grasp phases as training data. Besides, to figure out which phase played a better role model training, we studied another five cases. For cases 4–5, we used Firm Grasping Phase for training and reduced the testing data size. In cases 6–7, we used a combined Phases for training and sweet period for testing. In case 8, we used the data in the sweet period for both training and testing. It is worth mentioning again, the cross-validation method used for all the cases was leave-one-repetition-out cross-validation which used one repetition data for testing and the rest three repetitions for training the model, and repeated this process four times to cover all repetitions for testing, such that all testing data was excluded from training the model. For example, in one testing repetition of case 8, the data from the sweet period of three repetitions were used for training the model and the rest one for testing. The results are presented in Table 3.

As shown in the Table 3, we get the highest accuracy of 85.50% when we train with the all grasp phases and test with the only sweet period. Besides, from case 1 to 3, we find that if we keep the training data unchanged, the accuracy increases as the decrease of testing data size.

4. Discussion

Our hypothesis is supported by the results that there is a sweet period located in the Early Grasping Phase where sEMG signals can be used to achieve a similar or higher accuracy and lower delay of grasp classification than other time windows, which would help to improve the performance of robotic hand implementation in the real-life applications. This is important as the classifier can get the data much faster instead of waiting the muscle getting into the Firm Grasping Phase.

We found that during the Reaching Phase, the mean accuracy of this phase is only about 63%. This is because, in this period, the subjects moved their hands to reach the object and start to perform the grasp gesture, keeping the muscle status changing. Therefore, the sEMG signals in this period fluctuate very much, making it difficult in decoding the sEMG signals, see Figure 2.

When getting into the Early Grasping phase, the accuracy reaches approximately 85%, which is as high as that in the Firm Grasping phase. The possible reason for this is the hand has already fully formed into the target gesture during the Early Grasping phase. Although this formed gesture is slightly different to the final target gesture, it can provide sufficient information for the classification. Therefore, the accuracy reaches to a high level at the start of the Early Grasping phase. After the subject firmly grasps the object (getting into the Firm Grasping phase), the accuracy keeps stable at around 85% because the sEMG signals started to be stable, which also make the classification performance stable.

Notice that the sEMG signal is more active in the Reaching and Early Grasping phases with high amplitude of the sEMG waveform as shown in Figure 2. This is because the hand starts to perform the corresponding grasping gestures related activities such as hand aperture, where the sEMG signals from the forearm are usually active with higher amplitude than other phases [26], although the hand has not grasped to the object during the Reaching Phase. In contrast, starting from the mid-Early Grasping phase to the whole Firm Grasping Phase, the muscle status keeps relatively unchanged, which makes the amplitude sEMG signal slightly lower than that in the reaching and grasping phase; this is also why better grasp classification performance was achieved during the Early Grasping phase and the Firm Grasping Phase where the sEMG signal patterns are relatively similar.

Using all three grasp phases for training the model and only using sweet period for controlling is found to be the best strategy for Myoelectric prosthetic hand application in real-life condition, not only because the sweet period during Early Grasping phase is suitable for prosthesis control as discussed before, but also this strategy can also increase the recognition accuracy compared to other strategies. The possible reason of higher accuracy achieved by this strategy could be more variation data was included in the model training. From case 3 and 6 in the Table 3, we can see that if we remove the Firm Grasping Phase from training set, the accuracy decreases from 85.5% to 81.01%. This means that the Firm Grasping Phase is essential for training data because it may contain the information about the final target gesture. From case 3 and 7, we find that if we remove Reaching Phase from training set, the accuracy decreases from 85.5% to 82.51%. This means that Reaching Phase is also important for training data because it is the progress in which the gesture is formed.

For case 5, the accuracy is only 60.80% when only using the Firm Grasping Phase for training because this period lost much information about gesture formation in Reaching and Early Grasping Phases. For case 8, the accuracy reaches 74.99% only using sweet period for training because this training data also lost the part of information about the gesture in the Reaching Phase and the Firm Grasping Phase. However, using all phases data for training and the sweet period data for testing achieved the best accuracy, which can be the common practice in real-life situations where training a model is not time-sensitive.

5. Conclusions

In order to reduce the delay of myoprosthetic hand control in the real-life situation while maintaining a high recognition accuracy, we investigated the grasp classification performance during three grasping phases to identify the sweet period. We found that the sweet period located between 1.1 s and 1.4 s from the start of the hand grasping which happens in the Early Grasping phase before the hand is firmly grasped.

Furthermore, we found using sEMG from all three grasping phases (Reaching, Early Grasping, and Firm Grasping phases) for grasp classification model training achieved the best accuracy. Together with the identified sweet period for controlling, the grasp classification accuracy and the response speed of prosthetic hand can be balanced to achieve high performance.

Author Contributions: Conceptualization, S.W., X.J.; methodology, S.W., J.Z., X.J.; software, S.W.; validation, S.W., J.Z., B.Z., X.J.; formal analysis, S.W., J.Z., X.J.; investigation, S.W., J.Z.; resources, S.W., X.J.; data curation, S.W.; writing—original draft preparation, S.W., X.J.; writing—review and editing, J.Z., X.J., B.Z., S.W.; visualization, S.W.; supervision, X.J., B.Z.; project administration, X.J., B.Z,; funding acquisition, X.J., B.Z. All authors have read and agreed to the published version of the manuscript.

Funding: This research was funded by Natural Sciences and Engineering Research Council of Canada (NSERC) Discovery grant number RGPIN-2020-05525.

Institutional Review Board Statement: Ethical review and approval were waived for this study, due to the public available dataset we used has already approved.

Informed Consent Statement: Not applicable.

Data Availability Statement: The open source data used in this study is from MeganePro dataset 1 which is available at: https://dataverse.harvard.edu/dataverse/meganepro, accessed on 23 September 2020.

Conflicts of Interest: The authors declare no conflict of interest.

References

1. Niedernhuber, M.; Barone, D.G.; Lenggenhager, B. Prostheses as extensions of the body: Progress and challenges. *Neurosci. Biobehav. Rev.* **2018**, *92*, 1–6. [CrossRef] [PubMed]
2. Jang, C.H.; Yang, H.S.; Yang, H.E.; Lee, S.Y.; Kwon, J.W.; Yun, B.D.; Choi, J.Y.; Kim, S.N.; Jeong, H.W. A Survey on Activities of Daily Living and Occupations of Upper Extremity Amputees. *Ann. Rehabil. Med.* **2011**, *35*, 907–921. [CrossRef] [PubMed]
3. Burger, H.; Marinček, Č. Return to work after lower limb amputation. *Disabil. Rehabil.* **2007**, *29*, 1323–1329. [CrossRef]
4. Maat, B.; Smit, G.; Plettenburg, D.; Breedveld, P. Passive prosthetic hands and tools: A literature review. *Prosthetics Orthot. Int.* **2018**, *42*, 66–74. [CrossRef]
5. Castellini, C.; Smagt, P. Surface EMG in advanced hand prosthetics. *Biol. Cybern.* **2009**, *100*, 35–47. [CrossRef] [PubMed]
6. Sun, Y.; Li, C.; Li, G.; Jiang, G.; Jiang, D.; Liu, H.; Zheng, Z.; Shu, W. Gesture recognition based on kinect and sEMG signal fusion. *Mob. Netw. Appl.* **2018**, *23*, 797–805. [CrossRef]
7. Jiang, X.; Merhi, L.K.; Xiao, Z.G.; Menon, C. Exploration of Force Myography and surface Electromyography in hand gesture classification. *Med. Eng. Phys.* **2017**, *41*, 63–73. [CrossRef] [PubMed]
8. Jiang, X.; Merhi, L.K.; Menon, C. Force exertion affects grasp classification using force myography. *IEEE Trans.-Hum.-Mach. Syst.* **2017**, *48*, 219–226. [CrossRef]
9. Asfour, M.; Menon, C.; Jiang, X. A Machine Learning Processing Pipeline for Reliable Hand Gesture Classification of FMG Signals with Stochastic Variance. *Sensors* **2021**, *21*, 1504. [CrossRef]
10. Chen, X.; Zhang, X.; Zhao, Z.Y.; Yang, J.H.; Lantz, V.; Wang, K.Q. Multiple hand gesture recognition based on surface EMG signal. In Proceedings of the 2007 1st IEEE International Conference on Bioinformatics and Biomedical Engineering, Wuhan, China, 6–8 July 2007; pp. 506–509. [CrossRef]
11. Chen, X.; Zhang, X.; Zhao, Z.Y.; Yang, J.H.; Lantz, V.; Wang, K.Q. Hand gesture recognition research based on surface EMG sensors and 2D-accelerometers. In Proceedings of the 2007 11th IEEE International Symposium on Wearable Computers, Boston, MA, USA, 11–13 October 2007; pp. 11–14. [CrossRef]

12. Ummar, A.F.; Nisheena, V.I. The Upper Limb Invariant Myoelectric Prosthetic Control: A Review. In Proceedings of the 2020 International Conference on Futuristic Technologies in Control Systems Renewable Energy (ICFCR), Malappuram, India, 23–24 September 2020; pp. 1–4. [CrossRef]

13. Cognolato, M.; Gijsberts, A.; Gregori, V.; Saetta, G.; Giacomino, K.; Hager, A.G.M.; Gigli, A.; Faccio, D.; Tiengo, C.; Bassetto, F.; et al. Gaze, visual, myoelectric, and inertial data of grasps for intelligent prosthetics. *Sci. Data* **2020**, *7*, 43. [CrossRef]

14. Cutkosky, M. On grasp choice, grasp models, and the design of hands for manufacturing tasks. *IEEE Trans. Robot. Autom.* **1989**, *5*, 269–279. [CrossRef]

15. Sebelius, F.C.; Rosén, B.N.; Lundborg, G.N. Refined Myoelectric Control in Below-Elbow Amputees Using Artificial Neural Networks and a Data Glove. *J. Hand Surg.* **2005**, *30*, 780–789. [CrossRef]

16. Crawford, B.; Miller, K.; Shenoy, P.; Rao, R. Real-time classification of electromyographic signals for robotic control. *AAAI* **2005**, *5*, 523–528.

17. Feix, T.; Romero, J.; Schmiedmayer, H.B.; Dollar, A.M.; Kragic, D. The GRASP Taxonomy of Human Grasp Types. *IEEE Trans.-Hum.-Mach. Syst.* **2016**, *46*, 66–77. [CrossRef]

18. Bullock, I.M.; Zheng, J.Z.; De La Rosa, S.; Guertler, C.; Dollar, A.M. Grasp Frequency and Usage in Daily Household and Machine Shop Tasks. *IEEE Trans. Haptics* **2013**, *6*, 296–308. [CrossRef] [PubMed]

19. Kuzborskij, I.; Gijsberts, A.; Caputo, B. On the challenge of classifying 52 hand movements from surface electromyography. In Proceedings of the 2012 Annual International Conference of the IEEE Engineering in Medicine and Biology Society, San Diego, CA, USA, 28 August–1 September 2012; pp. 4931–4937. [CrossRef]

20. Englehart, K.; Hudgins, B. A robust, real-time control scheme for multifunction myoelectric control. *IEEE Trans. Biomed. Eng.* **2003**, *50*, 848–854. [CrossRef]

21. Phinyomark, A.; Phukpattaranont, P.; Limsakul, C. Feature reduction and selection for EMG signal classification. *Expert Syst. Appl.* **2012**, *39*, 7420–7431. [CrossRef]

22. Friedman, J.; Hastie, T.; Tibshirani, R. *The Elements of Statistical Learning*; Springer Series in Statistics New York; Springer: New York, NY, USA, 2001; Volume 1.

23. Ke, G.; Meng, Q.; Finley, T.; Wang, T.; Chen, W.; Ma, W.; Ye, Q.; Liu, T.Y. LightGBM: A Highly Efficient Gradient Boosting Decision Tree. *Adv. Neural Inf. Process. Syst.* **2017**, *30*, 3146–3154.

24. Mason, C.R.; Gomez, J.E.; Ebner, T.J. Primary motor cortex neuronal discharge during reach-to-grasp: Controlling the hand as a unit. *Arch. Ital. Biol.* **2002**, *140*, 229–236. [CrossRef]

25. Supuk, T.; Bajd, T.; Kurillo, G. Assessment of reach-to-grasp trajectories toward stationary objects. *Clin. Biomech.* **2011**, *26*, 811–818. [CrossRef] [PubMed]

26. Batzianoulis, I.; Krausz, N.E.; Simon, A.M.; Hargrove, L.; Billard, A. Decoding the grasping intention from electromyography during reaching motions. *J. Neuroeng. Rehabil.* **2018**, *15*, 57. [CrossRef] [PubMed]

Article

Atrial Fibrillation Prediction from Critically Ill Sepsis Patients

Syed Khairul Bashar [1], Eric Y. Ding [2], Allan J. Walkey [3], David D. McManus [2] and Ki H. Chon [1,*]

[1] Biomedical Engineering Department, University of Connecticut, Storrs, CT 06269, USA; syed.bashar@uconn.edu
[2] Division of Cardiology, University of Massachusetts Medical School, Worcester, MA 01655, USA; eric.ding@umassmed.edu (E.Y.D.); david.mcmanus@umassmed.edu (D.D.M.)
[3] Department of Medicine, Boston University School of Medicine, Boston, MA 02118, USA; alwalkey@bu.edu
* Correspondence: ki.chon@uconn.edu

Abstract: Sepsis is defined by life-threatening organ dysfunction during infection and is the leading cause of death in hospitals. During sepsis, there is a high risk that new onset of atrial fibrillation (AF) can occur, which is associated with significant morbidity and mortality. Consequently, early prediction of AF during sepsis would allow testing of interventions in the intensive care unit (ICU) to prevent AF and its severe complications. In this paper, we present a novel automated AF prediction algorithm for critically ill sepsis patients using electrocardiogram (ECG) signals. From the heart rate signal collected from 5-min ECG, feature extraction is performed using the traditional time, frequency, and nonlinear domain methods. Moreover, variable frequency complex demodulation and tunable Q-factor wavelet-transform-based time–frequency methods are applied to extract novel features from the heart rate signal. Using a selected feature subset, several machine learning classifiers, including support vector machine (SVM) and random forest (RF), were trained using only the 2001 Computers in Cardiology data set. For testing the proposed method, 50 critically ill ICU subjects from the Medical Information Mart for Intensive Care (MIMIC) III database were used in this study. Using distinct and independent testing data from MIMIC III, the SVM achieved 80% sensitivity, 100% specificity, 90% accuracy, 100% positive predictive value, and 83.33% negative predictive value for predicting AF immediately prior to the onset of AF, while the RF achieved 88% AF prediction accuracy. When we analyzed how much in advance we can predict AF events in critically ill sepsis patients, the algorithm achieved 80% accuracy for predicting AF events 10 min early. Our algorithm outperformed a state-of-the-art method for predicting AF in ICU patients, further demonstrating the efficacy of our proposed method. The annotations of patients' AF transition information will be made publicly available for other investigators. Our algorithm to predict AF onset is applicable for any ECG modality including patch electrodes and wearables, including Holter, loop recorder, and implantable devices.

Keywords: sepsis; atrial fibrillation; prediction; heart rate variability; feature extraction; random forest; annotations

Citation: Bashar, S.K.; Ding, E.Y.; Walkey, A.J.; McManus, D.D.; Chon, K.H. Atrial Fibrillation Prediction from Critically Ill Sepsis Patients. *Biosensors* **2021**, *11*, 269. https://doi.org/10.3390/bios11080269

Received: 16 July 2021
Accepted: 6 August 2021
Published: 9 August 2021

Publisher's Note: MDPI stays neutral with regard to jurisdictional claims in published maps and institutional affiliations.

1. Introduction

Sepsis is a life-threatening, dysregulated response to infection and is the leading cause of death in the hospitals of the United States. Sepsis affects more than 1.5 million Americans yearly at an annual cost of over $20 billion [1]. Atrial fibrillation (AF) is a common and deadly complication of sepsis; it is associated with poor outcomes during hospitalization and confers risk for significant adverse events long thereafter [2]. The mechanisms of AF during sepsis are unclear and may involve rapid remodeling from infection as well as triggers from autonomic nervous system activation, fluid shifts, and electrolyte disturbances [3]. Patients with sepsis have sixfold higher risk of new-onset AF as compared with hospitalized patients without sepsis and similar cardiovascular risk factors. New-onset AF during sepsis is a common and deadly dysrhythmia during sepsis, affecting nearly 1 in 5 septic patients [4,5] and is associated with significant morbidity and mortality [6]. As a result, early prediction of AF during sepsis could potentially lead to AF intervention strategies, thereby minimizing poor hospital outcomes during sepsis.

For the past two decades, there have been many studies of AF prediction using electrocardiogram (ECG) signals outside of the ICU setting. In [7], the frequent occurrence of atrial premature beats prior to the onset of premature atrial contraction (PAC) was reported to be predictive. PAC is characterized by analyzing the quantities of atrial and ventricular ectopic beats from the RR intervals; an increase in atrial ectopic beats is reported in subjects prior to AF episodes [8]. In [9], correlation coefficients, time domain, frequency domain, power spectral densities, and P waves were used to predict paroxysmal AF (PAF). Spectral, bispectral, and nonlinear measurements from 30-min heart rate variability data were used in [10] to predict PAF events. Time domain, frequency domain, nonlinear, and bispectrum features were calculated from 15-min heart rate data; genetic-algorithm-based optimization and a support vector machine classifier were used to predict PAF in [11]. In [12], time, frequency, and nonlinear domain heart rate variability (HRV) features were extracted first, which were then fed into an SVM classifier; feature subset and classifier tuning were performed using nondominated sorting genetic algorithm III. A predictor based on the number of premature atrial complexes not followed by a regular RR interval, runs of atrial bigeminy and trigeminy, and the length of any short run of paroxysmal atrial tachycardia was presented in [13]. In [14], short-term heart rate variability-based features were extracted first; then, genetic-algorithm-based feature selection and k-nearest neighbor classifier were applied to predict PAF. An AF prediction algorithm based on nonlinear features calculated from the return map and difference map of HRV signals was reported in [15]. A symbolic dynamic approach known as footprint analysis was presented in [16] to investigate heart rate dynamics before PAF episodes. In [17], a combination of linear, time–frequency, and nonlinear analysis were performed on heart rate variability and a mixture of experts classification was used for PAF prediction.

However, the common factor for most of the above-mentioned methods is that they were developed and validated using the 2001 Computing in Cardiology (CinC) Challenge data set, as this is the only publicly available data set so far for AF prediction. Thus, the AF prediction studies are limited by the available data sets. In the CinC data set, PAF prediction is performed within the PAF subjects using the two ECG records (pre-AF and distant from AF data segments) from the same subject. Moreover, none of these methods examined AF prediction in critically ill ICU patients. The mechanisms of AF during sepsis may differ from other clinical scenarios; therefore, AF prediction algorithms may differ during sepsis [3,5]. As a result, the above-mentioned methods lack a prospective head-to-head evaluation with clinically derived real life data [17].

In order to address the novel challenges of AF prediction during sepsis, in this study, we present a machine learning approach for AF prediction for ICU patients with sepsis. We used traditional HRV parameters as well as novel time–frequency-based features to identify pre-AF ECG recordings from critically ill sepsis patients.

The major contributions of this study are threefold. First, this is one of the first studies to propose an AF prediction method for critically ill sepsis patients. For this purpose, we use the CinC data for training and only the MIMIC III ICU data for testing; the previous methods used only the CinC data for both training and testing. Second, we not only predict AF immediately before its onset, but also analyze how much in advance we can predict the AF by using the prior 5 min of ECG data, thus allowing adequate time for potential clinical interventions prior to AF onset in real-world scenarios. Third, we provide valuable annotations for the normal sinus rhythm (NSR)-to-AF transition subjects (pre-AF recordings) collected from the MIMIC III ICU data, which will benefit other researchers and advance AF prediction research.

2. Description of the Database

Two different data sets were used in this study:

2.1. Mimic III Database (Used Only as Testing Data)

In this study, a subset of the Medical Information Mart for Intensive Care (MIMIC) III data set was used. MIMIC III is a large open source medical record database publicly available in PhysioNet [18] which contains deidentified health-related data from patients who stayed in critical care units of the Beth Israel Deaconess Medical Center between 2001 and 2012 [19]. It includes a variety of information such as patient demographics, laboratory test results, vital sign measurements, medications, nurse and physician notes, imaging reports, and out-of-hospital mortality, which are some of the notable parameters among many others that are available. In many patients, MIMIC III links continuous ECG waveforms to a wealth of time-varying clinical and hemodynamic data. The sampling frequency of the ECG recordings was 125 Hz and the measurement unit was millivolts (mV).

We have used a total of 50 critically ill ICU patients from the MIMIC III database. Twenty-five of these subjects had non-AF to AF transition, who are henceforth referred as "AF transition subjects." Additionally, these AF transition subjects had at least 1 h of non-AF rhythms before the AF onset. It is to be noted that the first onset of AF was adjudicated by two physicians (AW and DDM). The physicians at the University of Massachusetts Medical School and Boston University's Medical School were involved in finding AF transition subjects.

Finding AF subjects with the above-described requirements was a manually demanding task since it required searching through thousands of patients' ECG data records. Consequently, due to the above requirement, we found a limited number of patients for examining our algorithm's predictive capability. From the "MIMIC III waveform database matched subset" [20], 18 subjects were identified with non-AF to AF transition. ECG signals were annotated by board-certified physicians specializing in AF management (AW and DDM). These AF transition subjects had sepsis according to the International Classification of Diseases, Ninth Revision (ICD-9) codes.

Moreover, the physicians identified seven additional subjects with non-AF to AF transition who were not included in the MIMIC III matched subset, rather only from the MIMIC III database. These seven subjects were from the critically ill group; however, since these seven subjects were not from the MIMIC III matched subset, no clinical information about sepsis was available. Overall, a total of 25 (=18 + 7) subjects with non-AF to AF transition (i.e., pre-AF) were identified.

Similarly, in order to form the control group, 25 NSR subjects were chosen to match the number of non-AF to AF transition subjects. These control subjects were randomly chosen using the previous AF and NSR annotations provided by our group, and were adjudicated to not be in AF for the entire duration of the waveform recording [21]. These 25 NSR control subjects were from a critically ill group with sepsis. As a result, the total number of subjects in this study was 50, and they were used only as the test data set. The annotated data will be made publicly available at https://biosignal.uconn.edu/resources/ to facilitate further research.

2.2. AFPDB Database (Used Only as Training Data)

The AFPDB data set is a publicly available paroxysmal atrial fibrillation prediction database, which originated from the PAF prediction challenge administered by Computers in Cardiology in 2001 [18,22]. The training database contains 25 pairs of ECG recordings obtained from patients with paroxysmal AF where each pair is recorded from different PAF patients. Each pair of data contains one 30-min ECG segment that ends just prior to the onset of a PAF event and another 30-min ECG segment at least 45 min distant from the onset of PAF. Moreover, recordings from 25 normal subjects were provided; each recording is 30 min long and has two channels.

For this study, we used 25 control ECGs and 25 ECG segments which are just prior to the onset of PAF (referred to as pre-AF). As a result, the 50 recordings are from different subjects. Each ECG segment contained two-channel traces from Holter recordings with a sampling rate of 128 Hz and 12-bit resolution.

3. Proposed Method

The AF prediction scheme is illustrated in Figure 1. AF onset refers to the time point when AF started and the ECG recording prior to this onset is referred to as "pre-AF." The goal of our proposed method is to be able to predict the AF onset using this "pre-AF" data. For the control group, since there is no AF event, a random ECG portion is identified as the control. The aim is to discriminate these pre-AF segments from the NSR or normal segments.

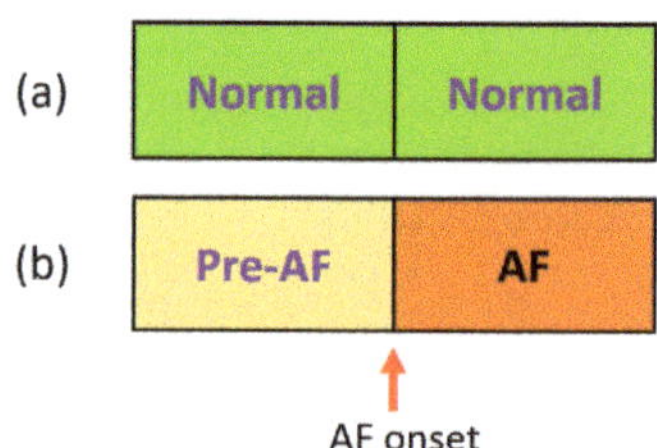

Figure 1. AF prediction schematic. (**a**) Normal recordings followed by normal recordings (control group). (**b**) Pre-AF (i.e., normal) recordings followed by AF onset.

Our method consisted of first preprocessing the ECG recordings, followed by feature extraction using several standard heart rate variability (HRV) analysis methods as well as time–frequency-based analysis of the heart rate signal. Finally, the pre-AF segments/ECG data are identified from the control group using the extracted features and machine learning classifiers.

3.1. Preprocessing

The first step of the preprocessing is the extraction of the heart rate data from ECG recordings. For short-term heart rate analysis, a 5-min ECG segment is recommended [14,23]. For the non-AF to AF transition subjects, a 5-min segment was taken from the ECG recordings immediately prior to the AF onset. Next, the R-peaks of the ECG segment were determined by a newly developed R-peak detection method which can reconstruct the ECG from the time-frequency-based sub-band decomposition [24]. After the R–R interval series was obtained, several preprocessing steps were performed depending on the feature extraction approaches. For calculating the frequency domain heart rate features, ectopic beats were first removed using the impulse rejection method described in [25] to obtain the corrected heart rates. Next, the corrected heart rate was resampled at 4 Hz by cubic spline, which was followed by trend removal.

For the time–frequency-domain-based analysis methods (variable frequency complex demodulation (VFCDM) and tunable Q-factor wavelet transform (TQWT)), the original R–R interval series was resampled at 4 Hz by cubic spline to make the samples evenly spaced; ectopic beat removal was not performed. For the time domain and nonlinear feature extraction methods, the original R–R interval was used without any further preprocessing.

3.2. Feature Extraction from RR Intervals

Figure 2A shows a representative 5-min heart rate signal, which is immediately prior to AF onset (from the CinC data set), whereas Figure 2B shows the same for a sample segment from the MIMIC data. Figure 2C,D show sample HRV segments from the CinC and MIMIC data, respectively, for the control group. From the pre-AF segments, it can be seen that there are several occurrences of PAC beats.

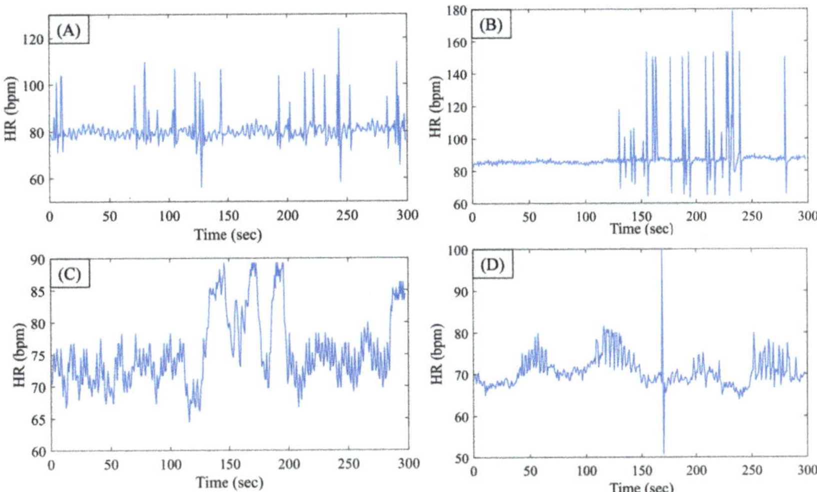

Figure 2. Representative 5-min heart rate signal, which is immediately prior to the AF onset: (**A**) from the CinC data set and (**B**) from the MIMIC III data set. Representative 5-min heart rate signal for control group: (**C**) from the CinC data set and (**D**) from the MIMIC III data set.

In order to predict AF, the following HR signal-based feature extraction methods were used in this study:

3.2.1. Time Domain Features

From the 5-min original heart rate signal, several standard time domain HRV features were calculated. The features include: standard deviation of the heart rate data (SDNN); total number of consecutive heart rate data differences greater than 50 ms (NN50); sum of NN50 divided by the total number of RR intervals (pNN50); skewness and kurtosis of the heart rate data; and root mean square of successive differences (RMSSD) of heart rate, which is divided by the mean heart rate of the corresponding segment to counter the variability among different subjects and segments. Finally, triangular index was also calculated as a geometric HRV feature, defined as the total number of RR intervals divided by the number of RR intervals that fall into a modal bin [11,23].

3.2.2. Nonlinear Features

In order to calculate the nonlinear features, the original heart rate signal was used. The extracted nonlinear HRV features include:

Poincaré Features

The Poincaré plot is a geometrical method that can be used to assess the dynamics of HRV. For HRV analysis, it is generated by plotting every RR interval against the prior interval, which creates a scatter plot [26]. For the Poincaré plot feature, an ellipse is fitted to the scattered points and the two following parameters are calculated for the quantification of the geometry.

SD_1 is the standard deviation of the projection of the Poincaré plot on the line perpendicular to the line of identity, which reflects the level of short-term variability. SD_2 is the standard deviation of the projection of the Poincaré plot on the line of identity, which is thought to indicate the level of long-term variability [27]. They are defined as follows, where SD is the standard deviation and RR_i is the ith RR interval [28]:

$$SD_1 = \frac{1}{\sqrt{2}} SD(RR_{i+1} - RR_i)$$

$$SD_2 = \sqrt{2 \times SD(RR_i)^2 - 0.5 \times SD(RR_{i+1} - RR_i)^2}$$

(1)

Moreover, the SD_1/SD_2 ratio was used as another Poincaré plot feature.

Sample Entropy

Sample entropy ($SampEn$) measures the randomness of the HRV signal. $SampEn$ is defined as the negative logarithm of the conditional probability that two sequences similar for m points remain similar at the next point, where self-matches are excluded [29,30]. $SampEn$ has two main parameters: template length 'm' and tolerance 'r'. A lower value of $SampEn$ indicates more self-similarity in the heart rate time series [30].

Multiscale Entropy

Multiscale entropy (MSE) analyzes the dynamic complexity of a system by quantifying its entropy over a range of temporal scales [31]. MSE is a two-step procedure: the first step consists of generating a coarse-grained time series by averaging the data points of the original HRV series while the second step consists of computing the sample entropy of each coarse-grained time series [32].

Approximate Entropy

Approximate entropy ($ApEn$) is the conditional probability of two segments of a time series of length N matching at a length $m + 1$ if they match at a length m [33]. $ApEn$ is a function of three parameters—N, m, and r—where N is the length of the HRV signal, m is the embedding dimension, and r is the tolerance/distance threshold, which is fixed to match segments when they are compared with each other [34].

Autoregressive (AR) Model

The RR interval time series can be described as the output of an AR model. By fitting an AR model, the fluctuations of the HRV series can be separated into those of the regulated component and the random component, which is the residual of the AR model [35].

$$RR(t) = \sum_{k=1}^{p} A(k)RR(t - k) + n(t)$$

(2)

Here, $A(k)$ is the AR model coefficient, $n(t)$ is the model error or residual component, and p is the model order. The variance of this residual component $n(t)$ is an estimate of the residual noise power (σ_{AR}^2), which is used as a feature in our work. For the pre-AF segments, this σ_{AR}^2 is expected to have high value due to the frequent occurrence of ectopic beats. In this study, the 12th order AR model was empirically selected.

3.2.3. Frequency Domain Features

Frequency domain parameters can provide useful information about the sympathetic and parasympathetic nervous activity and are shown to be effective for predicting PAF onset [12,14]. In order to calculate the frequency domain HRV features, ectopic beat removal was performed using the McNames impulse removal filter [25]. This corrected HRV was then resampled at 4 Hz by cubic spline and trend removal.

The power spectra of HRV data were calculated using Welch's periodogram method with 50% overlap. First, a Blackman window (length of 256) was applied to each segment, and then the fast Fourier transform was calculated for each windowed segment. Finally, the power spectra of the segments were averaged [36]. Figure 3a,b show a sample preprocessed heart rate signal obtained from a control subject and the corresponding PSD, respectively. Figure 3c,d show similar examples for a pre-AF segment. From the PSD, the very-low-frequency power (VLF) in the range 0–0.04 Hz, low-frequency power (LF) in the range 0–0.15 Hz, high-frequency power (HF) in the range 0.15–0.40 Hz, and total power were computed first [23]. Next, LF/HF, normalized LF (LFn = LF/total power), and normalized HF (HFn = HF/total power) were calculated and analyzed for pre-AF vs. NSR discrimination.

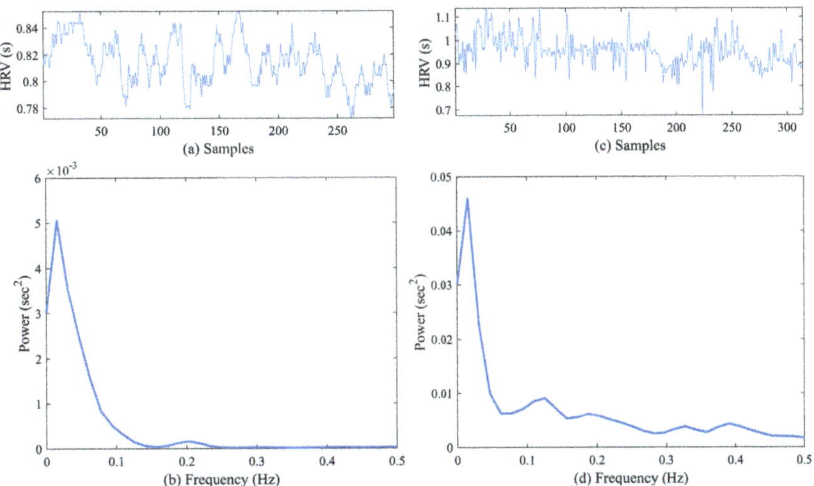

Figure 3. (**a**) Preprocessed heart rate signal from a control subject and (**b**) the corresponding PSD. (**c**) Preprocessed heart rate signal from a pre-AF segment and (**d**) the corresponding PSD.

3.2.4. VFCDM-Based Features

Variable frequency complex demodulation (VFCDM) is a high-resolution time–frequency domain method, which is widely used for various biosignal processing, including ECG [24,37], EDA [36], PPG [38,39] and other signals. First, the heart rate signal was resampled at 4 Hz to make the samples evenly spaced, which was followed by high-pass filtering (0.01 Hz) to remove any trends.

Using the VFCDM, the preprocessed heart rate signal was decomposed into K number of components or sub-bands [37]:

$$hrv(t) = \sum_{i=1}^{K} V_i(t) \tag{3}$$

where hrv is the input heart rate signal, $V_i(t)$ is the ith component or sub-band, and K is the number of sub-bands. In this study, by applying the VFCDM, the input $hrv(t)$ was divided into $K = 12$ sub-bands. These sub-bands were evenly spaced in the frequency range and their frequencies depend on the sampling rate. Since the heart rate data were resampled at 4 Hz, the spectral components (i.e., $V_i(t)$) were centered at 0.08, 0.24, 0.40, 0.56, 0.72, 0.88, 1.04, 1.20, 1.36, 1.52, 1.68, and 1.84 Hz.

Figure 4 shows a sample of preprocessed heart rate signal (the input) and the time–frequency representation obtained using the VFCDM.

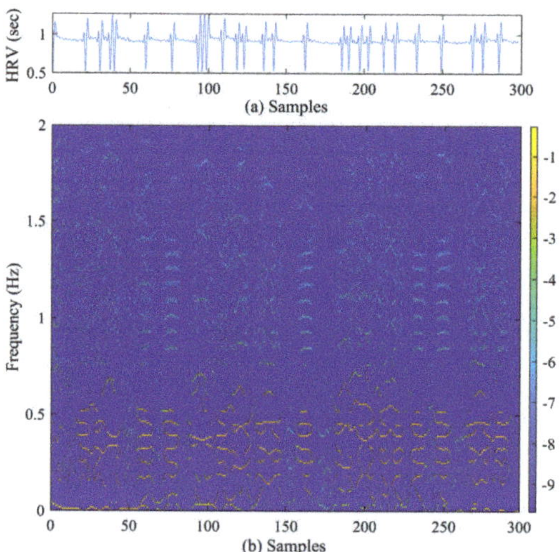

Figure 4. (**a**) Sample 5-min heart rate signal from a pre-AF ECG segment. (**b**) Time–frequency representation obtained using VFCDM.

From the 12 VFCDM components (3), only the third and fourth components were added to make a reconstructed heart rate time series, $hrv_{rec}(t) = V_3(t) + V_4(t)$.

This reconstructed heart rate time series (hrv_{rec}) contained the high-frequency components and was found to be highly useful for analyzing the heart rate variation due to frequent ectopic beats and subsequently, for AF prediction when compared to the control group. Figure 5a–d shows the third and fourth components obtained from the VFCDM decomposition of a sample heart rate signal (pre-AF group) and their respective power spectral density (PSD). The PSDs shows that third component is centered at 0.40 Hz, whereas the fourth component is centered at 0.56 Hz, which surrounds the HF part of HRV and represents the variation due to ectopic beats. Figure 5e shows the reconstructed heart rate signal (hrv_{rec}) and Figure 5f shows the PSD of the reconstructed HRV.

Using this reconstructed heart rate signal, we performed the Hilbert transform to obtain the signal envelope as follows [36]:

$$H(t) = \frac{1}{\pi} P \int_{-\infty}^{\infty} \frac{hrv_{rec}(\tau)}{t - \tau} d\tau \tag{4}$$

where P indicates the Cauchy principal value. $hrv_{rec}(t)$ and $H(t)$ form the complex conjugate pair, which can be used to define the analytic signal $A(t)$:

$$A(t) = hrv_{rec}(t) + iH(t) = a(t)e^{j\theta(t)} \tag{5}$$

where

$$a(t) = [hrv_{rec}^2(t) + H^2(t)]^{1/2}$$
$$\theta(t) = \arctan\left[H(t)/hrv_{rec}(t)\right] \tag{6}$$

The $a(t)$ is considered the instantaneous amplitude or envelope of $A(t)$. Figure 5e shows the reconstructed HRV (hrv_{rec}) and the Hilbert transform envelope or instantaneous amplitude ($a(t)$). From this instantaneous amplitude, mean, variance, and energy were calculated as features.

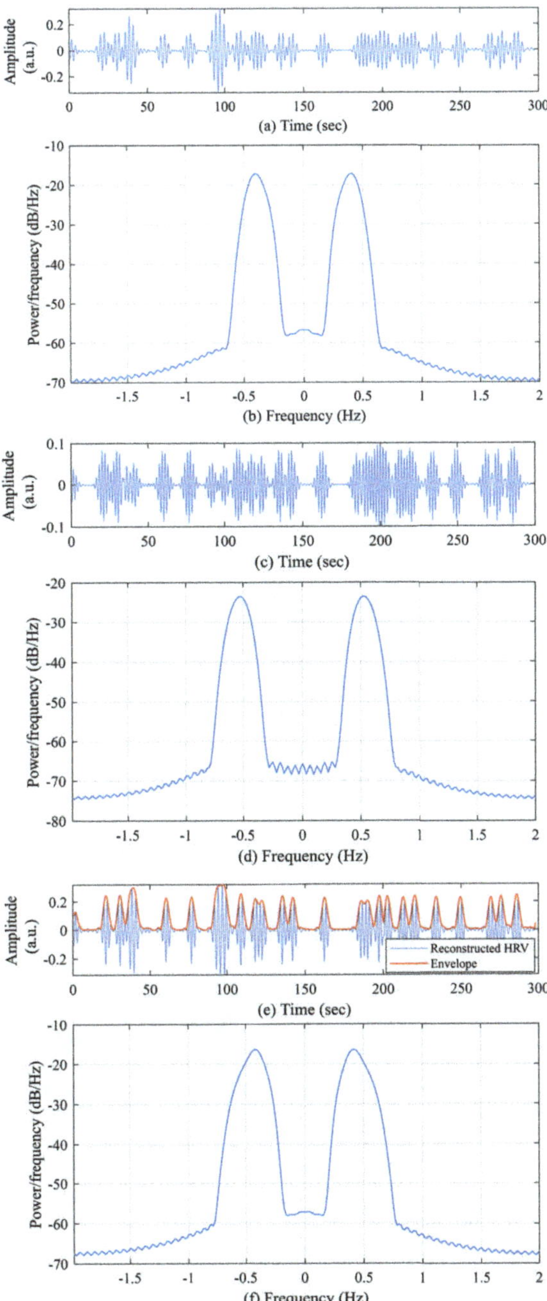

Figure 5. For a sample 5-min pre-AF heart rate signal: (**a**) third component of the VFCDM decomposition; (**b**) PSD of the third component; (**c**) fourth component of the VFCDM decomposition; (**d**) PSD of the fourth component; (**e**) reconstructed heart rate signal along with the Hilbert transform envelope; and (**f**) PSD of the reconstructed heart rate.

3.2.5. TQWT-Based Features

The tunable Q-factor wavelet transform (TQWT) is a flexible full-discrete wavelet transform, which is suitable for analyzing oscillatory signals [40]. TQWT facilitates analysis of oscillatory signals using three adjustable parameters: Q-factor (Q), redundancy or total oversampling rate (r), and the number of decomposition levels (J). Q controls the number of oscillations of the wavelet and affects the extent to which the oscillations of the wavelet are sustained [41]. r helps to localize the wavelet in the time domain without affecting its shape.

For a certain decomposition level J, TQWT decomposes an input signal into $J + 1$ sub-bands. It is performed by iteratively applying the two-channel filter bank on its low-pass channel. TQWT consists of a sequence of two-channel filter banks, with the low-pass output of each filter bank being used as the input to the successive filter bank [42].

For a low oscillatory signal, Q will be lower, whereas a higher Q value is required for high oscillatory signals. As a result, the wavelets will be more oscillatory with narrower frequency response. Unwanted excessive ringing of wavelets needs to be prevented while performing TQWT by appropriately choosing the value of r, which is recommended to be greater than or equal to 3. Details about TQWT can be found in [40,41]. In order to extract AF predicting features from the HRV signal using TQWT, $J = 17$, $Q = 3$, and $r = 4$ have been selected empirically in this study.

Figure 6A,B shows the input heart rate signal (from a pre-AF segment) and the TQWT coefficients, respectively, obtained from levels $J = 8$ to $J = 13$ where the resampled heart rate signal was used as the input. Figure 6C shows the frequency response of the TQWT transform for the selected parameters where the gain is normalized to have unity amplitude. Since $J = 17$ was used, the center frequencies of the TQWT sub-bands were (in descending order): 2 Hz, 1.31 Hz, 1.15, 1 Hz, 0.88 Hz, 0.77 Hz, 0.67 Hz, 0.59 Hz, 0.52 Hz, 0.45 Hz, 0.39 Hz, 0.35 Hz, 0.30 Hz, 0.26 Hz, 0.23 Hz, 0.20 Hz, and 0.18 Hz. The frequencies corresponding to $J = 8$ to $J = 13$ are marked in black in Figure 6C.

We analyzed the mean, variance, energy, entropy, and spectral entropy calculated from the coefficients of different sub-bands and found that energy as well as spectral entropy were the most useful ones as the discriminating features to be used for pre-AF and control segments.

Spectral entropy is a generalization of information entropy and it measures the distribution of frequencies. Spectral entropy treats the signal's normalized power distribution in the frequency domain as a probability distribution and calculates Shannon entropy from it [43,44]. For a given time–frequency spectrogram $S(t, f)$, the probability distribution at time t is given by:

$$P(t, m) = \frac{S(t, m)}{\sum_f S(t, f)} \tag{7}$$

The instantaneous spectral entropy at time t is calculated as [44]:

$$H(t) = -\sum_{k=1}^{N} P(t, k) \log_2 P(t, k) \tag{8}$$

In order to obtain a scalar feature value, L_2 norm of this instantaneous spectral entropy was used as the feature (referred to as "ENT").

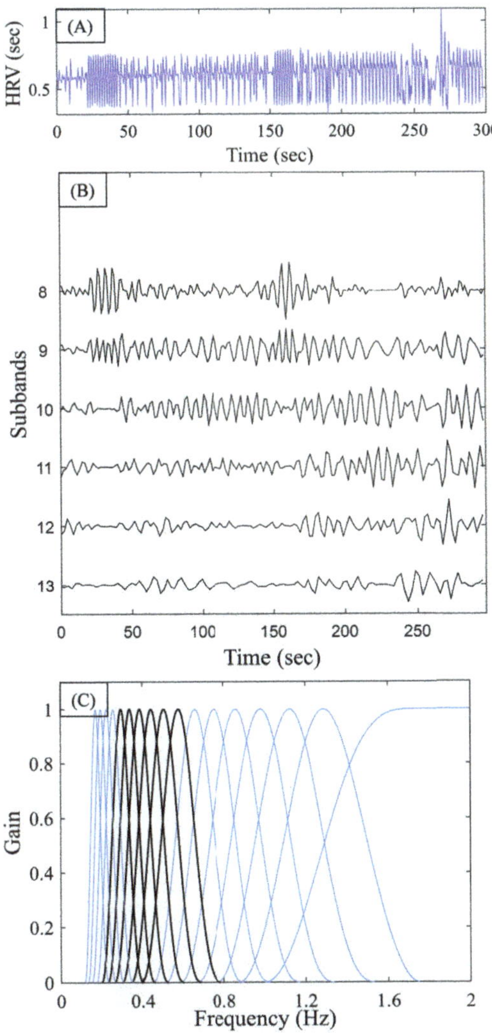

Figure 6. (**A**) Sample 5-min heart rate signal. (**B**) TQWT coefficients obtained from levels 8 to 13. (**C**) Frequency response of the TQWT transform with the selected parameters (normalized to have unity gain).

3.3. AF Prediction Framework

After several features were extracted from the five different domains, suitable features were selected by visual analysis (scatter plots and box plots) as well as cross-validation on the training data.

Based on the analysis performed using the training data, 14 features were selected. The selected features for the machine learning model include RMSSD; SD_1; AR residual noise; variance from VFCDM; LF/HF; LFn; TQWT spectral entropy from bands 8, 11, 12, and 13; and TQWT energy from sub-bands 9, 10, 11, and 12.

Several machine learning classifiers were analyzed using the selected feature subset and the performance of those classifiers is described in the Results section. Finally, support vector machine (SVM) and random forest (RF) were chosen for our AF prediction. SVM is a popular and well-established method for binary classification problems where a maximum margin between the training and test data is constructed [45]. RF classifier is formed by combining multiple randomly constructed tree models [46]. In the bagging (bootstrap aggregation) learning concept, many weak learners are trained over subsets drawn with replacements from the training set and their outputs are voted to determine a predictive estimate. This is shown to decrease the variance of the model without increasing the bias, thus resulting in diverse ensembles [47].

Figure 7A shows the scatter plot for the variance of VFCDM, whereas Figure 7B shows the 3D scatter plot for TQWT energy and spectral entropy. The scatter plots show that control and pre-AF samples have some visible separation for most of the cases. Figure 7C,D show the box plots for RMSSD and AR residual noise. The box plots have nonoverlapping medians, indicating the discriminatory property of the features. Figure 8 shows the complete flowchart of the proposed AF prediction method.

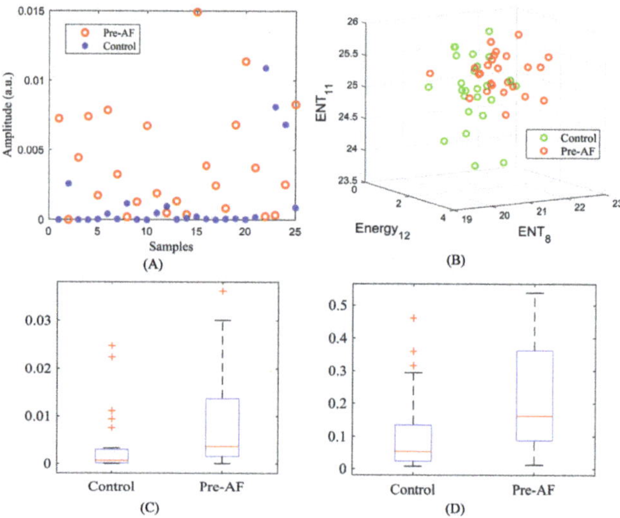

Figure 7. (**A**) VFCDM feature. (**B**) 3D scatter plot of spectral entropy (level 8, 11) and energy of level 12. (**C**) Box plots of RMSSD. (**D**) Box plots of AR residual noise.

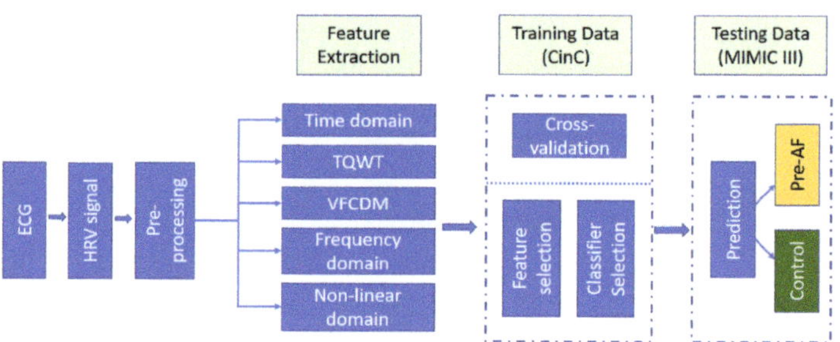

Figure 8. Overview of the proposed AF prediction method.

4. Results

In order to evaluate the prediction performance of our proposed method, commonly used binary classification accuracy measures were used. An ECG segment prior to the AF onset was denoted as a positive class, whereas an ECG segment from the control group was referred to as a negative class.

$$
\begin{aligned}
Sensitivity\,(SEN) &= TP/(TP+FN) \\
Specificity\,(SPE) &= TN/(TN+FP) \\
Accuracy\,(ACC) &= (TP+TN)/(TP+FN+TN+FP) \\
Positive\,predictive\,value\,(PPV) &= TP/(TP+FP) \\
Negative\,predictive\,value\,(NPV) &= TN/(TN+FN)
\end{aligned}
\tag{9}
$$

where TP denotes the number of true positives, TN is the number of true negatives, FP is a false positive, and FN is a false negative.

4.1. Results on Training Data (CinC Data)

From the CinC data set, we have 25 control and 25 pre-AF ECG segments. With these 50 segments, the well-established k-fold cross-validation was performed to select the classifier model and tune the hyperparameters. The training data were split into K disjoint partitions ($K = 5$) and each time ($K - 1$) folds were used for training while the last fold was treated as test data; the entire process was repeated k times [44]. For this study, we have explored several machine learning classifiers including support vector machine (SVM), discriminant analysis (DA), k-nearest neighbor (kNN) and random forest (RF). For the discriminant analysis, both linear and quadratic boundaries along with Mahalanobis distance were analyzed. Moreover, diagonal linear and diagonal quadratic discriminant functions were also used (referred to as "diaglinear" and "diagquadratic"), which are similar to linear and quadratic discriminant functions except the estimate of the covariance matrix is diagonal [48]. For SVM, both the linear and radial basis function (RBF) kernels were used. In kNN, both Euclidean and Cityblock (Manhattan) distance were used with the variation of "K" values, which denotes the number of the nearest neighbors to be used. For the RF, the hyperparameters were varied during the fivefold cross-validation and it was found that with the selected feature subset, 50 trees resulted in the best prediction performance. Table 1 shows the performance of several machine learning classifiers using the training data (CinC) for the fivefold cross-validation.

Table 1. Performance comparison of different classifiers.

Classifier Name	Type/ Hyperparameter	Sensitivity (%)	Specificity (%)	Accuracy (%)
	Linear	64	72	68
	Diaglinear	60	76	68
DA	Quadratic	88	60	74
	Diagquadratic	52	80	66
	Mahalanobis	92	52	72
SVM	Linear	64	72	68
	RBF	**76**	**76**	**76**
KNN	$K=5$, Euclidean	60	60	60
	$K=5$, Cityblock	64	60	62
RF	50 trees	**80**	**76**	**78**

It can be seen from Table 1 that the SVM and RF models resulted in better performance than the rest. The confusion matrices for both SVM and RF are shown in Table 2. With the fivefold cross-validation, the RF classifier achieved 80% sensitivity, 76% specificity, and

78% accuracy on the training data, whereas the SVM obtained 76% accuracy, sensitivity, and specificity.

Table 2. Confusion matrix on the training data.

		SVM				**RF**	
		Predicted Label				Predicted Label	
		Pre-AF	Control			Pre-AF	Control
True Label	Pre-AF	19	6	True Label	Pre-AF	20	5
	Control	6	19		Control	6	19

4.2. Results on Test Data (MIMIC III ICU)

Next, the trained model was tested using the critically ill ICU data from MIMIC III. It is to be noted that the feature subset and model parameters were fixed by doing the cross-validation on the training data (CinC); the trained model was blindly tested on the ICU data. The test data set contained 25 ECG recordings from the subjects with no-AF to AF transition (pre-AF segments) and 25 control ECG recordings.

4.2.1. Test Results on the Data Prior to AF Onset

First, the model was tested using the ECG data, which are immediately prior to the AF onset. These immediately prior data are expected to exhibit the most AF-predicting properties. Table 3 shows the confusion matrices for these test data using both the RF and SVM classifiers. The RF classifier identified 20 pre-AF segments correctly, resulting in sensitivity of 80%. Moreover, for the control group, RF detected 24 segments correctly, resulting in 96% specificity and an overall 88% accuracy. The radial basis SVM achieved 80% sensitivity, 100% specificity, and 90% accuracy.

Table 3. Confusion matrix on the test data (immediately before onset).

		SVM				**RF**	
		Predicted Label				Predicted Label	
		Pre-AF	Control			Pre-AF	Control
True Label	Pre-AF	20	5	True Label	Pre-AF	20	5
	Control	0	25		Control	1	24

In order to demonstrate the efficacy of our proposed method, we compared its performance with the Narin et al. method [14]. Narin et al. reported two different models: one is for PAF as well as the control subjects (model 1), whereas the other is only for the PAF subjects (model 2). Model 1 consisted of RMSSD, NN20, pNN20, FFT_{VLF}, and FFT_{HF} features; the kNN classifier is reported in Narin et al. [14]. The second model used RMSSD, FFT_{VLF}, FFT_{LF}, and total power of FFT along with the kNN classifier [14]. In order to compare the performance, both of these models were trained and tested using the same data that we used (CinC and MIMIC, respectively) and the resulting confusion matrices are presented in Table 4. From the table, it can be seen that both of the reported models of [14] have low sensitivity compared to ours (Table 3).

Table 4. Test results of the compared methods.

		Model 1 [14]				**Model 2** [14]	
		Predicted Label				Predicted Label	
		Pre-AF	Control			Pre-AF	Control
True Label	Pre-AF	15	10	True Label	Pre-AF	16	9
	Control	4	21		Control	3	22

4.2.2. Test Results for Moving Backward from AF Onset

In the next step, we analyzed how much in advance in time we can predict the AF onset. We analyzed how the prediction performs if we started far before AF onset. In order to study this, we took 5-min ECG segments and moved backward in a 50% overlap all the way up to 15 min prior to AF onset. As a result, the algorithm was tested using the ECG data from 2.5 min, 5 min, 7.5 min, and 10 min prior to AF onset. Figure 9 illustrates this testing scenario; for example, Figure 9E shows that one prediction was performed using the ECG data that was from 15 to 10 min prior to the AF onset.

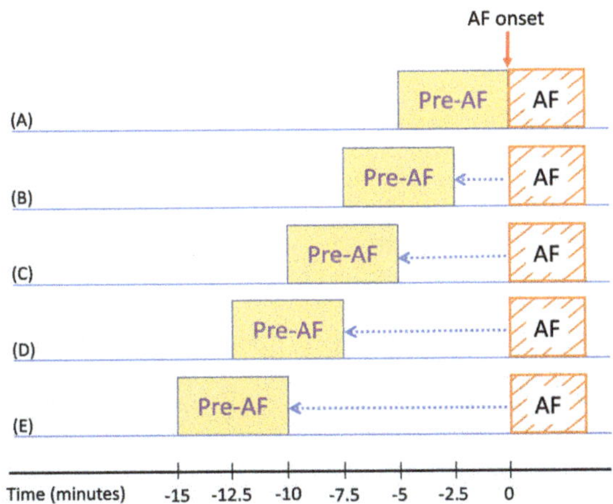

Figure 9. Illustration of AF prediction for moving backward in time from the onset.

For each of the testing scenarios demonstrated in Figure 9, we tested the already trained classifiers as mentioned in the previous subsection. Tables 5–8 show the confusion matrices from testing the critically ill ICU ECG data for the four different scenarios illustrated in Figure 9B–E. For each scenario, the results are presented using both RF and SVM classifiers.

Table 5. Confusion matrix on the test data (2.5 min before onset).

		SVM				RF	
		Predicted Label				Predicted Label	
		Pre-AF	Control			Pre-AF	Control
True Label	Pre-AF	16	9	True Label	Pre-AF	18	7
	Control	0	25		Control	1	24

Table 6. Confusion matrix on the test data (5 min before onset).

		SVM				RF	
		Predicted Label				Predicted Label	
		Pre-AF	Control			Pre-AF	Control
True Label	Pre-AF	15	10	True Label	Pre-AF	18	7
	Control	1	24		Control	3	22

Table 7. Confusion matrix on the test data (7.5 min before onset).

		SVM					RF	
		Predicted Label					Predicted Label	
		Pre-AF	Control				Pre-AF	Control
True Label	Pre-AF	13	12		True Label	Pre-AF	18	7
	Control	2	23			Control	4	21

Table 8. Confusion matrix on the test data (10 min before onset).

		SVM					RF	
		Predicted Label					Predicted Label	
		Pre-AF	Control				Pre-AF	Control
True Label	Pre-AF	16	9		True Label	Pre-AF	18	7
	Control	2	23			Control	3	22

It can be seen from Table 5 that when 2.5 min prior to AF onset ECG data were used for AF prediction, the RF model predicted 18 pre-AF segments correctly; the prediction was correct for 24 segments for the control class, resulting in 84% accuracy. However, the prediction performance slightly degraded as we moved farther from the AF onset. When we used the ECG data from 10 min before the AF onset, the AF prediction sensitivity, specificity, and accuracy were 72%, 88%, and 80%, respectively.

Finally, in Table 9 the sensitivity, specificity, accuracy, PPV, and NPV for different time durations are reported for both of SVM and RF classifiers. Although for the immediately prior to AF data SVM had slightly higher accuracy than did RF, for all other durations RF had better performance than did the SVM. Moreover, we compared the performance of our presented method with Narin et al. [14] for different time durations. For the comparison, we extracted the reported features described in [14], trained the *k*NN classifier on the CinC data, and tested the model using the MIMIC III ICU data. From the table, it is evident that our proposed method achieved better performance than the compared method for all cases.

Table 9. Confusion matrix on the test data (7.5 min before onset).

Prior Duration	Method	SEN (%)	SPE (%)	ACC (%)	PPV (%)	NPV (%)
0 min	SVM	80	100	**90**	100	83.33
	RF	80	96	88	95.24	82.76
	Method in [14]	60	84	72	78.95	67.74
2.5 min	SVM	64	100	82	100	73.53
	RF	72	96	**84**	94.74	77.42
	Method in [14]	56	84	70	77.78	65.63
5 min	SVM	60	96	78	93.75	70.59
	RF	72	88	**80**	85.71	75.86
	Method in [14]	52	88	70	81.25	64.71
7.5 min	SVM	52	92	72	86.67	65.71
	RF	72	84	**78**	81.82	75
	Method in [14]	36	84	60	69.23	56.76
10 min	SVM	64	92	78	88.89	71.88
	RF	72	88	**80**	85.71	75.86
	Method in [14]	68	80	74	77.27	71.43

5. Discussion

We presented a novel approach to predict AF during sepsis from critically ill ICU patients using the RR interval variability of ECG. Since the frequent occurrence of premature ectopic beats is shown to be a predictor of AF, the HRV-derived features were well suited for describing the variability.

In order to use this variability due to frequent occurrence of ectopic beats, ectopic beat removal was not used for preprocessing the HRV signal for the time domain, time–frequency domain, or nonlinear methods. Ectopic removal was used only for the frequency domain feature extraction, as this is the standard procedure for calculating frequency domain HRV features [17,23].

In this work, we extracted several features from 5-min heart rate signals using five different methods: time domain, VFCDM, TQWT, nonlinear, and frequency domain. With the extracted features, we trained machine learning models using the CinC data and performed cross-validation to select suitable features as well as the model parameters. Once we obtained the highest accuracy using the training data, we directly applied the trained model to the ICU data. For different combinations of the extracted features, we performed cross-validation using the training data and measured the prediction accuracy. Next, we chose the combination of classifier model and associated features which provided the best training performance. When other features were selected, the prediction accuracy on the training data was lower. Finally, our proposed method achieved reasonable performance on this blind test data, which shows the efficacy of our method.

For the performance comparison, the Narin et al. [14] method was chosen for a few reasons. First, unlike most other methods, the authors of [14] used normal subjects along with the PAF subjects. They performed the cross-validation performance using both the normal and PAF subjects, and not only the PAF subjects. Second, they analyzed how early they could predict AF by going backward in time. Finally, their method studied 5-min HRV signals to predict PAF. However, similar to most other reported AF prediction methods, no evaluation using an external test data set was performed. The fact that our method achieved higher performance than [14] for all the different time durations clearly shows the efficacy of the presented method. Moreover, this reflects that overfitting can be an issue when only the cross-validation results are reported using a small data set without doing an external test set.

There are three main contributions of our study: we tested AF prediction using a new and different data set, which consists of critically ill sepsis patients. After obtaining good prediction accuracy for the ECG data immediately prior to the AF onset, we analyzed how much in advance we could predict the AF. We achieved 80% overall accuracy for predicting AF 10 min prior to its onset. Currently, we do not have interventions to effectively prevent AF. Hence, the ability to predict AF will enable enrichment of trials of interventions to prevent AF. While 10 min of notice ahead of AF occurrence would be tight to institute an AF preventive strategy in practice, it may be enough time to give an intervention in an experimental setting. This work is foundational for predicting AF with longer duration. Most importantly, our work would help minimize the amount of time a patient spends in AF, as reducing the time burden of a patient's AF to only a few minutes may mitigate their risk for ischemic stroke.

Additionally, though the accuracy expectedly trends downward further away from AF onset, this decrease is largely a function of the algorithm's sensitivity, and the specificity actually remains high in all time windows examined. Therefore, this suggests that our approach can be especially useful for confirming true positive cases given a positive result. Finally, we provided new annotations for other researchers, which can be used as a valuable resource for future work in AF prediction.

Our study is different than the CinC 2001 data-based works [7–17]. In CinC-based works, AF prediction was analyzed using only the PAF subjects. In other words, within the PAF subjects, the AF prediction analysis was performed where each subject had two recordings: pre-AF and distant from AF. However, in this study, we performed AF prediction using the control subjects and the critically ill ICU sepsis subjects who had a transition from non-AF to AF. As a result, the pre-AF and control segments are from different subjects, which is distinct from the CinC data set.

Finally, our findings should be considered in light of study limitations. The main limitation is that we had a relatively small sample size from the MIMIC III ICU data.

Moreover, for the few MIMIC III transition subjects (which were not in the matched subset), we were unable to determine the sepsis status due to the lack of available clinical information. However, given the scarcity of the AF prediction data, it is understandable that getting this kind of rare data can be difficult. As a result, we provide our data annotations for other researchers so that people can use this data for advancing AF prediction research. Our work can be viewed as a preliminary study wherein we showed that by using the RR interval variation characteristics, we can achieve satisfactory AF prediction accuracy for critically ill ICU patients. Future works can focus on validation using a larger database when it becomes available and analyze whether the AF prediction performance differs between sepsis and nonsepsis ICU patients. Moreover, we aim to extend the prediction timeframe to further in advance in order to give an even more comfortable margin for taking action. Although our algorithm was validated on ECG data collected in the ICU with the standard ECG electrodes and leads, it is equally effective for any ECG modality including patch electrodes and wearables, including Holter, loop recorder, and implantable devices. Our algorithm uses the variability and morphology of the ECG to predict AF, hence, any ECG modality will suffice.

6. Conclusions

In this study, we have presented a novel approach to predict AF from critically ill sepsis patients using the MIMIC III ECG data. We have extracted various features from 5-min heart rate signals using time domain, frequency domain, nonlinear, VFCDM, and TQWT methods. With a subset of selected features, we have trained RF and SVM models using the CinC data; next, the trained models were directly applied to the MIMIC III ICU data without any further tuning. The proposed algorithm achieved good AF prediction performance on the test data and when compared with a state-of-the-art method, our method achieved better accuracy, thus showing the effectiveness of the presented method for real-life ICU data. Moreover, we analyzed how much in advance we can predict AF using the heart rate data. Since this is the first work to predict new onset AF in critically ill sepsis patients, we provide our annotations of the MIMIC III data to facilitate further AF prediction research. Future studies can explore how the AF prediction differs between sepsis and nonsepsis patients as well as validating the method using a larger number of AF subjects.

Author Contributions: Conceptualization, A.J.W., D.D.M. and K.H.C.; methodology, algorithm development, and validation, S.K.B.; data curation, S.K.B., E.Y.D., D.D.M. and A.J.W.; writing—original draft preparation, S.K.B.; writing—review and editing, E.Y.D., A.J.W., D.D.M. and K.H.C.; supervision, K.H.C.; project administration and funding acquisition, A.J.W., D.D.M. and K.H.C. All authors have read and agreed to the published version of the manuscript.

Funding: This research was supported by NIH grant R01 HL136660.

Data Availability Statement: The annotations of patients' AF transition information will be made publicly available at https://biosignal.uconn.edu/resources/.

Conflicts of Interest: The authors declare no conflict of interest.

References

1. Hershey, T.B.; Kahn, J.M. State sepsis mandates-a new era for regulation of hospital quality. *N. Engl. J. Med.* **2017**, *376*, 2311–2313. [CrossRef] [PubMed]
2. Walkey, A.J.; Greiner, M.A.; Heckbert, S.R.; Jensen, P.N.; Piccini, J.P.; Sinner, M.F.; Curtis, L.H.; Benjamin, E.J. Atrial fibrillation among Medicare beneficiaries hospitalized with sepsis: Incidence and risk factors. *Am. Heart J.* **2013**, *165*, 949–955. [CrossRef]
3. Bosch, N.A.; Cimini, J.; Walkey, A.J. Atrial fibrillation in the ICU. *Chest* **2018**, *154*, 1424–1434. [CrossRef] [PubMed]
4. Sibley, S.; Muscedere, J. New-onset atrial fibrillation in critically ill patients. *Can. Respir. J.* **2015**, *22*, 179–182. [CrossRef]
5. Walkey, A.J.; Wiener, R.S.; Ghobrial, J.M.; Curtis, L.H.; Benjamin, E.J. Incident stroke and mortality associated with new-onset atrial fibrillation in patients hospitalized with severe sepsis. *JAMA* **2011**, *306*, 2248–2254. [CrossRef] [PubMed]
6. Gandhi, S.; Litt, D.; Narula, N. New-onset atrial fibrillation in sepsis is associated with increased morbidity and mortality. *Neth. Heart J.* **2015**, *23*, 82–88. [CrossRef]

7. Zong, W.; Mukkamala, R.; Mark, R. A methodology for predicting paroxysmal atrial fibrillation based on ECG arrhythmia feature analysis. In *Computers in Cardiology 2001. Vol. 28 (Cat. No. 01CH37287)*; IEEE: Piscataway, NJ, USA, 2001; pp. 125–128.

8. Langley, P.; Di Bernardo, D.; Allen, J.; Bowers, E.; Smith, F.; Vecchietti, S.; Murray, A. Can paroxysmal atrial fibrillation be predicted? In *Computers in Cardiology 2001. Vol. 28 (Cat. No. 01CH37287)*; IEEE: Piscataway, NJ, USA, 2001; pp. 121–124.

9. De Chazal, P.; Heneghan, C. Automated assessment of atrial fibrillation. In *Computers in Cardiology 2001. Vol. 28 (Cat. No. 01CH37287)*; IEEE: Piscataway, NJ, USA, 2001; pp. 117–120.

10. Mohebbi, M.; Ghassemian, H. Prediction of paroxysmal atrial fibrillation based on non-linear analysis and spectrum and bispectrum features of the heart rate variability signal. *Comput. Methods Programs Biomed.* **2012**, *105*, 40–49. [CrossRef]

11. Boon, K.H.; Khalil-Hani, M.; Malarvili, M.; Sia, C.W. Paroxysmal atrial fibrillation prediction method with shorter HRV sequences. *Comput. Methods Programs Biomed.* **2016**, *134*, 187–196. [CrossRef] [PubMed]

12. Boon, K.H.; Khalil-Hani, M.; Malarvili, M. Paroxysmal atrial fibrillation prediction based on HRV analysis and non-dominated sorting genetic algorithm III. *Comput. Methods Programs Biomed.* **2018**, *153*, 171–184. [CrossRef]

13. Thong, T.; McNames, J.; Aboy, M.; Goldstein, B. Prediction of paroxysmal atrial fibrillation by analysis of atrial premature complexes. *IEEE Trans. Biomed. Eng.* **2004**, *51*, 561–569. [CrossRef]

14. Narin, A.; Isler, Y.; Ozer, M.; Perc, M. Early prediction of paroxysmal atrial fibrillation based on short-term heart rate variability. *Phys. Stat. Mech. Its Appl.* **2018**, *509*, 56–65. [CrossRef]

15. Lynn, K.; Chiang, H. A two-stage solution algorithm for paroxysmal atrial fibrillation prediction. In *Computers in Cardiology 2001. Vol. 28 (Cat. No. 01CH37287)*; IEEE: Piscataway, NJ, USA, 2001; pp. 405–407.

16. Yang, A.; Yin, H. Prediction of paroxysmal atrial fibrillation by footprint analysis. In *Computers in Cardiology 2001. Vol. 28 (Cat. No. 01CH37287)*; IEEE: Piscataway, NJ, USA, 2001; pp. 401–404.

17. Ebrahimzadeh, E.; Kalantari, M.; Joulani, M.; Shahraki, R.S.; Fayaz, F.; Ahmadi, F. Prediction of paroxysmal Atrial Fibrillation: A machine learning based approach using combined feature vector and mixture of expert classification on HRV signal. *Comput. Methods Programs Biomed.* **2018**, *165*, 53–67. [CrossRef]

18. Goldberger, A.L.; Amaral, L.A.; Glass, L.; Hausdorff, J.M.; Ivanov, P.C.; Mark, R.G.; Mietus, J.E.; Moody, G.B.; Peng, C.K.; Stanley, H.E. PhysioBank, PhysioToolkit, and PhysioNet: Components of a new research resource for complex physiologic signals. *Circulation* **2000**, *101*, e215–e220. [CrossRef]

19. Johnson, A.E.; Pollard, T.J.; Shen, L.; Li-Wei, H.L.; Feng, M.; Ghassemi, M.; Moody, B.; Szolovits, P.; Celi, L.A.; Mark, R.G. MIMIC-III, a freely accessible critical care database. *Sci. Data* **2016**, *3*, 1–9. [CrossRef]

20. Moody, B.; Moody, G.; Villarroel, M.; Clifford, G.; Silva, I. MIMIC-III Waveform Database Matched Subset. *PhysioNet* **2017**. [CrossRef]

21. Bashar, S.K.; Hossain, M.B.; Ding, E.; Walkey, A.J.; McManus, D.D.; Chon, K.H. Atrial Fibrillation Detection During Sepsis: Study on MIMIC III ICU Data. *IEEE J. Biomed. Health Inform.* **2020**, *24*, 3124–3135. [CrossRef] [PubMed]

22. Moody, G.; Goldberger, A.; McClennen, S.; Swiryn, S. Predicting the onset of paroxysmal atrial fibrillation: The Computers in Cardiology Challenge 2001. In *Computers in Cardiology 2001. Vol. 28 (Cat. No. 01CH37287)*; IEEE: Piscataway, NJ, USA, 2001; pp. 113–116.

23. Force, T. Standards of measurement, physiological interpretation and clinical use. Task Force of the European Society of Cardiology the North American Society of Pacing Electrophysiology. *Circulation* **1996**, *93*, 1043–1065. [CrossRef]

24. Bashar, S.K.; Noh, Y.; Walkey, A.J.; McManus, D.D.; Chon, K.H. VERB: VFCDM-based electrocardiogram reconstruction and beat detection algorithm. *IEEE Access* **2019**, *7*, 13856–13866. [CrossRef]

25. McNames, J.; Thong, T.; Aboy, M. Impulse rejection filter for artifact removal in spectral analysis of biomedical signals. In Proceedings of the The 26th Annual International Conference of the IEEE Engineering in Medicine and Biology Society, San Francisco, CA, USA, 1–5 September 2004; IEEE: Piscataway, NJ, USA, 2004; Volume 1, pp. 145–148.

26. Shaffer, F.; Ginsberg, J. An overview of heart rate variability metrics and norms. *Front. Public Health* **2017**, *5*, 258. [CrossRef] [PubMed]

27. Brennan, M.; Palaniswami, M.; Kamen, P. Do existing measures of Poincare plot geometry reflect nonlinear features of heart rate variability? *IEEE Trans. Biomed. Eng.* **2001**, *48*, 1342–1347. [CrossRef] [PubMed]

28. Karimi Moridani, M.; Setarehdan, S.K.; Motie Nasrabadi, A.; Hajinasrollah, E. Non-linear feature extraction from HRV signal for mortality prediction of ICU cardiovascular patient. *J. Med. Eng. Technol.* **2016**, *40*, 87–98. [CrossRef]

29. Lake, D.E.; Moorman, J.R. Accurate estimation of entropy in very short physiological time series: The problem of atrial fibrillation detection in implanted ventricular devices. *Am. J. Physiol.-Heart Circ. Physiol.* **2011**, *300*, H319–H325. [CrossRef]

30. Alcaraz, R.; Rieta, J.J. A review on sample entropy applications for the non-invasive analysis of atrial fibrillation electrocardiograms. *Biomed. Signal Process. Control.* **2010**, *5*, 1–14. [CrossRef]

31. Costa, M.; Goldberger, A.L.; Peng, C.K. Multiscale entropy analysis of biological signals. *Phys. Rev.* **2005**, *71*, 021906. [CrossRef]

32. Vest, A.N.; Da Poian, G.; Li, Q.; Liu, C.; Nemati, S.; Shah, A.J.; Clifford, G.D. An open source benchmarked toolbox for cardiovascular waveform and interval analysis. *Physiol. Meas.* **2018**, *39*, 105004. [CrossRef] [PubMed]

33. Richman, J.S.; Moorman, J.R. Physiological time-series analysis using approximate entropy and sample entropy. *Am. J. Physiol.-Heart Circ. Physiol.* **2000**, *278*, H2039–H2049. [CrossRef] [PubMed]

34. Udhayakumar, R.K.; Karmakar, C.; Palaniswami, M. Approximate entropy profile: A novel approach to comprehend irregularity of short-term HRV signal. *Nonlinear Dyn.* **2017**, *88*, 823–837. [CrossRef]

35. Hayano, J.; Ohashi, K.; Yoshida, Y.; Yuda, E.; Nakamura, T.; Kiyono, K.; Yamamoto, Y. Increase in random component of heart rate variability coinciding with developmental and degenerative stages of life. *Physiol. Meas.* **2018**, *39*, 054004. [CrossRef]
36. Posada-Quintero, H.F.; Florian, J.P.; Orjuela-Cañón, Á.D.; Chon, K.H. Highly sensitive index of sympathetic activity based on time-frequency spectral analysis of electrodermal activity. *Am. J. Physiol.-Regul. Integr. Comp. Physiol.* **2016**, *311*, R582–R591. [CrossRef] [PubMed]
37. Hossain, M.B.; Bashar, S.K.; Lazaro, J.; Reljin, N.; Noh, Y.; Chon, K.H. A robust ECG denoising technique using variable frequency complex demodulation. *Comput. Methods Programs Biomed.* **2021**, *200*, 105856. [CrossRef]
38. Chon, K.H.; Dash, S.; Ju, K. Estimation of respiratory rate from photoplethysmogram data using time–frequency spectral estimation. *IEEE Trans. Biomed. Eng.* **2009**, *56*, 2054–2063. [CrossRef]
39. Bashar, S.K.; Han, D.; Hajeb-Mohammadalipour, S.; Ding, E.; Whitcomb, C.; McManus, D.D.; Chon, K.H. Atrial fibrillation detection from wrist photoplethysmography signals using smartwatches. *Sci. Rep.* **2019**, *9*, 1–10. [CrossRef]
40. Selesnick, I.W. Wavelet transform with tunable Q-factor. *IEEE Trans. Signal Process.* **2011**, *59*, 3560–3575. [CrossRef]
41. Patidar, S.; Pachori, R.B.; Acharya, U.R. Automated diagnosis of coronary artery disease using tunable-Q wavelet transform applied on heart rate signals. *Knowl.-Based Syst.* **2015**, *82*, 1–10. [CrossRef]
42. Hassan, A.R.; Siuly, S.; Zhang, Y. Epileptic seizure detection in EEG signals using tunable-Q factor wavelet transform and bootstrap aggregating. *Comput. Methods Programs Biomed.* **2016**, *137*, 247–259. [CrossRef] [PubMed]
43. Pan, Y.; Chen, J.; Li, X. Spectral entropy: A complementary index for rolling element bearing performance degradation assessment. *Proc. Inst. Mech. Eng. Part J. Mech. Eng. Sci.* **2009**, *223*, 1223–1231. [CrossRef]
44. MathWorks—Makers of MATLAB and Simulink. Available online: https://in.mathworks.com/ (accessed on 15 October 2018). .
45. Yu, S.N.; Chou, K.T. Selection of significant independent components for ECG beat classification. *Expert Syst. Appl.* **2009**, *36*, 2088–2096. [CrossRef]
46. Breiman, L. Random forests. *Mach. Learn.* **2001**, *45*, 5–32. [CrossRef]
47. Elola, A.; Aramendi, E.; Irusta, U.; Del Ser, J.; Alonso, E.; Daya, M. ECG-based pulse detection during cardiac arrest using random forest classifier. *Med. Biol. Eng. Comput.* **2019**, *57*, 453–462. [CrossRef] [PubMed]
48. Alkan, A.; Günay, M. Identification of EMG signals using discriminant analysis and SVM classifier. *Expert Syst. Appl.* **2012**, *39*, 44–47. [CrossRef]

Article

Epileptic Seizure Detection on an Ultra-Low-Power Embedded RISC-V Processor Using a Convolutional Neural Network

Andreas Bahr [1,*,†], Matthias Schneider [1,†], Maria Avitha Francis [1,†], Hendrik M. Lehmann [2], Igor Barg [3], Anna-Sophia Buschhoff [4], Peer Wulff [4], Thomas Strunskus [3] and Franz Faupel [3]

1   Sensor System Electronics, Institute of Electrical Engineering and Information Technology, Kiel University, 24143 Kiel, Germany
2   CMOS Design, Technical University Braunschweig, 38106 Braunschweig, Germany
3   Multicomponent Materials, Institute for Material Science, Kiel University, 24143 Kiel, Germany
4   Institute of Physiology, Kiel University, 24118 Kiel, Germany
*   Correspondence: ab@tf.uni-kiel.de
†   These authors contributed equally to this work.

Abstract: The treatment of refractory epilepsy via closed-loop implantable devices that act on seizures either by drug release or electrostimulation is a highly attractive option. For such implantable medical devices, efficient and low energy consumption, small size, and efficient processing architectures are essential. To meet these requirements, epileptic seizure detection by analysis and classification of brain signals with a convolutional neural network (CNN) is an attractive approach. This work presents a CNN for epileptic seizure detection capable of running on an ultra-low-power microprocessor. The CNN is implemented and optimized in MATLAB. In addition, the CNN is also implemented on a GAP8 microprocessor with RISC-V architecture. The training, optimization, and evaluation of the proposed CNN are based on the CHB-MIT dataset. The CNN reaches a median sensitivity of 90% and a very high specificity over 99% corresponding to a median false positive rate of 6.8 s per hour. After implementation of the CNN on the microcontroller, a sensitivity of 85% is reached. The classification of 1 s of EEG data takes $t = 35\,\text{ms}$ and consumes an average power of $P \approx 140\,\mu\text{W}$. The proposed detector outperforms related approaches in terms of power consumption by a factor of 6. The universal applicability of the proposed CNN based detector is verified with recording of epileptic rats. This results enable the design of future medical devices for epilepsy treatment.

Keywords: convolutional neural network; EEG; epileptic seizure detection; RISC-V; ultra-low-power

Citation: Bahr, A.; Schneider, M.; Francis, M.A.; Lehmann, H.M.; Bard, I.; Buschhoff, A.-S.; Wulff, P.; Strunskus, T.; Faupel, F. Epileptic Seizure Detection on an Ultra-Low-Power Embedded RISC-V Processor Using a Convolutional Neural Network. *Biosensors* **2021**, *11*, 203. https://doi.org/10.3390/bios11070203

Received: 14 May 2021
Accepted: 18 June 2021
Published: 23 June 2021

Publisher's Note: MDPI stays neutral with regard to jurisdictional claims in published maps and institutional affiliations.

1. Introduction

With about 1% of the population affected, epilepsy is one of the most common neurological diseases globally [1]. Epilepsy requires ongoing medical attention and is associated with a decrease in the patients' quality of life and higher mortality rates [2,3]. Every year in the USA alone, the direct medical expenses including lost or reduced earnings associated with epilepsy are estimated to be $15.5 billion [4]. Despite ongoing research and development of new AEDs [5,6], the most common treatment in form of systemic administration of anti-epileptic drugs (AEDs) does not achieve sufficient long-term seizure suppression in ~30% of the patients. Therefore, alternative treatment methods to refractory epilepsy such as intracranial drug delivery [7] or neurostimulation [8] have been suggested. The pinnacle of development would be an implantable closed-loop system for on-demand intervention during ictal periods, which have to be identified sufficiently fast through an automated seizure classification system.

Since the beginning of research on automated classification of epileptic seizures in the 1970s, several algorithms to detect seizures have been developed [9,10]. The challenge in classical approaches of seizure detection is developing a model that is capable of dealing with the changing characteristics of seizures within the same subject. The different approaches rely on feature extraction coupled with a classification strategy. These features are for example wavelet-based filters, frequency band and spectral analysis, the slope, height, and duration characteristics of a seizure or cross-channel correlations [10]. Recently, with a view at hardware efficiency in order to enable implantable systems, convolutional neural networks have been analyzed for seizure detection algorithms. Lawhern et al. [11] showed with EEGNet that state-of-the-art seizure classification and interpretation is possible with a compact convolutional neural network. A CNN optimized for ultra-low power requirements was introduced in [12]. The detector, called SeizureNet, reaches a median sensitivity of 0.96 long-term for invasive intracranial EEG recordings. The efficiency of epileptic seizure prediction based on deep learning is analyzed and compared in [13,14].

Biomedical implantable and wearable devices are usually limited by size and energy restrictions. To meet the devices' high energy efficiency requirements and form factor budget, many functions are incorporated into the device by application-specific integrated circuits (ASICs) [15,16]. Applying this concept, several applications have been developed successfully, going far beyond standard applications like pacemakers and hearing aids. A very small and lightweight bioelectric recording system for flying insects has been shown in [17]. An implantable cortical microstimulator for Brain–Computer Interfaces was realized in [18]. Benabi et al. [19] demonstrated that a tetraplegic patient can control an exoskeleton by an implanted epidural wireless brain–machine interface. In [15,16], it has been shown that an electronic system can be miniaturized to such an extent that even neural recording from neonatal mice to monitor growing processes of the brain can be feasible. The integration of digital logic into systems with low-power microcontrollers based on RISC-V architecture is promising to further advance this field of research. The RISC-V architecture is based on the reduced instruction set computer principles introduced by the University of Berkeley, California. It is available under an open source license and thus, unlike most other microcontrollers, free to use.

In this paper, an epileptic seizure detector suitable for ultra-low-power RISC-V embedded processors and based on a CNN is presented. The implementation, training, and verification of the CNN is performed in MATLAB using the open source CHB-MIT dataset. The main requirement for the detection algorithm is the feasibility for an ultra-low-power hardware implementation. At the same time the detection algorithm has to achieve state-of-the-art detection performance. Low-power and low-complexity architectures call for dimensionality reduction and small memory usage. A multi-channel EEG is a high dimensional dataset and memory usage is thus a challenge. In this work, a CNN fulfilling the low-power and low-complexity requirements is presented. It consists of only a few layers and a manageable number of weights, thus fulfilling low memory requirements.

The paper is structured as follows. Section 2 presents the dataset and Section 3 the microcontroller hardware. In Section 4, the CNN architecture and its training and implementation in Matlab are presented. The implementation of the classifier on a RISC-V-based embedded microcontroller is presented in Section 4.4. The transferability of the proposed detector and its functionality is cross-validated and proven with EEG data from a rodent absence epilepsy model (Genetic Absence Epilepsy Rats from Strasbourg (GAERS)) in Section 4.5. In Section 4.6, the capability of the developed classifier to predict epileptic seizures is demonstrated. Section 5 presents the measurements and results of the performane of the detector for each of the presented chapters. The performance of the classifier is compared with similar state-of-the-art approaches in Section 6.

2. Dataset

In this work the open source CHB-MIT dataset, collected at the Children's Hospital Boston, is used [20,21]. It contains continuous scalp EEG recordings from 24 children with intractable seizures, which have been labeled by medical professionals. The type of epilepsy is not specified in [20,21]. However, this does not limit the performance of classification, as our network is independent of specific epilepsy types, but instead is trained patient specifically. The EEG sampling frequency for all patients was 256 Hz.

Most recordings contain 23 channels with EEG signals. For electrode positioning the international 10–20 system of EEG electrode positions and nomenclature is used. Overall, this dataset contains approximately 865 h of EEG signals with 198 seizures that usually last several seconds.

In addition, the presented algorithm and epilepsy detection method is verified with data from EEG recordings in a rodent model. We have used the Genetic Absence Epilepsy Rat from Strasbourg (GAERS), which represent one of the best established rodent models for generalized epilepsy. The rats show seizures with characteristic "spike and wave discharge" EEG patterns. For this study, male rats between 6 and 9 months were implanted with epidural electrodes. In total recordings with a length of more than 150 h are available. The data set has been made available via open access on the portal IEEE Dataport [22]. Experiments were performed in accordance with the German law on animal protection and were approved by the Animal Care and Ethics Committee of the University of Kiel.

3. Hardware Description

State-of-the-art ultra-low-power microcontrollers allow the execution of complex CNNs complying with real-time, power and size requirements of implantable systems. The microcontroller board chosen for implementing the CNN on hardware is the GAPuino Board developed by GreenWaves Technology [23]. The main processing unit is a GAP8, which is a multi-core RISC-V processor derived from the PULP platform. It is optimized through different approaches to run IoT applications on an ultra-low power base, especially CNNs. These approaches include a powerful programmable parallel processing unit, a hardware convolutional engine and an on-chip power management to reduce the component count while maximizing battery power down-conversion efficiency. The board is an Arduino Uno form factor board including several peripheral interfaces necessary for prototyping [24]. The open source RISC-V-based processor is chosen to enable complete customization and free use of the processor for implantable systems. In future, it is planed to adapt and optimize the RISC-V hardware, especially the hardware convolutional engine, to the specific needs of biosignal processing. In addition, the RISC-V processor architecture is forecasted to be an important processor architecture in industry and research within the next 5 years with over 60 billion processor cores fabricated [25].

4. Implementation

4.1. Dataset Preparation

Seizure detection can be modeled as a time-series classification problem that classifies the input data into ictal and inter-ictal parts. The CHB-MIT dataset is used as the input for a convolutional neural network. The EEG files of the dataset are preprocessed as described in detail in the Appendix A. The dataset contains a labeling of epileptic activity into ictal and inter-ictal. This labeled EEG data is used for further analysis. The CNN is trained for each patient individually. The datasets of each patients are processed separately and data from individual patients is not shared between the patients, neither in training or test nor in validation phase.

The holdout-method is used to split the dataset into training, validation, and test set. The ratio between these three parts is 60-20-20. The training set is used to train the neural network. On the basis of the validation set, the model optimizes its weights. The final model will be the model which maximizes the classification performance for the validation set. Consequently, the performance on the validation set is not a good estimation for the performance of unseen data. This problem is solved by using the third split: the data set. The data set is only used the test of the final model. It is therefore a good database to evaluate the performance of the model on unseen, new data. The 60-20-20 split is done in the same manner for the ictal data as well as for the inter-ictal data.

For most patients the files contain recordings of 23 channels. However, for almost every patient adjustments have to be conducted to provide a homogeneous input. Empty channel recordings or strongly alternating amplitudes for single channels have been excluded from the analysis.

4.2. Data Structuring

The CNN of the epileptic seizure detector processes time signals with a dedicated and fixed length. The dataset is split into parts with a length of 1 s. These blocks of data are the input for the neural network. The length of 1 s is selected to keep the time period of the runtime of a forward pass low while not losing valuable time-dependent information, which is necessary for real-time sensor and actuator systems [12]. The forward pass is the calculation process of traversing through all neurons from first to last layer. The procedure to get these short signal windows is to slide a window with the dimension $W = E \times (T \cdot f_s)$ over the data, where E are the number of channels, T is the time length of the window and f_s is the sampling frequency. A sampling frequency of $f_s = 256\,\text{Hz}$ and $E = 23$ channels leads to an input matrix of the size 23×256 for $T = 1\,\text{s}$.

For the inter-ictal data, the window is sliding with no overlap. As already stated, there is an imbalanced number of ictal and inter-ictal data, which is generally challenging for classifiers [26]. Truong et al. [27] propose an approach to solve this problem by generating additional ictal data for training. Similar to the windowing process for the inter-ictal data a window of the dimension 23×256 is shifted over the ictal recordings. The difference here is that the window is only shifted by one sample per iteration, compared to 256 samples for non-ictal data. This corresponds to an overlapping of 99.6 %, illustrated in Figure 1. To exemplify this, a epileptic seizure with a length of 10 s is considered. Without the overlapping techniques, this seizure would be cut into chunks of 1 s length generating 10 seizure events. Using the overlapping technique, 2304 seizures events with a respective length of 1 s (256 samples) are generated. This massive overlapping technique is only used for seizure data in the validation and training set. The test set is cut into samples without overlapping.

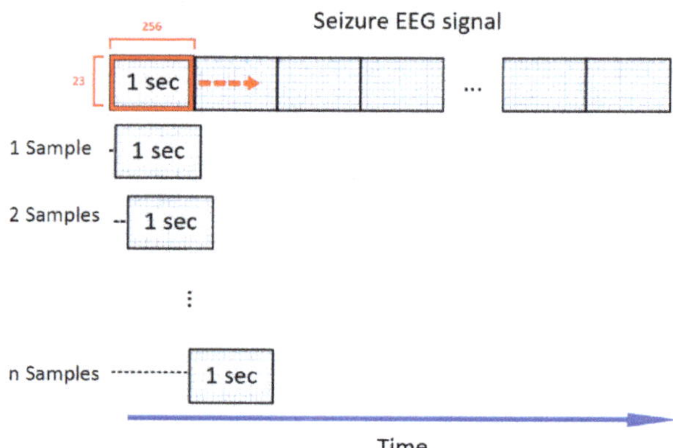

Figure 1. Sliding window technique [27]: A window with a length of 256 samples/1 s is sliding over seizure data with a step size of $S = 1$ sample to generate extra seizures for a balanced training set.

An EEG signal recorded at the head of a patient is easily corrupted by physiological and non-physiological artifacts such as action potentials from scalp muscle or motion of EEG cables, respectively [20,21]. The CHB-MIT dataset is partially corrupted by such artifacts. The recordings, e.g., contain 60 Hz noise, caused by the power supply. This 60 Hz noise differs between channels as well as between patients. To reduce the data processing and thus the hardware requirements, a signal preprocessing is not implemented in this work. It is not known to the authors, and not stated in the description of the dataset, if the dataset contains preprocessed data or if the recording equipment of the dataset performs data preprocessing of any kind. For future implantable systems it is assumed that the amount of distortions is reduced due to internal intracranial recordings compared to external recordings.

4.3. CNN Architecture

The architecture of the CNN used in this work is illustrated in Figure 2 and based on the SeizureNet CNN [12]. In [12], various architecture elements and functions are analyzed in order to evaluate the runtime and memory requirements for an energy-efficient seizure detecting classifier. This includes layer types like convolutions, dense layers (fully connected layer), pooling layers, and different activation functions. While the network in [12] was evaluated using the "Epilepsiae" dataset, especially the intracranial EEG dataset recorded at the University of Freiburg, our work is based on the CHB-MIT [21] dataset. The CNN in our work is optimized for ultra-low-power and energy consumption for future implantable systems. Training, verification and optimization of the CNN developed and presented in our work is performed in MATLAB using the MATLAB deep learning toolbox. The source code has been made available online under open access license [28] to enable rapid adoption in future research projects.

The implemented model architecture for this work is illustrated in Figure 2. The input is a 23×256 matrix. The first layer is a convolutional layer using a kernel with the dimensions 23×17. In the first layer, a convolution over all electrodes is chosen. The data from a multi-channel EEG recording are not spatially uncorrelated and with the selected size the kernel size can be reduced effectively. This approach is similar to spatial pattern recognition approaches and was also recommended in [11]. By adding a filtering over time, the first layer can efficiently learn spatial-time features. For the first layer, the number of kernels is given by $20 \times$ to provide a sufficient quantity of learnable patterns while keeping the amount of weights to be trained on a low level. By implementing a kernel over all electrodes, the output of the first layer is significantly smaller than the input. This is important when taking the needed memory into account. For all convolutional layers the kernel is sliding over the input with a stride of $S = 1$.

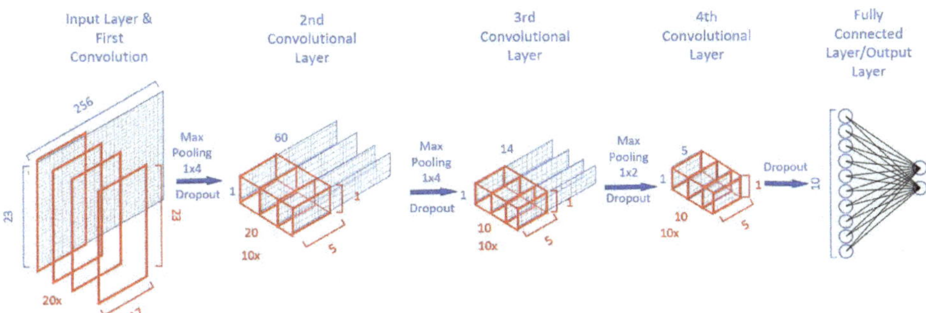

Figure 2. Schematic depiction of the CNN architecture showing the convolutional layers with their respective input matrix (blue rectangles) and kernels (red rectangles). Input is a 23×256 matrix. Between each convolutional layer, a dropout layer and max pooling layer is placed. Output are the two classes ictal and inter-ictal.

The next three convolutional layers extract key features and reduce the dimensionality of the network. The kernel size for the second layer is $10 \times 1 \times 5 \times 20$. For the third and fourth layer it is $10 \times 1 \times 5 \times 10$. A rectified linear activation function (ReLu) is used for each layer. The output layer is a 10×2 fully connected layer using a sigmoid activation function for each of the ten hidden neurons. As the classification task is to decide between ictal and inter-ictal recordings, each output neuron stands for a class. With a softmax function right at the end of the neural network the probability for the two output classes is calculated.

To reduce the output size for layer 1–2 even further, each convolutional layer is followed by a 1×4 max pooling layer with a stride of $S = 4$. The third convolutional layer is followed by a 1×2 max pooling layer with a stride of $S = 2$. To avoid overfitting, each layer except the output layer also contains a dropout layer with a dropout rate of 20%. The dropout layers are only used during the training phase.

4.4. CNN Hardware Implementation

The trained network is implemented on a RISC-V based GAP8 ultra-low-power microprocessor. To make use of its efficiency-increasing hardware convolutional engine and to implement the MATLAB trained CNN on the GAP8, processor optimizations and adjustments have to be made. This includes an adaptation of the network architecture, a transmission of the network architecture from MATLAB to C-Code and a quantization of all parameters including the input matrix. This is necessary, as GAP8 comes, for energy efficiency purposes, without a floating point unit. The quantization performed is an 16-bit Q1.14 fixed point quantization. The quantized parameters on the GAPuino Board are stored in integer form.

In total the CNN has a number of 10.162 trainable parameters and a memory requirement of 62.7 kB for 32-bit floats. The detailed structure, the number of trainable parameters and the required memory size per layer are presented in the Appendix A, Table A2.

To implement the CNN trained in MATLAB on the GAPuino Board the TensorFlow SDK used. The Greenwaves GAP8 SDK only supports quadratic convolutions. As the architecture developed in Matlab includes only non-quadratic kernels ($23 \times 17, 1 \times 5$ etc.) the number of layers and the filter dimensions have to be changed. The new architecture is given in Table A2. The network only contains two convolutional layers with a kernel size of 5×5 with a stride $S = 1$ and a fully connected layer of the size 2440×2. The Max Pooling layers are reduced to a 2×2 pooling with a stride $S = 2$.

4.5. Transferability of the CNN Based Classifier

The transferability of the presented algorithm and epilepsy detection method to classification tasks of the same structure is verified with data from EEG recordings in a rodent model. From the data set, random sample data sets are selected for further analysis. To meet the requirements of rat recordings, the CNN was adopted in such a way that only single channel recordings were used for training. The kernel matrix was adjusted to 1600 instead of 256 to meet the sampling frequency requirement of $f_s = 1.6\,\text{kHz}$ of the rats model recording. This ensures that a time frame with a length of 1 s is analyzed and equivalence to the CHB-MIT dataset is maintained. Measurements and results are presented in Section 5.5.

4.6. Seizure Prediction Based on Pre-Ictal Data

From a patient's perspective it would be highly desirable to detect an epileptic seizure before it occurs, instead of a detection during its occurrence. This would allow to issue warnings and take precautionary measures [29]. The ability to predict seizures with the developed CNN-based classification model is analyzed. While no clear unified definition of the length of the pre-ictal phase exists, various works define a period of up to 1 h before a seizure onset as pre-ictal [30–32]. In this section, only EEG data from non-seizure files are used for training, validation, and testing, and the models are trained in MATLAB. The seizure recordings are analyzed for different time periods defined as pre-ictal phases for 19 out of the 20 patients. It is assumed that the prediction quality depends on the length of the pre-ictal phase. If information about the upcoming seizure is available in the pre-ictal phase it can be expected that the prediction quality increases when longer time periods of the pre-ictal phase are taken into consideration by the classifier. The length of the pre-ictal phases are selected as 5, 10, 20, and 30 min. The performance of the prediction and the influence of the length of the pre-ictal phase is analyzed.

5. Measurements and Results

5.1. Evaluation Metrics

The performance of the CNN is evaluated with three metrics quantifying the quality of the binary classification task:

- Sensitivity, also called true positive rate (*TPR*): A measure for the proportion of ictal sequences (positives) that are correctly classified by the model as a seizure.

$$TPR = \frac{TP}{P};$$ (1)

  with *TP*: true positives, i.e., ictal sequences correctly classified as a seizure; *P*: positives, i.e., the total number of ictal sequences (positive cases) in the dataset.
- Specificity, also called true negative rate (*TNR*): Ratio of inter-ictal sequences correctly classified by the model as a non-seizure.

$$TNR = \frac{TN}{N};$$ (2)

  with *TN*: true negatives, i.e., inter-ictal sequences correctly classified as a non-seizure; *N*: negatives, i.e., the total number of inter-ictal sequences (negative cases) in the dataset. As the specificity of the classifier is very high with values of of 0.998 and higher, the specificity is measured in units of false positive rate for better comparability with other works.
- False positive rate (*FPR*) per hour (fp/h): Number of inter-ictal sequences (with a length of 1 s) wrongly classified by the model as a seizure per hour. The relation between these measures is given by:

$$TNR = 1 - FPR.$$ (3)

- AUC-score: Area under receiver operating characteristic (ROC) Curve—Measure of the model's ability to distinguish between the seizure and non-seizure classes.

These metrics ensure a comparability between different patients and results from related work.

5.2. MATLAB Classification Results

For each patient, an individual CNN is trained based on a personal dataset. The training phase of the neural network is limited to a length of 25 epochs. This makes the results comparable and provides equal conditions for all patients. All together the training is performed for 20 patients. All evaluation metrics are calculated for each patient separately.

Figures 3 and 4 illustrate the results of the classification in the time domain. The figures show the amplitude of a single channel EEG data in gray and the classification result as a probability in blue. A classified seizure event with a length of 101 s (marked in red) with a high detection probability throughout the event and low probability outside of the event is depicted in Figure 3. Figure 4 shows a classified seizure event with a length of 264 s (marked in red) and a highly fluctuating detection probability (blue). A possible reason for the poor detection probability in the event of Figure 4 could be the high background and low SNR in the signal (grey).

The detection sensitivity of all analyzed patients is shown in Figure 5 for a classification threshold of 0.5. The median sensitivity is 90% with a minimum outlier of 62.5% and a maximum of 100%. In this and the subsequent analysis, the median value instead of the average value is calculated in order to take into account possible outliers due to bad signal quality or corrupted (non-physiological) data in single patient's data sets.

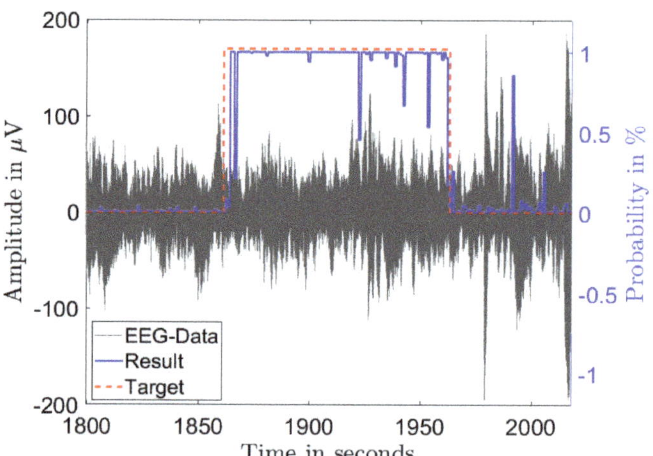

Figure 3. Single channel EEG data (gray) from a seizure record file of patient 1 showing 101 s of diagnosed seizure (red) with the output probability of the classification (blue).

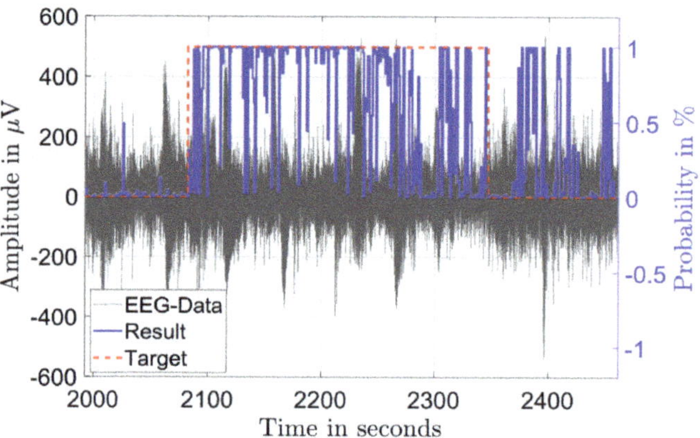

Figure 4. Single channel EEG data (gray) from a seizure record file of patient 8 showing 264 s of diagnosed seizure (red) and the output probability of the classification (blue).

The AUC-score is illustrated in Figure 5 showing a high median of 98% with the high quartile reaching 90% and the low quartile 95%. Overall, the sensitivity is the most important metric to evaluate the detection algorithm. Its importance is related to the fact that not detecting a seizure is worse than having a false positive alarm. The AUC is mainly an additional score to compare different neural network architectures since it is independent of the classifier threshold.

The distribution of the false positives per hour, which is according to Equation (3) a measure for the specificity, is shown in Figure 6. The median fp/h rate is 6.8 fp/h. Seventy-five percent of the results show less than 20 false positives per hour. For three patients, the fp/h rate is significantly higher (101, 95, and 65) than it is for the other patients. It is hypothesized and verified with a random sample test that this is due to a lower signal quality and higher noise level for these three patients. The best 5 patients stand out with a maximum of 1.7 fp/h. The minimum is 0.5 fp/h for patient 2.

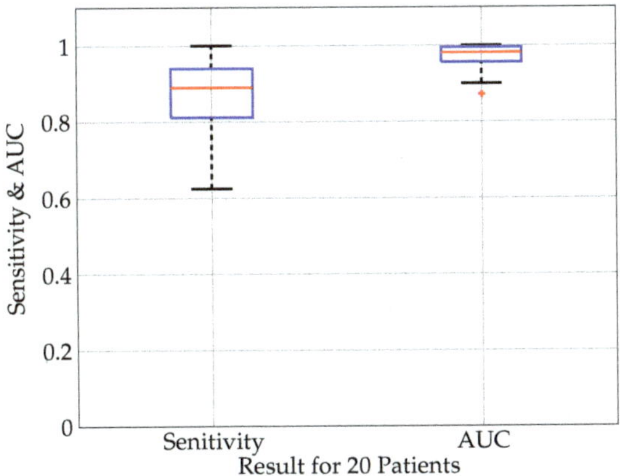

Figure 5. Boxplot (median value (red), lower and upper quartile (blue), min. and max. value (black), outlier (red cross)) of the evaluation measures sensitivity and AUC score for 20 patients. The median sensitivity is 0.90, 75 percentile: 0.94, 25 percentile: 0.81. The median AUC score is 0.98, 75 percentile: 0.99, 25 percentile: 0.98.

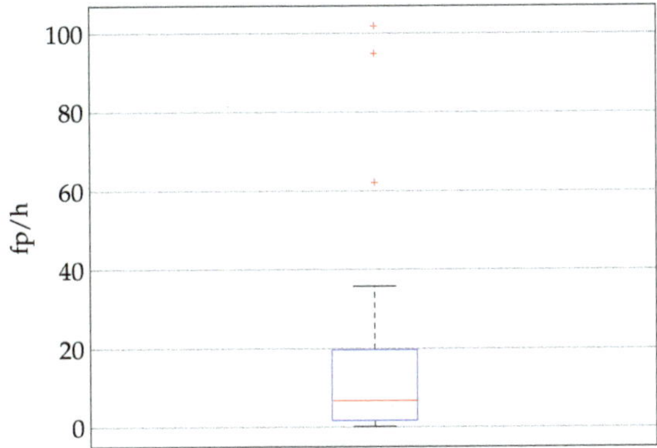

Figure 6. Illustration of the specificity of the classification showing a boxplot (median value (red), lower and upper quartile (blue), min. and max. value (black), outlier (red cross)) of the fp/h for 20 patients. The median fp/h is 6.8, 75 percentile: 19.8, 25 percentile: 1.75. The analysis is done on time signals with a length of 1 s. A false positive rate of 6.8 fp/h corresponds to a specificity of 0.998, this means that 99.8% of inter-ictal time frames of 1 s are classified correctly.

5.3. Classification Results for Hardware-Optimized CNN

The sensitivity and specificity for 10 EEG recordings classified in Python using the hardware optimized CNN structure are presented in Figure 7. The median sensitivity is 88.8%, the median specificity is 97.7%.

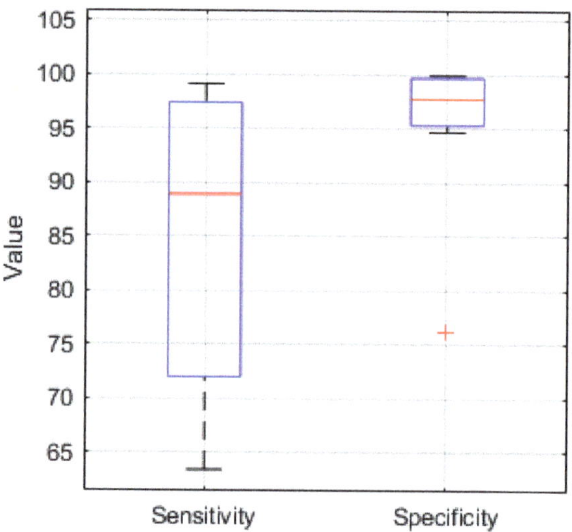

Figure 7. Sensitivity and Specificity boxplot (median value (red), lower and upper quartile (blue), min. and max. value (black), outlier (red cross)) for 10 EEG recordings classified in Python with a median sensitivity and specificity of 88.8% and 97.7%, respectively.

For comparison Figure 8 shows the boxplot of the sensitivity and specificity of the EEG recordings classified in MATLAB. For the 21 analyzed recordings the median values are 83.3% and 99.8% respectively.

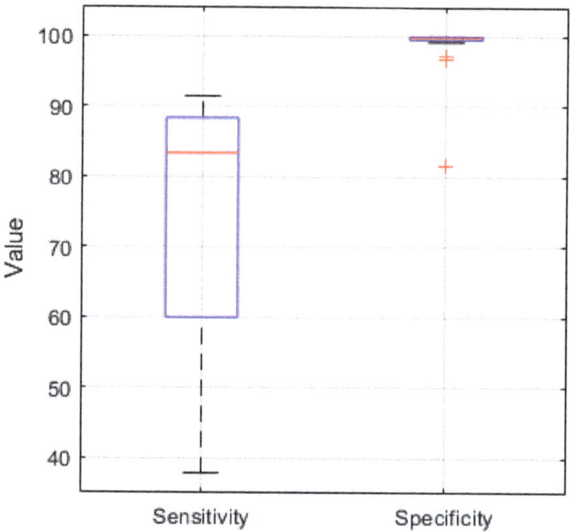

Figure 8. Sensitivity and Specificity boxplot (median value (red), lower and upper quartile (blue), min. and max. value (black), outlier (red cross)) for 21 EEG recordings classified in MATLAB with a median of 83.3% and 99.8%, respectively.

5.4. Power Consumption

The power consumption of the classification task is measured using a shunt resistor ($R = 1\,\Omega$) connected in series with the power supply of the processor of the GAPuino Board. While the processor performs the classification task, the voltage drop at the shunt resistor is measured. Based on the voltage and the value of the resistor, the power consumption is calculated. The measurement setup is depicted in Figure A1.

In order to have a trigger for the energy measurement, a digital I/O port is set to high before the GAP8 kernels are launched and set to low after the classification task is done. The voltage curve while classifying 1 s of EEG data is shown in Figure 9. The time period between on- and offset of the 8 processor kernels is marked by the trigger. The classification task of 1 s of EEG data takes a length of 33.5 ms. The maximum voltage reached is 15.44 mV. The trigger is scaled by a factor of 100 to simplify the representation. The maximum power consumption is 238.4 µW. The average consumption over the classification period is 140 µW. The energy required for one classification task is $E = 4.9$ µJ. The declared energy consumption is for classifying 1 s of EEG data and only for the GAP8 processor itself. The consumption of the whole board is not measured.

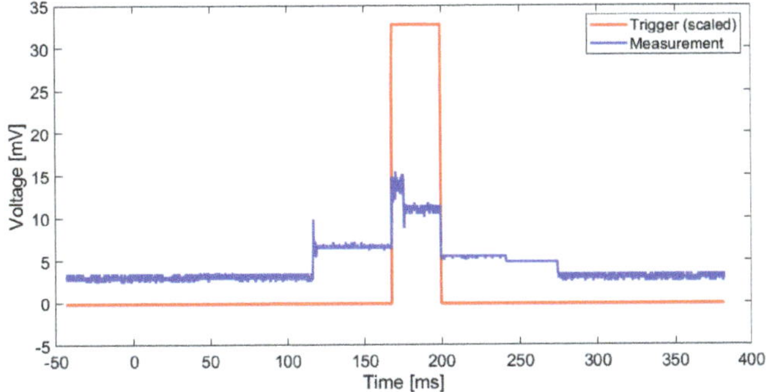

Figure 9. Measured voltage between TP5 and TP6 to measure the power consumption of GAP8 while classifying 1 s of EEG data (blue), trigger signal indicating the start and the end of the processing of 1 s of EEG data.

5.5. Verification with EEG Recordings in a Rodent Model

The presented algorithm and epilepsy detection method are verified with data from EEG recordings in a rodent model. Figure 10 shows one channel of the time signal of a recording of an epileptic rat with seizures and the classification result of the algorithm. The gray signal shows the recorded EEG signal in µV. The dashed red line shows the classification target indicating a seizure event with target value one and a non-seizure event with target value low, classified by neurological experts. The blue solid line shows the output probability of the classifier. As shown in the example in Figure 10, the four seizure events are detected by the classifier and the classification results of the CNN match very well with the classification of experts.

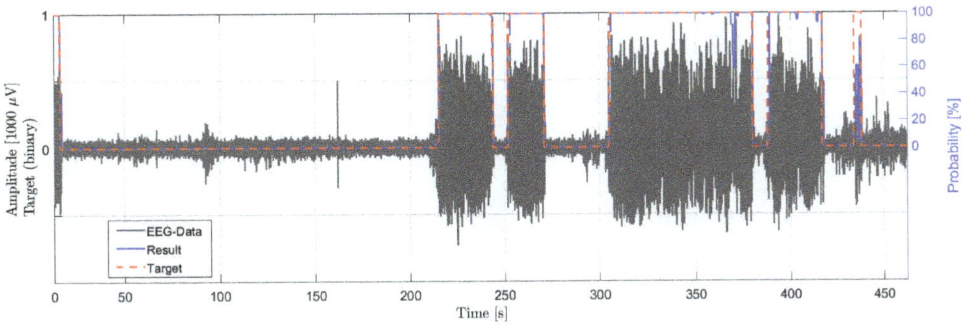

Figure 10. Single-channel recordings from a GAERS rat (gray) with a duration of 470 s showing a seizure event (red), as diagnosed by an expert, and the output probability of the classification (blue).

This exemplified result illustrates, that the developed CNN based seizure detection model is generally applicable and transferable to similarly structured classifications task. The performance of the adoption of the CNN is not evaluated in this work.

5.6. Seizure Prediction

For 19 out of the 20 patients, the seizure recordings are analyzed for different time periods defined as pre-ictal phases prior to seizure onset. Figure 11 shows the result of the analysis in the boxplot of fp/h for pre-ictal time periods. For this work the analysis is conducted for the time periods of 30, 20, 10, and 5 min. The classification results show a median false positive rate of 2.15 fp/h for a pre-ictal time period of 30 min, of 1.8 fp/h for 20 min, 3.0 fp/h for 10 min, and 3.9 fp/h for 5 min. While the shortest phase of 5 min shows the worst prediction, a clear trend for an increase in prediction quality for longer periods of pre-ictal phases can be seen. This strengthens the assumption that information about the upcoming seizure is available in the pre-ictal phase and that the prediction quality increases when more information is available or longer time periods are taken into consideration, respectively.

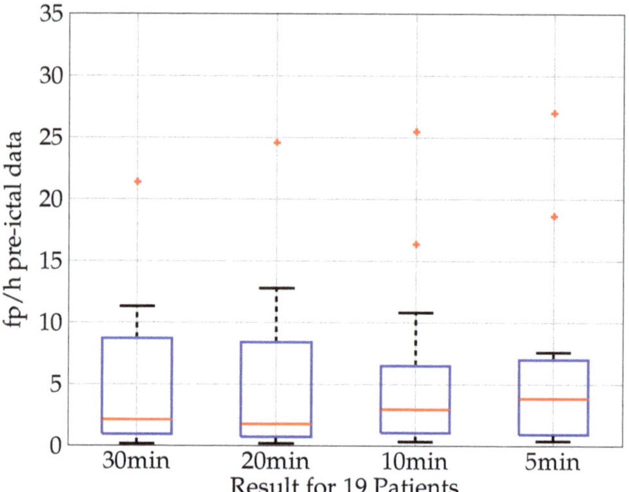

Figure 11. Seizure prediction based on pre-ictal data. Boxplot (median value (red), lower and upper quartile (blue), min. and max. value (black), outlier (red cross)) of the classification results in fp/h for 19 patients. The time period defined as "pre-ictal" varies from 30, 20, 10 to 5 min. The classification results show a median false positive rate of 2.15 fp/h for a pre-ictal time period of 30 min, of 1.8 fp/h for 20 min, 3.0 fp/h for 10 min and 3.9 fp/h for 5 min.

6. Comparison with State-of-the-Art

A comparison of the overall classification performance of the developed approach with recently published work is shown in Table 1. This comparison contains recent works that focus on future implantable systems. Thus, the approaches focus on low hardware complexity and low power requirement. Although it has to be admitted that it is difficult to compare the performance of classifiers based on classification results from different databases, it can be stated that all three classifiers show compatible performance.

Table 1. Comparison of the overall classification performance of the developed approach with published work.

	SeizureNet [12]	IntegerNet [27]	This Work
Database	iEEG dataset	CHB-MIT	CHB-MIT
Median sensitivity	96%	unspecified	90%
Median False positives per hour	10 fp/h	unspecified	6.8 fp/h
AUC-score	93%	94%	98%

A comparison of the power consumption with measurement results from state-of-the-art solutions is shown in Table 2. The average power consumption for a classification is more than 80% smaller than in [12]. With respect to energy consumption the presented work achieves a reduction by a factor of 6.9 and higher compared to that in [27], which consumes 34–90 µJ for each classification.

Table 2. Comparison of the energy consumption of a classification task for different chips and classifiers while running a seizure classification.

	SeizureNet [12]	IntegerNet [27]	This Work
Chip	unspecified	Microcontroller in 45 nm, 0.9 V CMOS process	GAP8 (8 core, RISC-V)
Power	850 µW	unspecified	140 µW
Energy	unspecified	34–90 µJ/classification	4.9 µJ/classification

The performance of different algorithms to detect epileptic seizures is compared in Table 3, based on the work in [33]. In this comparison, hardware requirements are not considered. The analysis are based on different data sets, thus the performance parameters cannot be compared directly. Nevertheless, it can be stated that it is possible to achieve comparable sensitivity and specificity with a variety of algorithms.

Table 3. Comparison of the performance of different algorithms for seizure detection independent of the hardware requirements, based on those in [33].

	Gabor [34]	Kelly [35]	Hopfengärtner [36]	This Work
Algorithm	Neural networks, CNET	Pattern-match regularity statistics, local max. frequency, amplitude variation	Power spectral analytical techniques, Short time Fourier transform	Convolutional Neural Network
EEG Sample [h]	528	1200	3248	865
Patients	22	55	19	20
Seizures	62	146	148	198
Sensitivity	90.3	79.5	90.9	90
Specificity [fp/h]	0.71	0.08	0.29	6.8

7. Conclusions

In this work, an epileptic seizure detection algorithm using a convolutional neural network has been presented, analyzed in MATLAB and implemented on an ultra-low-power RISC-V processor. In an implementation of the CNN on a RISC-V-based GAPuino microcontroller, a sensitivity of 85% is reached. The classification of 1 s of EEG data requires $E = 4.9\,\mu J$, which is suitable for low-power implantable systems. The specificity is higher than 99%. The classification of 1 s of input data takes 35 ms. Thus the low latency required for real-time applications is achieved. The proposed detector reduces the power consumption by the factor of 6 compared to related approaches. This is reached by the adoption of the CNN and by exploiting the hardware convolution engine of the of GAP8 microprocessor, which allows an energy efficient computation of the convolution operator. The CNN presented here is trained individually for each patient. Accordingly, this approach is not limited to a specific type of epilepsy. Instead it is generally applicable for epilepsy with recurring and comparable seizure events. This was confirmed with recordings from a rat model.

The classifiers, codes, and the data from recordings in the rodent model are made available to the public under open access license. This enables easy reuse and rapid adoption of the presented approach for future developments and applications.

Author Contributions: Conceptualization, A.B., P.W., T.S. and F.F.; methodology, A.B., M.S., M.A.F., H.M.L., I.B., A.-S.B., P.W., T.S. and F.F.; software, M.S., M.A.F. and H.M.L.; validation, A.B., M.S., M.A.F., H.M.L., I.B., A.-S.B., P.W., T.S. and F.F.; formal analysis, A.B., M.S., M.A.F., H.M.L., I.B. and A.-S.B.; investigation, A.B., M.S., M.A.F., H.M.L., I.B., A.-S.B. and T.S.; resources, A.B., A.-S.B., P.W., T.S. and F.F.; data curation, M.S., M.A.F., H.M.L., I.B. and A.-S.B.; writing—original draft preparation, M.S.; writing—review and editing, A.B., M.S., M.A.F., H.M.L., I.B., A.-S.B., P.W., T.S. and F.F.; visualization, A.B., M.S., M.A.F., H.M.L., I.B. and A.-S.B.; supervision, A.B., P.W., T.S. and F.F.; project administration, A.B., P.W., T.S. and F.F.; funding acquisition, A.B., P.W., T.S. and F.F. All authors have read and agreed to the published version of the manuscript.

Funding: This work was supported by the German Research Foundation, research training group RTG 2154 "Materials for Brain".

Institutional Review Board Statement: All experiments were performed in accordance with the German law on animal protection and approved by the Animal Care and Ethics Committee of the University of Kiel.

Informed Consent Statement: Not applicable.

Data Availability Statement: The GAERS data set is available via open access on the portal IEEE Dataport [22]. The MATLAB source code is available online under open access license [28] to enable rapid adoption in future research projects.

Conflicts of Interest: The authors declare no conflict of interest.

Appendix A

The Appendix A gives detailed information on the dataset, implementation, and the code in Matlab and TensorFlow.

Appendix A.1. Dataset

In the majority of cases the dataset is segmented in one-hour files, where files with a seizure occurring are called 'seizure records' and files without a seizure 'non-seizure records'. The recordings are formatted in the European Data Format (.edf). The information about used channels as well as starting and ending of a seizure period is stored in a separate text document.

Appendix A.2. Hardware Implementation

Appendix A.2.1. Dataset Preparation

The recordings of patients 11, 14, 19, 20, 21, and 22 contain empty channels. As these do not represent physiological data, they are not considered in the model. For patients 12, 13, and 18, the assigned scalp electrodes for each channel change massively throughout the recorded files. For this reason, the EEG data of these three patients is discarded from the analysis. The recordings of patient 17 contains only 22 channels, this patient is neglected too.

Appendix A.2.2. CNN Architecture

The Matlab code created for implementing the CNN in this work has been made available at Github [28]. The folder CHB_MIT includes all used functions, for data import of EEG data data preprocessing (e.g., removal of empty channels). The main function for constructing the CNN is located in the file CHB_MIT\seizure_detection_cnn\seizure_nn.m The folder Basis includes basic Matlab functionalities and auxiliary functions, e.g., functions for reading of data format EDF or storage of Matlab matrices in numpy arrays.

Appendix A.2.3. CNN Hardware Implementation

The architecture of the CNN, its dimensions and the number of filters is summarized in Table A1.

Table A1. Structure of the CNN, dimensions, and number of filters.

Layer	Name	Dimension	Number of Filters
	Input	$23 \times 256 \times 1$	-
1	Conv2D	23×17	20
	MaxPool2D (1×4) Dropout (0.2)		
2	Conv2D	1×5	10
	MaxPool2D (1×4) Dropout (0.2)		
3	Conv2D	1×5	10
	MaxPool2D (1×2) Dropout (0.2)		
4	Conv2D	1×5	10
	Dropout (0.2)		
5	Conv2D	1×1	1
6	Sigmoid	-	-

Appendix A.2.4. CNN Implementation in Tensorflow

The structure of the CNN is adopted and optimized to meet the requirements of the GAPuino board using the GAP8 software development kit. The detailed structure and the required memory size per layer is presented in Table A2.

Table A2. Structural adaptation of the CNN for implementation on GAPuino Board using the GAP8 SDK and memory size per layer. In total the CNN has a number of 10.162 trainable parameters and a memory requirement of 62.7 kB for 32-bit floats.

n-Layer	Operation	Output	Size
1	Input (23×256)	23×256	5888
2	$10 \times$ Convolution (5×5)	$19 \times 252 \times 10$	260
3	MaxPooling (2×2)	$9 \times 126 \times 20$	-
4	Dropout (0.2)	-	-
5	$20 \times$ Convolution (5×5)	$5 \times 122 \times 20$	5020

Table A2. *Cont.*

n-Layer	Operation	Output	Size
6	MaxPooling (2 × 2)	$2 \times 61 \times 20$	-
7	Dropout (0.2)	-	-
8	FullyConnected (2440 × 2)	2	4880
9	Softmax (2)	2	2
Total Number of trainable parameters		=	**10,162**
Input (23 × 256)		=	**5888**
Needed memory for 32-bit floats		=	**≈62.7 kB**

Appendix A.3. Measurements and Results

Power Consumption

The measurement setup for the power consumption during classification is depicted in Figure A1. The shunt resistor and the test points TP5 and TP6 are depicted at the bottom of the board.

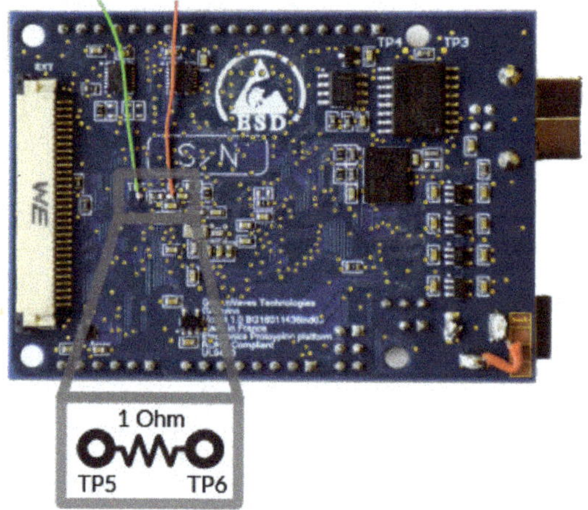

Figure A1. Measurement setup for power consumption showing a $R = 1\,\Omega$ shunt resistor at the bottom of the of the GAPuino evaluation [23] board with two measurement points (TP5 and TP6).

References

1. World Health Organization. Epilepsy: Key Facts. 2019. Available online: https://www.who.int/news-room/fact-sheets/detail/epilepsy (accessed on 21 June 2021).
2. Bishop, P.; Allen, C. The impact of epilepsy on quality of life: A qualitative analysis. *Epilepsy Behav.* **2003**, *4*, 226–233. [CrossRef]
3. Sperling, M.R. The Consequences of Uncontrolled Epilepsy. *CNS Spectrums* **2004**, *9*, 98–109. [CrossRef] [PubMed]
4. National Institute of Neurological Disorders and Stroke. *Epilepsy: Hope Through Research*; NIH Publication: Bethesda, MD, USA, 2015; Volume 15.
5. Löscher, W.; Potschka, H.; Sisodiya, S.M.; Vezzani, A. Drug Resistance in Epilepsy: Clinical Impact, Potential Mechanisms, and New Innovative Treatment Options. *Pharmacol. Rev.* **2020**, *72*, 606–638. [CrossRef]
6. Chen, Z.; Brodie, M.J.; Liew, D.; Kwan, P. Treatment Outcomes in Patients With Newly Diagnosed Epilepsy Treated With Established and New Antiepileptic Drugs: A 30-Year Longitudinal Cohort Study. *JAMA Neurol.* **2018**, *75*, 279–286. [CrossRef] [PubMed]
7. Gernert, M.; Feja, M. Bypassing the Blood–Brain Barrier: Direct Intracranial Drug Delivery in Epilepsies. *Pharmaceutics* **2020**, *12*, 1134. [CrossRef]

8. Trinka, E.; Boon, P.; Mertens, C. Neurostimulation for drug-resistant epilepsy: A systematic review of clinical evidence for efficacy, safety, contraindications and predictors for response. *Curr. Opin. Neurol.* **2018**, *31*, 198–210.

9. Astrakas, L.; Konitsiotis, S.; Tzaphlidou, M. Automated Epileptic Seizure Detection Methods: A Review Study. In *Epilepsy—Histological, Electroencephalographic and Psychological Aspects*; Stevanovic, D., Ed.; InTech: London, UK, 2012. [CrossRef]

10. Baldassano, S.N.; Brinkmann, B.H.; Ung, H.; Blevins, T.; Conrad, E.C.; Leyde, K.; Cook, M.J.; Khambhati, A.N.; Wagenaar, J.B.; Worrell, G.A.; et al. Crowdsourcing seizure detection: Algorithm development and validation on human implanted device recordings. *Brain A J. Neurol.* **2017**, *140*, 1680–1691. [CrossRef] [PubMed]

11. Lawhern, V.J.; Solon, A.J.; Waytowich, N.R.; Gordon, S.M.; Hung, C.P.; Lance, B.J. EEGNet: A Compact Convolutional Network for EEG-based Brain-Computer Interfaces. *J. Neural Eng.* **2016**, *15*, 056013.1–056013.17. [CrossRef]

12. Hügle, M.; Heller, S.; Watter, M.; Blum, M.; Manzouri, F.; Dümpelmann, M.; Schulze-Bonhage, A.; Woias, P.; Boedecker, J. Early Seizure Detection with an Energy-Efficient Convolutional Neural Network on an Implantable Microcontroller. *arXiv* **2018**, arXiv:1806.04549.

13. Daoud, H.; Bayoumi, M.A. Efficient Epileptic Seizure Prediction Based on Deep Learning. *IEEE Trans. Biomed. Circuits Syst.* **2019**, *13*, 804–813. [CrossRef]

14. Roy, Y.; Banville, H.; Albuquerque, I.; Gramfort, A.; Falk, T.H.; Faubert, J. Deep learning-based electroencephalography analysis: A systematic review. *J. Neural Eng.* **2019**, *16*, 051001. [CrossRef] [PubMed]

15. Bahr, A.; Saleh, L.A.; Hinsch, R.; Schroeder, D.; Isbrandt, D.; Krautschneider, W.H. Small area, low power neural recording integrated circuit in 130 nm CMOS technology for small mammalians. In Proceedings of the 2016 28th International Conference on Microelectronics (ICM), Giza, Egypt, 17–20 December 2016; pp. 349–352.

16. Bahr, A. *Aufnahme Von Hirnsignalen Mit Extrem Miniaturisierten Elektronischen Systemen Zur Untersuchung Von Lokalen Neuronalen Vernetzungen*; Logos Verlag Berlin: Berlin, Germany, 2017; ISBN: 3832545255.

17. Thomas, S.J.; Harrison, R.R.; Leonardo, A.; Reynolds, M.S. A Battery-Free Multichannel Digital Neural/EMG Telemetry System for Flying Insects. *IEEE Trans. Biomed. Circuits Syst.* **2012**, *6*, 424–436. [CrossRef] [PubMed]

18. Laiwalla, F.; Lee, J.; Lee, A.; Mok, E.; Leung, V.; Shellhammer, S.; Song, Y.; Larson, L.; Nurmikko, A. A Distributed Wireless Network of Implantable Sub-mm Cortical Microstimulators for Brain-Computer Interfaces. In Proceedings of the 2019 41st Annual International Conference of the IEEE Engineering in Medicine and Biology Society (EMBC), Berlin, Germany, 23–27 July 2019; pp. 6876–6879.

19. Benabid, A.L.; Costecalde, T.; Eliseyev, A.; Charvet, G.; Verney, A.; Karakas, S.; Foerster, M.; Lambert, A.; Morinière, B.; Abroug, N.; et al. An exoskeleton controlled by an epidural wireless brain–machine interface in a tetraplegic patient: A proof-of-concept demonstration. *Lancet Neurol.* **2019**, *18*, 1112–1122. [CrossRef]

20. Shoeb, A. Application of Machine Learning to Epileptic Seizure Onset Detection and Treatment. Ph.D. Thesis, Massachusetts Institute of Technology, Cambridge, MA, USA, 2009. Available online: http://hdl.handle.net/1721.1/54669 (accessed on 21 June 2021).

21. Shoeb, A.; Guttag, J. Application of Machine Learning to Epileptic Seizure Detection. In Proceedings of the 27th International Conference on International Conference on Machine Learning, ICML'10, Haifa, Israel, 21–25 June 2010; pp. 975–982.

22. Dataport, I. EEG of Genetic Absence Epilepsy Rats (GAERS). 2020. Available online: https://ieee-dataport.org/open-access/eeg-genetic-absence-epilepsy-rats-gaers (accessed on 21 June 2021).

23. Greenwaves Technologies. GAP8 Manual. 2018. Available online: https://greenwaves-technologies.com/manuals/BUILD/HOME/html/index.html (accessed on 21 June 2021).

24. Flamand, E.; Rossi, D.; Conti, F.; Loi, I.; Pullini, A.; Rotenberg, F.; Benini, L. GAP-8: A RISC-V SoC for AI at the Edge of the IoT. In Proceedings of the 2018 IEEE 29th International Conference on Application-specific Systems, Architectures and Processors (ASAP), Milan, Italy, 10–12 July 2018; pp. 1–4. [CrossRef]

25. SEMICO Research Corporation. *RISC-V Market Analysis: The New Kid on the Block*; Study No. CC315-19; November 2019. Available online: https://semico.com/sites/default/files/TOC_CC315-19.pdf (accessed on 21 June 2021).

26. Branco, P.; Torgo, L.; Ribeiro, R.P. A Survey of Predictive Modeling on Imbalanced Domains. *ACM Comput. Surv.* **2016**, *49*, 31:1–31:50. [CrossRef]

27. Truong, N.D.; Nguyen, A.D.; Kuhlmann, L.; Bonyadi, M.R.; Yang, J.; Ippolito, S.; Kavehei, O. Integer Convolutional Neural Network for Seizure Detection. *IEEE J. Emerg. Sel. Top. Circuits Syst.* **2018**, *8*, 849–857. [CrossRef]

28. Github. Matlab Source Code. 2020. Available online: https://github.com/cau-etit-sse/cnn-chb-mit (accessed on 21 June 2021).

29. Cook, M.J. Advancing seizure forecasting from cyclical activity data. *Lancet Neurol.* **2021**, *20*, 86–87. [CrossRef]

30. Eberlein, M.; Hildebrand, R.; Tetzlaff, R.; Hoffmann, N.; Kuhlmann, L.; Brinkmann, B.; Muller, J. Convolutional Neural Networks for Epileptic Seizure Prediction. In Proceedings of the 2018 IEEE International Conference on Bioinformatics and Biomedicine (BIBM), Madrid, Spain, 3–6 December 2018; pp. 2577–2582. [CrossRef]

31. Gagliano, L.; Bou Assi, E.; Nguyen, D.K.; Sawan, M. Bispectrum and Recurrent Neural Networks: Improved Classification of Interictal and Preictal States. *Sci. Rep.* **2019**, *9*, 15649. [CrossRef]

32. Ozcan, A.R.; Erturk, S. Seizure Prediction in Scalp EEG Using 3D Convolutional Neural Networks With an Image-Based Approach. *IEEE Trans. Neural Syst. Rehabil. Eng.* **2019**, *27*, 2284–2293. [CrossRef] [PubMed]

33. Baumgartner, C.; Koren, J.P. Seizure detection using scalp-EEG. *Epilepsia* **2018**, *59*, 14–22. [CrossRef]

34. Gabor, A.; Leach, R.; Dowla, F. Automated seizure detection using a self-organizing neural network. *Electroencephalogr. Clin. Neurophysiol.* **1996**, *99*, 257–266. [CrossRef]
35. Kelly, K.; Shiau, D.; Kern, R.; Chien, J.; Yang, M.; Yandora, K.; Valeriano, J.; Halford, J.; Sackellares, J. Assessment of a scalp EEG-based automated seizure detection system. *Clin. Neurophysiol.* **2010**, *121*, 1832–1843. [CrossRef] [PubMed]
36. Hopfengärtner, R.; Kerling, F.; Bauer, V.; Stefan, H. An efficient, robust and fast method for the offline detection of epileptic seizures in long-term scalp EEG recordings. *Clin. Neurophysiol.* **2007**, *118*, 2332–2343. [CrossRef] [PubMed]

biosensors

MDPI

Article

On the Classification of ECG and EEG Signals with Various Degrees of Dimensionality Reduction

Monica Fira [1], Hariton-Nicolae Costin [1,2,*] and Liviu Goraș [1,3]

1   Institute of Computer Science, Romanian Academy, 700481 Iasi, Romania;
    monica.fira@iit.academiaromana-is.ro (M.F.); lgoras@etti.tuiasi.ro (L.G.)
2   Faculty of Medical Bioengineering, Grigore T. Popa University of Medicine and Pharmacy of Iasi,
    700115 Iasi, Romania
3   Faculty of Electronics, Telecomunications & Information Technology, Gheorghe Asachi Technical University
    of Iasi, 700050 Iasi, Romania
*   Correspondence: hcostin@gmail.com

Abstract: Classification performances for some classes of electrocardiographic (ECG) and electroencephalographic (EEG) signals processed to dimensionality reduction with different degrees are investigated. Results got with various classification methods are given and discussed. So far we investigated three techniques for reducing dimensionality: Laplacian eigenmaps (LE), locality preserving projections (LPP) and compressed sensing (CS). The first two methods are related to manifold learning while the third addresses signal acquisition and reconstruction from random projections under the supposition of signal sparsity. Our aim is to evaluate the benefits and drawbacks of various methods and to find to what extent they can be considered remarkable. The assessment of the effect of dimensionality decrease was made by considering the classification rates for the processed biosignals in the new spaces. Besides, the classification accuracies of the initial input data were evaluated with respect to the corresponding accuracies in the new spaces using different classifiers.

Keywords: dimensionality reduction; classifications; Laplacian eigenmaps; locality preserving projections; compressed sensing

Citation: Fira, M.; Costin, H.-N.; Goraș, L. On the Classification of ECG and EEG Signals with Various Degrees of Dimensionality Reduction. *Biosensors* **2021**, *11*, 161. https://doi.org/10.3390/bios11050161

Received: 27 March 2021
Accepted: 14 May 2021
Published: 19 May 2021

Publisher's Note: MDPI stays neutral with regard to jurisdictional claims in published maps and institutional affiliations.

1. Introduction

Manifold learning [1] is a method for reducing dimensionality using the fact that essential information for many classes of high dimensional signals lies in much smaller dimensional spaces/manifolds. This is as the process of generating the data happens to have fewer degrees of independence thus permitting to the transformed data to belong to a low-dimensional subspace. Thus, even though data can't be represented in the initial space, when embedded in two or three dimensions, they can be easily represented and show, when possible some inherent structure. Therefore, to be able to visualize data dimension has to be decreased to one, two or three [2].

One possibility to get dimensionality reduction as well as compression is by taking projections of the data on a reduced number of random signals. However, using random projections, it is expected that some significant structure of the data might be lost since the signals are only approximately sparse and thus cannot be recovered with good accuracy [3].

Concerning geometry preserving, the techniques of manifold learning can be categorized into two classes:

(a)   Techniques that preserve the local arrangement: locally linear embedding (LLE), Laplacian eigenmaps (LE), manifold charting (MC), Hessian locally linear embedding (HLLE), and

(b)   Techniques that conserve global structure: isometric mapping (ISOMAP), diffusion map.

Several linear methods in manifold learning are principal component analysis (PCA), locality preserving projections (LPP) and multidimensional scaling (MDS), while among nonlinear ones are Isomap, Hessian eigenmaps, Laplacian eigenmaps, local linear embedding, and diffusion maps. From another point of view linear dimensionality reduction algorithms such as PCA, independent component analysis (ICA), linear discriminant analysis (LDA), and many others exhibit certain aspects to define an "interesting" way of linear data projection [4,5] at the price of possibly missing nonlinear structure of data. This is why non-linear methods are often stronger. The three steps of such algorithms are generally the following [6]:

- a nearest-neighbor search,
- defining of distances or affinities between elements,
- resolving a generalized eigenproblem to obtain the embedding of the initial space into a lower dimensional one.

The two main ingredients for dimensionality reduction are feature selection and feature extraction.

As mentioned above, we will discuss three methods for dimensionality reduction, two "standard" ones and the third, CS, which is not necessarily specific but interesting and useful as it will be shown.

In order to compare the methods we count on the fact that good dimensionality reduction will permit classification rates (usually smaller but) close to the initial ones.

We made use for testing, electrocardiographic (ECG) and electroencephalographic (EEG) signals downloaded from Internet databases and we compared the outcomes got with LE, LPP and CS using several standard classifiers aiming at getting an image about the compromise between dimensionality reduction and classification results.

In this paper we analyze the way the classifiers give good results for signals with various rates of dimensionality reduction. Thus, we present relevant information regarding the chosen method according to (a) the adopted rates of dimensionality reduction; (b) requirements such as reduced complexity (up to 2 or 3 dimensions), and (c) need for reconstruction. The advantages of each method are presented in the Section 4.

2. Materials and Methods

2.1. Laplacian Eigenmaps—LE

In the literature there are reported two similar techniques, in the sense that they consist each of three stages, the first two being common. The difference between the two is in the final stage, one of the algorithms keeping the local data arrangement, compared to the other that finds the optimal directions to project the data in a small space, so as to keep the data neighborhoods. These two techniques are Laplacian eigenmaps (LE) and locality preserving projections (LPP). Besides, for training data, Kernel LPP has the same significance as LE.

The basic assumption of the two methods is that data belong to a nonlinear subspace or nearly to it and in this way aim at discovering a low-dimensional modeling by retaining local characteristics. In LE the local properties are built on the keeping even distances between close neighbors.

The initial step in the LE algorithm [7] is to construct an adjacency graph G so that each data point xi is linked to its k nearest neighbors. In this way two things are important, namely, the number of neighbors as well as the weights of the graph branches which convey information about the distances between points.

The graph G will be constructed so that the weight wij is high if the points are close and wij is small if the nodes are far away. These weights are computed for all pairs of points xi and xj of the initial space; however, for points exterior the neighborhood k of a certain xm, the weights will have null value. In addition to the simplest weight assignment rule—one for neighboring points and null for outer points—a more exquisite rule is to use the Gaussian kernel [7–9]. After the calculation of the weights, follows the stage in which the calculation of the small dimensional representations is performed and on the manifold involves minimizing the cost function.

$$\varnothing(Y) = \sum_{ij} \| y_i - y_j \|^2 w_{ij},$$

where great weights wij strongly penalize distant points, thus nearly items in the initial space will be represented as near as possible in the new low-dimensional space.

Briefly, the LE algorithm [9] can be sketched in three main steps, namely:

(i.) *Nearest-neighbor search and adjacency graph construction*

Choose a number between K or a distance $\varepsilon > 0$ such that the vicinities of each data point are established: for a k-neighborhood nodes i and j are linked by a branch if i is through the k nearest neighbors of j or j is through the k nearest neighbors of i. On the other hand, nodes i and j are linked by a branch if $\| x_i - x_j \|^2 < \epsilon$, in which the Euclidean norm appears.

(ii.) *Weighted adjacency matrix (Choosing the weights)*

The weights w_{ij} of the symmetric ($n \times n$) vicinity matrix are computed as:

$$w_{ij} = w(x_i - x_j) = \begin{cases} \exp\left\{ -\frac{\|x_i - x_{ij}\|^2}{2\sigma^2} \right\}, & \text{if } x \in N_i; \\ 0, & \text{otherwise,} \end{cases}$$

according to the graph G that is assumed to be connected.

(iii.) *Eigenmaps*

In this stage, the eigenvalues and eigenvectors are calculated for the general eigenvector problem,

$$Lf = \lambda Df, \tag{1}$$

where $\mathbf{D} = (d_{ij})$ is an (n × n) diagonal matrix with

$$d_{ii} = \sum_{j \in N_i} w_{ij},$$

and $L = D - W$ is a Laplacian matrix which may be considered as an operator on functions applied on the nodes of G.

Ultimately, the eigenvector f_0 suitable to the 0 eigenvalue is discarded. The next m eigenvectors related to the next m eigenvalues in increasing gamut are utilized for embedding in a m-dimensional Euclidean space:

$$x_i \rightarrow (f_1(i), \ldots, f_m(i)), \tag{2}$$

where $f_0, \ldots, f_{k-1}$ are the solutions of (1).

2.2. Locality Preserving Projections—LPP

The locality preserving projections (LPP) method is established on the similarly variation rule as for the LE method. It has alike locality conserving attributes: the training data are utilized to learn a projection and the testing samples are embedded into the low-dimensional space [10].

Therefore, the first two stages of the LPP algorithm are alike as those of the LE while the final stage assumes calculating the eigenvectors and eigenvalues for the generalized eigenvector problem:

$$XLXTa = \lambda XDXTa, \tag{3}$$

in which $\mathbf{X}$ is the training data matrix and L, D have the same meaning as before.

Designating with $a_0, \ldots, a_{l-1}$ the column vectors related to the solutions of (2), ordering increasingly $\lambda_0 < \ldots < \lambda_{l-1}$, the mapping is defined as:

$$X_i \rightarrow y_i = A^T x_I, \ A = (A_0, \ A_1, \ldots, A_{l-1}), \tag{4}$$

in which y_i is l-dimensional, and $\mathbf{A}$ is a (nxl) matrix.

2.3. Compressed Sensing—CS

Compressed sensing is an acquisition technique that requires fewer samples than the Nyquist rate in the hypothesis of sparsity of signals [11]. Thus a signal x can be expressed by the projections:

$$y = \varnothing \, x, \tag{5}$$

where $x \in R^N$, $y \in R^M$ is the projection vector and $\varnothing \in R_{M, \, N}$ is the compressed sensing matrix whose entries are random i.i.d. (independent and identically distributed) signals. In this paper we will use the low dimensional projection vector y for signal classifications [12] and not for restoration signals.

2.4. Classifier Types

Since there are many methods of classification presented in the literature, it is difficult to decide which algorithm is superior to the others. The choice of one or the other depends on the type of application in which the classifier is incorporated but also on the specifics of the type of data used in the application. For example, for the classes linear separable, if the classes are linearly separable, the linear classifiers as logistic regression, Fisher's linear discriminant can surpass complex models as support vector machine (SVM) and artificial neural networks (ANN) and vice versa [13–15].

For the classification of ECG and EEG segments in the original space and in decreased dimensions, several classes of classifiers were used, namely: Decision Trees; Discriminant Analysis; Naive Bayes; SVM; Nearest Neighbor; Ensembles. Most of these classes have subclasses that have been used. In what follows several short descriptions of the main classifiers are given.

2.4.1. Decision Trees

Given data of attributes annotated with classes, a decision tree provides a series of rules that can be applied to classify new data. It utilizes an *if-then* command set which is reciprocally exclusive and exhaustive for classification. The commands are read sequentially utilizing the training data one at a time. Each time a rule is learned, the tuples incorporated by the rules are eliminated. This process is sustained on the training set until fulfilling a finish condition.

Advantages: Decision Tree is easy to comprehend and to view, the data does not require much preparation and the method can manage both numerical and qualitative data.

Drawback: This method can yield trees that do not generalize well and can be unstable i.e., small fluctuations in data could lead to the generation of a completely different tree.

2.4.2. Discriminant Analysis

This is a common primary classification method to test since it is quick, precise and simple to comprehend. Discriminant analysis is appropriate for voluminous datasets.

This technique presumes that particular categories provides data to whom they are assigned certain Gaussian distributions. In the training stage, the fitting function assesses the variables of a Gaussian law for every class.

2.4.3. Naive Bayes

Bayes' theorem is the source of this technique and it is based on the hypothesis of independence between every couple of attributes. Naive Bayes decision making behaves appropriately well in many real environments circumstances and applications, such as spam removal, document classification and person recognition. Naive Bayes is a simple method to apply and favorable outcomes have been acquired in the vast majority of situations. Additionally, it can be quickly used for voluminous datasets because it implies a linear function in time rather than by very time consuming iterative algorithms as in the case of a lot of other types of classifiers.

Advantages: Usually it needs a small number of training data to assess the necessary parameters. Naive Bayes decision making is very fast in contrast with more complex techniques.

Drawbacks: The big problem with this classifier is that it can manifest the so called "the zero probability problem". Thus, in the situation where the conditioned probability is zero for a certain attribute, the classifier is not able to offer a correct decision. This problem is usually solved by means of a Laplacian estimator.

2.4.4. Support Vector Machine—SVM

The support vector machine classifications consider the training data set as points divided into classes by an interval which is, ideally, as large as possible. The new data points are then embedded and estimated to belong to a certain class on one side or the other of the gap between the initial points.

In this way a SVM finds the most appropriate hyperplane that divides data points into two classes, in the sense that this hyperplane has the largest margin between the two classes. In other words, the SVM finds the maximal thickness of the area that is parallel to the hyperplane that has no inner data points [14].

Advantages: This classifier is efficient in high dimensional spaces and utilizes a subset of training data in the decision function that makes its memory very efficient.

Drawback: The SVM method does not directly give probability approximations. They are determined by applying usually an inefficient five-fold cross-validation.

2.4.5. Nearest Neighbor

The neighbors based classification is a type of slow training as it does not attempt to build a universal internal pattern, but simply stores cases of the training data. Classification is estimated from a simple majority vote of the k nearest neighbors of each point. Upper bound of the error rate approaches twice that of the ideal Bayes classifier.

Benefits: This method is easy to apply, powerful for noisy training sets, and efficient if the training set is huge.

Drawback: The main problem is the necessity to calculate k and the computation effort is great as it needs to compute the distance of each input point to all the training data.

2.4.6. Ensembles of Classifiers

The ensemble classifier combines a collection of classifiers that might perform superior classification performance compared to every single classifier. The principal rule behind the ensemble model is that a collection of poor learners join together to build a powerful learner. Qualities depend on the choice of the algorithm. Some techniques to perform ensemble decision trees are bagging and boosting.

Bagging (Bootstrap Aggregation) is applied when the object is to decrease the variance of a decision tree. The main idea is to create different data subsets from the training sample chosen randomly with replacement. Now, each group of subset data is utilized to train their decision trees. As a consequence, we end up with an ensemble of distinct models. Average of all the predictions from different trees are applied which is a more strong solution than a singular decision tree.

Boosting ensemble is another method to build a combination of classifiers. In this method, learners are determined sequentially with early learners applying uncomplicated models to the data and then evaluating data for errors. Hence, it fits consecutive trees (random sample) and, at all step, the object is to solve for net error from the previous tree.

Another type of ensemble of classifiers is the ensemble of nearest neighbor classifiers where each individual of the ensemble uses a random feature subset only and the decisions of these multiple classifiers are amalgamated for the ultimate decision.

Starting from the boosted trees ensemble, boosting being the most popular decision tree ensemble, Random under-sampling boosting (RUSBoost) has been introduced. Random under-sampling boosting (RUSBoost) is exceptionally successful at classifying irregular data. That means some classes with the training data have many more members than others. The method uses N, the number of members in the class with the fewest members in the training data, as the basic structure for sampling. In this way, by taking only N data points, classes with more members are under-sampled. If we have K classes, during the training stage, RUSBoost uses a smaller set of the data with N data points from each of those K classes. Then the method achieves the re-weighting and building the ensemble in Adaptive Boosting for Multiclass Classification [15].

3. Experimental Results and Discussions

3.1. ECG Signals

To analyze the feasibilities of dimension reduction utilizing LE, LPP and CS methods, we used for testing methods 44 ECG records from the MIT-BIH Arrhythmia database, including Holter data (so from wearable acquisition devices), collected at a sampling frequency of 360 Hz and on precision by 11 bits/sample [16]. Taking into account the annotations in the database, 7 pathological classes and the normal beating class were identified. The pathological classes included in this study are atrial premature beat (A), left bundle branch block beat (L), right bundle branch block beat (R), premature ventricular contraction (V), fusion of ventricular and normal beat (F), paced beat (/), fusion of paced and normal beat (f) and a class of normal beats (N).

For segmentation ECG signals we applied the segmentation method presented in a previous paper, namely, segmentation with centered R wave [17]. Our segmentation method begins with the precise determination of the R-wave, which has the maximum amplitude of ECG. Thus, the ECG signals are split in heartbeats cycles. An ECG cycle starts in the midst of a certain RR interval and finishes in the midst of the following RR interval. The R wave is placed in the center of the ECG cycle by resampling the signals on both parts of R. Thus cycles with the centered R waveform have been computed. Thereby, all ECG cycles are defined by 301 samples with the R wave being situated on the 150-th sample. Figure 1 shows an example of segmentation of the ECG signals belonging to each of the eight pattern categories.

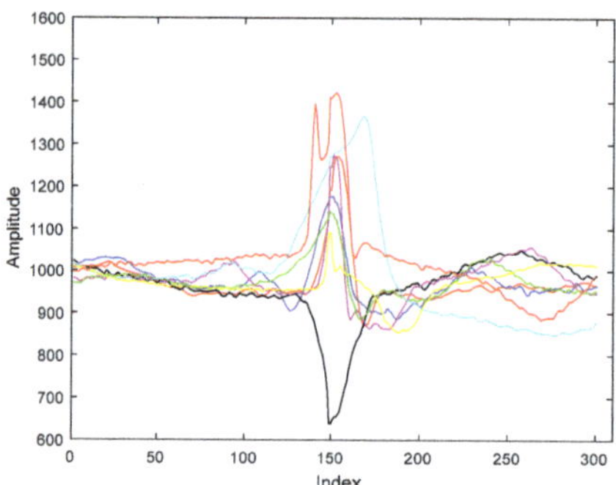

Figure 1. ECG patterns of the eight pattern classes used.

The database constructed is a data collection including 5608 ECG patterns, with 701 patterns for each of the eight considered types (seven pathological groups and a normal one).

A comparison of ECG behavior in the initial and reduced spaces implies first the classification of the ECG signals with the centered R-wave in the original space. The work was done in MATLAB® medium (MathWorks, Natick, MA, USA) and we used the next classifiers, each with different versions for tuning their key settings: Decision Trees (with fine, medium and coarse type classifier), Linear Discriminant and Quadratic Discriminant, Naive and Kernel Naive Bayes, Support Vector Machine (Linear, Quadratic, Cubic and Gaussian), *k*-nearest neighbors (fine, medium, coarse, Cosine, Cubic and Weighted KNN), besides different kinds of the ensemble of classifiers (Boosted and Bagged trees, discriminant and KNN Subspace and RUSBoosted Trees).

Figure 2 and Table 1 (its first column) show the classification accuracies for ECG signals with R-wave centered, in the initial space (raw data only). One can observe that good outcomes (over 90% classification accuracies) with SVM classifiers (Cubic, Quadratic and Medium Gaussian SVM), Fine KNN, and Ensemble Subspace KNN are got.

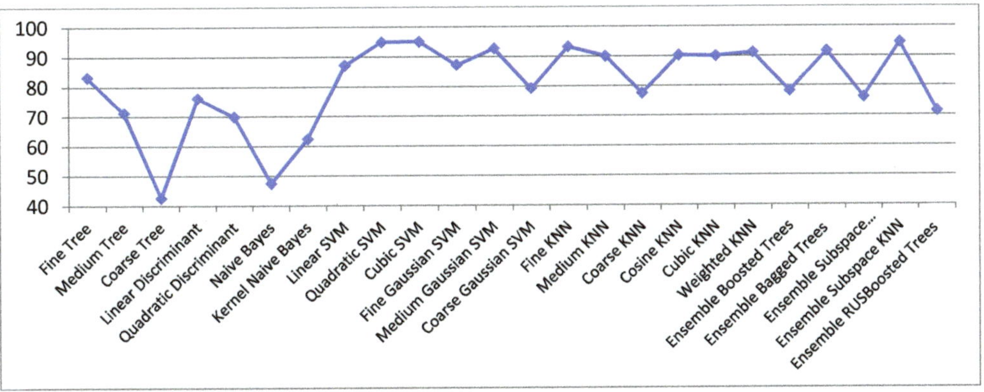

Figure 2. Classification rate in the original ECG space (centered 301 samples segments).

Table 1. Classification accuracies with CS, LE, LPP algorithms for 2, 3 and 25 dimensions respectively.

	ECG Original Centered	Compressed Sensed (CS)			Laplacian Eigenmaps (LE)			Locality Preserving Projections (LPP)		
	ECG Original	CS 2	CS 3	CS 25	LE 2	LE 3	LE 25	LPP 2	LPP 3	LPP 25
Fine Trees	83.44	49.41	55.34	79.81	76.25	77.32	86.73	54.00	66.65	81.15
Medium Trees	71.32	45.35	48.00	69.23	71.53	68.85	79.62	52.34	60.43	67.91
Coarse Trees	42.83	32.21	34.41	40.32	45.64	45.64	50.67	40.85	41.54	49.75
Linear Discriminant	76.32	24.23	33.72	73.94	34.77	38.81	77.44	30.42	35.41	73.64
Quadratic Discriminant	70.00	34.00	47.53	89.77	47.34	54.54	84.22	44.41	56.24	91.51
Naive Bayes	47.63	33.43	38.93	52.22	37.64	38.34	74.36	42.51	49.37	77.21
Kernel Naive Bayes	62.53	45.94	48.8	71.85	70.34	69.95	81.74	52.54	62.26	82.64
Linear SVM	87.34	29.52	38.9	85.14	49.08	61.37	85.62	37.52	47.72	85.92
Quadratic SVM	95.11	44.54	54.3	94.54	43.95	59.92	90.54	44.52	64.64	94.24
Cubic SVM	**95.24**	42.72	53.00	**94.50**	26.10	33.00	**91.20**	27.10	47.92	94.24
Fine Gaussian SVM	87.47	**51.80**	**62.90**	87.91	75.36	78.75	90.69	**54.40**	**70.10**	61.14
Medium Gaussian SVM	92.91	49.84	58.74	93.00	67.92	69.88	87.12	53.44	67.84	94.14
Coarse Gaussian SVM	79.47	32.85	43.65	80.97	54.36	55.41	80.92	44.45	57.82	83.82
Fine KNN	93.42	39.14	55.14	93.71	79.92	83.36	89.84	45.11	63.90	93.74
Medium KNN	90.27	48.72	60.82	90.82	80.76	83.92	89.65	52.42	68.00	91.32
Coarse KNN	77.62	50.47	57.71	77.44	74.00	75.35	80.12	53.63	65.74	78.34
Cosine KNN	90.54	29.64	47.15	90.74	61.25	81.42	89.55	32.80	54.62	92.76
Cubic KNN	90.22	48.81	60.81	90.81	80.88	83.95	89.72	52.38	68.34	90.77
Weighted KNN	91.47	43.60	59.44	92.34	**81.52**	**84.82**	90.32	48.51	67.42	92.35
Ensemble Boosted Trees	78.34	45.97	49.45	76.81	72.65	70.19	82.49	53.55	61.36	77.67
Ensemble Bagged Trees	91.81	43.94	59.45	90.4	80.00	83.91	90.91	48.86	68.31	91.84
Ensemble Subspace Discriminant	76.24	24.31	29.14	70.3	35	38.95	76.93	30.22	34.32	73.05
Ensemble Subspace KNN	94.71	23.34	44.00	94.04	51.24	80.82	89.98	24.14	56.10	**95.34**
Ensemble RUS Boosted Trees	71.54	45.34	47.94	69.31	71.54	68.84	79.64	52.84	60.67	67.97

The decision borders obtained with the KNN classifier are much more complex than for all Decision Trees, so getting an excellent classification for Fine KNN. The bad outcomes got with Bayes as opposed to KNN may have the following explanation: the fundamental distinction between KNN and Naive Bayes methods is that KNN is a discriminative classifier, and the Naive Bayes is a generative classifier. The Fine KNN classifier behaves better because it has the characteristic to be optimized locally. The great results achieved with Fine KNN were expected to be so. With an ensemble subspace KNN even better outcomes may be acquired.

In our approach the best accuracy is achieved with Cubic SVM, i.e., 95.2%. This parameter is valuable because the 8 classes studied are not easily distinguishable, and they are even intertwining.

In Table 1 and Figure 3 there are the classification outcomes: (a) in the original space with 301 samples; (b) results for ECG signals with dimensionality reduction by LE, LPP and CS methods for 2, 3 and 25 dimensions, respectively. We computed the classification accuracies for 2- and 3-dimensional cases because the signals with these dimensionalities can be easily illustrated graphically, which is very helpful and significant for comprehension the data spatial grouping. The graphic representation is very useful when we have many classes to handle and know nothing concerning their volumetric disposing. We also calculated the classification rate for dimensionality decrease to 25-space as we considered that a reduction from 301 to 25 dimensions is plausible both from the point of view of dimensionality reduction as well as in terms of classification accuracy.

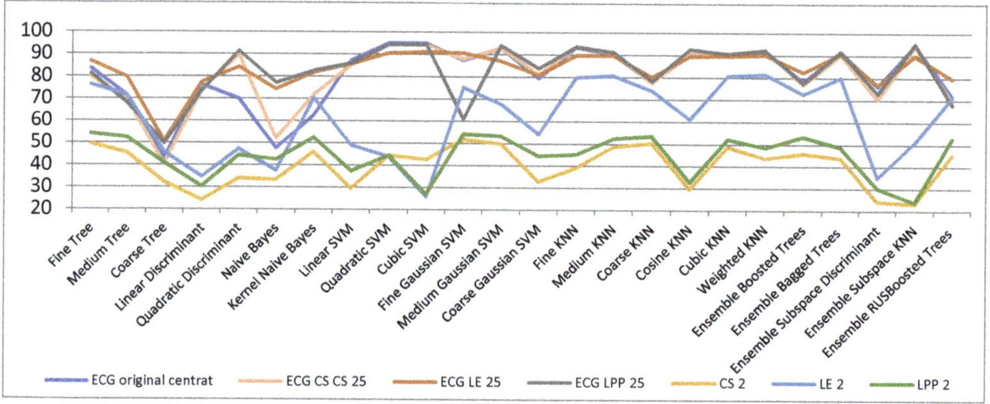

Figure 3. Classification results with CS, LE, LPP methods for 2, 3 and 25 dimensions, respectively.

Figure 4 and Table 2 show the results for various spatial dimensions for the Compressed Sensing (CS) method. It is observed that utilising Coarse Decision Tree very bad outcomes are got in the original space as well as in all other reduced spaces. Outcomes similar to those of the original space are achieved beginning with more than 10 dimensions in the projected space. Additionally, it can be observed the best outcomes hold with the SVM classifier. Depending on the degree of the dimensionality decrease they can be with cubic SVM or with fine Gaussian SVM. These classifiers achieve excellent classification rates, near to the medium Gaussian SVM. As a finding, for the dimensionality decrease with CS method, the SVM algorithm is best suited for that.

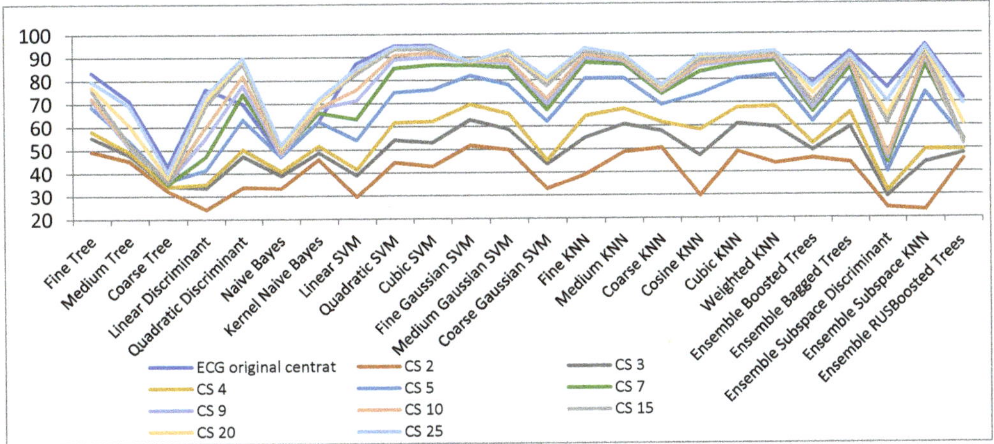

Figure 4. Classification results with CS method for dimensionality reduction.

Table 2. Classification results with CS method for dimensionality reduction.

	ECG Original Centered	CS 2	CS 3	CS 4	CS 5	CS 7	CS 9	CS 10	CS 15	CS 20	CS 25
Fine Tree	83.4	49.4	55.3	58.1	68.6	72.3	71.5	72.4	75.7	77.3	79.8
Medium Tree	71.3	45.3	48.0	49.3	54	52.8	51.6	52.3	52.7	60.6	69.2
Coarse Tree	42.8	32.2	34.4	34.2	36.5	35.2	36.2	36.7	35.9	38.0	40.3
Linear Discriminant	76.3	24.2	33.7	35.2	41.4	47.3	55.3	60.0	69.2	71.6	73.9
Quadratic Discriminant	70.0	34.0	47.5	50.3	63.2	74.1	77.8	82.0	87.6	*89.1*	*89.7*
Naïve Bayes	47.6	33.4	38.9	40.8	47.2	48.6	47.8	49.1	50.3	50.9	*52.2*
Kernel Naïve Bayes	62.5	45.9	48.8	51.7	62.4	*66.1*	*68.0*	*68.1*	*70.5*	*70.5*	*71.8*
Linear SVM	87.3	29.5	38.9	41.6	54.2	63.2	71.3	75.9	82.8	84.4	85.1
Quadratic SVM	95.1	44.5	54.3	61.7	74.7	85.2	88.9	90.8	93.3	94.2	94.5
Cubic SVM	**95.2**	42.7	53.0	62.2	75.9	86.6	**90.1**	**91.7**	**93.4**	**94.7**	**94.5**
Fine Gaussian SVM	87.4	**51.8**	**62.9**	**69.5**	**82.0**	86.4	*87.8*	*88.5*	*88.0*	*87.6*	*87.9*
Medium Gaussian SVM	92.9	49.8	58.7	65.4	78.0	85.4	87.3	88.6	91.2	92.0	*93.0*
Coarse Gaussian SVM	79.4	32.8	43.6	45.2	62.1	67.2	69.5	71.8	77.5	*79.5*	*80.9*
Fine KNN	93.4	39.1	55.1	64.4	80.7	87.6	89.4	91.0	92.4	*93.5*	*93.7*
Medium KNN	90.2	48.7	60.8	67.5	80.6	86.5	87.8	88.4	89.6	*90.3*	*90.8*
Coarse KNN	77.6	50.4	57.7	61.5	69.2	73.8	74.9	75.5	76.3	76.6	77.4
Cosine KNN	90.5	29.6	47.1	58.2	73.8	83.2	85.9	86.7	88.3	89.7	90.7
Cubic KNN	90.2	48.8	60.8	67.7	80.3	86.4	87.7	88.5	89.8	*90.5*	*90.8*
Weighted KNN	91.4	43.6	59.4	68.2	81.9	**88.1**	89.3	90.1	91.5	*92.1*	*92.3*
Ensemble Boosted Trees	78.3	45.9	49.4	52.2	61.8	66.1	67.5	70.6	69.5	73.8	76.8
Ensemble Bagged Trees	91.8	43.9	59.4	65.6	80.3	85.2	87.1	88.2	89.7	90.2	90.4
Ensemble Subspace Discriminant	76.2	24.3	29.1	31.5	40.0	43.9	45.6	47.0	61.1	64.4	70.3
Ensemble Subspace KNN	94.7	23.3	44.0	49.5	74.2	86.0	89.0	90.3	92.4	93.6	94.0
Ensemble RUSBoosted Trees	71.5	45.3	47.9	49.4	53.9	53.8	52.0	52.5	52.8	60.6	69.3

In the original 301-dimensional space the classification accuracy is 95.2%. In the case of decreasing to 10 and 25 dimensions, an accuracy of 91.7% and 93.4% were obtained, respectively. An interesting aspect that can be remarked in Table 2 (underlined numbers) is that for dimensionality reduction to 20 or 25 slightly improved results compared to those in the initial space have been obtained with some classifiers. A possible explanation is that through dimensionality reduction the classification problem complexity diminishes and thus the classification rate increases.

Figure 5 and Table 3 show the results obtained with LE, both for the initial and reduced ECG signals. In the original space the best outcomes are attained with cubic SVM classifier. On the contrary, in the case of very small dimensions (between 2 and 5) of the projected space with the LE algorithm very weak outcomes are achieved. For very small manifolds, the best outcomes are accomplished with the Weighted KNN classifier. This statement can be justified by maintaining the vicinities at the local level. Likewise, excellent outcomes for very small spaces are obtained by using the Fine Gaussian SVM classifier. Thus, for these small spaces, the classification of the test data is strongly dependant on the quality of the classifier. In other words, the classifier has to be able to draw very precise decision limits for very close data. It is the case of the Fine Gaussian SVM kernel range, that is establish to (1/4) sqrt(no. of features).

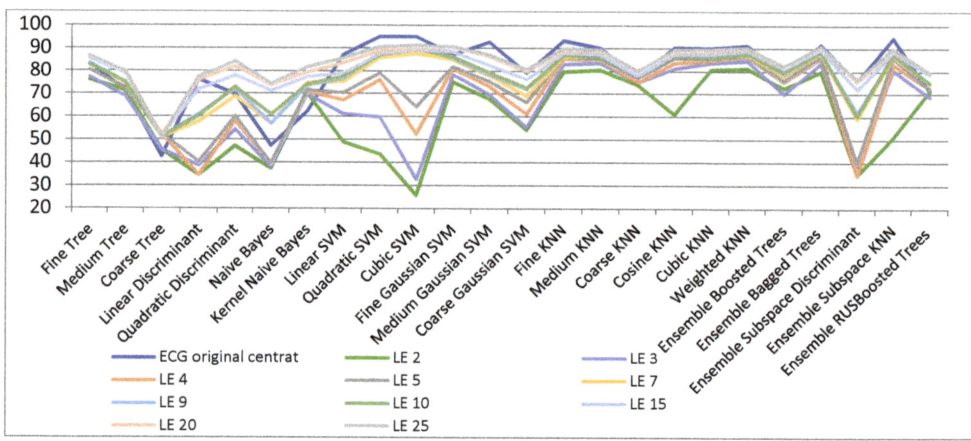

Figure 5. Classification results with LE method for dimensionality reduction.

Table 3. Classification results with LE method for dimensionality reduction.

	ECG Original Centred	LE 2	LE 3	LE 4	LE 5	LE 7	LE 9	LE 10	LE 15	LE 20	LE 25
Fine Tree	83.4	76.2	77.3	80.4	80.4	82.9	*83.7*	*82.8*	*85.8*	*86.5*	*86.7*
Medium Tree	71.3	71.5	68.8	*72.7*	*72.4*	*74.9*	*75*	*75.1*	*78.9*	*80.1*	*79.6*
Coarse Tree	42.8	*45.6*	*45.6*	*52.5*	*52.5*	*50.9*	*51.2*	*51.3*	*51.8*	*51.6*	*50.6*
Linear Discriminant	76.3	34.7	38.8	34.7	40.3	57.8	61.1	60.3	72.1	76.2	*77.4*
Quadratic Discriminant	70	47.3	54.5	58.3	60.1	69	*72.2*	*73*	*78.1*	*82.1*	*84.2*
Naive Bayes	47.6	37.6	38.3	39.8	39.5	*57*	*57.1*	*60.9*	*71.4*	*73.7*	*74.3*
Kernel Naive Bayes	62.5	*70.3*	*69.9*	*70.8*	*71.5*	*74.9*	*73.6*	*74*	*77.3*	*79.5*	*81.7*
Linear SVM	87.3	49	61.3	67.3	70.2	75.3	76.9	77.5	79.1	83.7	85.6
Quadratic SVM	95.1	43.9	59.9	76.2	79	86.1	87.6	87.3	87.7	89	90.5
Cubic SVM	**95.2**	26.1	33	52.5	64.2	87.9	**90.1**	**89.7**	89.6	90.4	**91.2**
Fine Gaussian SVM	87.4	75.3	78.7	81.1	82	85.2	85.9	86.5	*88.6*	*90.4*	*90.6*
Medium Gaussian SVM	92.9	67.9	69.8	73.4	75.4	78.3	78.6	79.5	82.8	86.6	87.1
Coarse Gaussian SVM	79.4	54.3	55.4	61.2	66.2	69.2	72.1	72.5	76.6	*80.1*	*80.9*
Fine KNN	93.4	79.9	83.3	85.7	86.2	86.2	87.2	87.1	88.1	88.9	89.8
Medium KNN	90.2	80.7	83.9	85	85.5	86.8	87	86.3	87.4	88.9	89.6
Coarse KNN	77.6	74	75.3	75.3	77.1	*79*	*78.6*	*78.5*	*78.3*	*80.6*	*80.1*
Cosine KNN	90.5	61.2	81.4	83.8	85.9	86.9	86.7	86.9	87.6	88.9	89.5
Cubic KNN	90.2	80.8	83.9	84.7	85.5	86.8	86.8	86.1	87.4	89	89.7
Weighted KNN	91.4	**81.5**	**84.8**	**86.6**	**86.9**	87.4	88.1	87.8	89.1	89.9	90.3
Ensemble Boosted Trees	78.3	72.6	70.1	75.5	76	78.3	79.2	79.9	*81.4*	*82.2*	*82.4*
Ensemble Bagged Trees	91.8	80	83.9	86.2	86.6	**88.2**	88.6	88.7	**89.9**	**90.9**	90.9
Ensemble Subspace Discriminant	76.2	35	38.9	34.7	40.2	59.2	61.9	60.5	72.2	75.9	76.9
Ensemble Subspace KNN	94.7	51.2	80.8	83.2	86.1	86.9	87.6	87.8	88.7	89.6	89.9
Ensemble RUSBoosted Trees	71.5	71.5	68.8	*72.7*	*72.4*	*74.9*	*75*	*75.1*	*79*	*80.1*	*79.6*

However, the Laplacian Eigenmaps technique for very small spaces, such as 2 and 3 dimensions, leads to very good classification results (81.5% and 84.5% classification accuracy, respectively) with Weighted KNN classifier. It is to remember here that the current classification problem is a difficult one, as there are 8 categories of ECG signals. We may state that a classification rate with only almost 10% under the original space versus a decrease in size from 301 to 2 is a remarkable result. The exceptional benefit of shrinking to 2 or 3 dimensions is the input data may be easily visualized graphically, allowing certain comprehension of the spatial arrangement. For a dimensionality reduction over 10, it can be observed that for some classifiers (results underlined in Table 3) higher classification accuracy than in the initial space has been obtained reminding of a kind of feature selection algorithm.

Figure 6 and Table 4 show the results of dimensionality reduction when using the LPP algorithm. As seen, the results are very similar to those achieved with the Laplacian Eigenmaps technique besides for very low dimensions (of 2, 3, and 4), when the classification measures achieved are much inferior (54%, 70.1%, and 77.3%, respectively). In the case of dimensions superior to 5, the classification measures are similar to those attained with the Laplacian Eigenmaps technique. For dimensions upper 20, classification measures very near to those in the original space are reached. As an example, for 20- and 25-dimensional spaces classification accuracies of above 95% are achieved by means of the Ensemble Subspace KNN classifier.

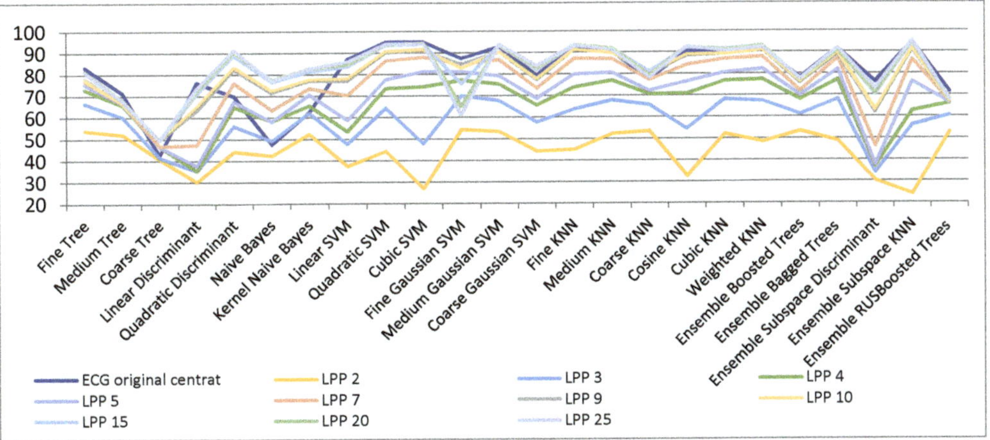

Figure 6. Classification results with LPP method for dimensionality reduction.

Table 4. Classification results with LPP method for dimensionality reduction.

	ECG Original Centered	LPP 2	LPP 3	LPP 4	LPP 5	LPP 7	LPP 9	LPP 10	LPP 15	LPP 20	LPP 25
Fine Tree	83.4	54	66.6	73	75.6	77.2	77.8	77.5	81.5	81.3	81.1
Medium Tree	71.3	52.3	60.4	65.9	66.5	66.8	66.9	67	68	68.1	67.9
Coarse Tree	42.8	40.8	41.5	*46.7*	*46.6*	*46.9*	*49.7*	*49.9*	*49.7*	*49.7*	*49.7*
Linear Discriminant	76.3	30.4	35.4	35.5	37.8	47.5	63.2	65.3	71.2	72.6	73.6
Quadratic Discriminant	70	44.4	56.2	65.1	67.6	*76.2*	*82.3*	*83.4*	*89.1*	*90.5*	*91.5*
Naive Bayes	47.6	42.5	49.3	*58.3*	*58.1*	*63.5*	*71.5*	*72.5*	*76.5*	*77.5*	*77.2*
Kernel Naive Bayes	62.5	52.5	62.2	*65.6*	*70.6*	*73.6*	*77*	*77.7*	*81.3*	*82.6*	*82.6*
Linear SVM	87.3	37.5	47.7	53.6	58.9	70.4	76.9	78.1	83.5	84.8	85.9
Quadratic SVM	95.1	44.5	64.6	73.5	77.6	86.4	90.2	90.9	93.7	94.1	94.2
Cubic SVM	**95.2**	27.1	47.9	74.3	81.2	**88.1**	91.2	**91.8**	94.3	94.5	94.2
Fine Gaussian SVM	87.4	**54.4**	**70.1**	**77.3**	81.2	84.8	84.4	82.9	75.8	65.2	61.1
Medium Gaussian SVM	92.9	53.4	67.8	75.4	79.2	86.7	90.2	90.4	*93.5*	*93.8*	*94.1*
Coarse Gaussian SVM	79.4	44.4	57.8	65.8	68.9	73.4	77.4	78	*82.1*	*83*	*83.8*
Fine KNN	93.4	45.1	63.9	73.9	80	87.3	**91.4**	91.5	*93.3*	*93.8*	*93.7*
Medium KNN	90.2	52.4	68	77	80.8	87	89.9	89.9	*91.9*	*92.1*	*91.3*
Coarse KNN	77.6	53.6	65.7	70.6	72.2	77.3	*80*	*80.3*	*81*	*79.3*	*78.3*
Cosine KNN	90.5	32.8	54.6	70.7	76.4	84.1	88.4	88.9	*92.2*	*92.7*	*92.7*
Cubic KNN	90.2	52.3	68.3	76.8	80.6	86.8	89.2	89.3	*91.6*	*91.1*	*90.7*
Weighted KNN	91.4	48.5	67.4	**77.3**	**82.3**	87.9	*91*	*91.1*	*93*	*92.9*	*92.3*
Ensemble Boosted Trees	78.3	53.5	61.3	68.1	70	72	75.8	76.5	77.6	77.3	77.6
Ensemble Bagged Trees	91.8	48.8	68.3	77.2	81.9	87.3	89.1	89.9	91.2	90.8	91.8
Ensemble Subspace Discriminant	76.2	30.2	34.3	37	37.7	46.3	62	63.2	70.3	70.9	73
Ensemble Subspace KNN	94.7	24.1	56.1	62.6	76.3	86.4	91.2	91.6	**94.5**	**95.4**	**95.3**
Ensemble RUSBoosted Trees	71.5	52.8	60.6	66	66.5	66.8	66.8	67.1	68	68.1	67.9

It has been observed again (underlined numbers in Table 4) that for dimensionality reduction over 10, in some cases improved results have been obtained.

In Figure 7 ECG signals with reduced dimensionality to 3D obtained with the 3 techniques are presented (each color corresponds to a different class) [18]; the great advantage of the possibility of data graphical visualization is obvious.

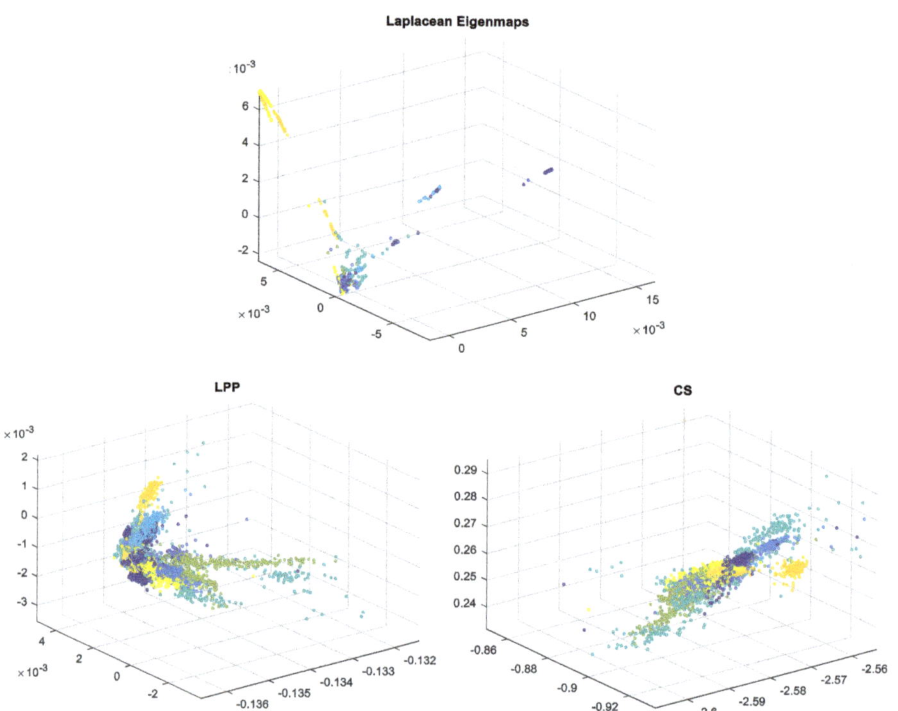

Figure 7. ECG data mapped into a 3-dimensional space with LE, LPP and CS techniques.

It can be observed that LE leads to a better data clustering/spatial separation than the other two methods for which, even though data are clustered, overlapping occurs. This is the reason why, when choosing dimensionality reduction to 3D, the classification ratio is better for LE compared to LPP and CS.

3.2. EEG Signals

For testing the dimensionality reduction methods, the EEG signals collected by Hoffmann and collaborators in their laboratory were used; a small database is free on the internet at [19]. This database includes EEG signals collected on the configuration with 32 channels, arranged in 942 vectors to be classified, lasting 1 sec. each [20,21]. The classification task is to detect the P300 waveform from a single EEG trial which has been used to build a P300 based spelling device for Brain-Computer Interface—BCI. We used configurations with 23, 8 and 4 channels for original EEGs for preprocessing and classifications tasks. The paradigm with P300 spelling device [22] that has been used is as follows.

One of the first examples for BCI is the algorithm proposed by Farwell and Donchin [22] that relies on the unconscious decision-making processes expressed via P300 in order to lead a computer. Another example, described in [23], refers to a real-time training of voted perceptron for classification of EEG data, also for a BCI application.

Now returning to the experiments proposed in [22], a (6 × 6) matrix containing (as in Figure 8) the letters of the alphabet and the numbers 1–9 were shown to the subjects on a computer display. The horizontal and vertical lines of the table were run at random for 100 ms with a 100 ms pause between sparkles i.e., after 12 sparkles every horizontal and vertical line was glowing once. Two datasets were acquired from every subject. During the first meeting subjects were requested to write the French words "lac", "nuage", "montagne", and "soleil", while for the second recording the subjects had to write the words "fromage", "chocolat", "pain", and "vin" [21].

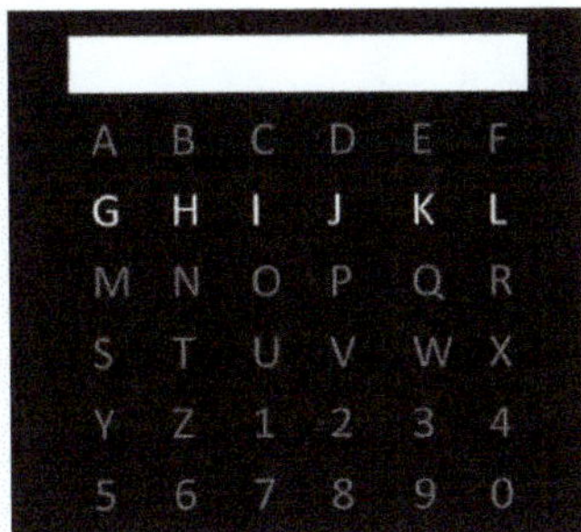

Figure 8. Classical P300 spelling paradigm described by Farwell–Donchin (1988).

As reported in [20] the EEG signals were registered from channels FP1, FP2, AF3, AF4, F7, F3, FZ, F4, F8, FC1, FC5, FC6, FC2, T7, C3, CZ, C4, T8, CP1, CP5, CP6, CP2, P7, P3, PZ, P4, P8, PO3, PO4, O1, OZ, O2 with a Biosemi Active 2 system (NEUROSPEC AG, Stans, Switzerland) at 2048 Hz. The signals were then referred to the average of channels O1, OZ, O2, low pass filtered (0 . . . 9) Hz with a 7th order Butterworth filter, and re-sampled with 128 Hz. The channels used as reference and channels T7, T8 were not used for EEG processing as they did not bring significant information for the P300s waveform detection. A more detailed explanation of the experimental work, i.e., EEG acquisition, preprocessing and artifact rejection is presented in [21].

In Figure 9 the electrodes configurations with 4, 8 and 23 channels are shown.

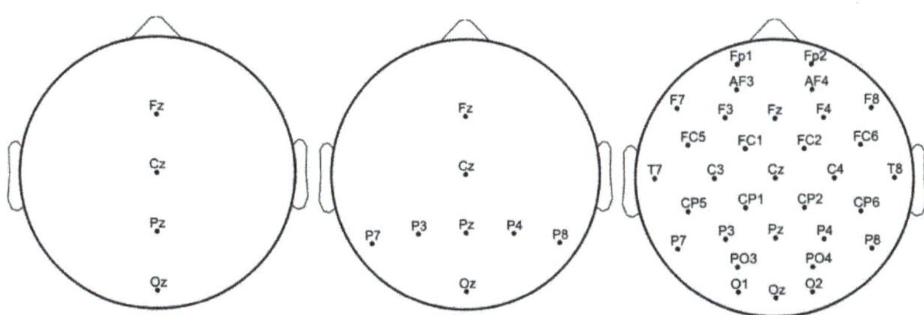

Figure 9. The electrodes configurations with 4, 8 and 23 channels.

Figure 10 shows the classification results for different channel configurations cases. It is observed that in general for the 8-channel version the best classification results of the original EEG signals are obtained. In general, good results are obtained for linear, quadratic and cubic SVM, but the best results are obtained with medium Gaussian SVM in the 8-channel configuration.

Because, in general, the configuration with 8 electrodes offers the best results, in the following we will present the results of this configuration for dimensionality reduction through the three analyzed methods. It should be mentioned that the initial EEG signals are segmented according to the stimulus applied to segments of 128 samples, i.e., we will consider that the space of the initial EEG signals is 128-dimensional.

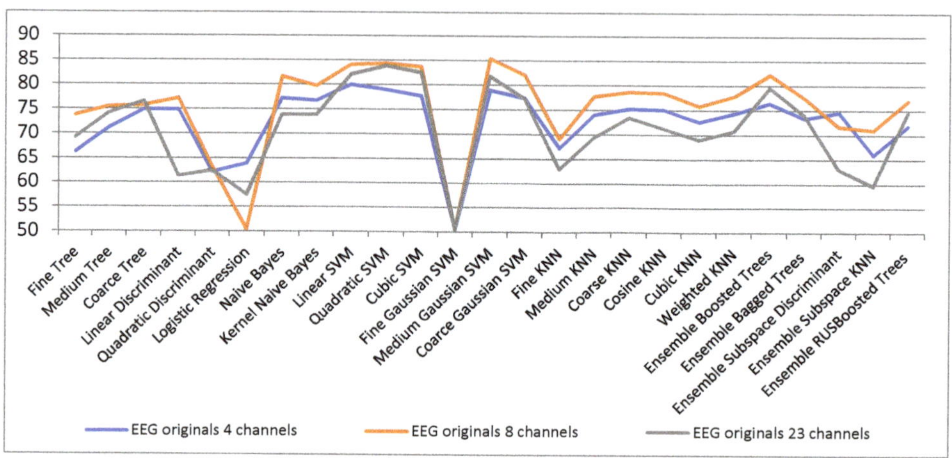

Figure 10. Classification results with original EEG signals for configurations with 4, 8 and 23 channels.

Figure 11 and Table 5 show the results for the dimensionality reduction with CS algorithm. It is found that there are classifiers with which better results are obtained in a space reduced to 15 dimensions compared to the initial space. This is the case of the discriminant linear classifier for which in the original space the classification rate is 77.2% and in a space reduced to 15 dimensions it classifies with a rate of 84.6%. Additionally, Quadratic Discriminant and Logistic Regression offers improved results for all spaces compared to the initial space. Additionally, in the case of Discriminant Subspace Ensembles the results in the reduced spaces are generally superior to the initial space. These results for which in spaces of reduced dimensionality improved results are obtained, compared to the initial spaces, are an example that the initial signals are in reality in a space of a much smaller dimensionality. It is much easier to classify data with a small dimension compared to the same data that is represented in a false large space.

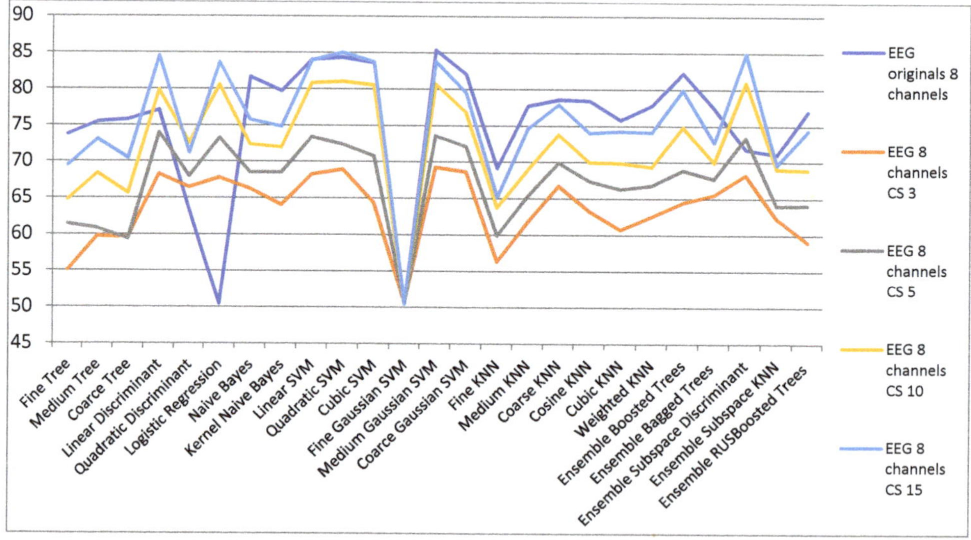

Figure 11. Results for the dimensionality reduction with CS algorithm for configurations with 8 channels.

Table 5. Classification results with CS method for configurations with 8 channels.

	ECG Orig.	EEG 8 Channels CS			
	8 Channels	CS 3	CS 5	CS 10	CS 15
Fine Tree	73.8	55.1	61.4	64.8	69.5
Medium Tree	75.5	59.8	60.8	68.4	73.1
Coarse Tree	75.8	59.6	59.4	65.7	70.5
Linear Discriminant	77.2	68.3	74	<u>79.9</u>	<u>84.6</u>
Quadratic Discriminant	63.4	<u>66.5</u>	<u>68</u>	<u>72.6</u>	<u>71.2</u>
Logistic Regression	50.5	<u>67.8</u>	<u>73.2</u>	<u>80.6</u>	<u>83.7</u>
Naive Bayes	81.7	66.3	68.5	72.4	75.8
Kernel Naive Bayes	79.8	64.1	68.5	72	74.9
Linear SVM	84.1	68.3	**73.4**	80.9	<u>84</u>
Quadratic SVM	84.4	69	72.4	**81.1**	**85.1**
Cubic SVM	83.7	64.4	70.8	80.6	<u>83.8</u>
Fine Gaussian SVM	50.5	50.7	50.5	50.5	50.5
Medium Gaussian SVM	**85.4**	**69.3**	73.6	80.8	83.9
Coarse Gaussian SVM	82.1	68.7	72.1	76.9	79.6
Fine KNN	69.2	56.4	59.9	63.8	65.2
Medium KNN	77.8	61.8	65.4	69	74.7
Coarse KNN	78.7	66.8	69.9	73.9	<u>78</u>
Cosine KNN	78.5	63.3	67.4	70	74.1
Cubic KNN	75.9	60.8	66.3	69.9	74.3
Weighted KNN	77.9	62.6	66.8	69.4	74.2
Ensemble Boosted Trees	82.3	64.5	68.9	74.9	80
Ensemble Bagged Trees	77.5	65.6	67.7	70	72.8
Ensemble Subspace Discriminant	71.8	68.3	73.4	<u>81</u>	<u>85</u>
Ensemble Subspace KNN	71.1	62.3	64	69.1	69.7
Ensemble RUSBoosted Trees	77	59.1	64.1	69	74.4

Figure 12 shows the results obtained with the LE algorithm to reduce the dimensionality of the space for EEG signals in the 8-channel configuration. It can be seen in Table 6 that in the case of the CS algorithm, the Linear and Quadratic Discriminant and Logistic Regression classifiers offer improved classification rates. Additionally, Discriminant Subspace Ensembles and KNN Subspace Ensembles classify better in reduced spaces with LE algorithm. The major difference from the CS method is that for very small spaces of dimensionality 3 and 5 the results are much better for the LE method compared to CS method. Hence the utility of the LE algorithm for data representation in 2 and 3 dimensional spaces for better visualization and understanding of spatial and geometric data arrangement.

Figure 13 shows the results obtained with the LPP algorithm to reduce the dimensionality of space for EEG signals in the 8-channel configuration. It is observed in Table 7 that the best results are obtained with all the classifiers for the initial space. These poor results are obtained both when applying LPP on each channel and then concatenating the signals with small spaces, or concatenating the initial EEG signals for the 8 channels and then applying the LPP method for dimensionality reduction.

In Figure 14 EEG signals with dimensionality reduced to 3D with all three techniques are represented. Signals containing the P300 wave have been plotted in blue and the others in red. It can be observed that for CS and LPP the two classes overlap, thus explaining the modest classification results for the 3D case. When using LE we get a better clustering of the two classes on the left laying non-P300 waves marked in red and on the right the P300 ones marked in blue. This is why LE leads to better results for 3D compared to LPP and CS.

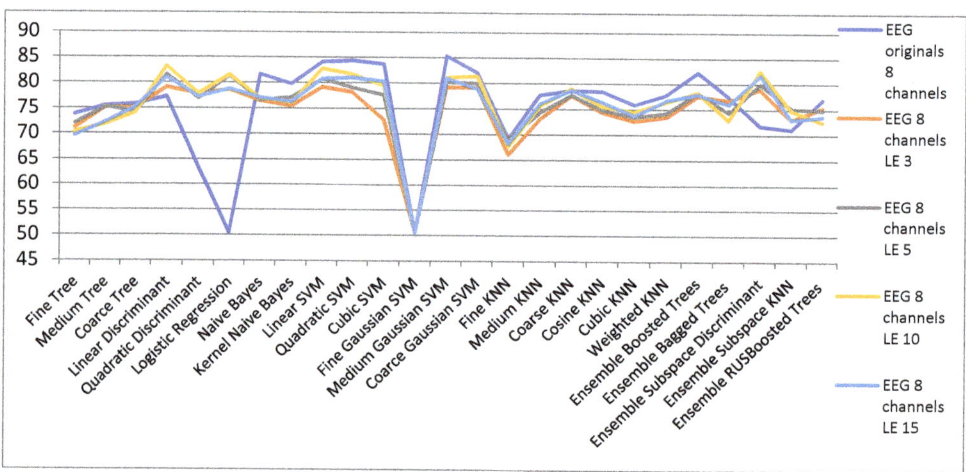

Figure 12. Results for the dimensionality reduction with LE algorithm for configurations with 8 channels.

Table 6. Classification results with LE algorithm for configurations with 8 channels.

	ECG Originals		EEG 8 Channels LE		
	8 Channels	LE 3	LE 5	LE 10	LE 15
Fine Tree	73.8	71.1	72	70.3	69.6
Medium Tree	75.5	75.1	75.3	71.8	72.3
Coarse Tree	75.8	75.1	74.3	74.1	75.2
Linear Discriminant	77.2	79.1	**81.6**	**83.2**	81.1
Quadratic Discriminant	63.4	77.8	76.9	77.9	77.2
Logistic Regression	50.5	78.7	81.4	81.6	78.8
Naive Bayes	81.7	76.5	76.6	77	77.1
Kernel Naive Bayes	79.8	75.5	77.1	76.1	76.3
Linear SVM	84.1	**79.2**	80.8	82.8	80.8
Quadratic SVM	84.4	78.2	79.1	81.7	81.1
Cubic SVM	83.7	72.9	77.7	79.5	80.4
Fine Gaussian SVM	50.5	50.7	50.5	50.5	50.5
Medium Gaussian SVM	**85.4**	**79.2**	80.3	81.1	81
Coarse Gaussian SVM	82.1	**79.2**	80	81.4	79
Fine KNN	69.2	66.1	69.1	67.6	68.2
Medium KNN	77.8	73.1	74.4	75.6	76.1
Coarse KNN	78.7	77.7	77.8	79.1	78.8
Cosine KNN	78.5	74.4	74.8	75.5	76.4
Cubic KNN	75.9	72.7	73.5	74.5	73.8
Weighted KNN	77.9	73.5	74.3	76.5	76.8
Ensemble Boosted Trees	82.3	77.7	78.3	78.3	78
Ensemble Bagged Trees	77.5	76.8	74.4	72.9	76
Ensemble Subspace Discriminant	71.8	79	80	82.5	**81.7**
Ensemble Subspace KNN	71.1	73	75.2	74.8	73
Ensemble RUSBoosted Trees	77	75.4	74.9	72.6	73.6

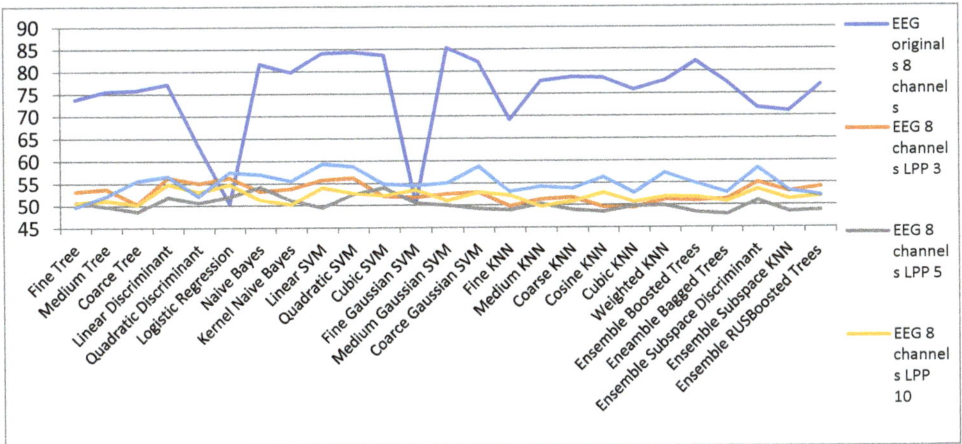

Figure 13. Results for dimensionality reduction with LPP algorithm for configurations with 8 channels.

Table 7. Classification results with LPP algorithm for configurations with 8 channels.

	EEG Orig.	EEG 8 Channels			
	8 Channels	LPP 3	LPP 5	LPP 10	LPP 15
Fine Tree	73.8	53.2	50.8	50.7	49.8
Medium Tree	75.5	53.8	49.8	51.2	52.2
Coarse Tree	75.8	50.4	48.6	50.3	55.6
Linear Discriminant	77.2	**56.3**	51.9	**54.9**	56.6
Quadratic Discriminant	63.4	55	50.7	53.1	52.1
Logistic Regression	50.5	**56.3**	52	**54.8**	57.5
Naïve Bayes	81.7	53.2	**54.2**	51.3	57
Kernel Naïve Bayes	79.8	53.8	51.2	50.2	55.6
Linear SVM	84.1	55.7	49.5	54	**59.4**
Quadratic SVM	84.4	56.2	52.5	52.7	58.8
Cubic SVM	83.7	52	54	52.1	54.9
Fine Gaussian SVM	50.5	51.8	50.5	53.5	54.5
Medium Gaussian SVM	**85.4**	52.5	50	51	55.1
Coarce Gaussian SVM	82.1	52.9	49.2	52.9	58.8
Fine KNN	69.2	49.8	48.9	52.1	53.1
Medium KNN	77.8	51.3	50.3	49.7	54.2
Coarse KNN	78.7	51.7	48.9	50.8	53.7
Cosine KNN	78.5	49.6	48.5	52.7	56.4
Cubic KNN	75.9	49.4	49.7	50.6	52.7
Weighted KNN	77.9	51.3	49.9	51.8	57.3
Ensemble Boosted Trees	82.3	51	48.3	51.7	54.9
Ensemble Bagged Trees	77.5	51.3	47.9	50.8	52.8
Ensemble Subspace Discriminant	71.8	55	51	53.5	58.4
Ensemble Subspace KNN	71.1	53	48.5	51.3	53.2
Ensemble RUSBoosted Trees	77	54	48.8	51.9	52.1

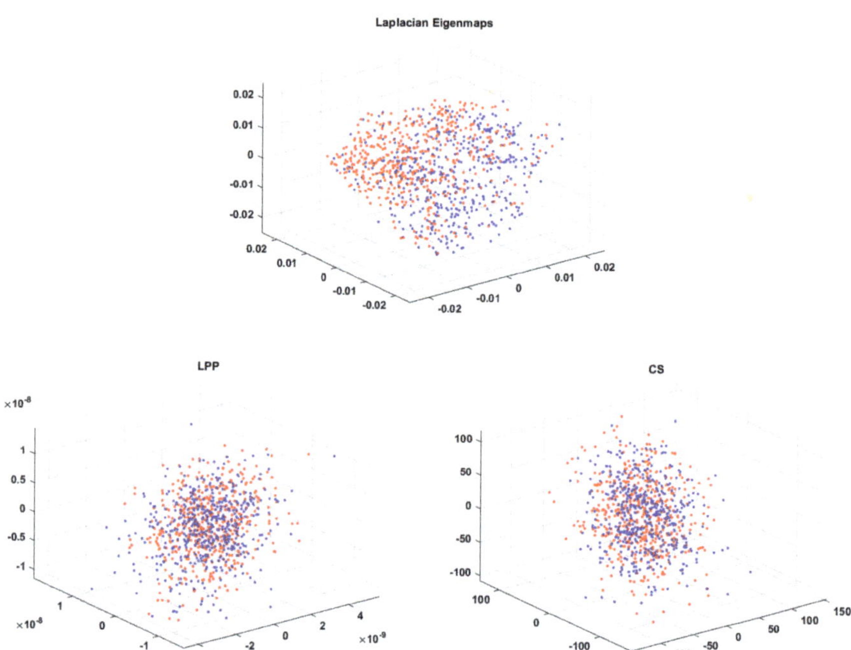

Figure 14. EEG data mapped into a 3-dimensional space with LE, LPP and CS techniques.

4. Conclusions

The aim of the paper was to offer a general view of the way the classifiers give good results for signals with various rates of dimensionality reduction.

Regarding ECG signals we stress the fact that they were preprocessed by aligning the R-wave. Our best results were obtained with SVM and KNN while for low dimensions (2 or 3), the best outcomes have been achieved with LE with the drawback that computations should be repeated for any new signal. Additionally, it has been found that in the case of CS for more than 10 dimensions the classification rate is near that obtained in the original space. Similar classification rates results have been achieved for dimensionality reduction larger than 10 with LPP for which the advantage for new testing signal is that no new calculations are necessary. Regarding CS, it is the most computationally advantageous compared to LE and LPP, which are much more computationally expensive.

For EEG signals, the CS and LE algorithms led to results similar to those obtained for ECG signals. The major difference that occurs in the case of EEG signals is for the LPP algorithm. This leads to much weaker results in reducing the dimensionality of the signals. To explain these results, we propose two hypotheses. A first one is that the LPP algorithm cannot find universal optimal projections for all 8 channels. The second hypothesis is that in the case of EEG signals the data are located on a manifold and the LPP algorithm fails to capture the local and at the same time general structure of the manifold, a situation encountered, for example, in the Swiss Roll manifold case.

The main conclusions of this work envisage the way dimensionality reduction and classification algorithms can be combined in order to obtain reasonable classification results even for (very) low dimensions both for ECG and a class of EEG signals. Choosing the rate of reduction of dimensionality is dependent on the motivation of the analysis. Thus, if we intend to reconstruct the initial signal, we will adopt CS, if we want intuition for 2 or 3 D we will choose LE while if we want to reduce dimensionality by about ten–twelve times and make classification in the reduced space without re-computation for new signals,

Biosensors **2021**, *11*, 161

we will use LPP. However, it seems LPP does not fit too well the global structure for EEG signals so that between LPP and LE the second one is better.

We assume these methods and outcomes might be extended in specific limits for more types of signals too, yet this concept should be attentively applied.

Author Contributions: Conceptualization, M.F., H.-N.C. and L.G.; methodology, M.F.; software, M.F.; validation, M.F., H.-N.C. and L.G.; formal analysis, M.F.; investigation, M.F.; resources, M.F.; data curation, M.F.; writing—original draft preparation, M.F., H.-N.C. and L.G.; writing—review and editing, M.F., H.-N.C. and L.G. All authors have read and agreed to the published version of the manuscript.

Funding: This research received no external funding.

Institutional Review Board Statement: Not applicable.

Informed Consent Statement: Not applicable.

Data Availability Statement: The data presented in this study are openly available in [physionet] at [10.1109/51.932724 and 10.1161/01.cir.101.23.e215], reference number [16] and [epfl], reference number [19]. The webpage of the MIT-BIH Arrhythmia Database is "https://www.physionet.org/content/mitdb/1.0.0/" (accessed on 17 May 2021) and "http://mmspg.epfl.ch/cms/page-58322.html" (accessed on 22 May 2017).

Conflicts of Interest: The authors declare no conflict of interest.

References

1. Mordohai, P.; Medioni, G. Dimensionality Estimation, Manifold Learning and Function Approximation using Tensor Voting. *J. Mach. Learn. Res.* **2010**, *11*, 411–450.
2. Boehmke, B.; Greenwell, B.M. Dimension Reduction. In *Hands-On Machine Learning with R*; Chapman & Hall, CRC Press: Boca Raton, FL, USA, 2019; pp. 343–396.
3. Bingham, E.; Mannila, H. Random projection in dimensionality reduction: Applications to image and text data. In Proceedings of the 7-th ACM SIGKDD International Conference on Knowledge Discovery and Data Mining, San Francisco, CA, USA, 26–29 August 2001; pp. 245–250.
4. Lee, J.A.; Verleysen, M. *Nonlinear Dimensionality Reduction*; Springer: Berlin/Heidelberg, Germany, 2007.
5. Bengio, Y.; Paiement, J.; Vincent, P.; Delalleau, O.; Le Roux, N.; Ouimet, M. Out-of-sample extensions for LLE, Isomap, MDS, eigenmaps, and spectral clustering. *Adv. Neural Inf. Process. Syst.* **2004**, *16*, 177–186.
6. Fodor, I. *A Survey of Dimension Reduction Techniques*; Technical Report; Center for Applied Scientific Computing, Lawrence Livermore National: Livermore, CA, USA, 2002.
7. Belkin, M.; Niyogi, P. Laplacian Eigenmaps and Spectral Techniques for Embedding and Clustering. *Adv. Neural Inf. Process. Syst.* **2001**, *14*, 586–691.
8. Belkin, M. Problems of Learning on Manifolds. Ph.D. Thesis, Department of Mathematics, The University of Chicago, Chicago, IL, USA, August 2003.
9. Belkin, M.; Niyogi, P. Laplacian eigenmaps for dimensionality reduction and data representation. *Neural Comput.* **2003**, *15*, 1373–1396. [CrossRef]
10. He, X.; Niyogi, P. Locality preserving projections. In Proceedings of the Conference on Advances in Neural Information Processing Systems, Vancouver, BC, Canada, 8–13 December 2003.
11. Donoho, D.L. Compressed sensing. *IEEE Trans. Inf. Theory* **2006**, *52*, 1289–1306. [CrossRef]
12. Candès, E.J.; Wakin, M.B. An Introduction to Compressive Sampling. *IEEE Signal Process. Mag.* **2008**, *25*, 21–30. [CrossRef]
13. Duda, R.; Hart, P. *Pattern Recognition and Scene Analysis*; Wiley Interscience: Hoboken, NJ, USA, 1973.
14. Bishop, C.M. *Pattern Recognition and Machine Learning*; Springer: Berlin/Heidelberg, Germany, 2006.
15. Alpaydin, E. *Introduction to Machine Learning*, 4th ed.; MIT Press: Cambridge, MA, USA, 2020.
16. MIT-BIH. Arrhythmia Database. Available online: http://www.physionet.org/physiobank/database/mitdb/ (accessed on 8 January 2021).
17. Fira, M.; Goraş, L.; Cleju, N.; Barabaşa, C. On the classification of compressed sensed signals. In Proceedings of the International Symposium on Signals, Circuits and Systems (ISSCS) 2011, Iasi, Romania, 30 June 2011.
18. Fira, M.; Goraş, L. On Some Methods for Dimensionality Reduction of ECG Signals. *Int. J. Adv. Comput. Sci. Appl.* **2019**, *10*, 326–607. [CrossRef]
19. EPFL. Available online: http://mmspg.epfl.ch/cms/page-58322.html (accessed on 22 May 2017).
20. Hoffmann, U.; Vesin, J.M.; Ebrahimi, T.; Diserens, K. An efficient P300-based brain-computer interface for disabled subjects. *J. Neurosci. Methods* **2008**, *167*, 115–125. [CrossRef] [PubMed]

21. Hoffmann, U.; Garcia, G.; Vesin, J.-M.; Diserens, K.; Ebrahimi, T. A Boosting Approach to P300 Detection with Application to Brain-Computer Interfaces. In Proceedings of the IEEE EMBS Conference on Neural Engineering, Arlington, VA, USA, 16–20 March 2005.
22. Farwell, L.A.; Donchin, E. Talking off the top of your head: A mental prosthesis utilizing event-related brain potentials. *Electroencephalogr. Clin. Neurophysiol.* **1988**, *70*, 510–523. [CrossRef]
23. Martišius, I.; Šidlauskas, K.; Damaševičius, R. Real-Time Training of Voted Perceptron for Classification of EEG Data. *Int. J. Artif. Intell.* **2013**, *10*, 207–217.

 *biosensors*

Article

Automatic Premature Ventricular Contraction Detection Using Deep Metric Learning and KNN

Junsheng Yu [1], Xiangqing Wang [1,*], Xiaodong Chen [2] and Jinglin Guo [1]

[1] School of Electronic Engineering, Beijing University of Posts and Telecommunications, Beijing 100876, China; jsyu@bupt.edu.cn (J.Y.); guojinglin@sgitg.sgcc.com.cn (J.G.)
[2] Queen Mary, University of London, London E1 4NS, UK; xiaodong.chen@qmul.ac.uk
* Correspondence: wangxiangqing@bupt.edu.cn

Abstract: Premature ventricular contractions (PVCs), common in the general and patient population, are irregular heartbeats that indicate potential heart diseases. Clinically, long-term electrocardiograms (ECG) collected from the wearable device is a non-invasive and inexpensive tool widely used to diagnose PVCs by physicians. However, analyzing these long-term ECG is time-consuming and labor-intensive for cardiologists. Therefore, this paper proposed a simplistic but powerful approach to detect PVC from long-term ECG. The suggested method utilized deep metric learning to extract features, with compact intra-product variance and separated inter-product differences, from the heartbeat. Subsequently, the k-nearest neighbors (KNN) classifier calculated the distance between samples based on these features to detect PVC. Unlike previous systems used to detect PVC, the proposed process can intelligently and automatically extract features by supervised deep metric learning, which can avoid the bias caused by manual feature engineering. As a generally available set of standard test material, the MIT-BIH (Massachusetts Institute of Technology-Beth Israel Hospital) Arrhythmia Database is used to evaluate the proposed method, and the experiment takes 99.7% accuracy, 97.45% sensitivity, and 99.87% specificity. The simulation events show that it is reliable to use deep metric learning and KNN for PVC recognition. More importantly, the overall way does not rely on complicated and cumbersome preprocessing.

Keywords: electrocardiogram; deep metric learning; k-nearest neighbors classifier; premature ventricular contraction

Citation: Yu, J.; Wang, X.; Chen, X.; Guo, J. Automatic Premature Ventricular Contraction Detection Using Deep Metric Learning and KNN. *Biosensors* **2021**, *11*, 69. https://doi.org/10.3390/bios11030069

Received: 26 January 2021
Accepted: 26 February 2021
Published: 4 March 2021

Publisher's Note: MDPI stays neutral with regard to jurisdictional claims in published maps and institutional affiliations.

1. Introduction

The heart is a vital part of the muscular system, which keeps blood circulating. Heart rhythm and heart rate are two fundamental indicators to assess whether the heart is working orderly [1]. Heart rhythm is usually rhythmic, and its clinical significance is more important than the heart rate. However, suppose the heart's four chambers, including the right atrium (RA), right ventricle (RV), left atrium (LA), and left ventricle (LV), cannot alternately contract and relax to pump blood through the heart. In that case, the heartbeat will be abnormal in speed and rhythm. The irregular heartbeat typifies arrhythmia and harms the body's organs and tissues, such as the lungs and brain [2]. Table 1 lists the most common types of arrhythmia.

Arrhythmias are closely related to electrical irregulars of the pumping heart [3]. Precisely, the heart's electrical system controls the heartbeat by the electrical signal. However, when these electrical signals that should have traveled on a fixed path change or the heart tissue changes, arrhythmias occur. For most arrhythmias, the electrocardiogram (ECG) is a handy and visual tool and has the advantages of being simple, fast, and accurate [4]. ECG can record the heart's electrical signals and is non-invasive and affordable for ordinary people. Moreover, a normal heartbeat in ECG has four main entities: A P wave, a QRS complex (a combination of the Q wave, R wave and S wave), a T wave, and a U wave, as shown in Figure 1. Table 2 shows the cause of generating these waves.

Table 1. The most common types of arrhythmia.

Type	Characteristic
Tachycardia	Heart rate over 100 beats per minute
Bradycardia	Heart rate below 60 beats per minute
Supraventricular arrhythmias	Arrhythmias that begin in the heart's upper chambers (atrium)
Ventricular arrhythmias	Arrhythmias that begin in the heart's lower chambers (ventricles)
Bradyarrhythmias	Arrhythmias that caused by a dysfunction in the cardiac conduction system

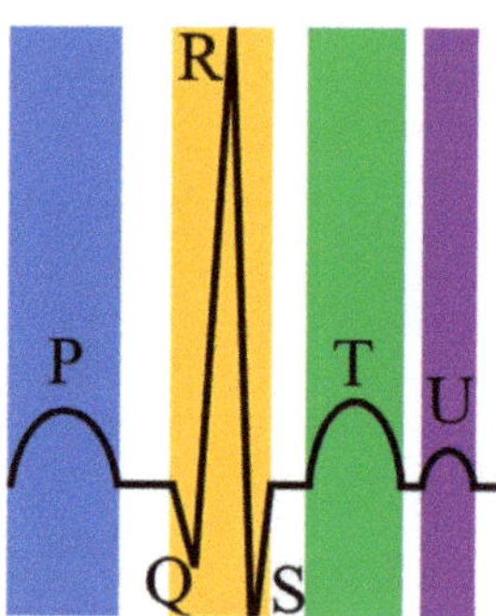

Figure 1. A normal heartbeat in an electrocardiogram (ECG).

Table 2. The cause of generating each wave in ECG.

Wave	Cause
P wave	Depolarization of the atrium
QRS complex	Depolarization of the ventricles
T wave	Repolarization of the ventricles
U wave	Repolarization of the Purkinje fibers

However, ECG is powerless for some particular arrhythmias, such as premature ventricular contraction (PVC), because the patient has a limited time for testing on the ECG machine during a standard ECG recording. PVC is a common arrhythmia initiated in the ventricles and often occurs in repeating patterns, as stated in Table 3. Specifically, PVC is ubiquitous in healthy individuals and patients and is associated with many diseases. There is a study evaluating the prevalence of frequent PVCs in Guangzhou, China [5]. Above 1.5% of the residents who received 12-lead ECG had PVCs, and nearly 1/6 of subjects who received 24-h Holter ECG were diagnosed with PVCs. According to the report provided by the American College of Cardiology Electrophysiology Council, PVC is related to left ventricular dysfunction and cardiomyopathy [6].

Table 3. The patterns of premature ventricular contraction (PVC) occurrence.

Patterns	Description
Bigeminy	Every other beat is a PVC
Trigeminy	Every third beat is a PVC
Quadrigeminy	Every fourth beat is a PVC
Couplet	Two consecutive PVCs
NSVT	Three-thirty consecutive PVCs

Furthermore, PVC is also associated with some disorders, such as ventricular tachycardia (VT), ventricular fibrillation (VF), underlying coronary artery disease, hypertension, and myocardial infarction (MI) [7–9]. Because PVC usually causes few or no symptoms, self-diagnosis is not accessible. Most people go to the hospital for help only after they notice severe symptoms.

Since the Holter monitor is a small wearable device and can record the heart's behavior in the patient's everyday life, cardiologists usually use the Holter monitor as a medium to obtain long-term ECG and diagnose PVC in clinical practice. However, analyzing so many long-term ECGs takes a lot of time and energy for cardiologists. Therefore, it is crucial to improve the efficiency of cardiologists regarding reliable and automatic searching for PVC from the long-term ECG.

With the continuous advancement of technology for collecting and processing physiological signals in recent years, many researchers have developed various algorithms to detect PVC from the long-term ECG automatically, as summarized in Table 4. In general, these algorithms are mainly of two types: Morphology-based methods and deep learning-based methods. In these morphology-based methods, extracting features relies on strong expertise, and most researchers have to manually design each feature to ensure that the features are practical. In the deep learning-based methods, extracting features is automatic, which is the most significant difference between the two methods.

Specifically, the morphology-based method's core is designing a series of trustworthy features manually with professional knowledge and experience. Compared with the normal heartbeat, PVC's waveform usually has three main characteristics, as shown in Figure 2: The QRS complex is broad and has an abnormal morphology (QRS-N and QRS-V); it occurs earlier than expected for the next sinus impulse ($T_1 < T_3 < T_2$); full compensatory pause ($T_1 + T_2 = T_3 + T_4$). Therefore, in the morphology-based methods, some classic features mostly come from the time-domain or frequency-domain of the ECG. Due to the continuous development of machine learning algorithms and the advancement of professional knowledge related to signal processing and ECG, most researchers have favored the morphology-based methods. Moreover, these approaches have occupied an unshakable status for a long time.

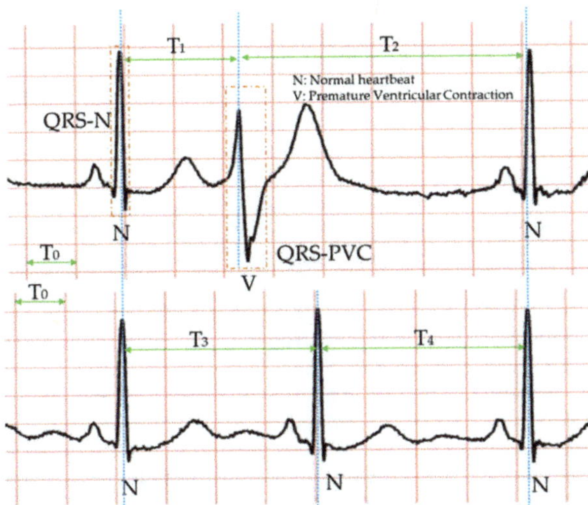

Figure 2. The waveforms of PVC and normal heartbeat. The two ECGs in this picture are from the same person. Each symbol is defined as follows. N (normal heartbeat); V (premature ventricular contraction); T_0 (0.20 s); T_1 (R-R interval); T_2 (R-R interval); T_3 (R-R interval); T_4 (R-R interval); QRS-N (QRS complex of normal heartbeat); QRS-V (QRS complex of PVC). The important thing is that T_3 and T_4 are usually equal, and the sum of them is generally similar to the sum of T_1 and T_2. The blue dotted line indicates the location of the R wave peak in each heartbeat.

The signals, collected directly from wearable devices, are always noisy. These noises mainly include baseline wander, 60 Hz power-line noise, electromagnetic interference, 100 Hz interference from fluorescence lights, and motion artifacts. Therefore, many morphology-based methods usually denoise the long-term ECG to extract features more accurately. These popular denoising algorithms are usually based on filters [10–12] or wavelet transforms [13,14].

Secondly, the morphology-based methods design and extract a series of features according to the expertise related to ECG and signal processing. Adnane et al. proposed a vital feature based on the Haar wavelet transform coefficients [15]. Du et al. also recommended an essential feature obtained by the chaotic analysis and Lyapunov exponent, named the chaotic feature [16]. Lek-uthai et al. extracted the four features based on cardiac electrophysiology: R-R interval, pattern of QRS complex, width of QRS complex, and ST-segment (the end of the QRS complex to the beginning of the T wave) level [17]. Jenny et al. suggested using the independent component analysis (ICA) algorithm to extract features and applying *t*-test analysis to evaluate these features [18]. Nuryani et al. redefine the width and the gradient of the QRS wave and regarded them as features [19].

Another factor determining the PVC detection method's performance is the classifier, which classifies samples with these extracted features. The essence of the classifier is a hypothesis or discrete-valued function. There are some popular classifiers used to distinguish regular and PVC beats: Artificial neural networks (ANN) [20–22], learning vector quantization neural network (LVQNN) [23], k-nearest neighbours (k-NN) algorithm [24,25], discrete hidden Markov model (DHMM) [26], support vector machine (SVM) [27,28], Bayesian classification algorithms [29], and random forest (RF) [30].

In summary, the morphology-based methods include three essential components: Denoising, designed features, and classifiers. Noise reduction is a prerequisite for accurately extracting features. Feature extraction is the core. The classifier directly plays a decisive role in the performance of these methods. Although the morphology-based methods have achieved significant success on this project after many researchers' efforts, these methods still have some limitations. First, the process of feature extraction relies heavily on preprocessing, such as wavelet transform and QRS detection. Preprocessing undoubtedly increases computational overhead. Further, extracting features is a complex and professional process. In this process, features are not imagined out of thin air but based on knowledge and experience. The features in each literature are often different from person to person, which makes it biased. Therefore, some scholars have proposed deep learning-based methods, which can detect PVC without manually designing features.

Deep learning-based methods are also inseparable from denoising, designed features, and classifiers. Compared with the morphology-based methods, the deep learning-based methods usually do not require professional knowledge and experience related to ECG or signal processing to design features automatically. Although these features are challenging to understand intuitively, these features are useful. That is to say, in most cases, we do not know the meaning of these features, but these features can be used to distinguish between a normal heartbeat and PVC.

Conway et al. used an ANN to detect PVC without manually extracting features [31]. The ANN's input corresponds to the 30 points of the QRS complex. Yang et al. proposed an innovative algorithm based on sparse auto-encoder (SAE) to extract features [32]. SAE is an unsupervised learning algorithm, including two processes of encoding and decoding. The encoding process performs the features' extraction, and the decoding process ensures the effectiveness of the features. Zhou et al. suggested an approach based on the lead convolutional neural network (LCNN) and long short-term memory (LSTM) network to extract features [33]. Liu et al. proposed a PVC detection method, which can directly analyze and process the ECG waveform images [34]. The finetuned Inception V3 model, developed by Google, is the core component of the method [35].

It is worth noting that feature extraction and classification are closely connected and inseparable. Liu et al. also recommend using a one-dimensional convolutional neural network (1D CNN) to classify the ECG time-series data obtained from ECG waveform images. Zhou et al. reported a PVC detection method based on the recurrent neural network (RNN) [36], which has natural and inherent advantages in processing time-series signals because of its internal memory. Hoang et al. proposed a PVC detection model deployed in wearable devices [37]. The model is based on a CNN and can be scalable from 3-lead to 16-lead ECG systems.

The deep learning-based methods alleviate the limitations of morphology-based methods and have the following three advantages. (1) The deep learning-based methods can use specific network structures to extract features, such as the convolutional kernel. This process does not require human intervention. (2) In extracting features, the deep learning-based methods can continuously optimize features to ensure that the features are practical and non-redundant, such as pooling operation. (3) The deep learning-based methods are less affected by preprocessing, such as detecting and locating the QRS waveform.

However, these existing deep learning-based methods are not without flaws. Most of the features extracted by deep learning algorithms are difficult to understand intuitively. The performance of the deep learning-based methods is slightly inferior to the morphology-based methods, as shown in Table 4. Some deep learning-based methods need to preprocess the ECG. In the literature [36], much preprocessing is required before the model training, such as resampling, signature detection, and normalization. In addition, the research [37] takes 2D time-frequency images obtained by wavelet transform on the ECG as the proposed network's input. No doubt preprocessing increases the computational overhead.

Table 4. Some algorithms for detecting PVC.

Reference	Features	Classifier	Accuracy	Sensitivity	Specificity
[10]	Eight features based on RQA	KNN and PNN	92.25%	73.33%	94.74%
[11]	Template-matching procedures	Threshold method	98.2%	93.1%	81.4%
[12]	12 features based on FT	ANN	98.54%	99.93%	98.3%
[13]	8 generalized wavelets transformed coefficients	FNN	99.8%	99.21%	99.93%
[14]	10 ECG morphological features and one interval feature	MLP	95.4%	-	-
[15]	Wavelet detail coefficients	Threshold method	98.48%	97.21%	98.67%
[16]	Chaotic feature	Threshold method	99.1%	93.6%	-
[17]	R-R interval, pattern of QRS complex, width of QRS complex, and ST-segment level	Main parameters algorithm	-	97.75%	98.8%
[18]	Using the ICA algorithm to extracts features	K-means and fuzzy C-means	80.94%	81.1%	80.1%
[19]	The width and gradient of the QRS wave	SSVM	99.46%	98.94%	99.99%
[20]	R-R interval and QRS width	ANN	96.29%	94.58%	96.59%
[21]	R-peak, R-R, QRS, VAT, Q-peak, and S-peak	ANN	99.41%	96.08%	-
[22]	R-R interval, QS interval, QR amplitude, and RS amplitude	ANN	97.34%	-	-
[23]	Feature extraction of Lyapunov exponent curve	LVQNN	98.9%	90.26%	92.31%
[24]	Using the PCA, SOM, ICA algorithm to extracts features	KNN	99.63%	99.29%	99.89%
[25]	Four morphological characteristics	DHMM	96.59%	97.57%	96.85%
[26]	Feature selection with GA	KNN	99.69%	99.46%	99.91%
[27]	Form factor and R-R interval	SVM	95%	-	-
[28]	A set of geometrical features	SVM	99%	98.5%	99.5%
[29]	80 features based on DFT	BCM	98.3%	100%	
[30]	R-R interval, R amplitude, and QRS area	RF	96.38%	97.88%	97.56%
[31]	Resampled QRS waveform	ANN	95%	-	-
[32]	20-dimensional feature vector obtained by using SAE	ANN	99.4%	97.9%	91.8%
[33]	Learned features automatically	LCNN, LSTM, and rules inference	99.41%	97.59%	99.54%
[34]	Learned features automatically	1D CNN and 2D CNN	88.5%	-	-
[36]	Learned features automatically	RNN	96–99%	99–100%	94–96%
[37]	Learned features automatically	2D CNN	90.84%	78.6%	99.86%

Abbreviations: Recurrence quantification analysis (RQA), Fourier transform (FT), independent component analysis (ICA), principal component analysis (PCA), self-organizing maps (SOM), genetic algorithm (GA), discrete Fourier transform (DFT), sparse autoencoder SAEK-nearest neighbor (KNN), probabilistic neural network (PNN), artificial neural networks (ANN), fuzzy neural network (FNN), multilayer perceptron (MLP), support vector machine (SVM), swarm-based support vector machine (SSVM), learning vector quantization neural network (LVQNN), discrete hidden Markov model (DHMM), Bayesian classification models (BCM), random forest (RF), lead convolutional neural network (LCNN), long short-term memory network (LSTM), one-dimensional convolutional neural network (1D CNN), two-dimensional convolutional neural network (2D CNN), recurrent neural network (RNN). Further, "-" means that relevant information is not mentioned in the literature.

In summary, we can quickly draw the following conclusions according to the above discussion and Table 4. (1) Most of the methods mentioned in the literature are based on morphology. Table 4 lists 27 references, of which 22 belong to the morphology-based method, and only five belong to the deep learning-based method. (2) Most researchers prefer to use ANN, KNN, and SVM to identify PVC after completing the feature extraction. Six pieces of literature in Table 4 use ANN as a classifier. (3) The R-R interval is an excellent feature, which has been recognized by the majority of researchers. Nearly one-third of morphology-based methods have used this feature. (4) In terms of accuracy, sensitivity, and specificity, these three classifiers, FNN, BCM, and SSVM, achieved the best results, respectively. Overall, the morphology-based method's performances were slightly better than deep learning, due to the expert's knowledge and experience.

Consider the following: On the one hand, it is easy to understand the features extracted by the morphology-based methods, but feature engineering is the most significant limitation of this method; on the other hand, it is very difficult or even impossible to understand intuitively the features extracted by the deep learning-based methods, but deep learning algorithms can automatically extract and optimize features. This research proposed a novel approach based on deep metric learning and KNN to ensure that the features used to detect PVC can be extracted automatically and understood intuitively.

Specifically, the proposed method introduced deep metric learning into PVC inspection projects for the first time. It is worth mentioning that deep metric learning can automatically extract features, and these features are usually in the high-dimensional embedding space. In this case, the KNN classifier is undoubtedly an optimal choice. Second, the proposed method did not rely on expert knowledge and experience related to ECG, significantly reducing the threshold for studying physiological signals. In theory, the proposed method is suitable for the most physiological signals. Third, to improve the efficiency of detecting PVC from long-term ECG, this method can directly classify heartbeats. Preprocessing, such as denoising, is unnecessary. Finally, clinical ECG from the MIT-BIH (Massachusetts Institute of Technology-Beth Israel Hospital) Arrhythmia Database [38,39] evaluated and verified the proposed method's performance and effectiveness. The following is the remainder's arrangement: Section 2 describes the dataset, proposed framework, and evaluation measures; Section 3 presents and discusses the results; Section 4 gives the conclusion and directions.

2. Materials and Methods

2.1. Materials

In this paper, all ECG came from the MIT-BIH Arrhythmia Database, which plays an essential role as a referee in verifying arrhythmia detectors. The MIT-BIH Arrhythmia Database was first publicly released in 1980 and has been updated three times in 1988, 1992, and 1997. Its public release is a landmark event. Nearly one hundred research groups worldwide have used the MIT-BIH Arrhythmia Database in the eight years from the first release. Today, many academic and industrial researchers have affirmed the effectiveness of this database. Specifically, the MIT-BIH Arrhythmia Database contains 48 long-term Holter recordings obtained from 47 subjects: 25 men and 22 women. Every record is numbered from 100 to 234, with some numbers missing. Only records 201 and 202 are from the same male subject, and the remaining records corresponded to the other subjects one by one. Furthermore, each record contains two signals with a sampling rate of 360 Hz and a sampling duration of slightly over half an hour.

In most records, the first signal is a modified limb lead II (MLII), and the second signal is usually a modified lead V1 (occasionally V2, V5, and V4). It is worth noting that at least two cardiologists independently annotate all signals in this database. Undoubtedly, free access to a large number of ECGs and beat-by-beat annotations through the internet at any time and anywhere has improved the efficiency of the development of arrhythmia detectors, which has been beneficial to numerous researchers. The ECGs used in this study were from the MLII, which appeared in almost all records. Considering the suggestion proposed by the Association for the Advancement of Medical Instrumentation (AAMI), this study discarded records 102, 104, 107, and 217 because of the paced beats. Furthermore, this research divided ECGs in the MIT-BIH Arrhythmia Database into the training set and test set, as shown in Table 5.

Table 5. Dividing ECG into a training set and test set.

Dataset	Records	Normal Heartbeat	PVC
Training set	101, 106, 108, 109, 112, 114, 115, 116, 118, 119, 122, 124, 201, 203, 205, 207, 208, 209, 215, 220, 223, 230	35,640	2851
Test set	100, 103, 105, 111, 113, 117, 121, 123, 200, 202, 210, 212, 213, 214, 219, 221, 222, 228, 231, 232, 233, 234	33,868	2548

In this table, "Records" represents ECG recordings in the training set or test set. The "Normal heartbeat" and "PVC" represent the numbers of regular heartbeats and PVC in the training set or test set.

Notably, many datasets have adopted cross-validation to divide the training set and test set. However, applying cross-validation is unreasonable and may cause label leakage in this experiment. The reason is that the heartbeat of the subjects in the resting state hardly changes during a period of time. A reasonable division method should ensure that the same person's ECG can not appear in both the training and test sets. Therefore, like most other studies, this study adopted the division method shown in Table 5, ensuring a reasonable comparison.

2.2. Methodology

Figure 3 shows the proposed method's flow, namely, ECG collection, signal preprocessing, feature extraction, and classification. First, collecting long-term ECG is inseparable from wearable devices, such as Holter. Secondly, the proposed method extracted the single heartbeat from the MLII using a fixed time window and the R-peak detection algorithm. Then, the deep metric learning model could extract features of the heartbeat automatically. Finally, the KNN classifier predicted the category of the heartbeats based on the distance between the heartbeats. Since this research focused more on signal processing and analysis, the long-term ECGs and annotations came from the MIT-BIH Arrhythmia Database.

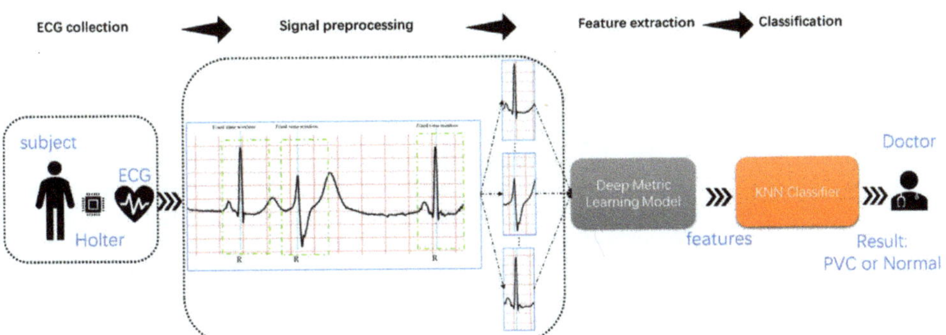

Figure 3. Block diagram of the proposed study.

2.2.1. Signal Preprocessing

Since the long-term ECG collected from the wearable device contained some noise, most existing research literature would use software algorithms to remove noise and baseline wander, such as the bandpass filter and wavelet transform. Considering that denoising increases the system's computational load, the deep metric learning model can automatically extract features indicating the difference between the normal heartbeats and

PVC heartbeats. Therefore, this study did not perform any operations related to denoising the signal but only segmenting the ECG.

The segmentation of ECG involves R-peak detection and a fixed time window. Specifically, the proposed method first applies the R-peak detection algorithm to locate the R-peak on the ECG. Because the existing R-peak detection algorithm [40–43] performs very well in accuracy and real-time, for example, Pan et al. designed an algorithm that can correctly detect 99.3% of the R-peak for the MIT-BIH Arrhythmia Database. This study directly used the MIT-BIH Arrhythmia Database's R-peak position.

Moreover, sliding a fixed time window on the ECG is a simple and straightforward way to obtain the same size's heartbeats. In this research, the window's length was 433. Each sliding should make the window's vertical centerline coincide with each heartbeat's R-peak. After these two steps, we could extract the normal heartbeats and PVCs from the ECG in each record.

2.2.2. Feature Extraction

Feature extraction is an essential step for the development of PVC detectors. It is no exaggeration to say that the feature extraction defines the upper limit of the PVC detector. The classifier bounds how close the PVC detector is to its upper limit. For existing morphology-based methods, feature extraction is a complicated process. It relies heavily on feature designers' knowledge and experience and reduces the efficiency of developing PVC detectors, because a set of excellent and efficient features often requires many researchers' concerted efforts and a large number of experiments.

Although deep learning-based methods can automatically extract features and avoid these limitations, the features, extracted through the classic network structures and optimization algorithms, are difficult to understand intuitively in these deep learning-based methods. Moreover, according to the existing literature, the deep learning-based methods' overall performance is not significantly better than the morphology-based methods. It is particularly noteworthy that most of the methods suggested in the current literature have inadvertently ignored a severe issue that the number of normal heartbeats is much greater than PVC heartbeats in the MIT-BIH Arrhythmia Database.

Fortunately, the metric learning model can entirely solve the above problems. Metric learning is a type of mechanism to combine features to compare observations effectively. There are many types of metric learning models, such as stochastic neighbor embedding (SNE) [44], locally linear embeddings (LLE) [45], mahalanobis metric for clustering (MMC) [46], and neighborhood component analysis (NCA) [47]. The first two are unsupervised, and the latter two are supervised. Specifically, the metric learning model predicts the samples' categories by measuring the similarity among samples [48]. Moreover, the model's core is to establish a mapping function to represent the optimal distance metric.

Distinguishing features makes the classifier perform better. Metric learning is very good at extracting distinguishing features. Metric learning aims to make objects with the same label behave closer in the feature while increasing the distance between objects with different labels. To deal with various classification or clustering problems, we can select appropriate features through prior knowledge and experience on specific tasks. However, this method is very time-consuming, labor-intensive, and may also be unrobust to data changes. As an ideal alternative, metric learning can independently learn the metric distance function for a specific task according to different studies.

Due to deep learning technology and activation functions, deep metric learning, as a combination of deep learning and metric learning, has provided excellent solutions in many classification tasks and attracted researchers' attention in academia and industry. In the Humpback Whale Identification competition held on the Kaggle platform, which is the world's largest data science community [49], the top five participating teams' solutions all applied deep metric learning models: Triplet neural network [50] and siamese neural network [51]. The most conspicuous characteristic of these networks is the sharing weights, which makes the samples related because the triplet neural network can simultaneously learn both positive and negative distances and the number of training data combinations increases significantly to avoid overfitting. This study intended to use the triplet neural network as the deep metric learning model's basic architecture, as shown in Figure 4.

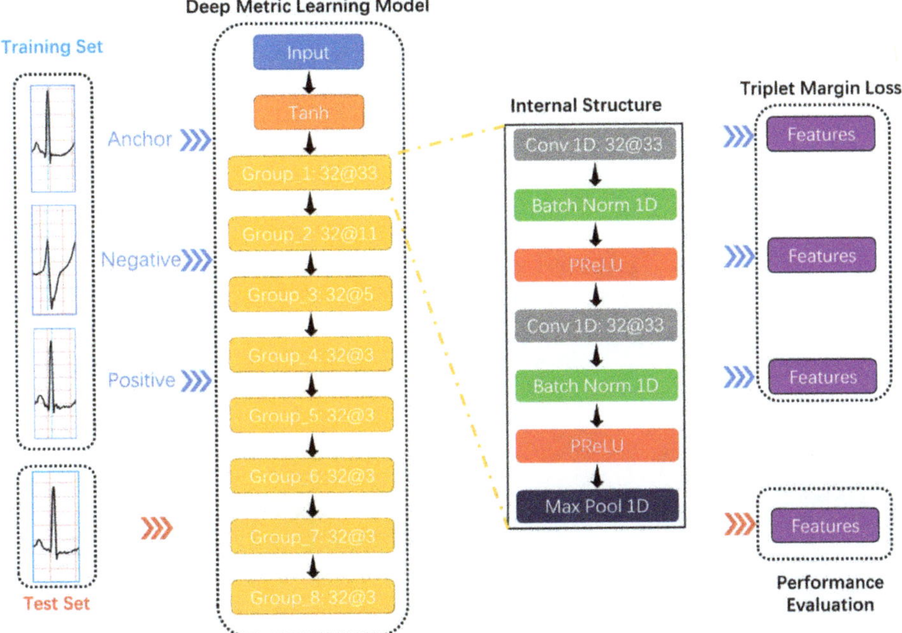

Figure 4. The proposed deep metric learning model's basic architecture. Take "Group_1 32@33" as an example to comprehend the convolution group. "Group _1" is the convolution group's name; "32@33" represents the number and size of the one-dimensional convolutional layer's convolution kernels in the convolution group. Each convolutional group contains two 1D convolutional layers, two batch normalization layers, two activation functions, and one max-pooling layer.

Considering that the R wave peak is much larger than other points in the whole heartbeat, normalizing the heartbeat was beneficial to the deep metric learning model's training. The Tanh function can normalize the input data between −1 and 1. Further, the Tanh function has little effect on real numbers close to 0 and has a more significant impact on real numbers far away from 0, especially these real numbers greater than one or less than −1. Equations (1) and (2) are the definitions of the Tanh function and its derivatives, respectively.

$$\text{Tanh}(x) \; = \; \frac{Sinh(x)}{Cosh(x)} \; = \; \frac{e^x - e^{-x}}{e^x + e^{-x}} \tag{1}$$

$$\frac{dTanh(x)}{dx} \; = \; sech^2 x = 1 - Tanh^2 x \tag{2}$$

Secondly, the proposed deep metric learning model had eight convolutional groups that resulted in a feature vector representing a detected feature's positions and intensity in the input data, as shown in Figure 4. Each convolutional group contained two 1D convolutional layers, two batch normalization layers, two activation functions, and one max-pooling layer.

The 1D convolutional layer was the necessary component of automatic feature extraction. The purpose of the convolution operation was to extract different features of the input of this layer. In the entire network, the first few convolutional layers can usually only extract some low-level features. In contrast, the last layers can iteratively extract more complex features from the low-level features. The calculation of convolution was not complicated. The generated sequence could be obtained by repeating the following process: Move the convolution kernel in fixed steps along the input vector and calculate the dot product of the horizontally flipped convolution kernel and the input vector. The convolution definition is expressed as Equation (3), where x, h, y, respectively, represent the input vector, convolution kernel, and generated sequence.

$$y_j = \sum_{i=-\infty}^{\infty} x_i \times h_{j-i} \tag{3}$$

Adding the batch normalization layer to the proposed deep metric learning model could improve the training efficiency by normalizing the convolutional layer's feature map. When training the model, the batch normalization layer would sequentially perform the following operations [52]:

1. Calculate the mean and variance of the input vector;

$$Batch\ mean \quad \mu_B = \frac{1}{m}\sum_{i=1}^{m} x_i \tag{4}$$

$$Batch\ variance \quad \sigma_B^2 = \frac{1}{m}\sum_{i=1}^{m} (x_i - \mu_B)^2 \tag{5}$$

2. Normalize the input using the mean and variance;

$$\overline{x_i} = \frac{x_i - \mu_B}{\sqrt{\sigma_B^2 + \epsilon}} \tag{6}$$

3. Attain the output with scaling and shifting;

$$y_i = \gamma\overline{x_i} + \beta \tag{7}$$

In the Equations (4) and (7), m and ϵ, respectively, represent the number of samples per batch and a small constant for numerical stability. Further, γ and β are learnable parameters.

The rectified linear unit (ReLU) dramatically promoted the development of deep learning. Its use provided a better solution than that of the sigmoid function. The parametric rectified linear unit (PReLU) has improved ReLU and become the default activation function in many classification tasks [53]. Although PReLU introduces slope parameters, PReLU can better adapt to the other parameters like weights, and the increase in training costs is negligible. The mathematical definition of PReLU is Equation (8), where y_i and a_i, respectively, represent the input on channel i and the negative slope which is a learnable parameter.

$$f(y_i) = \max(0,\ y_i) + a_i \times \min(0, y_i) \tag{8}$$

Adding the pooling layer to the proposed deep metric learning model could reduce the computational cost and effectively cope with the over-fitting by down-sampling and summarizing in the feature map. In addition, the pooling layer made the feature position change more robust, referred to by the "local translation invariance." Three types of pooling operations have been widely used: Max-pooling, min-pooling, and average-pooling, as described in Table 6. However, the simultaneous use of min-pooling and PReLU would make each layer's output results in the model almost all 0. Considering that the R wave waveform is sharp and high in a complete heartbeat, the max-pooling operation was applied in this study's pooling layer.

Table 6. Three types of pooling operations.

Type	Operation
Max-pooling	The maximum pixel value of the batch is selected
Min-pooling	The minimum pixel value of the batch is selected
Average-pooling	The average value of all the pixels in the batch is selected

Here, the "batch" means a group of features that are the overlapping parts of these two vectors: The pooling layer's kernel and the input vector.

Thirdly, training neural networks are inseparable from the loss function. The loss function can evaluate neural networks' performance and play an essential part during training. The triplet margin loss [54] is used for measuring a relative similarity between samples. In this study, the triplet margin loss based on the cosine similarity calculated the model error required in an optimization process used to train the proposed deep metric learning model. Furthermore, the loss function for each sample in the mini-batch is:

$$L(a, p, n) = \max\{d(a_i, p_i) - d(a_i, n_i) + margin,\ 0\} \tag{9}$$

where

$$d\left(\vec{x}, \vec{y}\right) = \vec{x} \cdot \vec{y} = \left|\vec{x}\right|\left|\vec{y}\right| \cos\theta \tag{10}$$

The anchor, positive example, and negative example were three feature vectors and composed a triplet. Further, to make the model's training process faster and more stable, applying the miner based on multi-similarity [55] could generate more valuable triplets. The multi-similarity contained three similarities in the general pair weighting (GPW) framework: Self-similarity, negative relative similarity, and positive relative similarity. In this study, the miner based on multi-similarity implemented the following process: Select a negative pair for the anchor if its similarity satisfies Equation (11); select a positive pair for the same anchor if its similarity satisfies Equation (12). Repeat the above steps with the feature vector obtained from each heartbeat as an anchor to obtain the index sets of its selected positive and negative pairs. These index sets are the basis of triples.

$$S_{ij}^- > \min_{y_k = y_j} S_{ik} - \epsilon \tag{11}$$

$$S_{ij}^+ < \max_{y_k \neq y_j} S_{ik} + \epsilon \tag{12}$$

$$S_{ij} = f(x_i; \theta) \cdot f(x_j; \theta) \tag{13}$$

In Equations (11)–(13), assume x_i is an anchor, y_i is the corresponding label, f is a neural network parameterized by θ, and $\cdot$ denotes the dot product, where S_{ij} and ϵ, respectively, represent the similarity of two samples and a given margin.

2.2.3. Classification

The classifier is the last link of the method proposed in this article and directly determines the classification system's performance. In other research projects, the choice of classifier often depends on the results of multiple experiments. In other words, choosing a classifier requires many repeated experiments and costs much time. Many researchers often do experiments on several commonly used classifiers, such as SVM and ANN. Further, there is no reliable theoretical basis or clear direction to determine which type of classifier to use in most cases. Even if the researcher has determined which specific classifier to use, it is a huge challenge to adjust this classifier's parameters.

However, in this article, since the features extracted by the deep metric model contain distance information, the KNN classification algorithm was the most suitable classifier. KNN classification algorithm is a type of non-generalizing learning. Unlike other classifiers that try to train a general model, the KNN classifier focuses on the distance. Moreover, the classification basis of the KNN is intuitive. The KNN classifier has only one parameter to control the number of votes, called K. The KNN classification algorithm first calculates the distance between the test data and each training data. If K is 1, the training data label with the closest distance is regarded as the predicted label. If K is greater than 1, the KNN classification algorithm votes according to the the top K training data labels with the smallest distance and finally determines the predicted label.

2.3. Evaluation Measures

The confusion matrix is a standard format for evaluating classification performance, and it usually appears in the form of a matrix. In most classification tasks, the confusion matrix summarizes the number of correctly and incorrectly predicted samples and those broken down by each class, providing researchers with a global perspective to comprehensively and efficiently evaluate the classifier's performance, especially in imbalanced datasets.

This study used the confusion matrix to measure the recognition performance of the proposed method. Further, this study used five evaluation indicators: Accuracy (ACC), sensitivity (Se), specificity (Sp), positive prediction (P_+), and negative prediction (P_-), based on the confusion matrix to compare more conveniently with experimental results in other literature. The confusion matrix and other five indicators, which also have been used in the literature [28], can be expressed as Equations (14)–(19). TN, FN, TP, and FP represent true negatives, false negatives, true positives, and false positives.

$$\text{Confusion Matrix} = \begin{bmatrix} \text{TN} & \text{FP} \\ \text{FN} & \text{TP} \end{bmatrix} \tag{14}$$

$$\text{Accuracy Acc} = \frac{\text{TP} + \text{TN}}{\text{TP} + \text{TN} + \text{FN} + \text{FP}} \tag{15}$$

$$\text{Sensitivity Se} = \frac{\text{TP}}{\text{TP} + \text{FN}} \tag{16}$$

$$\text{Specificity Sp} = \frac{\text{TN}}{\text{TN} + \text{FP}} \tag{17}$$

$$\text{Positive prediction } P_+ = \frac{\text{TP}}{\text{TP} + \text{FP}} \tag{18}$$

$$\text{Negative prediction } P_- = \frac{\text{TN}}{\text{TN} + \text{FN}} \tag{19}$$

3. Results and Discussion

In this study, the main factors affecting the proposed system's performance were as follows: The denoising method, the number of features, type of pooling layer, the loss function configuration, and type of classifier. First, denoising is a double-edged sword in the signal preprocessing stage. Denoising can improve the signal's quality, reducing the difficulty of training a deep metric learning model. However, the signal may also lose some valuable information because of denoising.

Second, as a bridge between the deep metric learning model and classifier, the number of features is an essential hyper-parameter. This value cannot be too large or too small. The greater the number of features, the easier the features become redundant. Conversely, if there are too few features, the less information the features contain cause the classifier's performance to deteriorate. Third, the type of pooling layer determines how features are summarized and retained and has the effect of de-redundancy. A proper pooling layer can select the most practical features to speed up the deep metric learning model's training speed.

Fourth, the loss function configuration is the top priority of training the deep metric learning model. The loss function and the miner based on multi-similarity cooperated in the proposed system. In the loss function, the margin should be within a reasonable range. The larger the margin, the more valuable the feature, but the harder it is to train the deep metric learning model. Conversely, the smaller the margin, the easier it is to train the model, but the less practical the features. Finally, the KNN classifier is hugely suitable for processing the deep metric learning model's features. However, the choice of K value is highly dependent on the distribution of features.

In this section, this study strictly divided the training set and the test set according to Table 5 and used them in each experiment. Before anything else, we evaluated the necessity of signal denoising. Secondly, we assessed the impact of the number of features on the proposed model. Immediately afterward, we tested pooling layers' influence on the feature extraction of deep metric learning models. To improve the proposed system's performance, we have adjusted the loss function and the miner parameters many times. Subsequently, we checked the performance of the KNN classifier and further optimized the classifier. Finally, we compared the proposed method with other research literature on multiple evaluation indicators, such as accuracy, sensitivity, and specificity. We carried out the simulation process on a Linux server with an Nvidia GeForce RTX 2070 GPU.

3.1. Experiment 1: Evaluation of the Necessity for Signal Denoising

In collecting ECG, wearable devices also collect noises. These noises can affect the quality of the signal and even distort the signal. The analog-to-digital conversion chip is a critical hardware component in wearable devices, directly determining the signal quality. Therefore, in the signal acquisition phase, researchers usually improve the sensor's hardware equipment to suppress noise as much as possible. On the other hand, most scholars use software algorithms in the signal preprocessing stage to remove noise further. However, it is worth mentioning that the noise reduction algorithm inevitably changes the signal more or less. For the metric learning model used in this paper, the convolutional layer can automatically extract useful features and ignore useless information, such as noise. Therefore, the necessity of denoising the signal in the preprocessing stage is worth exploring.

Considering that the data used in this article were all from the MIT-BIH Arrhythmia Database and the method proposed in this paper focused on signal analysis, the denoising methods only involve software algorithms in the signal preprocessing. Expressly, we set up a set of comparative experiments to evaluate the necessity of signal denoising. This comparative experiment first processes and classifies the ECG directly according to the method proposed in this article, without applying any denoising means. Secondly, based on the first experiment, we only added some denoising algorithms in the signal preprocessing stage.

These denoising algorithms include two finite impulse response (FIR) filters with a sampling rate of 1000 Hz and two median filters. Figure 5 shows the denoising effect of the ECG. The former can filter 60 Hz power-line noise and 100 Hz interference from fluorescence lights, and the latter can remove the baseline of the signal and some noise. It is worth noting that the sizes of these two median filters window are 71 and 215, respectively, which is the same as the setting in literature [56]. Tables 7 and 8 record the parameters and results of the comparative experiment in detail. In Table 7, the LR, WD, and K refer to the learning rate, weight decay, and the KNN classifier's parameter.

Table 7. The parameters related to the experiment.

Batch Size	Number of Features	Margin	Distance	Epsilon	Optimizer	LR	WD	K
32	32	0.2	Cosine Similarity	0	Adam	0.0001	0	1

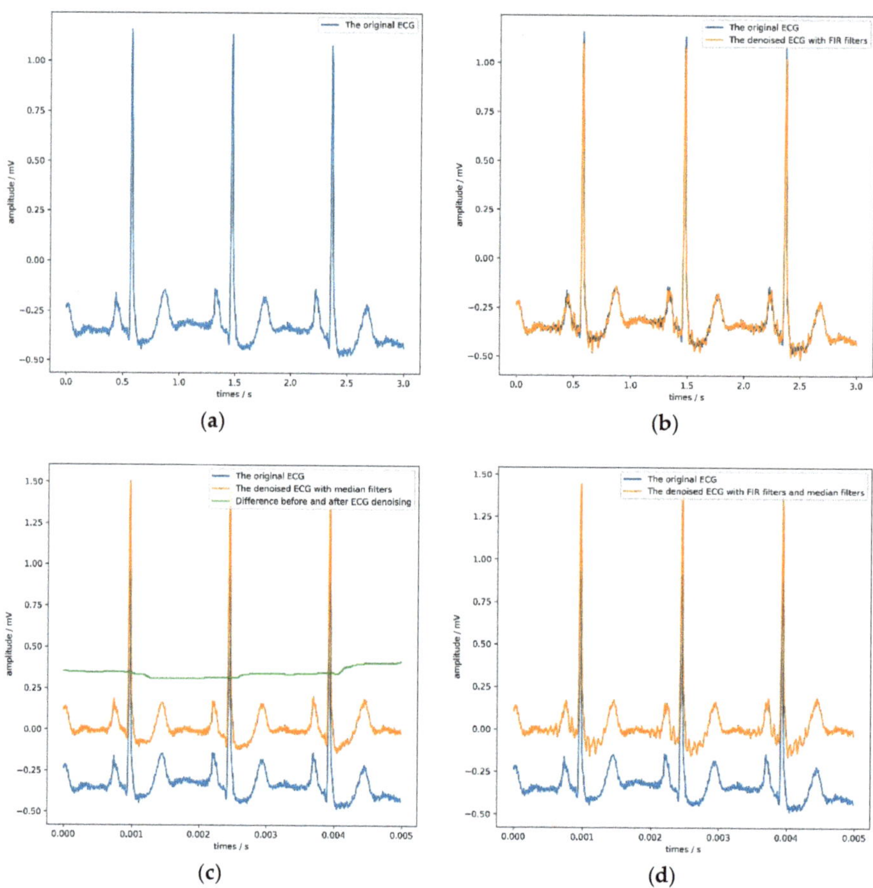

Figure 5. The result of applying different denoising algorithms on the ECG. (**a**) shows a 3-s ECG without denoising; (**b**) and (**c**) illustrate the effect of using finite impulse response (FIR) filters and median filters on the ECG, respectively; (**d**) shows the impact of using FIR filters and median filters on the ECG.

Table 8. The performance of applying different noise reduction algorithms on the proposed method.

Noise Reduction Algorithms	Acc (%)	Se (%)	Sp (%)	P_+ (%)	P_- (%)	Time
None	99.63	96.74	99.85	97.97	99.76	0.00
FIR filters	99.56	96.66	99.78	97.04	99.75	0.23
Median filters	99.53	96.9	99.73	96.37	99.77	6.58
FIR filters and median filters	99.46	96.15	99.71	96.12	99.71	7.04

Here, the "Time" means the time it takes to denoise a half-hour ECG.

It is not difficult to find from Table 8 that both the FIR filter and the median filter interfered with the model's judgment to a certain extent, especially when applying both filters at the same time. Adding FIR filters and median filters in the signal preprocessing stage reduces each evaluation index of the model. The median filter can maximize the model's sensitivity, but the model's accuracy would drop slightly. According to the model's overall performance, the most appropriate choice was not to use FIR filters or median filters. Figure 5 directly confirms this conclusion.

By observing the four sub-pictures in Figure 5, we can quickly and intuitively discover two phenomena. First of all, the FIR filters could filter out specific frequency components but make the ECG show more obvious glitches simultaneously, which would be counterproductive. Second, the median filters could effectively remove the baseline but slightly change the ECG's contour, which would be hidden danger for the model's judgment.

According to Table 8 another thing worth noting is that the time required to process a half-hour-long ECG using the FIR filter and the median filter was 0.23 and 6.58 s, respectively, in this experiment. If this experiment used the computer hardware equipment with a lower frequency, the time spent on noise removal would become longer.

Considering the actual situation and experimental results, the method proposed in this paper had a particular anti-noise ability. Therefore, denoising was an option in this experiment's signal preprocessing stage, though not a necessary option. Since this article focused on the classification of electrocardiograms, no more detailed research was done on noise reduction methods.

3.2. Experiment 2: The Choice of the Number of Features

There is no doubt that features are essential and directly determine the performance of the classifier. In theory, practical features should be informative, differentiated, and independent. The deep metric learning model can automatically extract features. In the process of producing high-quality features, the number of features is a critical parameter.

Suppose the number of features is too small. In that case, the deep metric learning model's training process would be challenging. The acquired features are indistinguishable, and the information contained in the features is not enough to smoothly train the classifier.

On the contrary, too many features are redundant and increase the deep metric learning model's training time. Further, the excessive features have the following shortcomings for the classifier: Expanding the classifier's complexity, causing the dimensional disaster, and resulting in ill-posed problems and sparse features problems. These disadvantages eventually lead to a decline in the performance of the classifier.

Considering the above points, we conducted a series of experiments to find the appropriate number of features. We set different values for the number of kernels in the last convolutional layer to adjust the features. Table 9 provides the detailed results. Moreover, this experiment also adopted the basic configuration in Table 7.

Table 9. The results of the varying number of features.

The Number of Features	TN	FP	FN	TP	Acc (%)	Se (%)	Sp (%)	P_+ (%)	P_- (%)
2	33,808	60	91	2457	99.59	96.43	99.82	97.62	99.73
8	33,800	68	89	2459	99.57	96.51	99.8	97.31	99.74
32	33,817	51	83	2465	99.63	96.74	99.85	97.97	99.76
64	33,802	66	84	2464	99.59	96.7	99.81	97.39	99.75

According to the experimental results in Table 9, we found that the features extracted by the deep metric learning model could make the KNN classifier perform best when the number of features was 32. Further, the number of features and sensitivity were positively correlated. In other words, the more features, the more confident the proposed system was in PVC predictions. To better analyze these results, we used *t*-distributed stochastic neighbor embedding (t-SNE) [57] to reduce the features' dimension and then visualize the features in Figure 6. The t-SNE is a machine learning algorithm for dimension reduction, which is very suitable for reducing high-dimensional data to 2 or 3 dimensions for visualization.

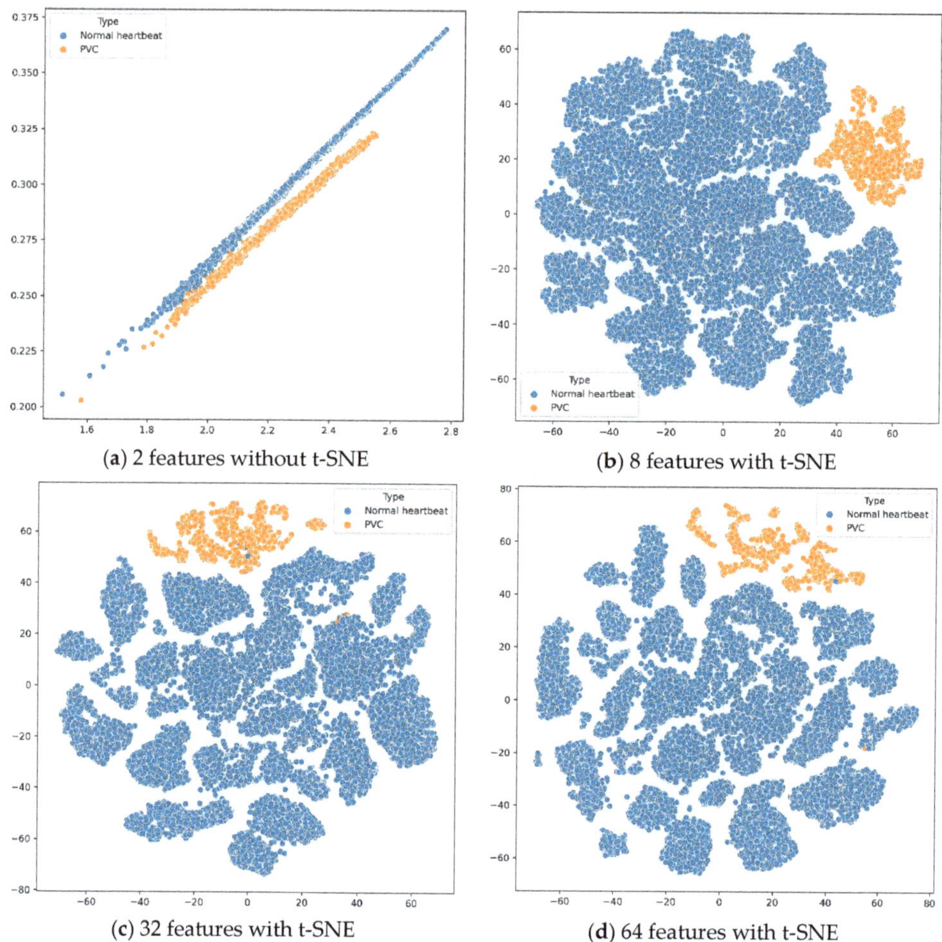

(a) 2 features without t-SNE

(b) 8 features with t-SNE

(c) 32 features with t-SNE

(d) 64 features with t-SNE

Figure 6. Visualizing the features of training data.

Suppose we used the deep metric learning model to extract only two features. In that case, we could directly draw the features in a two-dimensional coordinate system without dimension reduction by t-SNE. When the number of features was greater than 2, we would use the t-SNE algorithm to reduce the features' dimensions and display them on a two-dimensional plane. The four sub-images in Figure 6 show the distribution of different quantity features, extracted from all training data through the deep metric learning model, on a two-dimensional plane.

First, as shown in subfigure (a), the normal heartbeats and PVC were distributed on two parallel straight lines. However, when the first feature was around 2.1 and the second feature was around 0.26, the boundary between the normal heartbeat and PVC was not stark. Secondly, the other three subfigures showed that these features had obvious boundaries on the two-dimensional plane, distinguishing between the normal heartbeats and PVC. Finally, although the results in Table 9 are not much different, it is better to use the deep metric learning model to extract 32 features after comparing evaluation indicators such as accuracy and sensitivity.

3.3. Experiment 3: Assess the Impact of Pooling Type

In the CNN architecture, most researchers tend to insert a pooling layer in-between consecutive convolutional layers periodically. On the one hand, the pooling layer reduces the number of parameters to learn, avoiding over-fitting, and accelerating the deep metric learning model's training speed. On the other hand, unlike the convolutional layer that extracts features with precise positioning, the pooling layer summarizes the features generated by a convolution layer, making the deep metric learning model more robust to variations in the position of the features in the input ECG. In other words, the pooling layer has a natural advantage in analyzing heartbeats of different cycles, even if these heartbeats come from different people.

Generally speaking, the core of the pooling layer is a fixed-shaped window. According to a set stride, this window slid overall feature regions and computed a single output for each location. It is worth noting that the way the pooling layer computes the output has no kernel and is deterministic, typically based on the maximum or average value of the features in the pooling window.

Specifically, the output after the max-pooling layer would contain the previous feature map's most prominent features, which guarantees that each feature used to transmit to the next layer is practical. The average-pooling gives the average of features, taking into account global features in the pooling window. Therefore, in this experiment, we tested these two pooling layers' performances in feature extraction with the configuration in Table 7. Figure 7 shows the results of this experiment in the form of a confusion matrix. Table 10 illustrates the detailed results in each evaluation index.

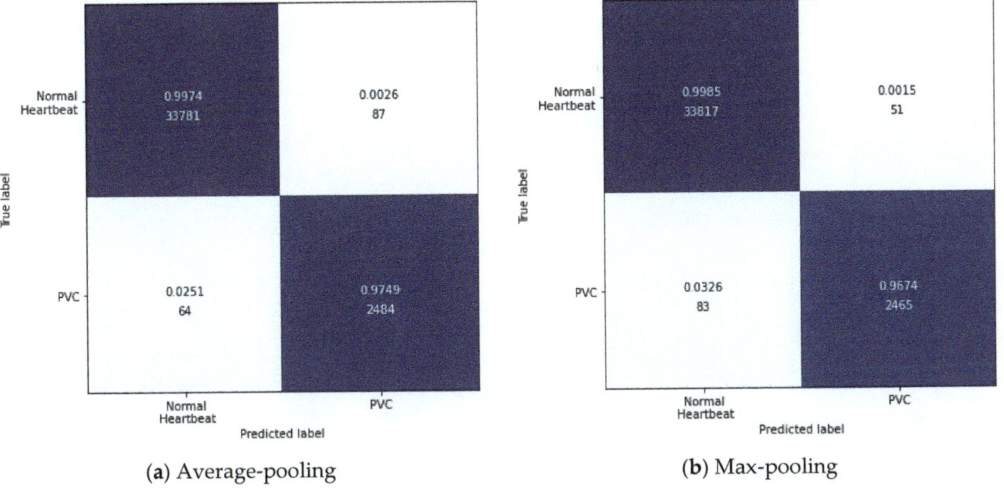

(a) Average-pooling   (b) Max-pooling

Figure 7. The confusion matrix about testing the pooling layer.

Table 10. The detailed results of testing the pooling layer.

Pooling Type	Acc (%)	Se (%)	Sp (%)	P_+ (%)	P_- (%)
Max-pooling	99.63	96.74	99.85	97.97	99.76
Average-pooling	99.59	97.49	99.74	96.62	99.81

According to Figure 7, it can be found intuitively that the deep metric learning model with the max-pooling layer misjudged 134 test data, 17 fewer than the model with the average-pooling layer. Although the two models' performances were similar, the max-pooling layer model was better at predicting normal heartbeats. The model with the average-pooling layer was more confident in predicting PVC, as shown in Table 10.

In the proposed method, feature extraction's error mainly came from two aspects: The pooling window size and the feature shift caused by convolutional layer parameters. Generally speaking, the average-pooling operation could reduce the former error to preserve more information in the pooling window. The max-pooling operation can reduce the latter error to focus on the highest intensity information.

Since the loss function was based on cosine similarity, the desired model used to extract features should make the cosine similarity between samples of different classes as small as possible. Suppose the number of features was 2. The PVC and normal heartbeat features should be as close as possible to the two coordinate axes, respectively, in a two-dimensional coordinate system. Under careful consideration, the max-pooling layer was better than the average-pooling layer.

3.4. Experiment 4: Configure the Parameters of the Loss Function and Miner

In the triplet margin loss, the margin is an indispensable parameter that directly affects training the deep metric learning model. The definition of margin is the desired difference between the anchor-positive distance and the anchor-negative distance. Generally speaking, the larger the margin, the higher the quality of the extracted features. However, a large margin makes the model's training process very unstable, and the loss makes it challenging to approach zero.

Secondly, in this paper, when we trained the model using the triples format's training data, there were countless triples. However, since some triples met the margin requirements in the loss function, these triples did not contribute to the training model. There is no doubt that blindly and directly using all triples is time-consuming and inefficient for training models.

Fortunately, the miner based on multi-similarity can solve this problem. In this miner, epsilon is an important parameter that controls which triples are selected to train the model. Generally speaking, the larger the epsilon, the more triples are involved in training the model. To maximize the deep metric learning model's performance, we conducted a series of experiments on margin and epsilon values with the configuration in Table 7. Table 11 lists the results for different values of margin and epsilon.

Table 11. The experiment results about the margin and epsilon.

Margin	Epsilon	TN	FP	FN	TP	Acc (%)	Se (%)	Sp (%)	P_+ (%)	P_- (%)
0.1	0.0	33,824	44	65	2483	99.70	97.45	99.87	98.26	99.81
0.2	0.0	33,817	51	83	2465	99.63	96.74	99.85	97.97	99.76
0.4	0.0	33,812	56	69	2479	99.66	97.29	99.83	97.79	99.80
0.8	0.0	33,786	82	49	2499	99.64	98.08	99.76	96.82	99.86
0.1	0.1	33,808	60	78	2470	99.62	96.94	99.82	97.63	99.77
0.1	0.2	33,787	81	64	2484	99.60	97.49	99.76	96.84	99.81
0.1	0.3	33,795	73	70	2478	99.61	97.25	99.78	97.14	99.79

First of all, Table 11 shows that specificity and margin are negatively correlated, provide epsilon is 0. When the margins were 0.2, 0.4, 0.8, the proposed PVC detection system reached an accuracy of about 99.64% in these three experiments. However, when the margin was 0.1, the proposed PVC detection system performed best in the following indicators: Accuracy, specificity, and positive prediction. Secondly, increasing epsilon made the system's overall performance worse, especially accuracy and positive prediction.

For the same batch of training data, the greater the margin, the greater the loss. In the case of a fixed learning rate, an enormous loss makes it difficult for the optimizer to find the best point, which leads to a decline in the quality of the extracted features. On the other hand, epsilon determines the number of triples involved in training. The larger the epsilon, the greater the number of triples in the same batch of training data, which undoubtedly increases the computational load. Furthermore, although the larger epsilon increases the number of triples, most of the triples can only produce a minimal loss, which leads to a reduction in the batch loss. A small loss may cause the optimizer to fall into a local optimum. Therefore, according to the experimental results, it is suitable to set the margin and epsilon to 0.1 and 0, respectively.

3.5. Experiment 5: Optimization of KNN Classifier and Comparison with Other Literature

In this article, the KNN classifier is suitable thanks to the spatiality of the features extracted by the deep metric learning model. Nevertheless, the performance of the KNN classifier is very dependent on the K value. A small K value is likely to cause overfitting, while an immense K value is likely to overlook some useful information in the training data. Therefore, it is necessary to test the K value. Table 12 lists the performance of the KNN classifier under different K values.

Table 12. The performance of the KNN classifier with different K values.

K	TN	FP	FN	TP	Acc (%)	Se (%)	Sp (%)	P_+ (%)	P_- (%)
1	33,824	44	65	2483	99.7007	97.449	99.8701	98.2588	99.8082
3	33,824	44	66	2482	99.6979	97.4097	99.8701	98.2581	99.8053
5	33,825	43	68	2480	99.6952	97.3312	99.873	98.2957	99.7994
9	33,822	46	69	2479	99.6842	97.292	99.8642	98.1782	99.7964
11	33,822	46	70	2478	99.6815	97.2527	99.8642	98.1775	99.7935

Overall, the best value of K was 1, which made the classifier obtain the highest accuracy. Secondly, as the K value continued to increase, the number of misjudgments by the KNN classifier for PVC was rising since the number of normal heartbeats was much larger than that of PVC. Finally, all the experimental results in Table 12 confirmed the effectiveness of the PVC detection method proposed in this article. Finally, we compared the proposed method with other literature, as shown in Figure 8.

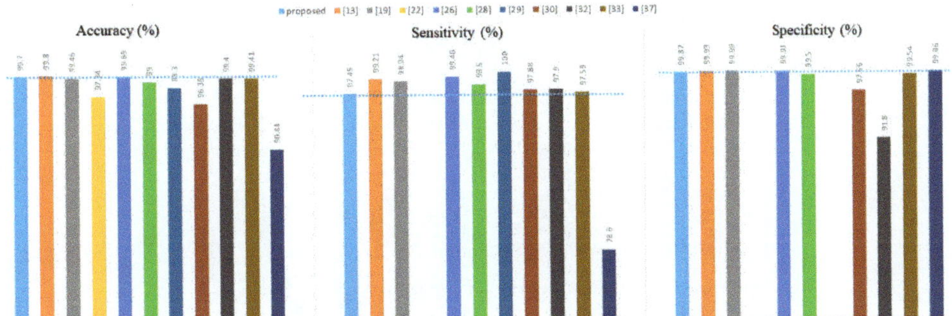

Figure 8. Comparison with other literature.

As a whole, the proposed method was not superior in terms of accuracy, specificity, or sensitivity compared to the references [13,19,22,26]. However, they used long-term ECGs with no more than ten records from the MIT-BIH Arrhythmia Database to experiment. For example, references [19,22] used only six and five patient ECGs, respectively. In addition to this, randomly dividing the training set and test set should attract our attention and vigilance. For example, reference [26] randomly divided the training set and the test set at a ratio of 2:1. References [19,22] are no exception to this problem. It is particularly noteworthy that the training set and the test set were the same in reference [13], making their results unconvincing.

Second, the proposed method was only 0.1% lower than the reference [13] in accuracy and outperformed the others. In terms of specificity, the proposed method was also only inferior to reference [13,19,22,26]. However, this paper's proposed PVC detection system did not perform exceptionally well in terms of sensitivity.

It is worth mentioning that the results of reference [28] were based on five-fold cross-validation. Suppose the division scheme of the training and test sets mentioned in this paper were used in reference [28]. In that case, the accuracy, sensitivity, and specificity of reference [28] would be 97.6%, 72.1%, and 99.9%, respectively. Therefore, our proposed method was superior to reference [28,29] used ten-fold cross-validation to select the classifier. Cross-validation is not suitable in the PVC detection task because it lays a hidden danger for label leakage. Further, our method did not rely on complex preprocessing and was superior to reference [37] in all metrics. Finally, the proposed system's sensitivity was similar to that of reference [30,32,33]. Our method was superior to the methods presented in these three literature pieces in terms of accuracy and specificity.

In summary, our method outperformed other studies. Further, applying deep metric learning can automatically extract features and ensure that the features are spatially informative. Finally, the PVC detection system proposed in this paper was highly portable. The system could be directly applied to analyze many other physiological signals.

4. Conclusions

This study successfully applied a deep metric learning model to extract spatial features from heartbeats. These features were useful and practical. Moreover, the KNN classifier could directly classify heartbeats based on the distance between features. This paper's series of experimental results showed that the proposed method achieved significantly better classification results than the existing morphology-based and deep learning-based methods. It was also practical and easy to migrate the proposed method to other physiological signals, such as heart sounds and pulses. Third, in this paper, we developed cosine similarity-based features. There were many other types of distance features to be developed. We plan to develop deep metric learning models based on different types of distances in future work to extract features. Combining multiple features helped to improve the performance of the proposed system. Finally, deploying the proposed method on cloud servers is in our plan, which will be of great help to patients and physicians in remote areas.

Author Contributions: Conceptualization, X.C.; data curation, X.W. and J.G.; formal analysis, X.W.; methodology, J.Y., X.C., and J.G.; project administration, X.C.; resources, J.Y.; software, J.G.; visualization, J.G.; writing—original draft, X.W.; writing—review and editing, J.Y. and X.C. All authors have read and agreed to the published version of the manuscript.

Funding: This research received no external funding.

Institutional Review Board Statement: Not applicable.

Informed Consent Statement: Not applicable.

Data Availability Statement: The data presented in this study are openly available in [physionet] at [10.1109/51.932724 and 10.1161/01.cir.101.23.e215], reference number [38,39]. The webpage of the MIT-BIH Arrhythmia Database is "https://www.physionet.org/content/mitdb/1.0.0/" (accessed on 22 February 2021).

Conflicts of Interest: The authors declare no conflict of interest.

References

1. Heerdt, P.M.; Dickstein, M.L. Regulation and assessment of cardiac function. In *Foundations of Anesthesia*; Mosby: St. Louis, MO, USA, 2006; pp. 511–523.
2. Annam, J.R.; Kalyanapu, S.; Ch, S.; Somala, J.; Raju, S.B. Classification of ECG Heartbeat Arrhythmia: A Review. *Proced. Comput. Sci.* **2020**, *171*, 679–688. [CrossRef]
3. Franco, D. Molecular Determinants of Cardiac Arrhythmias. *Hearts* **2020**, *1*, 146–148. [CrossRef]
4. Bae, T.W.; Lee, S.H.; Kwon, K.K. An Adaptive Median Filter Based on Sampling Rate for R-Peak Detection and Major-Arrhythmia Analysis. *Sensors* **2020**, *20*, 6144. [CrossRef]
5. Huang, W.; Liu, Y.; Liu, F.; Deng, H.; Xue, Y.; ShuLin, W. Epidemiological study of premature ventricular contraction in Guangzhou communities. *South China J. Cardiol.* **2018**, *19*, 80–88.

6. Babayiğit, E. Important tips reflected in our daily practice from the American College of Cardiology Electrophysiology Council report on premature ventricular contractions. *Anatol. J. Cardiol.* **2020**, *23*, 196–203. [CrossRef] [PubMed]

7. Gerstenfeld, E.P.; De Marco, T. Premature Ventricular Contractions. *Circulation* **2019**, *140*, 624–626. [CrossRef] [PubMed]

8. Park, K.-M.; Im, S.I.; Chun, K.J.; Hwang, J.K.; Park, S.-J.; Kim, J.S.; On, Y.K. Asymptomatic ventricular premature depolarizations are not necessarily benign. *Europace* **2016**, *18*, 881–887. [CrossRef]

9. Pramudita, B.A.; Nugraha, A.F.; Nugroho, H.A.; Setiawan, N.A. Premature Ventricular Contraction (PVC) Detection Using R Signals. *KnE Life Sci.* **2019**, *4*, 1–7. [CrossRef]

10. Hock, T.S.; Faust, O.; Lim, T.C.; Yu, W. Automated detection of premature ventricular contraction using recurrence quanti-fication analysis on heart rate signals. *J. Med. Imaging Health Inform.* **2013**, *3*, 462–469. [CrossRef]

11. Li, P.; Liu, C.; Wang, X.; Zheng, D.; Li, Y.; Liu, C. A low-complexity data-adaptive approach for premature ventricular contraction recognition. *Signal Image Video Process.* **2013**, *8*, 111–120. [CrossRef]

12. Chikh, M. Application of artificial neural networks to identify the premature ventricular contraction (PVC) beats. *Electron. J. Comb.* **2004**, *4*, 8–21. [CrossRef]

13. Lim, J.S. Finding Features for Real-Time Premature Ventricular Contraction Detection Using a Fuzzy Neural Network System. *IEEE Trans. Neural Netw.* **2009**, *20*, 522–527. [CrossRef] [PubMed]

14. Ebrahimzadeh, A.; Khazaee, A. Detection of premature ventricular contractions using MLP neural networks: A comparative study. *Measurement* **2010**, *43*, 103–112. [CrossRef]

15. Adnane, M.; Belouchrani, A.; Adnane, M. Premature ventricular contraction arrhythmia detection using wavelet coefficients. In Proceedings of the 2013 8th International Workshop on Systems, Signal Processing and their Applications (WoSSPA), Algiers, Algeria, 12–15 May 2013; pp. 170–173.

16. Du, H.; Bai, Y.; Zhou, S.; Wang, H.; Liu, X. A novel method for diagnosing premature ventricular contraction beat based on chaos theory. In Proceedings of the 2014 11th International Conference on Fuzzy Systems and Knowledge Discovery (FSKD), Xiamen, China, 19–21 August 2014; pp. 497–501.

17. Lek-Uthai, A.; Ittatirut, S.; Teeramongkonrasmee, A. Algorithm development for real-time detection of premature ventricular contraction. In Proceedings of the TENCON 2014-2014 IEEE Region 10 Conference, Bangkok, Thailand, 22–25 October 2014; pp. 1–5.

18. Jenny, N.Z.N.; Faust, O.; Yu, W. Automated Classification of Normal and Premature Ventricular Contractions in Electrocardiogram Signals. *J. Med. Imaging Health Inform.* **2014**, *4*, 886–892. [CrossRef]

19. Nuryani, N.; Yahya, I.; Lestari, A. Premature ventricular contraction detection using swarm-based support vector machine and QRS wave features. *Int. J. Biomed. Eng. Technol.* **2014**, *16*, 306. [CrossRef]

20. Nugroho, A.A.; Nuryani, N.; Yahya, I.; Sutomo, A.D.; Haijito, B.; Lestari, A. Premature ventricular contraction detection using artificial neural network developed in android application. In Proceedings of the Joint International Conference on Electric Vehicular Technology and Industrial, Mechanical, Electrical and Chemical Engineering (ICEVT & IMECE), Surakarta, Indonesia, 4–5 November 2015; pp. 212–214.

21. Jun, T.J.; Park, H.J.; Minh, N.H.; Kim, D.; Kim, Y.-H. Premature Ventricular Contraction Beat Detection with Deep Neural Networks. In Proceedings of the 2016 15th IEEE International Conference on Machine Learning and Applications (ICMLA), Anaheim, CA, USA, 18–20 December 2016; pp. 859–864.

22. Jeon, E.; Jung, B.-K.; Nam, Y.; Lee, H. Classification of Premature Ventricular Contraction using Error Back-Propagation. *KSII Trans. Internet Inf. Syst.* **2018**, *12*, 988–1001. [CrossRef]

23. Liu, X.; Du, H.; Wang, G.; Zhou, S.; Zhang, H. Automatic diagnosis of premature ventricular contraction based on Lyapunov exponents and LVQ neural network. *Comput. Methods Programs Biomed.* **2015**, *122*, 47–55. [CrossRef]

24. Kaya, Y.; Pehlivan, H. Feature selection using genetic algorithms for premature ventricular contraction classification. In Proceed-ings of the 2015 9th International Conference on Electrical and Electronics Engineering (ELECO), Bursa, Turkey, 26–28 November 2015; pp. 1229–1232.

25. Kaya, Y.; Pehlivan, H. Classification of Premature Ventricular Contraction in ECG. *Int. J. Adv. Comput. Sci. Appl.* **2015**, *6*, 34–40. [CrossRef]

26. Bouchikhi, S.; Boublenza, A.; Chikh, M.A. Discrete hidden Markov model classifier for premature ventricular contraction detection. *Int. J. Biomed. Eng. Technol.* **2015**, *17*, 371. [CrossRef]

27. Gonzalez, L.; Walker, K.; Challa, S.; Bent, B. Monitoring a skipped heartbeat: A real-time premature ventricular contraction (PVC) monitor. In Proceedings of the 2016 IEEE Virtual Conference on Applications of Commercial Sensors (VCACS), Piscataway, NJ, USA, 15 June 2016–15 January 2017; pp. 1–7.

28. De Oliveira, B.R.; De Abreu, C.C.E.; Duarte, M.A.Q.; Filho, J.V. Geometrical features for premature ventricular contraction recognition with analytic hierarchy process based machine learning algorithms selection. *Comput. Methods Programs Biomed.* **2019**, *169*, 59–69. [CrossRef]

29. Casas, M.M.; Avitia, R.L.; González-Navarro, F.F.; Cardenas-Haro, J.A.; Reyna, M.A. Bayesian Classification Models for Premature Ventricular Contraction Detection on ECG Traces. *J. Health Eng.* **2018**, *2018*, 2694768. [CrossRef]

30. Xie, T.; Li, R.; Shen, S.; Zhang, X.; Zhou, B.; Wang, Z. Intelligent Analysis of Premature Ventricular Contraction Based on Features and Random Forest. *J. Health Eng.* **2019**, *2019*. [CrossRef]

31. Conway, J.C.D.; Raposo, C.A.; Contreras, S.D.; Belchior, J.C. Identification of Premature Ventricular Contraction (PVC) Caused by Disturbances in Calcium and Potassium Ion Concentrations Using Artificial Neural Networks. *Health* **2014**, *6*, 1322–1332. [CrossRef]

32. Yang, J.; Bai, Y.; Li, G.; Liu, M.; Liu, X. A novel method of diagnosing premature ventricular contraction based on sparse auto-encoder and softmax regression. *Bio-Med. Mater. Eng.* **2015**, *26*, S1549–S1558. [CrossRef]

33. Zhou, F.-Y.; Jin, L.-P.; Dong, J. Premature ventricular contraction detection combining deep neural networks and rules inference. *Artif. Intell. Med.* **2017**, *79*, 42–51. [CrossRef] [PubMed]

34. Liu, Y.; Huang, Y.; Wang, J.; Liu, L.; Luo, J. Detecting Premature Ventricular Contraction in Children with Deep Learning. *J. Shanghai Jiaotong Univ. Sci.* **2018**, *23*, 66–73. [CrossRef]

35. Szegedy, C.; Vanhoucke, V.; Ioffe, S.; Shlens, J.; Wojna, Z. Rethinking the inception architecture for computer vision. *Conf. Proc.* **2016**, 2818–2826. [CrossRef]

36. Zhou, X.; Zhu, X.; Nakamura, K.; Mahito, N. Premature Ventricular Contraction Detection from Ambulatory ECG Using Recurrent Neural Networks. In Proceedings of the 2018 40th Annual International Conference of the IEEE Engineering in Medicine and Biology Society (EMBC), Honolulu, HI, USA, 18–21 July 2018; pp. 2551–2554.

37. Hoang, T.; Fahier, N.; Fang, W.-C. Multi-Leads ECG Premature Ventricular Contraction Detection using Tensor Decomposition and Convolutional Neural Network. In Proceedings of the 2019 IEEE Biomedical Circuits and Systems Conference (BioCAS), Nara, Japan, 17–19 October 2019; pp. 1–4.

38. Moody, G.; Mark, R. The impact of the MIT-BIH Arrhythmia Database. *IEEE Eng. Med. Biol. Mag.* **2001**, *20*, 45–50. [CrossRef] [PubMed]

39. Goldberger, A.L.; Amaral, L.A.N.; Glass, L.; Hausdorff, J.M.; Ivanov, P.C.; Mark, R.G.; Mietus, J.E.; Moody, G.B.; Peng, C.-K.; Stanley, H.E. PhysioBank, PhysioToolkit, and PhysioNet: Components of a new research resource for complex physiologic signals. *Circulation* **2000**, *101*, e215–e220. [CrossRef]

40. Pan, J.; Tompkins, W.J. A Real-Time QRS Detection Algorithm. *IEEE Trans. Biomed. Eng.* **1985**, *32*, 230–236. [CrossRef] [PubMed]

41. Lin, C.-C.; Chang, H.-Y.; Huang, Y.-H.; Yeh, C.-Y. A Novel Wavelet-Based Algorithm for Detection of QRS Complex. *Appl. Sci.* **2019**, *9*, 2142. [CrossRef]

42. Chen, C.-L.; Chuang, C.-T. A QRS Detection and R Point Recognition Method for Wearable Single-Lead ECG Devices. *Sensors* **2017**, *17*, 1969. [CrossRef]

43. Chen, A.; Zhang, Y.; Zhang, M.; Liu, W.; Chang, S.; Wang, H.; He, J.; Huang, Q. A Real Time QRS Detection Algorithm Based on ET and PD Controlled Threshold Strategy. *Sensors* **2020**, *20*, 4003. [CrossRef]

44. Wu, H.; Dai, D.; Wang, X. A Novel Radar HRRP Recognition Method with Accelerated T-Distributed Stochastic Neighbor Embedding and Density-Based Clustering. *Sensors* **2019**, *19*, 5112. [CrossRef] [PubMed]

45. Jia, M.; Li, T.; Wang, J. Audio Fingerprint Extraction Based on Locally Linear Embedding for Audio Retrieval System. *Electronics* **2020**, *9*, 1483. [CrossRef]

46. Siomos, N.; Fountoulakis, I.; Natsis, A.; Drosoglou, T.; Bais, A. Automated Aerosol Classification from Spectral UV Measurements Using Machine Learning Clustering. *Remote Sens.* **2020**, *12*, 965. [CrossRef]

47. Zhang, L.; Rao, Z.; Ji, H. NIR Hyperspectral Imaging Technology Combined with Multivariate Methods to Study the Residues of Different Concentrations of Omethoate on Wheat Grain Surface. *Sensors* **2019**, *19*, 3147. [CrossRef] [PubMed]

48. Kaya, M.; Bilge, H. Şakir Deep Metric Learning: A Survey. *Symmetry* **2019**, *11*, 1066. [CrossRef]

49. Humpback Whale Identification—Kaggle. Available online: https://www.kaggle.com/c/humpback-whale-identification (accessed on 16 November 2020).

50. Bozo, M.; Aptoula, E.; Cataltepe, Z. A Discriminative Long Short Term Memory Network with Metric Learning Applied to Multispectral Time Series Classification. *J. Imaging* **2020**, *6*, 68. [CrossRef]

51. Xie, C.; Wang, X.; Qian, C.; Wang, M. A Source Code Similarity Based on Siamese Neural Network. *Appl. Sci.* **2020**, *10*, 7519. [CrossRef]

52. Garbin, C.; Zhu, X.; Marques, O. Dropout vs. batch normalization: An empirical study of their impact to deep learning. *Multimed. Tools Appl.* **2020**, *79*, 12777–12815. [CrossRef]

53. He, K.; Zhang, X.; Ren, S.; Sun, J. Delving deep into rectifiers: Surpassing human-level performance on imagenet classification. In Proceedings of the IEEE International Conference on Computer Vision, Santiago, Chile, 7–13 December 2015; pp. 1026–1034.

54. Özcan, U.G.; Toklu, B. A new hybrid improvement heuristic approach to simple straight and U-type assembly line balancing problems. *J. Intell. Manuf.* **2009**, *20*, 123–136. [CrossRef]

55. Wang, X.; Han, X.; Huang, W.; Dong, D.; Scott, M.R. Multi-Similarity Loss with General Pair Weighting for Deep Metric Learning. In Proceedings of the IEEE Conference on Computer Vision and Pattern Recognition, Long Beach, CA, USA, 16–20 June 2019; pp. 5022–5030.

56. Mondéjar-Guerra, V.; Novo, J.; Rouco, J.; Penedo, M.; Ortega, M. Heartbeat classification fusing temporal and morphological information of ECGs via ensemble of classifiers. *Biomed. Signal Process. Control* **2019**, *47*, 41–48. [CrossRef]

57. Maaten, L.; Hinton, G. Visualizing High-Dimensional Data using t-SNE. *J. Mach. Learn. Res.* **2008**, *9*, 2579–2605.

MDPI
St. Alban-Anlage 66
4052 Basel
Switzerland
Tel. +41 61 683 77 34
Fax +41 61 302 89 18
www.mdpi.com

Biosensors Editorial Office
E-mail: biosensors@mdpi.com
www.mdpi.com/journal/biosensors